FUNCTIONS OF A COMPLEX VARIABLE

Functions of a Complex Variable

CONSTRUCTIVE THEORY

V. I. SMIRNOV and N. A. LEBEDEV

Translated by Scripta Technica Ltd

THE M.I.T. PRESS

Massachusetts Institute of Technology
Cambridge, Massachusetts

Originally published in Moscow by the State Scientific Publishing House, under the title
Konstruktivnaya teoriya funktsii kompleksogo peremennogo

Translated and prepared for press by Scripta Technica Ltd
139a New Bond Street, London W.1

English edition first published 1968 by
Iliffe Books Ltd, Dorset House, Stamford Street, London S.E.1

Printed and bound in Great Britain by
Butler & Tanner Ltd, Frome and London

Published in the U.S.A. by
The M.I.T. Press, Massachusetts Institute of Technology
Cambridge, Massachusetts

Library of Congress Catalog Card Number: 68-20049

Contents

Introduction

One of the basic problems of a constructive theory of functions is that of approximate representation of functions of a certain class by functions of a sub-class which are in some sense simpler. The present book basically deals with this problem for regular functions of a complex variable. It also covers a different type of question relating to estimations of the given class of functions. Questions such as this are of independent interest and are also connected with the inverse problems of the theory of best approximation.

The greater part (the beginning) of the first chapter is devoted to the question of an approximate representation of functions which are regular in closed sets by means of polynomials and rational fractions obtainable by interpolation. The approximation of functions which are continuous in a closed disc and regular within this circle by means of polynomials is also investigated. The end of the first chapter is devoted to the well-known results of Mergelian, which relate to the approximation of functions that are continuous in bounded, closed sets and regular at internal points of these sets by using polynomials and rational functions, and to the question of evaluating the best polynomial approximation of functions of the type given above. In this section we also introduce Vitushkin's theorem for approximating functions which are continuous on a closed, bounded set without internal points by means of rational functions.

The second chapter sets out the theory of the fundamental and generalised Faber polynomials and some applications of these polynomials. The end of the chapter gives an account of the theory of biorthogonal functions which is indirectly connected with the contents of the chapter.

The third and fourth chapters deal with the mean square approximation with respect to a domain or its boundary.

Due to the nature of the material, the fifth chapter is divided into two parts. The first part sets out (for the case of a disc) the solution to a series of concrete problems connected with polynomials deviating least from zero, with Mendeleev's problem and the well-known results of A.A. Markov and Bernstein. Analogous results are obtained for the intersection of the real axis and ellipse by means of an elementary transformation of the independent variable.

The second part of Chapter 5 contains a series of general theorems relating to the theory of best approximation. The examination is carried out in linear metric spaces and leads in particular to theorems for the existence and uniqueness of the best approximation and to the establishment of necessary and sufficient conditions for the generalised polynomial to be the polynomial of best approximation.

At the end of the chapter there are some applications of the general theorems previously set out.

The present book is based on a special course on the constructive theory of functions of a complex variable given for some years in Leningrad University, firstly by Smirnov and subsequently by Lebedev. The material in the book has been considerably expanded in comparison with the lectures.

In drawing up this work we have confined ourselves to the simplest and, in our view, fundamental problems and the methods applied to them are classical in the familiar meaning of the word.

To our profound regret we were unable to include a whole series of extremely interesting new studies being carried out mainly in the Moscow and Yerevan Schools.

In particular, we have not touched at all on questions relating to a constructive theory for entire functions, inverse theorems for best approximation and the theory of weighted uniform approximation nor to the theory of bases of different types of systems of functions. In the matter of evaluating the best approximation we have confined ourselves to theorems involving the use of external contour lines.

Questions on the approximation of functions originate from

the famous work of Chebyshev. The works of Bernstein were of fundamental importance in producing the current constructive theory of functions. His numerous works and those of his pupils deal particularly with functions of real variables. More recently works on analogous topics for the case of complex variables have also begun to appear.

Khavin carefully studied the manuscript and pointed out various mistakes in the presentation. We should like to extend to him our deepest gratitude. We should also like to thank Vidensky who read through Chapter 5 and gave us valuable information about literature relating to the subject.

V.I. Smirnov and N.A. Lebedev

CHAPTER ONE

Uniform approximation of functions by polynomials and rational functions

This chapter is devoted to the uniform approximation of functions by polynomials and rational functions.

Let B be some infinite set of points of the extended z plane. The closure of the set B will always be denoted by $\bar{B}$ and it assumed that $\bar{B}$ is not empty and is not the extended plane unless otherwise stated. The complement of any set B is the set of all points $z \notin B$ of the extended plane and this complement is denoted by CB. The boundary of a set B is the set of all those points whose every neighbourhood contains points of B and of CB. The boundary of B will be denoted by L.

Let $\mathscr{P}$ be a class of functions regular on $\bar{B}$, e.g. each function $p(z) \in \mathscr{P}$ is regular on some open set D_p containing $\bar{B}$ (the set D_p can be different for different functions $p(z) \in \mathscr{P}$). We shall first ascertain those conditions which must be applied to a function $f(z)$ given on B in order that it can be uniformly approximated on B as well as is desired by means of functions $p(z) \in \mathscr{P}$, e.g. so that there shall be a sequence of functions $p_n(z) \in \mathscr{P}$, $n=1, 2, \ldots,$ such that the measure of uniform approximation of $f(z)$ by means of functions $p_n(z)$ on B

$$E(f\ \ p_n\ \ B) = \sup_{z \in B} |f(z) - p_n(z)| \to 0 \text{ as } n \to \infty$$

Let us show that in this case the function $f(z)$ is completely defined on the closed set $\bar{B}$ and the sequence $p_n(z)$ uniformly converges on $\bar{B}$ to $f(z)$. Let a be an arbitrary point which is a limit point of B but does not belong to B, and z_k, $k=1, 2, \ldots$, be some sequence of points from B converging to a such that there exists a finite or infinite limit

$$f_r(a) = \lim_{k\to\infty} f(z_k)$$

The limit $f_r(a)$ is called the partial limit of the function $f(z)$ as $z \to a$. For $\varepsilon > 0$ there is N_ε such that for $n > N_\varepsilon$

$$|f(z) - p_n(z)| < \varepsilon$$

for all $z \in B$ and, in particular,

$$|f(z_k) - p_n(z_k)| < \varepsilon, \quad k = 1, 2, \ldots$$

From the fact that $p_n(z_k) \to p_n(a)$ as $k \to \infty$ it follows that the limit $f_r(a)$ is a finite number, and in the limit we have

$$|f_r(a) - p_n(a)| \leqslant \varepsilon, \quad n > N_\varepsilon$$

Clearly all partial limits of $f(z)$ coincide as $z \to a$, as there is a $\lim\limits_{z\to a} f(z) = A$. Let us define the function $f(z)$ at point a, assuming that $f(a) = A$. Thus, let us then define the function $f(z)$ on B, and, in such a case, where $n > N_\varepsilon$

$$|f(z) - p_n(z)| \leqslant \varepsilon, \ z \in \bar{B}$$

e.g. the sequence $p_n(z)$ uniformly converges on $\bar{B}$ to $f(z)$. Since the functions $p_n(z)$ are regular on $\bar{B}$, because of the uniform convergence of the sequence $p_n(z)$, the function $f(z)$ should be continuous on B and regular at every internal point of the set $\bar{B}$.

Note that we can confine ourselves to considering only bounded sets $\bar{B}$. In fact, if $\bar{B}$ is an unbounded set, then taking an arbitrary point a outside $\bar{B}$, let us map the set $\bar{B}$ with the help of the function $\zeta = \dfrac{1}{z-a}$ on to a bounded set $\bar{B}$ of the plane ζ. The question of approximating $f(z)$ on $\bar{B}$ by functions $p(z)$ will then become that of approximating $f_*(\zeta) = f\left(a + \frac{1}{\zeta}\right)$ on $\bar{B}_*$ by functions $p_*(\zeta) = p\left(a + \frac{1}{\zeta}\right)$.

Let $\bar{B}$ be an arbitrary closed bounded set of the z plane and L be the boundary of $\bar{B}$. The complement $C\bar{B}$ of the set B consists of a finite or enumerable number of domains B^j, $j=1, 2, \ldots$ These domains B^j are called complementary for $\bar{B}$. The boundary of the domain B^j is denoted by L^j. If $C\bar{B}$ consists of a finite number of domains B^j, $j=1, 2, \ldots, s$, then $L=\bigcup\limits_{j=1}^{s} L^j$. Let the function $f(z)$ be continuous on $\bar{B}$ and regular at every internal point of the set $\bar{B}$, and $p_n(z)$, $n=1, 2, \ldots,$ be a sequence of functions, regular on $\bar{B}$, uniformly converging on $\bar{B}$ to $f(z)$. Let us assume that all functions $p_n(z)$ are regular in some domain B^j Since the sequence $p_n(z)$ uniformly converges on L^j, it converges uniformly on $\bar{B}^j$ to some function which is regular on B^j and continuous on $\bar{B}^j$. Thus, with the given assumption, the function $f(z)$ can be defined on B^j such that the function obtained is regular at internal points of the set $\bar{B} \cup B^j$ and continuous on $\bar{B} \cup B^j$. Since our aim is the uniform approximation with an unspecified degree of accuracy of the arbitrary function $f(z)$, continuous on $\bar{B}$ and regular at internal points of this set, by functions $p(z) \in \mathscr{P}$, then from the above it follows that the class $\mathscr{P}$ should have the following property: given some finite set K of domains B^j, there exists a function $p(z) \in \mathscr{P}$ having at least one singular point in each of the domains B^j of the set K. The simplest class $\mathscr{P}$ having this property is the class of all rational functions

$$p(z)=\frac{b_0+b_1z+\ldots+b_mz^m}{(z-a_1)^{m_1}(z-a_2)^{m_2}\ldots(z-a_s)^{m_s}}$$

where $a_1, a_2, \ldots, a_s$ are fixed points of $C\bar{B}$. As we shall see we can restrict ourselves by including in the class $\mathscr{P}$ only such rational functions that in each of the subsidiary domains B^j for $\bar{B}$ there is not more than one previously given pole of the function. However, we cannot limit ourselves by including in the class $\mathscr{P}$ only rational functions with a limited order of pole even in one of the subsidiary domains B^j. In fact, let all the functions of the class $\mathscr{P}$ have, for example, a single pole $a \neq \infty$ of an order not higher than m in the subsidiary domain B^1, and let us assume that for a function $f(z)=\frac{1}{(z-a)^{m+1}}$ regular on $\bar{B}$ there exists a sequence $p_n(z)$, $n=1, 2, \ldots,$ of rational functions from $\mathscr{P}$, which uniformly converge on $\bar{B}$ to $f(z)$. The functions of the sequence $\varphi_n(z) = p_n(z)(z-a)^m$ are regular on B^1, and this sequence uniformly converges on $\bar{B}$ to the function $\frac{1}{z-a}$. On the basis of what has been said previously this function considered on $\bar{B}$ can be defined on B^1 such that the function obtained will be regular on $\bar{B} \cup B^1$, which cannot be so in view of the theorem of the uniqueness of an analytic continuation. Analogously we can consider the case where $a = \infty$, putting $f(z) = z^{m+1}$.

We immediately come up against the question of the best uniform approximation by polynomials. Let $\bar{B}$ be a closed bounded set of points of the z plane; $f(z)$ a bounded function given on $\bar{B}$; $p_n(z)$ an arbitrary polynomial of a degree not higher than n. Let us assume

$$\rho_n(f, \bar{B}) = \inf_{p_n(z)} \sup_{z \in \bar{B}} |f(z) - p_n(z)|$$

It will be shown in Chapter 5 that there is a polynomial $\pi_n(z)$ of a degree not higher than n such that

$$\sup_{z \in \bar{B}} |f(z) - \pi_n(z)| = \rho_n(f, \bar{B})$$

The polynomial $\pi_n(z)$ is called the polynomial of best uniform approximation to $f(z)$ on $\bar{B}$, and $\rho_n(f, \bar{B})$ the measure of best approximation to $f(z)$ by means of polynomials of a degree not higher than n.

If for some polynomial $p_n^*(z)$ we find an upper estimate for $|f(z) - p_n^*(z)|$ on $\bar{B}$, then clearly we obtain also an upper estimate for the value of $\rho_n(f, \bar{B})$.

Analogous statements can be made about the best uniform approximation by rational functions.

Henceforward we shall make use of the following definition: if the sequence of functions $\varphi_n(z)$, $n = 1, 2, \ldots$, converges uniformly to $f(z)$ on every compact set contained in the open set B, then we shall say that the sequence $\varphi_n(z)$ converges to $f(z)$ uniformly on the compact sub-sets of B and we shall write $\varphi_n(z) \rightrightarrows f(z)$ in B. For a compact set $\bar{B}$, the symbol $\varphi_n(z) \rightrightarrows f(z)$ on $\bar{B}$ denotes that $\varphi_n(z)$ converges to $f(z)$ uniformly on $\bar{B}$.

Note that Section 1.7, devoted to presenting the results of Mergelian and Vitushkin, could usefully be read before the intervening sections.

1.1 ELEMENTARY RESULTS FROM THE THEORY OF INTERPOLATION

1.1.1 The problem of simple and multiple interpolation

We wish to construct a polynomial* $p(z)$ which at various given points

$$w_1, w_2, \ldots, w_{n+1}$$

has corresponding given values

$$z_1, z_2, \ldots, z_{n+1}$$

There are infinitely many such polynomials. If we restrict ourselves to finding a polynomial $p_n(z)$ of a degree not higher than n, then the polynomial is unique. In fact, the difference between two polynomials with the given properties is a polynomial of a degree not higher than n becoming zero at $(n+1)$ different points z_k, $k=1, 2, \ldots, n+1$. But such a polynomial must be identically zero. We call a polynomial of a degree not higher than n, having given values w_k at different points z_k, an interpolating polynomial.

Lagrange proposed the following form for writing an interpolating polynomial:

$$p_n(z) = \sum_{k=1}^{n+1} \frac{(z-z_1)(z-z_2)\ldots(z-z_{k-1})(z-z_{k+1})\ldots(z-z_{n+1})}{(z_k-z_1)(z_k-z_2)\ldots(z_k-z_{k-1})(z_k-z_{k+1})\ldots(z_k-z_{n+1})} w_k \tag{1}$$

It can be verified directly that this polynomial is of a degree not higher than n and that $p_n(z_k)=w_k$, $k=1, 2, \ldots, n$. Formula (1) is called Lagrange's interpolating polynomial. Quantities z_k are called basic points of interpolation.

The interpolating polynomial (1) can be written in the form

$$p_n(z) = \sum_{k=1}^{n+1} \frac{\omega(z)}{(z-z_k)\,\omega'(z_k)} w_k, \qquad \omega(z) = \prod_{k=1}^{n+1} (z-z_k) \tag{2}$$

Let us now define the problem of multiple interpolation. Let there be different points z_k, $k=1, 2, \ldots, m$, and let

*Here and in everything that follows, the index shown when designating the polynomial signifies that the polynomial in question is of a degree not higher than the given index.

each point z_k be associated with quantities

$$w_k^{(\nu)} \quad \nu = 0, 1, \ldots, \alpha_k - 1$$

where α_k are given natural numbers. It is required to construct a polynomial $p_n(z)$ of the lowest degree satisfying the conditions

$$p_n^{(\nu)}(z_k) = w_k^{(\nu)}, \quad k = 1, 2, \ldots, m; \quad \nu = 0, 1, \ldots, \alpha_k - 1 \tag{3}$$

We shall also call such a polynomial an interpolating polynomial (for multiple interpolation). The quantities z_k, as before, will be called the basic points of interpolation, and the quantity α_k will be called the multiplicity of interpolation for the basic point z_k. If all $\alpha_k = 1$, then we come back to the problem of simple interpolation discussed above.

To find a polynomial $p_n(z)$ satisfying conditions (3) we have in all $\sum\limits_{k=1}^{m} \alpha_k$ conditions. Since the number of coefficients of a polynomial is one more than its degree, we have to try to find a polynomial with a degree $n = \sum\limits_{k=1}^{m} \alpha_k - 1$.

Assuming that the desired interpolating polynomial with a degree not higher than n and satisfying conditions (3) exists, we shall denote it by $p_n(z)$. Let us introduce the polynomial

$$\omega(z) = \prod_{k=1}^{m} (z - z_k)^{\alpha_k} \tag{4}$$

of degree $n + 1 = \sum\limits_{k=1}^{m} \alpha_k$. Let us expand the rational function $\frac{p_n(z)}{\omega(z)}$ into partial functions:

$$\frac{p_n(z)}{\omega(z)} = \sum_{k=1}^{m} \sum_{l=0}^{\alpha_k - 1} \frac{A_{k,l}}{(z - z_k)^{\alpha_k - l}} \tag{5}$$

To find the coefficient $A_{k,l}$ we multiply this equation by $(z - z_k)^{\alpha_k}$. Then attached to the coefficient $A_{k,l}$ there will be a factor $(z - z_k)^l$. Differentiating the equation obtained l times and putting $z = z_k$, we have

$$\frac{d^l}{dz^l} \left[p_n(z) \frac{(z - z_k)^{\alpha_k}}{\omega(z)} \right]_{z = z_k} = l!\, A_{k,l}$$

Hence, by using conditions (3), we easily find

$$A_{k,l} = \sum_{\nu=0}^{l} \frac{1}{\nu!\,(l-\nu)!}\, w_k^{(\nu)} \left[\frac{d^{l-\nu}}{dz^{l-\nu}} \frac{(z-z_k)^{\alpha_k}}{\omega(z)} \right]_{z=z_k} \tag{6}$$

$$k=1, 2, \ldots, m; \; l=0, 1, \ldots, \alpha_k - 1$$

and from Equation (5) we find that

$$p_n(z) = \sum_{k=1}^{m} \frac{\omega(z)}{(z-z_k)^{\alpha_k}} \sum_{l=0}^{\alpha_k - 1} A_{k,l}(z-z_k)^l \tag{7}$$

where the coefficients $A_{k,l}$ are defined according to Formula (6). Formula (7) in fact defines a polynomial with a degree not higher than n. This polynomial satisfies conditions (3) and, consequently, is the required one. To confirm this, differentiate polynomial (7) ν times, $\nu=0, 1, \ldots, \alpha_k-1$, and then put $z=z_k$, $k=1, 2, \ldots, m$. The uniqueness of the interpolating polynomial follows from Formula (6). Polynomial (7) is called Hermite's interpolating polynomial.

NOTE 1 On some set B let there be given a function $f(z)$ and various points z_k, $k=1, 2, \ldots, m$, of this set. The interpolating polynomial $p_n(z)$ for the function $f(z)$, set up in accordance with these data, should satisfy the conditions

$$w_k^{(\nu)} = p_n^{(\nu)}(z_k) = f^{(\nu)}(z_k), \; k=1, 2, \ldots, m; \; \nu=0, 1, \ldots, \alpha_k - 1$$

at points z_k.

It is to be expected that, when certain conditions are imposed on the set B, the function $f(z)$, the choice of points z_k and the choice of quantities α_k, the interpolating polynomial $p_n(z)$ will be a sufficiently good approximation of the function $f(z)$. Later we shall confirm that this is so.

NOTE 2 It is more convenient to consider that among the basic points of interpolation z_k, $k=1, 2, \ldots, n+1$, there may be some which are identical; in this case here and everywhere in the following we shall consider that, if some value z_k is found amongst the points α times, this value should be a point of interpolation of multiplicity α. In this scheme both simple and multiple interpolation are included.

Let us now investigate the problem of multiple interpolation by generalised polynomials (rational functions):

$$p_n(z) = b_0 + b_1^{(0)} z + \ldots + b_{m_0}^{(0)} z^{m_0} + \frac{b_1^{(1)}}{z - a_1} + \frac{b_2^{(1)}}{(z - a_1)^2} + \ldots + \frac{b_{m_1}^{(1)}}{(z - a_1)^{m_1}} + \ldots + \frac{b_1^{(s)}}{z - a_s} + \frac{b_2^{(s)}}{(z - a_s)^2} + \ldots + \frac{b_{m_s}^{(s)}}{(z - a_s)^{m_s}}$$

We shall call the number $n = \sum_{j=0}^{s} m_j$ the order of this generalised polynomial.

Let there be various given points z_k, $k = 1, 2, \ldots, m$, differing from points $a_1, a_2, \ldots, a_s$, which are considered to be fixed, and let there be associated with each point z_k quantities $w_k^{(\nu)}$, $\nu = 0, 1, \ldots, \alpha_k - 1$, where α_k, $\sum_{k=1}^{m} \alpha_k = n + 1$ are given natural numbers. It is required to construct a generalised polynomial $p_n(z)$ of the lowest order n satisfying the conditions

$$p_n^{(\nu)}(z_k) = w_k^{(\nu)}, \quad k = 1, 2, \ldots, m; \quad \nu = 0, 1, \ldots, \alpha_k - 1 \tag{8}$$

The quantities z_k are the basic points of interpolation, α_k is the multiplicity of interpolation for point z_k. We shall call such a generalised polynomial the generalised interpolating polynomial.

Let us assume that the required generalised interpolating polynomial exists, and let us denote it by $p_n(z)$. Following the same procedure as for multiple interpolation we now assume that

$$\omega(z) = \frac{\prod_{k=1}^{m} (z - z_k)^{\alpha_k}}{\prod_{j=1}^{s} (z - a_j)^{m_j}} \tag{9}$$

In this case we obtain Formulae (7) and (6) for the required polynomial where $\omega(z)$ is defined according to Formula (9). If all the points of interpolation are simple, then Formula (7) has the form of (2).

The uniqueness of the generalised polynomial follows, as before, from Formula (6).

We shall assume that the point ∞ is included amongst the points of interpolation. To be specific, let $z_m = \infty$. In this case we have $m_0 = 0$,

$$p_n^{(\nu)}(\infty) = \left[\frac{d^\nu}{d\zeta^\nu} p_n\left(\frac{1}{\zeta}\right) \right]_{\zeta=0}$$

and in Formula (9) for $\omega(z)$ we drop the factor $(z-z_m)^{\alpha_m}$. The expansion of the rational function $\frac{p_n(z)}{\omega(z)}$ then has the form

$$\frac{p_n(z)}{\omega(z)}=\sum_{k=1}^{m-1}\sum_{l=0}^{\alpha_k-1}\frac{A_{k,l}}{(z-z_k)^{\alpha_k-l}}+\sum_{l=0}^{\alpha_m-1}A_{m,l}z^{\alpha_m-1-l} \tag{10}$$

The coefficients $A_{k,l}$ for $k\neq m$ are defined as before by Formula (6). To find the coefficient $A_{m,l}$, $l=0,1,\ldots,\alpha_m-1$, we divide the equality (10) by z^{α_m-1}, change z to $\frac{1}{\zeta}$, differentiate the equality obtained l times with respect to ζ and finally assume that $\zeta=0$. We then have

$$A_{m,l}=\sum_{\nu=0}^{l}\frac{1}{\nu!(l-\nu)!}\,\varpi_m^{(\nu)}\left[\frac{d^{l-\nu}}{d\zeta^{l-\nu}}\frac{\zeta^{\alpha_m-1}}{\omega\left(\frac{1}{\zeta}\right)}\right]_{|\zeta=0},\quad l=0,1,\ldots,\alpha_m-1$$

Formula (7) will take the form

$$p_n(z)=\sum_{k=1}^{m-1}\frac{\omega(z)}{(z-z_k)^{\alpha_k}}\sum_{l=0}^{\alpha_k-1}A_{k,l}(z-z_k)^l+\omega(z)\sum_{l=0}^{\alpha_m-1}A_{m,l}z^{\alpha_m-1-l} \tag{11}$$

The observations made with respect to interpolating polynomials for functions relate entirely to generalised interpolating polynomials if we put $f^{(\nu)}(\infty)=\left[\frac{d^\nu}{d\zeta^\nu}f\left(\frac{1}{\zeta}\right)\right]_{|\zeta=0}$.

Interpolating polynomials are often investigated which have the more general form:

$$p_n(z)=\sum_{k=0}^{n}a_k\varphi_k(z)$$

where $\varphi_k(z)$ are given functions, regular on a set B of the complex plane, and a_k are fixed numbers - the coefficients of the polynomial. The problem of simple and multiple interpolation can also be applied to such generalised polynomials. In particular, amongst these generalised polynomials are the trigonometrical polynomials:

$$s_n(\zeta)=\sum_{k=0}^{n}(a_k\cos k\zeta+b_k\sin k\zeta)$$

where a_0, a_k, b_k, $k=1, 2, \ldots, n$ are constants. If we have $|a_n|^2+|b_n|^2 \neq 0$, then the number n will be called the order of the trigonometrical polynomial. On the basis of Euler's formula,

$$\cos k\zeta = \frac{1}{2}(e^{ik\zeta}+e^{-ik\zeta}), \quad \sin k\zeta = \frac{1}{2i}(e^{ik\zeta}-e^{-ik\zeta})$$

the trigonometrical polynomial $s_n(\zeta)$ can be written in the form

$$s_n(\zeta) = \sum_{k=-n}^{n} A_k e^{ik\zeta}$$

where

$$A_0 = a_0;\ A_k = \frac{1}{2}(a_k - ib_k),\ k=1, 2, \ldots, n;\ A_{-k} = \frac{1}{2}(a_k+ib_k)$$
$$k=1, 2, \ldots, n \quad (a_0 = A_0;\ a_k = A_k + A_{-k},\ k=1, 2, \ldots, n$$
$$b_k = i(A_k - A_{-k}),\ k=1, 2, \ldots, n)$$

Trigonometrical polynomials have a period of 2π and, consequently, if some trigonometrical polynomial $s_n(\zeta)$ gives a uniform approximation to a function $g(\zeta)$ on a set B_0 with a certain degree of approximation ρ, then on the set B_k, $k=\pm 1, \pm 2, \ldots$, obtained by the parallel displacement of B_0 by the vector $2k\pi$, the trigonometrical polynomial $s_n(\zeta)$ gives uniform approximation to the function $g(\zeta-2k\pi)$ with the same degree of approximation ρ. Hence it follows that when studying approximations by trigonometrical polynomials, it is reasonable to consider functions having a period of 2π and given on a set Q possessing the property that the set Q', obtained by parallel displacement by the vector 2π, coincides with Q. Such sets will be called 2π-strips.

It is clear how the problem of multiple interpolation is formulated for trigonometrical polynomials and we shall not stop to produce the corresponding formulae.

1.1.2 Representation of interpolating and generalised interpolating polynomials in the form of a contour integral (Hermite's formula)

Theorem 1 Let B be the interior of a closed, partially smooth Jordan curve L. Let the function $f(z)$ be regular in B and continuous in $\bar{B}$, the interpolation points z_k, $k=1, 2, \ldots, m$, lie in B and α_k be natural numbers. Then the interpolating polynomial $p_n(z)$ satisfying the conditions

$$p_n^{(\nu)}(z) = f^{(\nu)}(z_k), \quad k = 1, 2, \ldots, m; \quad \nu = 0, 1, \ldots, \alpha_k - 1$$
$$n = \sum_{k=1}^{m} \alpha_k - 1$$

can be represented by the formula*

$$p_n(z) = \frac{1}{2\pi i} \int_L \frac{f(\zeta)[\omega(\zeta) - \omega(z)]}{(\zeta - z)\,\omega(\zeta)} d\zeta, \quad \omega(z) = \prod_{k=1}^{m} (z - z_k)^{\alpha_k} \quad (1)$$

PROOF Under the integral sign in Formula (1) is a polynomial of z of degree $n = \sum_{k=1}^{m} \alpha_k - 1$, and after integrating with respect to ζ we obtain a polynomial with a degree not greater than n. By using Cauchy's formula and the fact that $\omega(z_k) = 0$, we have

$$p_n(z_k) = \frac{1}{2\pi i} \int_L \frac{f(\zeta)[\omega(\zeta) - \omega(z_k)]}{(\zeta - z_k)\,\omega(\zeta)} d\zeta = \frac{1}{2\pi i} \int_L \frac{f(\zeta)}{\zeta - z_k} d\zeta = f(z_k)$$

and, differentiating equality (1) with respect to z, ν times, $\nu = 0, 1, \ldots, \alpha_k - 1$, we have

$$p_n^{(\nu)}(z) = \frac{\nu!}{2\pi i} \int_L \frac{f(\zeta)}{(\zeta - z)^{\nu+1}} \frac{\omega(\zeta) - \omega(z)}{\omega(\zeta)} d\zeta$$
$$- \sum_{l=1}^{\nu} C_\nu^l \frac{(\nu - l)!}{2\pi i} \int_L \frac{f(\zeta)}{(\zeta - z)^{\nu - l + 1}} \frac{\omega^{(l)}(z)}{\omega(\zeta)} d\zeta$$

Assuming here that $z = z_k$ and taking $\omega^{(l)}(z_k) = 0$, $l = 0, 1, \ldots, \alpha_k - 1$, we obtain

$$p_n^{(\nu)}(z_k) = \frac{\nu!}{2\pi i} \int_L \frac{f(\zeta)}{(\zeta - z)^{\nu+1}} d\zeta = f^{(\nu)}(z_k), \quad \nu = 0, 1, \ldots, \alpha_k - 1$$

Thus

$$p_n^{(\nu)}(z_k) = f^{(\nu)}(z_k), \quad k = 1, 2, \ldots, m; \quad \nu = 0, 1, \ldots, \alpha_k - 1$$

e.g. $p_n(z)$ is in fact the desired interpolating polynomial and the theorem is proved.

*When integrating along the contour of a domain the integration is carried out in the positive direction. In other cases integration takes place in an anti-clockwise direction.

Formula (1) is called Hermite's formula. From it there follows that

$$r_n(z) = f(z) - p_n(z) = \frac{1}{2\pi i} \int_L \frac{f(\zeta)}{\zeta - z} \frac{\omega(z)}{\omega(\zeta)} d\zeta, \quad z \in B \tag{2}$$

The difference $r_n(z)$ is called the residue term of interpolation.

We now let $p_n(z)$ represent the generalised interpolating polynomial:

$$p_n(z) = \sum_{k=0}^{m_0} b_k^{(0)} z^k + \sum_{j=1}^{s} \sum_{k=1}^{m_j} \frac{b_k^{(j)}}{(z - a_j)^k}$$

where a_j are fixed numbers m_j, $\sum_{j=1}^{s} m_j = n$ are natural numbers, and b_0, b_k^j are the coefficients of the polynomials. There now follows:

Theorem 2 Let B be a bounded domain, the boundary L of which consists of a finite number of closed, partially smooth Jordan curves, and let the function $f(z)$ be regular in B and continuous in $\bar{B}$. If the points of interpolation z_k, $k = 1, 2, \ldots, m$, lie inside B, α_k, $k = 1, 2, \ldots, m$, are natural numbers and the points a_j, $j = 1, 2, \ldots, s$, lie outside $\bar{B}$, then the generalised interpolating polynomial satisfying the conditions

$$p_n^{(\nu)}(z_k) = f^{(\nu)}(z_k), \quad k = 1, 2, \ldots, m; \quad \nu = 0, 1, \ldots, \alpha_k - 1$$

$$n = \sum_{k=1}^{m} \alpha_k - 1$$

can be represented by Formula (1), where $\omega(z)$ is defined by Formula (9), Section 1.1.1.

The proof is analogous to the proof of Theorem 1.

If under the same conditions the domain B contains the point ∞, then in the formula for $p_n(z)$ $m_0 = 0$ and the term $f(\infty)$ must be added to the right-hand side of Formula (1). We assume that the point ∞ is also amongst the points of interpolation $z_k^{(n)}$ (and we alter the formula for $\omega(z)$, as indicated in Section 1.1.1).

As before we find that the residue term of interpolation $r_n(z) = f(z) - p_n(z)$ is defined by Formula (2), where $\omega(z)$ is defined according to Formula (9), Section 1.1.1.

1.2 CONVERGENCE CRITERIA OF INTERPOLATING PROCESSES FOR FUNCTIONS REGULAR IN BOUNDED, SIMPLY-CONNECTED SETS OF TYPE $\mathfrak{M}$

Let there be a function $f(z)$ regular on a closed bounded set $\bar{B}$ of the z plane, and on $\bar{B}$ let there be given a sequence of interpolating points

$$z_1^{(n)},\ z_2^{(n)},\ \ldots,\ z_{n+1}^{(n)},\qquad n=0,\ 1,\ 2,\ \ldots,\ z_k^{(n)}\in\bar{B} \tag{1}$$

amongst which, for fixed n, there may be some which are identical (see Note 2, Section 1.1.1). Let us construct for the points (1) an interpolating polynomial $p_n(z)$, $n=0, 1, 2, \ldots,$ for the function $f(z)$.

The set we shall call type $\mathfrak{M}$ is a closed set containing not fewer than two points, the complement of which with respect to the extended plane is a simply connected domain containing the point ∞.

Let us suppose that $\bar{B}$ is a set of type $\mathfrak{M}$ and let us denote its complement $C\bar{B}$ by B^1. We shall denote the common boundary of B^1 and $\bar{B}$ by L. The uniform convergence of polynomials $p_n(z)$ to the function $f(z)$ on $\bar{B}$ essentially depends on the behaviour of the sequence of the polynomials

$$\omega_n(z)=\prod_{k=1}^{n+1}\left(z-z_k^{(n)}\right)$$

in the domain B^1.

Let us now introduce the notation which we shall use in the following for the case where $\bar{B}$ is a set of type $\mathfrak{M}$: $w=\varphi(z)$ is a function conformally mapping B^1 on to the domain $|w|>1$, where $\varphi(\infty)=\infty$ and $\varphi'(\infty)>0$; $L_\rho\ (\rho>1)$ are the contour lines of the domain B^1, e.g. the lines defined by the equation $|\varphi(z)|=\rho$; B_ρ is the bounded domain with the boundary L_ρ; B_ρ^1 is the unbounded domain with boundary L_ρ; by the symbols $\bar{B}_1$, B_1^1 and L_1 we shall understand respectively $\bar{B}$, B^1 and L; $c=[\varphi'(\infty)]^{-1}$, and consequently in the neighbourhood of the point $z=\infty$ the expansion of $\varphi(z)$ has the form

$$c\varphi(z)=z+a_0+\frac{a_1}{z}+\frac{a_2}{z^2}+\ \ldots \tag{2}$$

The inverse function to $w=\varphi(z)$, is denoted by $z=\psi(w)$. The function $w=c\varphi(z)$ maps the domain B^1 on to the domain $|w|>c$. The quantity c is called the conformal radius of the domain B^1 with respect to the point $z=\infty$ or the capacity of the set $\bar{B}$. A more general concept for the capacity of a closed set will be given in Section 1.3.4. We shall also make use of the following lemma.

Lemma 1 If $\theta_n(z)$ is a sequence of functions regular in some domain B and uniformly bounded inside B, and is such that from any of its sub-sequences $\theta_{n_k}(z)$ a sub-sequence can be extracted which converges in B to some function $\varphi(z)$, then the whole sequence $\theta_n(z)$ uniformly converges inside B to $\varphi(z)$ and $\varphi(z)$ is regular in B.

PROOF Let us assume that at some point $z_0 \in B$ the sequence $\theta_n(z)$ does not converge to $\varphi(z_0)$. Let us extract from the sequence $\theta_n(z)$ a sub-sequence $\theta_{n_k}(z)$ such that $\theta_{n_k}(z_0) \to A \neq \varphi(z_0)$ as $k \to \infty$. But according to the premises we can extract from the sequence $\theta_{n_k}(z)$ a sub-sequence converging in B to $\varphi(z)$, and, in particular, at the point z_0 this subsequence tends to $\varphi(z_0)$. This contradiction shows that the sequence $\theta_n(z)$ converges in B to $\varphi(z)$. The uniform convergence of the sequence $\theta_n(z)$ inside B now follows from Vitali's theorem, and the lemma is proved.

NOTE 1 An analogous lemma is valid for a sequence $\vartheta_n(z)$ of functions which are harmonic in the domain B.

In the following an essential part will be played by the following functions defined in the domain B^1

$$\theta_n(z) = \frac{\sqrt[n+1]{\omega_n(z)}}{c\varphi(z)}, \qquad n = 0,\ 1,\ \ldots \tag{3}$$

We shall first ascertain the properties of these functions for any distribution of points (1) in $\bar{B}$. For the root in (3) we shall take that branch which in the neighbourhood of the point $z=\infty$ has the form

$$\sqrt[n+1]{\omega_n(z)} = \sqrt[n+1]{\prod_{k=1}^{n+1}\left(z - z_k^{(n)}\right)}$$
$$= z\sqrt[n+1]{\prod_{k=1}^{n+1}\left(1 - \frac{z_k^{(n)}}{z}\right)} = z + \alpha_0^{(n)} + \frac{\alpha_1^{(n)}}{z} + \ldots$$

The functions $\theta_n(z)$ are single-valued, regular in B^1 and do not become zero in B^1. This follows from the principle of single-valuedness (B^1 is a simply connected domain) and the fact that the zeros of the function $\omega_n(z)$ lie outside B^1. Besides $\theta_n(\infty)=1$.

Let us now introduce the notation:

$$M_n(\rho)=\sup_{z\in L_\rho}|\omega_n(z)|,\quad \rho>1,\quad M_n=\sup_{z\in L}|\omega_n(z)| \tag{4}$$

It is easy to see that $M_n(\rho)\to M_n$ as $\rho\to 1$. We have

$$\sup_{z\in L_\rho}|\theta_n(z)|=\frac{\sqrt[n+1]{M_n(\rho)}}{c\rho}$$

From the maximum modulus principle it follows that when ρ decreases the quantity $\dfrac{\sqrt[n+1]{M_n(\rho)}}{c\rho}$ does not diminish and consequently,

$$|\theta_n(z)|\leqslant\frac{\sqrt[n+1]{M_n}}{c},\quad z\in B^1 \tag{5}$$

We note certain properties of the functions $\theta_n(z)$ and the numbers M_n which are used later.

The quantity

$$d=d(\bar B)=\sup_{z_1,\, z_2\in\bar B}|z_1-z_2|$$

is called the diameter of the bounded closed set $\bar B$.

1. If d is the diameter of the set $\bar B$, then $|\omega_n(z)|\leqslant d^{n+1}$ for $z\in\bar B$ and, consequently, $M_n\leqslant d^{n+1}$. Hence, and from (5), we conclude that

$$|\theta_n(z)|\leqslant\frac{d}{c},\quad z\in B^1,\quad n=1,\,2,\,\ldots$$

e.g. the sequence of functions $\theta_n(z)$ is uniformly bounded in B^1.

2. Since $\theta_n(\infty)=1$, it follows, from (5), the maximum principle and the inequality $M_n\leqslant d^{n+1}$, that

$$1\leqslant\frac{\sqrt[n+1]{M_n}}{c}\leqslant\frac{d}{c}$$

$$n=1,\,2,\,\ldots$$

Remember that all the above is true for any distribution of points (1) on $\bar B$.

We shall also require the following two lemmas.

Lemma 2 The following inequalities hold

$$\left|\frac{t\psi'(t)}{\psi(t)-\psi(w)}\right| \leqslant \frac{|t|+|w|}{|t|-|w|}, \quad 1<|w|<|t|$$

$$\left|\frac{t\psi'(t)}{\psi(t)-z}\right| \leqslant \frac{|t|+1}{|t|-1}, \quad |t|>1, \quad z \in \bar{B}$$

PROOF The function

$$f(\zeta)=\frac{w}{c\left[\psi\left(\frac{w}{\zeta}\right)-\psi(w)\right]}$$

is regular and schlicht in the circle $|\xi|<1$ and $f(0)=0$, $f'(0)=1$, and therefore (see Goluzin [1] p. 146)

$$\left|\frac{\zeta f'(\zeta)}{f(\zeta)}\right| = \left|\frac{\frac{w}{\zeta}\psi'\left(\frac{w}{\zeta}\right)}{\psi\left(\frac{w}{\zeta}\right)-\psi(w)}\right| \leqslant \frac{1+|\zeta|}{1-|\zeta|}, \quad |\zeta|<1$$

Assuming that here $\zeta=\frac{w}{t}$, we obtain the first inequality. The second inequality is proved in an analogous fashion.

Lemma 3 If $\bar{B}$ is a set of type $\mathfrak{M}$ and $f(z)$ is a function regular in $\bar{B}_{\rho_0}$, $\rho_0>1$ then the residue term of interpolation has the value

$$|f(z)-p_n(z)| \leqslant M(\rho_0)\frac{\rho_0+r}{\rho_0-r}\,\frac{\sup\limits_{z\in L_r}|\omega_n(z)|}{\inf\limits_{z\in L_{\rho_0}}|\omega_n(z)|}, \quad z\in\bar{B}_r, \quad 1\leqslant r<\rho_0 \tag{6}$$

where

$$M(\rho_0)=\frac{1}{2\pi}\int_0^{2\pi}|f(\psi(\rho_0e^{i\theta}))|\,d\theta$$

PROOF From Equation 2, Section 1.1.2, it follows that

$$|f(z)-p_n(z)| \leqslant \frac{\sup\limits_{z\in L_r}|\omega_n(z)|}{\inf\limits_{z\in L_{\rho_0}}|\omega_n(z)|}\,\frac{1}{2\pi}\int_0^{2\pi}|f(\psi(\rho_0e^{i\theta}))|\left|\frac{\rho_0e^{i\theta}\psi'(\rho_0e^{i\theta})}{\psi(\rho_0e^{i\theta})-z}\right|d\theta$$

and Lemma 2 leads to value (6). The lemma is proved.

NOTE 2 In inequality (6) we can put

$$M(\rho_0)=\sup_{z\in L_{\rho_0}}|f(z)|$$

*Theorem 1 (basic convergence principle)** In order that a sequence of interpolating polynomials $p_n(z)$ for any function $f(z)$, regular on a set $\bar{B}$ of type $\mathfrak{M}$, shall uniformly converge to $f(z)$, it is necessary and sufficient that as $n \to \infty$

$$\left| \frac{\sqrt[n+1]{\omega_n(z)}}{c\varphi(z)} \right| = |\theta_n(z)| \rightrightarrows 1 \tag{7}$$

in B^1, or that

$$\lim_{n\to\infty} \frac{\sqrt[n+1]{M_n}}{c} = 1 \tag{8}$$

SUPPLEMENT If the function $f(z)$ is regular in B_{ρ_0}, $\rho_0 > 1$, and fulfils condition (7) or (8), then for any ρ_1, $1 < \rho_1 < \rho_0$,

$$|f(z) - p_n(z)| \leqslant \frac{M_{\rho_1}}{\rho_1^{n+1}}, \quad z \in \bar{B} \tag{9}$$

where M_{ρ_1} is a constant independent of n.

PROOF: NECESSITY Let $z_0\,(z_0 \neq \infty)$ be a point from B^1 and $\rho > 1$ so that $z_0 \in B^1_\rho$. The interpolating polynomial for the function $\frac{1}{z_0 - z}$, regular on $\bar{B}$, will be

$$p_n(z) = \frac{1}{2\pi i} \int_{L_\rho} \frac{\omega_n(\zeta) - \omega_n(z)}{(z_0 - \zeta)(\zeta - z)\,\omega_n(\zeta)}\, d\zeta$$

$$= \sum_{\zeta \in B_\rho} \operatorname{Res} \frac{\omega_n(\zeta) - \omega_n(z)}{(z_0 - \zeta)(\zeta - z)\,\omega_n(\zeta)} = -\sum_{\zeta \in B^1_\rho} \operatorname{Res} \frac{\omega_n(\zeta) - \omega_n(z)}{(z_0 - \zeta)(\zeta - z)\,\omega_n(\zeta)}$$

$$= \frac{\omega_n(z_0) - \omega_n(z)}{(z_0 - z)\,\omega_n(z_0)}$$

and the residue term of interpolation is

$$r_n(z) = \frac{1}{z_0 - z} - p_n(z) = \frac{\omega_n(z)}{(z_0 - z)\,\omega_n(z_0)} \tag{10}$$

We are proving the necessity for conditions (7) and (8), e.g.

* That condition (7) was necessary was proved by Kalmar [1], and that it was sufficient (in a more general formulation) by Walsch [2] (see also [1] page 192). That condition (8) was necessary and sufficient was proved by Fekete in unpublished work and later (in a more general formulation) by Walsch [3] (see also [1] page 203). Condition (8) was studied in detail by Geronimus [3].

let us suppose that $r_n(z) \rightrightarrows 0$ as $n \to \infty$ on $\bar{B}$, and from (10) it follows that

$$\lim_{n\to\infty} \frac{M_n}{|\omega_n(z_0)|} = 0$$

for any $z_0 \in B^1$, where M_n is defined by (4). Consequently, for every $z \in L_\rho$, $\rho > 1$, there is a number $N_z(\rho)$ such that for $n > N_z(\rho)$, $|\omega_n(z)| > M_n$ and this means

$$|\theta_n(z)| \geqslant \frac{\sqrt[n+1]{M_n}}{c\rho}, \quad z \in L_\rho, \quad n > N_z(\rho) \tag{11}$$

Let us now assume that (8) is not fulfilled and consequently there is a sub-sequence

$$\frac{\sqrt[n_k+1]{M_{n_k}}}{c} \to q > 1$$

as $k \to \infty$. From the sequence $\theta_{n_k}(z)$ we will choose a subsequence, uniformly converging inside B^1 to some function $\theta(z)$ regular in B^1. Without loss of generality we can assume that the sequence itself $\theta_{n_k}(z) \rightrightarrows \theta(z)$ as $k \to \infty$. Note that $\theta(\infty) = 1$. We choose $\rho > 1$ such that $q/\rho > 1$. Assuming that in (11) $n = n_k$ and that k tends to ∞, in the limit we have $|\theta(z)| \geqslant \frac{q}{\rho} > 1$, $z \in L_\rho$. But, on the strength of the minimum modulus principle, this contradicts the condition $\theta(\infty) = 1$ and (8) is proved.

Let us at this stage prove (7). If a sub-sequence $\theta_{n_k}(z)$ uniformly converges inside B^1 to some function $\theta(z)$ regular in B^1, then from (5) and (8) we conclude that $|\theta(z)| \leqslant 1$, $z \in B^1$. But $\theta(\infty) = 1$ and, consequently, $\theta(z) \equiv 1$ in B^1. Hence, from the fact that the sequence $\theta_n(z)$ is normal and from Lemma 1, we conclude that the whole sequence $\theta_n(z) \rightrightarrows 1$ in B^1, e.g. we have (7). The necessity is proved.

SUFFICIENCY Let (7) be fulfilled and the function $f(z)$ be regular in a set B and at the same time regular in some domain B_{ρ_0}, $\rho_0 > 1$. Let us arbitrarily set ρ_1 such that $1 < \rho_1 < \rho_0$, and choose ρ_2 and ρ_3 such that we have

$$1 < \rho_3 < \rho_1 < \rho_2 < \rho_0, \quad \rho_1\rho_3 < \rho_2$$

In the closed domain $\bar{D}$, bounded by the lines L_{ρ_2} and L_{ρ_3}, (7) holds, whence

$$|\omega_n(z)| = [c\,|\varphi(z)| + \varepsilon_n(z)]^{n+1}, \quad z \in \bar{D}$$

where $\varepsilon_n(z) \rightrightarrows 0$ as $n \to \infty$ on $\bar{D}$. Hence for $z \in L_{\rho_3}$ we have

$$|\omega_n(z)| \leqslant (c\rho_3 + \varepsilon_n)^{n+1}, \quad \varepsilon_n = \sup_{z \in L_{\rho_3}} |\varepsilon_n(z)|$$

and for $z \in L_{\rho_2}$

$$|\omega_n(z)| \geqslant (c\rho_2 - \varepsilon'_n)^{n+1}, \quad \varepsilon'_n = \sup_{z \in L_{\rho_2}} |\varepsilon_n(z)|$$

where ε_n and ε'_n tend to zero as $n \to \infty$.

From (6) on the strength of Lemma 3 and the maximum principle we have

$$|f(z) - p_n(z)| \leqslant \left[M(\rho_2) \frac{\rho_2 + 1}{\rho_2 - 1} \left(\frac{c\rho_3 + \varepsilon_n}{c\rho_2 - \varepsilon'_n} \rho_1 \right)^{n+1} \right] \frac{1}{\rho_1^{n+1}}, \quad z \in \bar{B} \qquad (12)$$

The expression in parentheses in Formula (12) tends to $\frac{\rho_3}{\rho_2}\rho_1$ as $n \to \infty$, and since $\frac{\rho_3}{\rho_2}\rho_1 < 1$, then the expression in the square bracket tends to zero as $n \to \infty$. At the same time for all n it is bounded by some quantity M_{ρ_1}, dependent on the choice of ρ_1 in the interval $1 < \rho_1 < \rho_0$, and from (12) we obtain inequality (9), from which it follows that $p_n(z)$ uniformly converges to $f(z)$ on $\bar{B}$ and that the supplement to the theorem is valid.

Let (8) now be fulfilled. As in the proof of the sufficiency we can show that as $n \to \infty$

$$\theta_n(z) \rightrightarrows 1 \qquad (13)$$

in B^1 which means $|\theta_n(z)| \rightrightarrows 1$ as $n \to \infty$ in B^1. Hence, as already shown, it follows that for any function $f(z)$, regular on $\bar{B}$, the sequence of interpolating polynomials $p_n(z)$ uniformly converges to $f(z)$ and the supplement to the theorem is valid. The sufficiency is proved.

NOTE 3 From the proof of the theorem it follows that conditions (7) and (8) are of equal strength. We shall make use of this in the future. Note also that (13) follows from (7). In fact, from (7) it follows that the sequence $\theta_n(z)$ is uniformly bounded on the compact sub-sets of B^1 and that every sub-sequence $\theta_{n_k}(z)$ converging uniformly on the compact sub-sets of B^1 converges to a function $\theta(z)$ regular in B^1 such that $|\theta(z)| \equiv 1$ in B^1. But $\theta(\infty) = 1$ and, consequently, $\theta(z) \equiv 1$. Hence we conclude that $\theta_n(z) \rightrightarrows 1$ in B^1.

NOTE 4 The whole proof of the theorem could have been carried out by using instead of sequence $\theta_n(z)$ the sequence $\vartheta_n(z) = \ln|\theta_n(z)|$ of functions harmonic in the domain B^1, and corresponding theorems on harmonic functions.

NOTE 5 From the proof of the theorem it follows that if the function $f(z)$ is regular in B_{ρ_0}, $\rho_0 > 1$, and bounded $|f(z)| \leqslant M, z \in B_{\rho_0}$, then in Formula (9) we can put $M_{\rho_1} = MC$, where C is a constant depending only on ρ_0 and ρ_1.

In the following paragraph we shall formulate a principle for the choice of interpolating points for which (7) is fulfilled and, consequently, (9) holds. To deal with this we require the following lemma.

Lemma 4 (Bernstein [1] page 21) We let $P(z) = a_n z^n + a_{n\ 1} z^{n-1} + \ldots$ be a function regular in a simply-connected domain B^1, containing the point $z = \infty$, except for the pole of order n at point ∞. If

$$\lim_{\rho \to 1+0} \sup_{z \in L_\rho} |P(z)| = M < \infty \tag{14}$$

where L_ρ is a contour line of B^1, then

$$|P(z)| \leqslant M\rho^n, \quad z \in L_\rho, \quad \rho > 1 \tag{15}$$

and the equal sign holds in the case, and only in the case, where $P(z) = M\eta\varphi(z)^n$, where η is a constant, $|\eta| = 1$.

PROOF Let us consider a function

$$f(w) = \frac{P(\psi(w))}{w^n} = \sum_{k=0}^{\infty} \frac{\delta_k}{w^k}$$

noting that $\psi(w) = cw + \alpha_0 + \frac{\alpha_1}{w} + \ldots$ is a function mapping $|w| > 1$ on to B^1 (inverse of $w = \varphi(z)$). This function f is regular in $|w| > 1$, and, on the strength of the maximum principle, the finite or infinite limit written below will exist. By virtue of (14) we have

$$\lim_{\rho \to 1+0} \sup_{|w|=\rho} |f(w)| = \lim_{\rho \to 1+0} \frac{\sup\limits_{|w|=\rho} |P(\psi(w))|}{\rho^n} = M$$

By virtue of the maximum principle for regular functions

$$|f(w)| = \left| \frac{P(\psi(w))}{w^n} \right| \leqslant M, \quad |w| > 1$$

and the equal sign holds in the case, and only in the case, when

$$\frac{P(\psi(w))}{w^n} \equiv M\eta, \quad |w| > 1$$

Equation (15) follows from this and the equal sign holds in the case, and only in the case, when

$$P(z) \equiv M\eta\,\varphi(z)^n, \quad z \in B^1$$

and the lemma is proved.

Theorem 2 (Bernstein [1] *)* If for a function $f(z)$, given on a set $\bar{B}$ of type $\mathfrak{M}$, for every natural n there is a polynomial $p_n(z)$ of a degree not greater than n such that

$$|f(z) - p_n(z)| \leqslant \frac{M}{\rho^{n+1}}, \quad z \in \bar{B}, \quad \rho > 1 \tag{16}$$

where M does not depend on z and n, then the function $f(z)$ is regular in the domain B_ρ.*

PROOF Let us denote $\tau_n(z) = f(z) - p_n(z)$. Then

$$|\tau_n(z)| \leqslant M\rho^{-n-1}, \; z \in \bar{B}$$

and so for $z \in \bar{B}$ we have

$$|p_{n+1}(z) - p_n(z)| = |\tau_n(z) - \tau_{n+1}(z)| \leqslant \frac{M}{\rho^{n+1}} + \frac{M}{\rho^{n+2}} \leqslant \frac{2M}{\rho^{n+1}}$$

In accordance with Lemma 4 we have

$$|p_{n+1}(z) - p_n(z)| \leqslant \frac{2M}{\rho^{n+1}}\,\rho_1^{n+1} = 2M\left(\frac{\rho_1}{\rho}\right)^{n+1}, \quad z \in \bar{B}_{\rho_1}, \; \rho_1 > 1$$

Hence it follows that for any ρ_1, $1 < \rho_1 < \rho$, the series

$$p_1(z) + (p_2(z) - p_1(z)) + \ldots + (p_{n+1}(z) - p_n(z)) + \ldots$$

uniformly converges in the domain $\bar{B}_{\rho_1}$. In accordance with Weirstrass's theorem, the sequence of polynomials $p_n(z)$ $n = 1, 2, \ldots$ uniformly converges to some function $f_*(z)$ regular in B_{ρ_1}. But on the strength of (16) the sequence of polynomials $p_n(z)$ converges in B to $f(z)$, e.g. $f(z) \equiv f_*(z)$

* More exactly, there is a function $f_*(z)$ regular in B_ρ and such that $f_*(z) = f(z)$ for $z \in B$.

in $\bar{B}$ and $f_*(z)$ is the analytical continuation of function $f(z)$ from $\bar{B}$ to B_{ρ_1}. In view of the arbitrariness of ρ_1, $1 < \rho_1 < \rho$, it follows that $f(z)$ is analytically continued in B_ρ. Such a continuation is unique. The theorem is proved.

From this theorem it follows that if the function $f(z)$ is regular in B and has on L_{ρ_0} a singular point the inequality (9) cannot hold for $\rho_1 > \rho_0$ not only for the interpolating polynomials but also for the polynomials of best approximation on $\bar{B}$.

1.3 PRINCIPLES OF CHOICE FOR INTERPOLATION POINTS

In this subsection we shall indicate some of the rules for choosing the points of interpolation ensuring that the conditions (7) and (8) obtained in the previous subsection are fulfilled.

1.3.1 First principle of choice (Fejer points)

Let $\theta_k^{(n)}$, $k = 1, 2, \ldots, n+1$; $n = 0, 1, \ldots$, be a sequence of points dividing the finite interval $[a, b]$:

$$a \leqslant \theta_1^{(n)} \leqslant \theta_2^{(n)} \leqslant \ldots \leqslant \theta_{n+1}^{(n)} \leqslant b$$

Let us denote for a fixed n, the number of points $\theta_k^{(n)}$ lying in the interval $[\alpha, \beta]$, $a \leqslant \alpha \leqslant \beta \leqslant b$, by $N^{(n)}(\alpha, \beta)$. If

$$\lim_{n\to\infty} \frac{N^{(n)}(a, c)}{n+1} = \frac{c-a}{b-a} \tag{1}$$

for every c from $[a, b]$, then we shall say that the points $\theta_k^{(n)}$ are uniformly distributed over the interval $[a, b]$. In particular, the points $\theta_k^{(n)}$ are uniformly distributed over $[a, b]$ if

$$\theta_{k+1}^{(n)} - \theta_k^{(n)} = \frac{b-a}{n+1}, \; k = 1, 2, \ldots, n; \; n = 1, 2, \ldots$$

Such points will be called equidistant on $[a, b]$.

It is easy to see that if points $\theta_k^{(n)}$ are equally distributed over the interval $[a, b]$, then for $a \leqslant \alpha \leqslant \beta \leqslant b$

$$\lim_{n\to\infty} \frac{N^{(n)}(\alpha, \beta)}{n+1} = \frac{\beta-\alpha}{b-a} \tag{2}$$

In fact, first let $\alpha = \beta$ and for the sake of being specific $\alpha > a$. Assuming that $\varepsilon > 0$ is sufficiently small, let us divide the obvious inequality

$$0 \leqslant N^{(n)}(\alpha, \beta) \leqslant N^{(n)}(a, \beta) - N^{(n)}(a, \alpha - \varepsilon)$$

by $n+1$ and let n tend to ∞:

$$0 \leqslant \overline{\lim_{n\to\infty}} \frac{N^{(n)}(\alpha, \beta)}{n+1} \leqslant \frac{\beta - a}{b-a} - \frac{\alpha - \varepsilon - a}{b-a} = \frac{\varepsilon}{b-a}$$

In view of the arbitrariness $\varepsilon > 0$ we conclude that equality (2) is true for $\alpha = \beta$. For $\beta > \alpha$ we have

$$N^{(n)}(\alpha, \beta) = N^{(n)}(a, \beta) - N^{(n)}(a, \alpha) + N^{(n)}(\alpha, \alpha)$$

Dividing this equality by $n+1$ and letting n tend to ∞, in the limit we obtain (2).

Clearly if (2) is fulfilled, then (1) is fulfilled, e.g. the points $\theta_k^{(n)}$ are uniformly distributed in the interval $[a, b]$.

Let us introduce points

$$\tilde{\theta}_k^{(n)} = a + \frac{b-a}{n+1}\left(k - \frac{1}{2}\right), \quad k = 1, 2, \ldots, n+1; n = 0, 1, \ldots,$$

equidistant on $[a, b]$.

Lemma 1 In order that the points $\theta_k^{(n)}$ shall be distributed equally over the interval $[a, b]$, it is necessary and sufficient that the difference $\theta_k^{(n)} - \tilde{\theta}_k^{(n)} \to 0$ as $n \to \infty$ uniformly with respect to k.

PROOF: NECESSITY Let us assume that the statement of the lemma is false. This means that while (1) holds, there is a fixed $\varepsilon > 0$ for which there exists an increasing sequence of natural numbers n_l, $l = 1, 2, \ldots,$ and corresponding numbers k_l such that

$$\left|\theta_{k_l}^{(n_l)} - \tilde{\theta}_{k_l}^{(n_l)}\right| \geqslant \varepsilon$$

From the sequence of numbers $\tilde{\theta}_{k_l}^{(n_l)}$ it is possible to extract a converging sub-sequence. To simplify the notation, we shall assume that the sequence itself converges to some quantity $\tilde{\theta}_*$, $a \leqslant \tilde{\theta}_* \leqslant b$. A converging sub-sequence can also be extracted from the sequence of numbers $\theta_{k_l}^{(n_l)}$. Let the sequence itself converge to some quantity θ_*, $a \leqslant \theta_* \leqslant b$.

It is clear that $|\theta_* - \tilde{\theta}_*| \geqslant \varepsilon$. To be specific, let us put $\theta_* - \tilde{\theta}_* \geqslant \varepsilon$. There exists $R > 0$ such that for $l > R$

$$\theta_{k_l}^{(n_l)} > \theta_* - \frac{\varepsilon}{3}, \quad \tilde{\theta}_{k_l}^{(n_l)} < \tilde{\theta}_* + \frac{\varepsilon}{3}$$

From the first inequality it follows that in the interval $\left[a, \theta_* - \frac{\varepsilon}{3}\right]$ there are fewer than k_l of the numbers $\theta_m^{(n_l)}, m = 1, \ldots, n_l + 1$, and we have

$$\frac{N^{(n_l)}\left(a, \theta_* - \frac{\varepsilon}{3}\right)}{n_l + 1} < \frac{k_l}{n_l + 1} = \frac{\tilde{\theta}_{k_l}^{(n_l)} - a}{b - a} + \frac{1}{2(n_l + 1)}$$

$$< \frac{\tilde{\theta}_* + \frac{\varepsilon}{3} - a}{b - a} + \frac{1}{2(n_l + 1)}$$

Letting l tend to ∞ in this inequality, by virtue of (1), in the limit we obtain $\theta_* - \frac{\varepsilon}{3} \leqslant \tilde{\theta}_* + \frac{\varepsilon}{3}$ or $\theta_* - \tilde{\theta}_* \leqslant \frac{2}{3}\varepsilon$, and at the same time $\theta_* - \tilde{\theta}_* \geqslant \varepsilon$. This contradiction proves the necessity.

SUFFICIENCY According to the premises, the difference $\theta_k^{(n)} - \tilde{\theta}_k^{(n)} \to 0$ uniformly with respect to k as $n \to \infty$. Consequently, the difference

$$\theta_{k+1}^{(n)} - \theta_k^{(n)} = \left(\theta_{k+1}^{(n)} - \tilde{\theta}_{k+1}^{(n)}\right) + \left(\tilde{\theta}_{k+1}^{(n)} - \tilde{\theta}_k^{(n)}\right) - \left(\theta_k^{(n)} - \tilde{\theta}_k^{(n)}\right), \quad k \leqslant n$$

also uniformly tends to zero with respect to k as $n \to \infty$. Let θ be any number from $[a, b]$, and for given n let the number of points $\theta_k^{(n)}$ not to the right of θ be k_n. Since $\theta_{k_n}^{(n)} \leqslant \theta < \theta_{k_n+1}^{(n)}$, $\theta_{k_n}^{(n)} \to \theta$ as $n \to \infty$. Further, since $N^{(n)}(a, \theta) = k_n$,

$$\frac{N^{(n)}(a, \theta)}{n + 1} = \frac{k_n}{n + 1} = \frac{\tilde{\theta}_{k_n}^{(n)} - a}{b - a} + \frac{1}{2(n + 1)}$$

$$= \frac{\tilde{\theta}_{k_n}^{(n)} - \theta_{k_n}^{(n)} + \theta_{k_n} - a}{b - a} + \frac{1}{2(n + 1)}$$

Letting n tend to ∞, in the limit we have (1), consequently the points $\theta_k^{(n)}$ are uniformly distributed over $[a, b]$, and the sufficiency is proved. It is easy to see that the lemma remains valid if it is formulated for the different points $\tilde{\theta}_k^{(n)}$, $k = 1, 2, \ldots, n + 1$, equidistant on $[a, b]$.

Lemma 2 Let $\theta_k^{(n)}$ be a sequence of points dividing the interval $[a, b]$. In order that $\theta_k^{(n)}$ shall be uniformly distributed over the interval $[a, b]$ it is necessary and sufficient that for any function $f(\theta)$, continuous in the interval $[a, b]$,

$$\lim_{n\to\infty} \frac{1}{n+1}\sum_{k=1}^{n+1} f\left(\theta_k^{(n)}\right) = \frac{1}{b-a}\int_a^b f(\theta)\,d\theta \tag{3}$$

PROOF: NECESSITY Let us introduce the quantities

$$\vartheta_k^{(n)} = a + \frac{b-a}{n+1}k, \qquad k = 0, 1, \ldots, n+1$$

We have

$$I_n = \frac{1}{n+1}\sum_{k=1}^{n+1} f\left(\theta^{(n)}\right) = \sum_{k=1}^{n+1} \frac{1}{b-a}\int_{\vartheta_{k-1}^{(n)}}^{\vartheta_k^{(n)}} f\left(\theta_k^{(n)}\right) d\theta$$

and

$$\frac{1}{b-a}\int_a^b f(\theta)\,d\theta - I_n = \sum_{k=1}^{n+1}\frac{1}{b-a}\int_{\vartheta_{k-1}^{(n)}}^{\vartheta_k^{(n)}} \left[f(\theta) - f\left(\theta_k^{(n)}\right)\right] d\theta \tag{4}$$

Since the function $f(\theta)$ is continuous in $[a, b]$, for any $\varepsilon > 0$ there is $\delta > 0$ such that, if $|\theta' - \theta''| < \delta$, θ' and $\theta'' \in [a, b]$, then

$$|f(\theta') - f(\theta'')| < \varepsilon$$

Since we are proving the necessity of condition (3) it is assumed that the points $\theta_k^{(n)}$ are uniformly distributed over $[a, b]$, and in accordance with Lemma 1 there exists N such that for $n > N$

$$\left|\theta_k^{(n)} - \tilde{\theta}_k^{(n)}\right| < \delta, \quad k = 1, 2, \ldots, n+1, \text{ and } \frac{b-a}{2(n+1)} < \delta$$

Then for any θ from $\left[\vartheta_{k-1}^{(n)}, \vartheta_k^{(n)}\right]$ we have

$$\left|f(\theta) - f\left(\theta_k^{(n)}\right)\right| \leqslant \left|f(\theta) - f\left(\tilde{\theta}_k^{(n)}\right)\right| + \left|f\left(\theta_k^{(n)}\right) - f\left(\tilde{\theta}_k^{(n)}\right)\right| < 2\varepsilon$$

Whence, and from (4), we obtain

$$\left|\frac{1}{b-a}\int_a^b f(\theta)\,d\theta - \frac{1}{n+1}\sum_{k=1}^{n+1} f\left(\theta_k^{(n)}\right)\right| \leqslant \sum_{k=1}^{n+1}\frac{1}{b-a}\int_{\vartheta_{k-1}^{(n)}}^{\vartheta_k^{(n)}} 2\varepsilon\,d\theta = 2\varepsilon$$

In view of the arbitrariness of ε the necessity is proved.

SUFFICIENCY Let us take any λ, $a \leqslant \lambda \leqslant b$, and sufficiently small $\varepsilon > 0$. Let us consider the function $y = f(\theta)$, whose graph is made up of straight-line segments connecting in succession the points

$$(a,\ 0),\ \left(a + \frac{\varepsilon}{2},\ 1\right),\ \left(\lambda - \frac{\varepsilon}{2},\ 1\right)(\lambda,\ 0),\ (b,\ 0)\ (\varepsilon < \lambda - a)$$

for $\lambda > a$, and $f(\theta) \equiv 0$ for $\lambda = a$. We have

$$I = \frac{1}{b-a} \int_a^b f(\theta)\, d\theta = \frac{\lambda - a - \frac{\varepsilon}{2}}{b-a}$$

$$I_n = \frac{1}{n+1} \sum_{k=1}^{n+1} f\left(\theta_k^{(n)}\right) \leqslant \frac{N^{(n)}(a,\ \lambda)}{n+1}$$

The latter inequality follows from the fact that $f\left(\theta_k^{(n)}\right) = 0$ at $\theta_k^{(n)} \geqslant \lambda$ and $f\left(\theta_k^{(n)}\right) \leqslant 1$ for $\theta_k^{(n)} < \lambda$. From the formulae obtained we conclude that

$$\frac{N^{(n)}(a,\ \lambda)}{n+1} - \frac{\lambda - a}{b-a} \geqslant (I_n - I) - \frac{\varepsilon}{2(b-a)}$$

Since according to the premises $\lim_{n \to \infty} I_n = I$ for sufficiently large n this inequality gives

$$\frac{N^{(n)}(a,\ \lambda)}{n+1} - \frac{\lambda - a}{b-a} \geqslant - \frac{\varepsilon}{b-a} \tag{5}$$

Let us now consider the function $y = f(\theta)$, the graph for which is made up of straight-line segments connecting in succession the points

$$(a,\ 1),\ (\lambda,\ 1),\ \left(\lambda + \frac{\varepsilon}{2},\ 0\right),\ \left(b - \frac{\varepsilon}{2},\ 0\right),\ (b,\ 1),\ (\varepsilon \leqslant b - \lambda)$$

for $\lambda < b$ and $f(\theta) \equiv 1$ for $\lambda = b$. We have

$$I = \frac{1}{b-a} \int_a^b f(\theta)\, d\theta = \frac{\lambda - a + \frac{\varepsilon}{2}}{b-a}$$

$$I_n = \frac{1}{n+1} \sum_{k=1}^{n+1} f\left(\theta_k^{(n)}\right) \geqslant \frac{N^{(n)}(a,\ \lambda)}{n+1}$$

The latter inequality follows from the fact that $f\left(\theta_k^{(n)}\right) = 1$ for $\theta_k^{(n)} \leqslant \lambda$ and $f\left(\theta_k^{(n)}\right) \geqslant 0$ for $\theta_k^{(n)} > \lambda$. From the formulae obtained it follows that

$$\frac{N^{(n)}(a, \lambda)}{n+1} - \frac{\lambda - a}{b - a} \leqslant I_n - I + \frac{\varepsilon}{2(b-a)}$$

and for sufficiently large n we obtain

$$\frac{N^{(n)}(a, \lambda)}{n+1} - \frac{\lambda - a}{b - a} \leqslant \frac{\varepsilon}{b-a}$$

Whence, and from (5), on the strength of the arbitrariness of ε, we conclude that (1) holds and, consequently, the points $\theta_k^{(n)}$ are uniformly distributed on $[a, b]$. The sufficiency is proved.

NOTE 1 The points $\theta_k^{(n)}$ are uniformly distributed on $[a, b]$ if (3) is fulfilled only for function $f(\theta)$, continuous on $[a, b]$ and satisfying the condition $f(a) = f(b)$ and such that every function $f_0(\theta)$, continuous on $[a, b]$ and satisfying the condition $f_0(a) = f_0(b)$, can be uniformly approximated as well as is desired on $[a, b]$ by functions of the type indicated.

Let D be an arbitrary, closed set of points of the extended plane z. The function $f(z, \theta)$, $z \in \bar{D}$, $a \leqslant \theta \leqslant b$, is said to be equicontinuous with respect to θ on $[a, b]$ for $z \in \bar{D}$, if for any $\varepsilon > 0$ there exists $\delta > 0$ such that, for any θ' and θ'' from $[a, b]$ such that $|\theta' - \theta''| < \delta$,

$$|f(z, \theta') - f(z, \theta'')| < \varepsilon$$

for all $z \in \bar{D}$. If the function $f(z, \theta)$ is continuous in (z, θ), $z \in \bar{D}$, $a \leqslant \theta \leqslant b$, then it is equicontinuous with respect to θ on $[a, b]$ for $z \in \bar{D}$.

From the proof of the necessity we obtain a corollary.

COROLLARY If the function $f(z, \theta)$ is equicontinuous with respect to θ on $[a, b]$ for $z \in \bar{D}$ and the points $\theta_k^{(n)}$ are uniformly distributed on $[a, b]$, then as $n \to \infty$

$$\frac{1}{n+1} \sum_{k=1}^{n+1} f\left(z, \theta_k^{(n)}\right) \rightrightarrows \frac{1}{b-a} \int_a^b f(z, \theta)\, d\theta, \quad z \in \bar{D} \tag{6}$$

Let the points $w_k^{(n)}$, $k = 1, 2, \ldots, n+1$; $n = 0, 1, \ldots$, lie on the circle $|w| = \rho$, $\rho \geqslant 1$. Let us introduce the points $\theta_k^{(n)} = \arg w_k^{(n)}$, $0 \leqslant \theta_k^{(n)} < 2\pi$, and let us repeat them at periods of 2π on the whole axis $-\infty < \theta < +\infty$. We can enumerate these latter in ascending order: $\vartheta_k^{(n)}$, $k = 0, \pm 1, \pm 2, \ldots$ and we shall say that the points $w_k^{(n)}$ are uniformly distributed (or equidistant) on the circle $|w|=\rho$, if the points $\vartheta_k^{(n)}$ (lying in the interval $[0, 2\pi]$) are uniformly distributed (respectively equidistant) over $[0, 2\pi]$. Note that if the points $\vartheta_k^{(n)}$ are equally distributed on $[0, 2\pi]$, then they are equally distributed on every interval $[a, a+2\pi]$, $-\infty < a < +\infty$, of length 2π. This follows easily from the fact that conditions (1) and (2) are of equal strength.

We shall call a set of type $\mathfrak{M}$ a set of type $\tilde{\mathfrak{M}}$ if all the points of the boundary of the domain B^1, complementary for $\bar{B}$ with respect to the extended plane, are accessible from B^1 (see Goluzin [1] pp. 40-43, 49, 50). In this case the function $z=\psi(w)$ introduced above is continuous in the domain $|w|\geqslant 1$, except for the point $w=\infty$.

We shall call a set of type $\mathfrak{M}$ a set of type $\overline{\mathfrak{M}}$ if the boundary of B^1 is a closed Jordan curve. It is known that if $\bar{B}$ is a set of type $\overline{\mathfrak{M}}$, then $\bar{B}$ is also a set of type $\tilde{\mathfrak{M}}$, but not always the reverse (see Goluzin [1] p. 41).

Let B be a set of type $\mathfrak{M}$ and let the points $w_k^{(n)}$ lying on the circle $|w|=\rho$, $\rho>1$, be equally distributed (equidistant) on $|w|=\rho$. In this case we shall say that the points $z_k^{(n)} =\psi(w_k^{(n)})$, lying on L_ρ are equally distributed on L_ρ (respectively uniformly divided). If $\bar{B}$ is a set of type $\tilde{\mathfrak{M}}$, then this definition will also hold for $\rho=1$.

Theorem 1 Let $\bar{B}$ be a set of type $\overline{\mathfrak{M}}$ and let the points

$$z_k^{(n)},\ k=1,\ 2,\ \ldots,\ n+1;\ n=1,\ 2,\ \ldots$$

lie on the boundary L of the domain B^1. In order that the points shall be uniformly distributed on L it is necessary and sufficient that as $n\to\infty$

$$|\theta_n(z)|\rightrightarrows 1 \tag{7}$$

in B^1 (the functions $\theta_n(z)$ are defined by Formula (3), Section 1.2).

PROOF: NECESSITY (Fejer [1]) Let $w_k^{(n)}=\varphi(z_k^{(n)})=e^{i\theta_k^{(n)}}$. Let us introduce the function

$$f(z,\ \theta)=\ln|z-\psi(e^{i\theta})|,\quad 0\leqslant\theta\leqslant 2\pi;\quad z\in L_\rho;\quad \rho>1$$

Since this function is equicontinuous with respect to θ on $[0, 2\pi]$ for $z\in L_\rho$, $\rho>1$, and since in addition the points $\theta_k^{(n)}$

are uniformly distributed on $[0, 2\pi]$ on the basis of the corollary to Lemma 2 we conclude that, as $n \to \infty$

$$\frac{1}{n+1} \sum_{k=1}^{n+1} \ln \left| z - \psi\left(e^{i\theta_k^{(n)}}\right) \right| \rightrightarrows \frac{1}{2\pi} \int_0^{2\pi} \ln | z - \psi(e^{i\theta}) | \, d\theta \tag{8}$$

on L_ρ, $\rho > 1$. We have

$$\frac{1}{n+1} \sum_{k=1}^{n+1} \ln \left| z - \psi\left(e^{i\theta_k^{(n)}}\right) \right| = \frac{1}{n+1} \sum_{k=1}^{n+1} \ln | z - z_k^{(n)} | = \ln \sqrt[n+1]{|\omega_n(z)|}$$

and putting $z = \psi(w)$, we obtain

$$\frac{1}{2\pi} \int_0^{2\pi} \ln | z - \psi(e^{i\theta}) | \, d\theta = \operatorname{Re} \left\{ \frac{1}{2\pi i} \int_{|\zeta|=1} \ln [\psi(w) - \psi(\zeta)] \frac{d\zeta}{\zeta} \right\}$$

$$= \operatorname{Re} \left\{ \frac{1}{2\pi i} \int_{|\zeta|=1} \ln \frac{\psi(w) - \psi(\zeta)}{w - \zeta} \frac{d\zeta}{\zeta} + \frac{1}{2\pi i} \int_{|\zeta|=1} \ln (w - \zeta) \frac{d\zeta}{\zeta} \right\}$$

$$= \operatorname{Re} \{ \ln \psi'(\infty) + \ln w \} = \ln [c | \varphi(z) |], \quad c = \psi'(\infty) = \frac{1}{\varphi'(\infty)} > 0$$

The last two integrals are evaluated by means of the residue theorem applied to the domains $|\zeta| > 1$ and $|\zeta| < 1$. From this it follows that condition (8) is equivalent to the fact that, as $n \to \infty$

$$\ln \frac{\sqrt[n+1]{|\omega_n(z)|}}{c | \varphi(z) |} \rightrightarrows 0, \quad \text{i. e.} \quad |\theta_n(z)| \rightrightarrows 1$$

on L_ρ and consequently on $\bar{B}_\rho^1$ for any $\rho > 1$ e.g. in B^1. The necessity of condition (7) is proved.

SUFFICIENCY (Kalmar [1]) According to the premises, the points $z_k^{(n)}$ are distributed on L and fulfil condition (7). Whence, as we saw above, it follows that, as $n \to \infty$

$$\theta_n(z) \rightrightarrows 1 \tag{9}$$

in B^1. The functions $\theta_n(z)$ do not become zero in the simply connected domain B^1 and as a result the functions $\ln \theta_n(z)$, normalised by the condition $\ln \theta_n(\infty) = 0$, are single-valued and regular in B^1 and, on the basis of (9), as $n \to \infty$

$$\ln \theta_n(z) = \ln \frac{\sqrt[n+1]{\omega_n(z)}}{c\varphi(z)} \rightrightarrows 0$$

in B^1. Whence by differentiation we find that, as $n \to \infty$

$$\frac{1}{n+1} \sum_{k=1}^{n+1} \frac{1}{z - z_k^{(n)}} \rightrightarrows \frac{\varphi'(z)}{\varphi(z)} \tag{10}$$

in B^1. Now let $f(z)$ be an arbitrary function regular in the domain $\bar{B}_\rho$, $\rho > 1$. Multiplying (10) by the function $f(z)$ and integrating with respect to L_ρ, we have

$$\lim_{n\to\infty} \frac{1}{n+1} \sum_{k=1}^{n+1} f(z_k^{(n)}) = \frac{1}{2\pi i} \int_{L_\rho} f(z) \frac{\varphi'(z)}{\varphi(z)} dz$$

Taking into consideration that $z_k^{(n)} = \psi\left(e^{i\theta_k^{(n)}}\right)$ and assuming that in the integral $z = \psi(\rho e^{i\theta})$, we have

$$\lim_{n\to\infty} \frac{1}{n+1} \sum_{k=1}^{n+1} f\left(\psi\left(e^{i\theta_k^{(n)}}\right)\right) = \frac{1}{2\pi} \int_0^{2\pi} f(\psi(\rho e^{i\theta}))\, d\theta$$

Whence it follows that, as $n \to \infty$

$$\frac{1}{n+1} \sum_{k=1}^{n+1} f\left(\psi\left(e^{i\theta_k^{(n)}}\right)\right) \to \frac{1}{2\pi} \int_0^{2\pi} f(\psi(e^{i\theta}))\, d\theta \tag{11}$$

for every function $f(z)$ regular on $\bar{B}$.

Note that the set $\bar{B}$ of type $\mathfrak{M}$ satisfies the conditions of Mergelian's Lemma 2, Section 1.7.3, and therefore every function which is regular in $B (B = C\bar{B^1})$ and continuous on $\bar{B}$ can be uniformly approximated on $\bar{B}$ as well as is desired by means of functions regular on $\bar{B}$, whence we can clearly conclude that Formula (11) holds for every function regular in B and continuous on $\bar{B}$ (the same result can be obtained by means of Rado's theorem, see Goluzin [1] p. 67).

Let us now introduce a sub-class $\mathscr{K}$ of such functions $f(z)$. Let the function $\xi = \lambda(z)$ be schlicht and let it conformally map the domain B on to the circle $|\xi| < 1$. Note that the function $\xi = \lambda(z)$ is regular in B and continuous on $\bar{B}$, and the inverse function $z = \mu(\xi)$ is regular in the circle $|\xi| < 1$ and continuous on $|\xi| \leqslant 1$ (see Goluzin [1] p. 50). It is known (see Gakhov [1] pp. 374-378)* that there is a function $g(\xi) = u(\xi) + iv(\xi)$ which is regular in $|\xi| < 1$

*The indicated monograph formulates the continuity only for $u(\xi)$ in $\bar{B}$, but from the proof it follows directly that under the given circumstances $v(\xi)$ is also continuous in $\bar{B}$.

and continuous on $|\xi| \leqslant 1$ such that its real part $u(\xi)$ is an arbitrary analytical function on the circle $|\xi|=1$ given in advance. From this it follows that the function $f(z)=g(\lambda(z)) = u(\lambda(z))+iv(\lambda(z))$ is regular in B and continuous on $\bar{B}$. Functions $f(z)$ of this type also form a class $\mathscr{K}$. Since by using analytic functions on the circle $|\xi|=1$ it is possible to approximate uniformly as well as is desired every function continuous on the circle $|\xi|=1$ (see Natanson [1] p. 26), by using functions

$$\chi(\theta)=\operatorname{Re} f(\psi(e^{i\theta}))=u(\lambda(\psi(e^{i\theta}))), \quad f(z)\in\mathscr{K}$$

it is possible to approximate uniformly as well as is desired any real function continuous in the interval $[0, 2\pi]$ $\chi_0(\theta)$ satisfying the condition $\chi_0(0)=\chi_0(2\pi)$. By writing (11) for $f(z)$ and taking the real section of both parts we obtain

$$\frac{1}{n+1}\sum_{k=1}^{n+1}\chi(\theta)\to\frac{1}{2\pi}\int_0^{2\pi}\chi(\theta)\,d\theta$$

Whence, from Lemma 2 and the associated Note 1, we conclude that the points $\theta_k^{(n)}$ are uniformly distributed over the interval $[0, 2\pi]$. This means that the points $z_k^{(n)}=\psi\left(e^{i\theta_k^{(n)}}\right)$ are uniformly distributed on L and the theorem is proved.

NOTE 2 It is obvious that the necessity for the conditions in the theorem still holds in the case where $\bar{B}$ is a set of type $\mathfrak{M}$. The sufficiency similarly holds for certain more special concepts of 'points uniformly distributed on L', but we shall not dwell on this.

COROLLARY If a function $f(z)$ regular in the closed set $\bar{B}$ of the type $\mathfrak{M}$ is regular in the domain B_{ρ_0}, $\rho_0>1$, then for the sequence of interpolating polynomials $p_n(z)$, constructed for the Fejer points for any ρ_1, $1<\rho_1<\rho_0$, we have the estimate

$$|f(z)-p_n(z)|\leqslant\frac{M_{\rho_1}}{\rho_1^{n+1}}, \quad z\in\bar{B} \tag{12}$$

where M_{ρ_1} is a constant independent of n.

Let us consider a particular case of this corollary. Let $\bar{B}$ be a segment of the real axis $-1\leqslant x\leqslant 1$. The Zhoukovskiy function $z=\frac{1}{2}\left(w+\frac{1}{w}\right)$ maps the domain $|w|>1$ on to the exterior of the segment $\bar{B}$ and $w=\infty$ corresponds with $z=\infty$. The external contour L_ρ, $\rho>1$, will be

$$z=\frac{1}{2}\left(\rho e^{i\varphi}+\frac{1}{\rho e^{i\varphi}}\right)$$

or, putting $z=x+iy$, we have

$$L_\rho:\begin{cases} x=\left(\rho+\frac{1}{\rho}\right)\cos\varphi, \\ y=\left(\rho-\frac{1}{\rho}\right)\sin\varphi, \end{cases} \qquad 0\leqslant\varphi<2\pi$$

Thus the contour line L_ρ is an ellipse with semi-axes $a=\rho+\frac{1}{\rho}$ and $b=\rho-\frac{1}{\rho}$ and foci at the points -1 and 1.

To the points

$$w_k^{(n)}=e^{\frac{4k+1}{2n}\pi i},\ k=0,1,\dots,n-1;\ n=1,2,\dots$$

equidistant on the circle $|w|=1$ there correspond the points

$$x_k^{(n)}=\frac{1}{2}\left(e^{\frac{4k+1}{2n}\pi i}+e^{-\frac{4k+1}{2n}\pi i}\right)=\cos\frac{4k+1}{2n}\pi$$

on the segment $\bar{B}$. They can be enumerated more easily by putting

$$x_k^{(n)}=\cos\frac{2k+1}{2n}\pi, \qquad k=0,1,\dots,n-1;\ n=1,2,\dots$$

From the corollary we find that if the function $f(z)$ regular on the closed segment $-1\leqslant x\leqslant 1$, is in fact regular inside the ellipse L_{ρ_0}, $\rho_0>1$, then for the sequence of interpolating polynomials $p_{n-1}(z)$ with interpolating points at $x_k^{(n)}$, $k=0,1,\dots,n-1$,

$$|f(z)-p_{n-1}(z)|\leqslant\frac{M_{\rho_1}}{\rho_1^n}, \qquad -1\leqslant x\leqslant 1$$

where ρ_1 and M_{ρ_1} have the same meaning as in the corollary.

For the solution of several problems the value of the quantity M_{ρ_1}, appearing in (12), is important. This can be found if the boundary L of the set $\bar{B}$ of type $\mathfrak{M}$ satisfies certain conditions and the interpolating points are uniformly distributed on L. For this we require several lemmas.

Lemma 3 Let

$$F(t, w) = \ln \frac{\psi(t) - \psi(w)}{c(t-w)}, \quad |t| > 1; \quad |w| > 1$$

Then for $r > 1$ we have

$$\left| \frac{\partial^m}{\partial \theta^m} F(re^{i\theta}, w) \right| \leqslant \frac{\sqrt{(2m-1)!}\, r^{m-1}}{(r^2-1)^m} \sqrt{\ln \frac{|w|^2}{|w|^2-1}} \tag{13}$$
$$m = 1, 2, \ldots$$

PROOF Any branch of the function $F(t, w)$ is regular in the domain $|t| > 1$ for every $w, |w| > 1$, and for the coefficients of its expansion

$$F(t, w) = \sum_{k=1}^{\infty} \frac{a_k(w)}{t^k}, \quad |w| > 1, \quad |t| > 1$$

there is the inequality (see Goluzin [1] p. 155)

$$\sum_{k=1}^{\infty} k\, |a_k(w)|^2 \leqslant \ln \frac{|w|^2}{|w|^2-1} \tag{14}$$

By using Cauchy's inequality and the well-known inequality

$$\sum_{k=1}^{\infty} k^m x^k \leqslant \frac{m!\, x}{(1-x)^{m+1}}, \quad 0 < x < 1, \quad m = 0, 1, \ldots$$

(easily proved by induction) we have

$$\left| \frac{\partial^m}{\partial \theta^m} F(re^{i\theta}, w) \right| \leqslant \sum_{k=1}^{\infty} \frac{k^m |a_k(w)|}{r^k} = \sum_{k=1}^{\infty} \frac{k^{m-\frac{1}{2}}}{r^k} \sqrt{k}\, |a_k(w)|$$
$$\leqslant \sqrt{\sum_{k=1}^{\infty} \frac{k^{2m-1}}{r^{2k}} \sum_{k=1}^{\infty} k\, |a_k(w)|^2} \leqslant \sqrt{\frac{(2m-1)!\, r^{2m-2}}{(r^2-1)^{2m}} \ln \frac{|w|^2}{|w|^2-1}}$$

The lemma is proved.

Lemma 4 If the function $f(x)$ has continuous derivatives up to the m th order, $(m = 0, 1, \ldots)$ for $-h \leqslant x \leqslant h$, $h > 0$, then

$$f(0) = \int_{-h}^{h} f(x)\, dx + \sum_{\nu=1}^{m} \frac{2^\nu B_\nu\left(\frac{1}{2}\right)}{\nu!} \left[f^{(\nu-1)}(h) - f^{(\nu-1)}(-h)\right] + R_m \tag{15}$$

where $B_\nu(x)$ is a Bernoulli polynomial of degree ν (see Stefenson [1]) and

$$R_m = -h^{m+1} \int_0^1 \frac{2^m B_m\left(\frac{1+\tau}{2}\right)}{m!} [f(h(1-\tau)) + (-1)^m f(-h(1-\tau))]\, d\tau \tag{16}$$

PROOF Let us make use of the general formula of Euler-Maclaurin (see Steffensen [1] p. 133):

$$\varphi(0) = \int_{-\frac{1}{2}}^{\frac{1}{2}} \varphi(x)\, dx + \sum_{\nu=1}^{m} \frac{B_\nu\left(\frac{1}{2}\right)}{\nu!} \left[\varphi^{(\nu-1)}\left(\frac{1}{2}\right) - \varphi^{(\nu-1)}\left(-\frac{1}{2}\right)\right] + R_m^*$$

$$R_m^* = -\int_0^1 \frac{\bar{B}_m\left(\frac{1}{2} - t\right)}{m!} \varphi^{(m)}\left(-\frac{1}{2} + t\right) dt$$

where $\bar{B}_m(x)$ is the periodic Bernoulli function of order m. Let us investigate the integral determining R_m^* as the sum of two integrals in the intervals $\left[0, \frac{1}{2}\right]$ and $\left[\frac{1}{2}, 1\right]$ and in the first put $t = \frac{1}{2}\tau$, and in the second $t = 1 - \frac{1}{2}\tau$:

$$R_m^* = -\frac{1}{2} \int_0^1 \left[\frac{B_m\left(\frac{1+\tau}{2}\right)}{m!} \varphi\left(\frac{1}{2}(1-\tau)\right) + \frac{B_m\left(\frac{1-\tau}{2}\right)}{m!} \varphi\left(-\frac{1}{2}(1-\tau)\right)\right] d\tau = -\frac{1}{2} \int_0^1 \frac{B_m\left(\frac{1+\tau}{2}\right)}{m!} \times \left[\varphi\left(\frac{1}{2}(1-\tau)\right) + (-1)^m \varphi\left(-\frac{1}{2}(1-\tau)\right)\right] d\tau$$

Now assuming $\varphi(x) = f(2hx)$, we reach the conclusion of the lemma.

NOTE 3 If m is even, $(m = 2, 4, \ldots)$,, then

$$\int_0^1 B_m\left(\frac{1+\tau}{2}\right) d\tau = \frac{2}{m+1} B_{m+1}\left(\frac{1+\tau}{2}\right)\Big|_0^1 = \frac{2}{m+1} [B_{m+1}(1) - B_{m+1}(0)] = 0$$

and therefore in this case we can add the term $-2f(0)$ to the expression in the square brackets in Formula (16).

This means it is possible to obtain the following upper estimates for $|R_m|$:

(a) if $f^{(m)}(x)$ has a bounded variation $\bigvee_{-h}^{h} f^{(m)}(x)$ occurring in the interval $[-h, h]$, then

$$|R_m| \leqslant \alpha_m (2h)^{m+1} \bigvee_{-h}^{h} f^{(m)}(x)$$

(b) if at all points of the interval $(-h, h)$ there exists the $(m+1)$th derivative and $M(f^{(m+1)}) = \sup_{-h \leqslant x \leqslant h} |f^{(m+1)}(x)|$

$$|R_m| \leqslant \beta_m (2h)^{m+2} M(f^{(m+1)})$$

where

$$\alpha_m = \int_0^1 \frac{1}{m!} \left| B_m\left(\frac{1+\tau}{2}\right) \right| d\tau, \quad \beta_m = \int_0^1 \frac{1}{m!} \left| B_m\left(\frac{1+\tau}{2}\right) \right| (1-\tau)\, d\tau \quad (17)$$

If $m = 2k-1$ is odd, then

$$\alpha_{2k-1} = \frac{4}{(2k)!}\left(1 - \frac{1}{2^{2k}}\right)|B_{2k}| = 8\left(1 - \frac{1}{2^{2k}}\right)\frac{1}{(2\pi)^{2k}} \sum_{\nu=1}^{\infty} \frac{1}{\nu^{2k}} \leqslant \frac{4}{3}\pi^2 \frac{1}{(2\pi)^{2k}}$$

$$\beta_{2k-1} = \frac{2}{(2k)!}\left(1 - \frac{1}{2^{2k-1}}\right)|B_{2k}|$$

$$= 4\left(1 - \frac{1}{2^{2k-1}}\right)\frac{1}{(2\pi)^{2k}} \sum_{\nu=1}^{\infty} \frac{1}{\nu^{2k}} \leqslant \frac{2}{3}\pi^2 \frac{1}{(2\pi)^{2k}}$$

From Lemma 4 and Note 3 we easily obtain the following lemma.

Lemma 5 Let the points θ_k, $k = 1, 2, \ldots, n$, $n = 1, 2, \ldots$ be equidistant in the interval $[a, b]$, i.e. $\theta_k = a + \frac{b-a}{2n}(2k-1)$, and let the function $f(\theta)$ have continuous derivatives of the mth order, $m = 0, 1, \ldots$ on $[a, b]$ and $f^{(2\nu-1)}(b) = f^{(2\nu-1)}(a)$ for all whole ν, $1 \leqslant 2\nu - 1 < m$. Then:

1. If $f^{(m)}(\theta)$ has a bounded variation $\bigvee_a^b f^{(m)}(x)$ on $[a, b]$, then

$$\left| \frac{1}{n} \sum_{k=1}^{n} f(\theta_k) - \frac{1}{b-a} \int_a^b f(\theta)\, d\theta \right| \leqslant \alpha_m \frac{(b-a)^m}{2n^{m+1}} \bigvee_a^b f^{(m)}(\theta)$$

2. If at all points of the interval (a, b) there exists the

$(m+1)$th derivative and $M(f^{(m+1)}) = \sup_{a \leqslant x \leqslant b} |f^{(m+1)}(x)|$, then

$$\left| \frac{1}{n} \sum_{k=1}^{n} f(\theta_k) - \frac{1}{b-a} \int_a^b f(\theta)\, d\theta \right| \leqslant \beta_m \frac{(b-a)^{m+1}}{2n^{m+1}} M(f^{(m+1)})$$

Here α and β are defined by Formulae (17).

Lemma 6 Let $\bar{B}$ be a set of type $\mathfrak{M}$ and the points

$$z_n^{(k)},\ k = 1, 2, \ldots, n+1;\ n = 1, 2, \ldots$$

be uniformly distributed on L_r, $r > 1$. Then for $\rho > r$ we have

$$\frac{\sup\limits_{z \in L_r} |\omega_n(z)|}{\inf\limits_{z \in L_\rho} |\omega_n(z)|} \leqslant \frac{2r^{n+1}}{\rho^{n+1} - r^{n+1}} e^{\frac{A_m(r)}{(n+1)^{2m-1}}}, \quad m = 1, 2, \ldots$$

where

$$A_m(r) = \frac{2\pi^2}{3} \frac{\sqrt{(4m-1)!}\, r^{2m-1}}{(r^2-1)^{2m}} \sqrt{\ln \frac{r^2}{r^2-1}} \tag{18}$$

PROOF Let $w_k^{(n)} = \varphi(z_k^{(n)}) = re^{i\theta_k^{(n)}}$ and a be the half-sum of any two successive numbers $\theta_k^{(n)}$. Writing $F(re^{i\theta}, w) = \ln \dfrac{\psi(re^{i\theta}) - \psi(w)}{c(re^{i\theta} - w)}$, $|w| > 1$, by virtue of Lemmas 3 and 4 and Note 3, we have

$$\left| \frac{1}{n+1} \sum_{k=1}^{n+1} \ln \frac{\psi\left(re^{i\theta_k^{(n)}}\right) - \psi(w)}{c\left(re^{i\theta_k^{(n)}} - w\right)} - \frac{1}{2\pi} \int_a^{a+2\pi} \ln \frac{\psi(re^{i\theta}) - \psi(w)}{c(re^{i\theta} - w)}\, d\theta \right|$$

$$\leqslant \frac{2}{3} \pi^2 \frac{1}{(2\pi)^{2m}} \frac{(2\pi)^{2m}}{2(n+1)^{2m}} \frac{\sqrt{(4m-1)!}\, r^{2m-1}}{(r^2-1)^{2m}} \sqrt{\ln \frac{|w|^2}{|w|^2 - 1}}$$

By arguing in the same manner as in the proof of Theorem 1, we can show that the integral within the absolute value sign is equal to zero, and since the function inside the absolute value sign is regular in the domain $|w| > 1$, we find

$$\left|\ln \frac{\omega_n(\psi(w))}{c^{n+1}(w^{n+1}-\eta r^{n+1})}\right| \leqslant \frac{A_m(r)}{2(n+1)^{2m-1}}, \quad |w| \geqslant r, \quad \eta = e^{i\sum_{k=1}^{n+1}\theta_k^{(n)}}$$

Whence the conclusion of the lemma easily follows.

Analogously we can prove the following.

Lemma 7 Let $\bar{B}$ be a set of type $\tilde{\mathfrak{M}}$ such that the function $F_*(t, w) = \ln\left|\frac{\psi(t)-\psi(w)}{c(t-w)}\right|$ is continuous in the domain $|t| \geqslant 1$ for every fixed w, $|w| > 1$, and the variation

$$\bigvee_0^{2\pi} F_*(re^{i\theta}, w) \leqslant V$$

for $|w| > 1$. If the interpolating points $z_k^{(n)}$, $k = 1, 2, \ldots, n+1$; $n = 1, 2, \ldots,$ are uniformly distributed on the boundary L of the set B, then for every $\rho > 1$ we have

$$\frac{\sup\limits_{z \in L} |\omega_n(z)|}{\inf\limits_{z \in L_\rho} |\omega_n(z)|} \leqslant 2e^V \frac{1}{\rho^{n+1}-1}$$

Whence (see inequality (6), Section 1.2 and Note 2 attached to it) follows

Theorem 2 Let the conditions of Lemma 7 be fulfilled. If the function $f(z)$ is regular in the domain B_{ρ_0}, $\rho_0 > 1$, and

$$\frac{1}{2\pi}\int_0^{2\pi} |f(\psi(\rho e^{i\theta}))|\, d\theta \leqslant M, \; 1 < \rho < \rho_0$$

then for the interpolating polynomials $p_n(z)$, $n = 1$, constructed for points uniformly distributed on L,

$$|f(z) - p_n(z)| \leqslant 2Me^V \frac{\rho_0+1}{\rho_0-1}\, \frac{1}{\rho_0^{n+1}-1}, \qquad z \in \bar{B}$$

Analogously on the basis of Lemma 6 we can show that the following is valid

Theorem 3 Let $\bar{B}$ be a set of type $\mathfrak{M}$. If then the function $f(z)$ is regular in the domain B_{ρ_0}, $\rho_0 > 1$, and if in addition

$$\frac{1}{2\pi}\int_0^{2\pi}|f(\psi(\rho e^{i\theta}))|\,d\theta\leqslant M,\ 1<\rho<\rho_0$$

then for the interpolating polynomials $p_n(z)$, $n=1,2,\ldots$ constructed for uniformly distributed points on L_r, $1<r<\rho_0$,

$$|f(z)-p_n(z)|\leqslant M\frac{\rho_0+r}{\rho_0-r}e^{\frac{A_m(r)}{(n+1)^{2m-1}}}\frac{2r^{n+1}}{\rho_0^{n+1}-r^{n+1}},\quad z\in\bar{B}_r \tag{19}$$
$$m=1,\ 2,\ \ldots,$$

where $A_m(r)$ is defined by Formula (18).

Theorem 4 Let B be a domain the boundary L of which is a closed analytical Jordan curve. If the function $f(z)$ is regular in the domain B_{ρ_0}, $\rho_0>1$, and

$$\frac{1}{2\pi}\int_0^{2\pi}|f(\psi(\rho e^{i\theta}))|\,d\theta\leqslant M,\ 1<\rho<\rho_0$$

then for the interpolating polynomials $p_n(z)$, $n=1,2,\ldots$, constructed for points uniformly distributed on L,

$$|f(z)-p_n(z)|\leqslant M\frac{\rho_0+1}{\rho_0-1}e^{\frac{A_m}{(n+1)^{2m-1}}}\frac{2}{\rho_0^{n+1}-1},\quad z\in\bar{B} \tag{20}$$
$$m=1,\ 2,\ \ldots,$$

where $\underline{A_m}$ depend only on m and the domain B. If the function $z=\psi(w)$ is regular and schlicht in a domain $\frac{1}{r}<|w|<+\infty$, $r>1$, then we can put $A_m=A_m(r)$, where $A_m(r)$ is defined by Formula (18).

PROOF The function $z=\psi(w)$ is regular and schlicht in a domain $\frac{1}{r}<|w|<+\infty$, $r>1$ (see Goluzin [1] p. 51), and maps the domain $|w|>\frac{1}{r}$ on to the complement B_*^1 of some set $\bar{B}_*\subset B$ of type $\mathfrak{M}$, and, further, the function $z=\psi_*(w)=\psi\left(\frac{w}{r}\right)$ represents a schlicht mapping of the domain $|w|>1$ on to B_*^1 Moreover, the contour line L_ρ^* for $\bar{B}_*$ coincides with the contour line $L_{\frac{\rho}{r}}$ for $\bar{B}$, $\rho>r$. Applying Theorem 3 to the set $\bar{B}_*$, we easily reach the conclusion of Theorem 4.

NOTE 4 In Formulae (19) and (20) for any given n we can choose m such that the right-hand side has the smallest possible value.

In the last three theorems we estimated $|f(z)-p_n(z)|$ in domains with fairly good boundaries. If the conditions of Theorem 3 are fulfilled and we wish to estimate $|f(z)-p_n(z)|$ only on $\bar{B}$, then the factor $\frac{\rho_0+r}{\rho_0-r}$ in the inequality (19) can be altered to $\frac{\rho_0+1}{\rho_0-1}$ and, for every n, points $z_k^{(n)}$ can be given on some curve L_{r_n}, $1<r_n<\rho_0$, chosen such that the right-hand side of the inequality (19) is a minimum or is close to a minimum. Let r_n be the positive root of the equation

$$r_n^2 = 1 + r_n^{\frac{2m-1}{2m+1}} \frac{1}{n+1} a^{\frac{1}{2m+1}}, \quad a = \frac{8}{3}\pi^2 m(n+1)\sqrt{(4m-1)!\ln(n+1)} \tag{21}$$

where m is a natural number which we shall define below. Clearly there is a unique root of this equation for which $r_n>1, r_n \to 1$ as $n\to\infty$ for any fixed m. We shall consider an n such that

$$r_n < \sqrt{\rho_0}, \quad r_n \leqslant a^{\frac{1}{2m+1}} \tag{22}$$

From (19) for $z \in \bar{B}$ we have

$$|f(z)-p_n(z)| \leqslant 2M \frac{\rho_0+1}{\rho_0-1} \frac{\rho_0^{\frac{n+1}{2}}}{\rho_0^{\frac{n+1}{2}}-1} \exp\left[\frac{2m+1}{4m} r_n^{\frac{2m-1}{2m+1}} a^{\frac{1}{2m+1}}\right] \frac{1}{\rho_0^{n+1}} \tag{23}$$

By using Stirling's formula it is easy to show that

$$a^{\frac{1}{2m+1}} \leqslant \left[2\,\frac{(2\pi)^2}{3}(n+1)\sqrt{(4m+1)!\ln(n+1)}\right]^{\frac{1}{2m+1}}$$
$$\leqslant \frac{4m+1}{e} \alpha^{\frac{1}{2m+1}} \leqslant \frac{2}{e}(2m+1)\,\alpha^{\frac{1}{2m+1}}$$

where

$$\alpha = \frac{2}{3}\sqrt{e}\,(2\pi)^{5/2}(n+1)\sqrt{\ln(n+1)}$$

The quantity $r_n^{\frac{2m-1}{2m+1}} \frac{(2m+1)^2}{2em} \alpha^{\frac{1}{2m+1}}$ has its smallest value for

$$m=\frac{1}{4}\left(\ln\alpha+\sqrt{\ln^2\alpha+4}\right)\leqslant\frac{1}{2}\ln\alpha+\frac{1}{2}$$

But m must be an integer and consequently, taking into account the behaviour of the function

$$\frac{(2m+1)^2}{m}\alpha^{\frac{1}{2m+1}}$$

we can put

$$m=\frac{1}{2}\ln\alpha+\frac{3}{2}$$

Then from (23) we have

$$|f(z)-p_n(z)|\leqslant 2M\frac{\rho_0+1}{\rho_0-1}\frac{\rho_0^{\frac{n+1}{2}}}{\rho_0^{\frac{n+1}{2}}-1}\beta^{r_n}\frac{1}{\rho_0^{n+1}}$$

$$\beta=\frac{2}{3}(2\pi)^{5/2}e^6(n+1)\sqrt{\ln(n+1)},\qquad z\in\bar{B}$$

for all sufficiently large n. This estimate holds for every n satisfying the inequality (22), a and r_n then being defined by Formula (21), where we put

$$m=\left[\frac{1}{2}\ln\left(\frac{2}{3}(2\pi)^{5/2}(n+1)\sqrt{\ln(n+1)}\right)+\frac{7}{4}\right]$$

For such m, $r_n\to 1$ as $n\to\infty$.

Other investigations of this problem and of a whole series of others are reviewed in the article by Mergelian [2].

1.3.2 Second principle of choice (Van der Monde points)

On a set $\bar{B}$ of type $\mathfrak{M}$ let us take $n+1$ points $\xi_k^{(n)}$, $k=1, 2, \ldots,$ $n+1$, $n=1, 2, \ldots,$ and let us form the Van der Monde determinant

$$v_n=v_n\left(\xi_1^{(n)}, \xi_2^{(n)}, \ldots, \xi_{n+1}^{(n)}\right)=\prod_{1\leqslant k<j\leqslant n+1}\left(\xi_k^{(n)}-\xi_j^{(n)}\right)$$

This determinant is a continuous, bounded function of the $n+1$ variables $\xi_k\in\bar{B}$, $k=1, 2, \ldots, n+1$. The function $|v_n|$

reaches a maximum, but possibly not for a unique system of points. Let $|v_n|$ attain a maximum for the system of points $z_k^{(n)}$, $k=1, 2, \ldots, n+1$. Clearly all the points $z_k^{(n)}$ are different. Let us show that they all lie on the boundary L of the set B. To do this let us consider the function v_n for fixed $\xi_j^{(n)} = z_j^{(n)}$, $j \neq k$, and variable $\xi_k^{(n)} = z$. In the determinant v_n the variable part is the factor

$$\sigma_k(z) = (z - z_1^{(n)})(z - z_2^{(n)}) \cdots (z - z_{k-1}^{(n)})(z_{k+1}^{(n)} - z) \cdots (z_{n+1}^{(n)} - z)$$

From the way of choosing the points $z_j^{(n)}$, $j = 1, 2, \ldots, n+1$, it follows that $|\sigma_k(z)|$ reaches its greatest value on B for $z = z_k^{(n)}$, and by virtue of the maximum principle the point $z_k^{(n)} \in L$, $k = 1, 2, \ldots, n+1$, and

$$\left| \frac{\sigma_k(z)}{\sigma_k(z_k^{(n)})} \right| \leqslant 1, \quad z \in \bar{B} \tag{1}$$

We consider the points $z_k^{(n)}$, $k = 1, 2, \ldots, n+1$, $n = 1, 2, \ldots$ as a system of interpolating points. Such interpolating points are called Van der Monde points. All these points are simple.

Let the function $F(z)$ be regular on set $\bar{B}$. Let us consider the interpolating polynomial

$$p_n(z) = \sum_{k=1}^{n+1} F(z_k^{(n)}) \frac{\omega_n(z)}{(z - z_k^{(n)})\, \omega_n'(z_k^{(n)})}$$

Since $\dfrac{\omega_n(z)}{z - z_k^{(n)}}$ only differs in sign from $\sigma_k(z)$, the inequality (1) gives

$$\left| \frac{\omega_n(z)}{(z - z_k^{(n)})\, \omega_n'(z_k^{(n)})} \right| \leqslant 1, \quad z \in \bar{B}$$

and for the interpolating polynomial constructed, we have the following estimate

$$|p_n(z)| \leqslant \sum_{k=1}^{n+1} |F(z_k^{(n)})|, \quad z \in \bar{B}$$

Theorem 1 (Fekete [2]*)* Let $\bar{B}$ be a set of type $\mathfrak{M}$ and $\xi_k^{(n)}$ be Van der Monde interpolation points for $\bar{B}$. Then $|\theta_n(z)| \rightrightarrows 1$ as $n \to \infty$ in B^1 $\left(\text{and } \lim\limits_{n \to \infty} \sqrt[n+1]{M_n} = c\right)$.

PROOF Let us take an arbitrary function $f(z)$ regular in $\bar{B}$, in B_{ρ_0} let $\rho_0 > 1$, and let us construct an interpolation polynomial $p_n^*(z)$ with Fejer-type interpolating points lying on L_{ρ_2}, $1 < \rho_2 < \rho_0$. We note that the function $w = \frac{1}{\rho_2}\varphi(z)$ maps domain $B_{\rho_2}^1$ on to domain $|w| > 1$ and the contour line L_ρ, $\rho > \rho_2$, on to the circumference $|w| = \frac{\rho}{\rho_2}$. Consequently the contour L_{ρ_0} for $\bar{B}$ will be the contour line $L_{\frac{\rho_0}{\rho_2}}$ for domain $\bar{B}_{\rho_2}$. Hence and from Theorem 1, Section 1.3.1, it follows that

$$|f(z) - p_n^*(z)| \leqslant \frac{M_1}{\rho_3^{n+1}}, \quad z \in \bar{B}_{\rho_2}, \quad 1 < \rho_3 < \frac{\rho_0}{\rho_2}$$

Let us introduce the function $F(z) = f(z) - p_n^*(z)$ and construct for it an interpolating polynomial $p_n^{**}(z)$ with interpolating points at the $z_k^{(n)}$ Van der Monde points.

We shall then have

$$|p_n^{**}(z)| \leqslant \sum_{k=1}^{n+1} |F(z_k^{(n)})| \leqslant (n+1)\frac{M_1}{\rho_3^{n+1}}, \quad z \in \bar{B}$$

The polynomials $p_n(z) = p_n^*(z) + p_n^{**}(z)$ will be interpolating polynomials for the function $f(z)$ with interpolated points in the Van der Monde points. In fact

$$p_n(z_k^{(n)}) = p_n^*(z_k^{(n)}) + p_n^{**}(z_k^{(n)}) = p_n^*(z_k^{(n)}) + f(z_k^{(n)}) - p_n^*(z_k^{(n)}) = f(z_k^{(n)})$$

Then

$$\begin{aligned} |f(z) - p_n(z)| &= |f(z) - p_n^*(z) - p_n^{**}(z)| \leqslant |f(z) - p_n^*(z)| + |p_n^{**}(z)| \\ &\leqslant \frac{M_1}{\rho_3^{n+1}} + (n+1)\frac{M_1}{\rho_3^{n+1}} = (n+2)\frac{M_1}{\rho_3^{n+1}}, \quad z \in \bar{B} \end{aligned}$$

i.e.

$$|f(z) - p_n(z)| \leqslant (n+2)\frac{M_1}{\rho_3^{n+1}}, \quad z \in \bar{B}$$

and from Theorem 1, Section 1.2 the theorem is proved.

NOTE 1 From the corollary to Theorem 1 in Section 1.3.1, it is directly consequent that we have the estimate

$$|f(z)-p_n(z)| \leqslant \frac{M_{\rho_1}}{\rho_1^{n+1}}, \quad z \in \bar{B}, \quad 1 < \rho_1 < \rho_0$$

NOTE 2 From the theorem just proved and Theorem 1, Section 3.1, follows that if $\bar{B}$ is a set of type $\mathfrak{M}$, then the Van der Monde points for the domain $\bar{B}$ are distributed on the contour L of the set B^1.

NOTE 3 Let the points $z_k^{(n)} \in \bar{B}$, $k=1, 2, \ldots, n+1$; $n=1, 2, \ldots$, and

$$U_n = \sup_{z \in \bar{B}} \sum_{k=1}^{n+1} \left| \frac{(z-z_1^{(n)}) \ldots (z-z_{k-1}^{(n)})(z-z_{k+1}^{(n)}) \ldots (z-z_{n+1}^{(n)})}{(z_k^{(n)}-z_1^{(n)}) \ldots (z_k^{(n)}-z_{k-1}^{(n)})(z_k^{(n)}-z_{k+1}^{(n)}) \ldots (z_k^{(n)}-z_{n+1}^{(n)})} \right|$$

(For Van der Monde points $U_n \leqslant n+1$.) If the points $z_k^{(n)}$ are such that for any $\rho > 1$

$$\lim_{n \to \infty} \frac{U_n}{\rho^n} = 0 \tag{2}$$

then the conclusion of the theorem remains the same if in its formulation the Van der Monde points are replaced by points $z_k^{(n)}$ satisfying condition (2); in this case the proof of the theorem does not alter significantly and analogous Notes 1 and 2 are valid (if we take $z_k^{(n)}$ on L). Note that condition (2) is satisfied when, and only when,

$$\overline{\lim_{n \to \infty}} \sqrt[n]{U_n} \leqslant 1$$

1.3.3 *Third principle of choice (Chebyshev points)*

Let $\bar{B}$ be an arbitrary, bounded, closed set. Let us take in $\bar{B}$ arbitrary points ξ_k $(k=1, 2, \ldots, n+1)$ and construct the polynomial of degree $n+1$

$$p(z) = \prod_{k=1}^{n+1} (z-\xi_k) = z^{n+1} + a_1 z^n + \ldots + a_{n+1} \tag{1}$$

Let us fix n and put

$$\tilde{\mu}_{n+1} = \inf_{\xi_k^{(n)}} \sup_{z \in \bar{B}} |p(z)| \tag{2}$$

D

There is a sequence $p_l(z)$ of polynomials of type (1) such that as $l \to \infty$ $\sup\limits_{z \in \overline{B}} |p_l(z)| \to \tilde{\mu}_{n+1}$. From this series of polynomials we choose a sub-sequence such that the zeros of the polynomials converge to certain limiting values, the chosen sub-sequence uniformly converging on $\overline{B}$ to some limiting polynomial which we shall denote by $\tilde{\tau}_{n+1}(z)$. This limiting polynomial $\tilde{\tau}_{n+1}(z)$ will be of degree $n+1$ with leading coefficient one, and all its zeros will lie in $\overline{B}$. Let us show that

$$\sup_{z \in \overline{B}} |\tilde{\tau}_{n+1}(z)| = \tilde{\mu}_{n+1} \tag{3}$$

In fact, letting l tend to infinity in the inequality

$$\sup_{z \in \overline{B}} |\tilde{\tau}_{n+1}(z)| \leqslant \sup_{z \in \overline{B}} |p_l(z) - \tilde{\tau}_{n+1}(z)| + \sup_{z \in \overline{B}} |p_l(z)|$$

in the limit we have

$$\sup_{z \in \overline{B}} |\tilde{\tau}_{n+1}(z)| \leqslant \tilde{\mu}_{n+1}$$

On the other hand, from the definition for $\tilde{\mu}_{n+1}$ it follows that

$$\sup_{z \in \overline{B}} |\tilde{\tau}_{n+1}(z)| \geqslant \tilde{\mu}_{n+1}$$

Comparing the last two inequalities we conclude that Formula (3) is valid.

The polynomial $\tilde{\tau}_{n+1}(z)$ is called the polynomial deviating the least from zero on $\overline{B}$ (or Chebyshev polynomial for $\overline{B}$) with zeros on $\overline{B}$. The points $z_k^{(n)}$, $k = 1, 2, \ldots, n+1$, $n = 0, 1\ldots$ of the polynomials $\tilde{\tau}_{n+1}(z)$ can be taken as the points of interpolation, and will be called the Chebyshev points for $\overline{B}$. For the Chebyshev points

$$\omega_n(z) = \prod_{k=1}^{n+1} (z - z_k^{(n)}) = \tilde{\tau}_{n+1}(z)$$

and

$$M_n = \sup_{z \in \overline{B}} |\omega_n(z)| = \tilde{\mu}_{n+1}$$

Theorem 1 Let $\overline{B}$ be a set of type $\mathfrak{M}$ and $z_k^{(n)}$ be the Chebyshev interpolation points for $\overline{B}$. Then $|\theta_n(z)| \rightrightarrows 1$ in B^1 as $n \to \infty$ (and $\lim_{n\to\infty} \sqrt[n+1]{M_n} = c$).

PROOF Let $z_k^{*(n)}$ be the Van der Monde points for $\overline{B}$ and

$$\omega_n^*(z) = \prod_{k=1}^{n+1} \left(z - z_k^{*(n)}\right), \quad M_n^* = \sup_{z \in \overline{B}} |\omega_n^*(z)|$$

It is clear that $M_n \leqslant M_n^*$ and therefore

$$1 \leqslant \frac{\sqrt[n+1]{M_n}}{c} \leqslant \frac{\sqrt[n+1]{M_n^*}}{c}$$

Hence, and from Theorem 1, Section 3.2, it follows that

$$\lim_{n\to\infty} \frac{\sqrt[n+1]{M_n}}{c} = 1$$

which is equivalent to $|\theta_n(z)| \rightrightarrows 1$ in B^1 and the theorem is proved.

NOTE The equality $\lim_{n\to\infty} \sqrt[n+1]{M_n} = c$ can be written:

$$\lim_{n\to\infty} \sqrt[n+1]{\tilde{\mu}_n} = c$$

1.3.4 *Chebyshev constant and transfinite diameter of a closed set*

We shall now establish a deeper connection between the Chebyshev points and the Van der Monde points and shall also introduce the concept of the transfinite diameter which plays an important role in approximation theory.

Let $\overline{B}$ be an arbitrary, bounded, closed set of points of the z plane. In an analogous way to Section 1.3.3, we introduce a polynomial $\tilde{\tau}_n(z) = z^n + a_1^{(n)} z^{n-1} + \ldots + a_n^{(n)}$ with zeros on $\overline{B}$ and deviating least from zero on $\overline{B}$, and as before we put

$$\tilde{\mu}_n = \sup_{z \in \overline{B}} |\tilde{\tau}_n(z)|$$

Let $p(z)=z^n+b_1z^{n-1}+\ldots+b_n$ be an arbitrary polynomial of degree n with a leading coefficient one. Putting

$$\mu_n=\inf_p \sup_{z\in\bar{B}} |p(z)|$$

Let $\bar{B}^*$ be the convex hull of the set $\bar{B}$.* To determine μ_n we only have to consider polynomials $p(z)$ with zeros on $\bar{B}^*$ since if ξ is some zero of the polynomial $p(z)$ not belonging to B^* then (making use of the convexity of the set $\bar{B}^*$) the modulus of the factor $(z-\xi)$ appearing in $p(z)$ can be reduced for all points of $\bar{B}$ putting ξ on the boundary of $\bar{B}^*$ with the shortest distance from ξ to $\bar{B}^*$. In a similar manner to Section 3.3 we can show that there is a polynomial $\tau_n(z)=z^n+\alpha_1^n z^{n-1}+\ldots+\alpha_n^{(n)}$, for which

$$\sup_{z\in\bar{B}} |\tau_n(z)|=\mu_n$$

The polynomial $\tau_n(z)$ is called the polynomial deviating least from zero on $\bar{B}$ (or Chebyshev polynomial for $\bar{B}^*$).

NOTE:

1. all zeros of the polynomial $\tau_n(z)$ belong to $\bar{B}^*$;
2. there is a unique Chebyshev polynomial of degree n for a closed set $\bar{B}$. We shall give a proof of this later (see Section 5.3.4);
3. $\mu_n\leqslant\tilde{\mu}_n$.

We can show that the following limits exist and are finite

$$\lim_{n\to\infty}\sqrt[n]{\mu_n}=\mu \text{ and } \lim_{n\to\infty}\sqrt{\tilde{\mu}_n}=\tilde{\mu} \tag{1}$$

Let us first consider the left-hand limit of (1). If the set $\bar{B}$ contains a finite number of points only, then for sufficiently large n, $\mu_n=0$ and consequently $\mu=0$. Assuming the set $\bar{B}$ contains an infinite number of points, let us denote the diameter of the set $\bar{B}^*$ by D and evaluate μ_n. As

$$\tau_n(z)=\prod_{k=1}^{n+1}(z-\xi_k),\ \xi_k\in\bar{B}^*$$

*A set G of points of the plane z are said to be convex if for any two points z_1 and z_2 of the set G, the segment $\overline{z_1z_2}$ joining the two points also belongs to the set G. The convex hull of a closed set $\bar{D}$ is the section of all closed convex sets containing $\bar{D}$. The convex hull of a closed set is a closed convex set.

then

$$0 \leqslant \mu_n \leqslant \inf_{\xi_k \in \bar{B}^*} \sup_{z \in \bar{B}} \left| \prod_{k=1}^{n} (z - \xi_k) \right| \leqslant D^n$$

and consequently $0 \leqslant \sqrt[n]{\mu_n} \leqslant D$. Let us put

$$\varliminf_{n \to \infty} \sqrt[n]{\mu_n} = \mu', \quad \varlimsup_{n \to \infty} \sqrt[n]{\mu_n} = \mu''$$

and show that $\mu' = \mu''$. Clearly $\mu' \leqslant \mu''$. All that is required is to show that $\mu'' \leqslant \mu'$. Let $\varepsilon > 0$. We choose n such that $\sqrt[n]{\mu_n} < \mu' + \varepsilon$. Then $|\tau_n(z)| < (\mu' + \varepsilon)^n$, $z \in \bar{B}$. If k and l are any natural numbers, then for $\xi \in \bar{B}^*$ and $z \in \bar{B}$

$$|(z - \xi)^l (\tau_n(z))^k| < D^l (\mu' + \varepsilon)^{nk}$$

and consequently $\mu_{nk+l} < D^l (\mu' + \varepsilon)^{nk}$ or

$$\sqrt[nk+l]{\mu_{nk+l}} < D^{\frac{l}{nk+l}} (\mu' + \varepsilon)^{\frac{nk}{nk+l}}$$

If $\sqrt[n_\nu]{\mu_{n_\nu}} \to \mu''$ as $\nu \to \infty$, then, assuming

$$n_\nu = nk_\nu + l_\nu, \quad 0 \leqslant l_\nu < n$$

from the last inequality, we have $\mu'' \leqslant \mu' + \varepsilon$, and in view of the arbitrary smallness of $\varepsilon > 0$ we conclude that $\mu'' \leqslant \mu'$. Therefore $\mu'' = \mu'$ and the first of the limits (1) exists. Analogously we can show that the second limit of (1) also exists. Clearly $\mu \leqslant \tilde{\mu}$. The quantity μ is called the Chebyshev constant of the set $\bar{B}$.

Now let $z_1, z_2, \ldots, z_n$, $n \geqslant 2$, be arbitrary points of a set $\bar{B}$. Let us now investigate, as in Section 1.3.2, the Van der Monde determinant of the points $z_{k,j}$, $k = 1, 2, \ldots, n$

$$v(z_1, z_2, \ldots, z_n) = \prod_{1 \leqslant k < l \leqslant n} (z_k - z_l)$$

and denote by v_n the least upper bound of $|v(z_1, z_2, \ldots, z_n)|$, where the points $z_1, z_2, \ldots, z_n$ run through different systems with n points (from $\bar{B}$). Let us assume $d_n = v_n^{\frac{2}{n-1}}$ and show that there exists a finite limit

$$\lim_{n\to\infty} \sqrt[n]{d_n} = d \tag{2}$$

If the set $\overline{B}$ contains a finite number of points only, then if n is sufficiently large $v_n = 0$ and consequently $d = 0$. Assuming the set $\overline{B}$ contains an infinite number of points we assume that $|v(z_1, \ldots, z_n)|$ reaches its highest value for $z_k = \xi_k^{(n)}$, $k = 1, 2, \ldots, n$, $n = 2, 3, \ldots$, ($\xi_k^{(n)}$ being the Van der Monde points), e.g.

$$v_n = \left| v\left(\xi_1^{(n)}, \ldots, \xi_n^{(n)}\right) \right|$$

We have

$$v_{n+1}^2 = \left| \prod_{1 \leqslant k < l \leqslant n+1} \left(\xi_k^{(n+1)} - \xi_l^{(n+1)}\right) \right|^2 = \left| \prod_{\substack{k,\, l=1 \\ k \neq l}}^{n+1} \left(\xi_k^{(n+1)} - \xi_l^{(n+1)}\right) \right|$$

$$= \left| \prod_{\substack{l=1 \\ l \neq k_0}}^{n+1} \left(\xi_{k_0}^{(n+1)} - \xi_l^{(n+1)}\right) \right|^2 \left| \prod_{\substack{k,\, l=1 \\ k \neq l \\ k,\, l \neq k_0}}^{n+1} \left(\xi_k^{(n+1)} - \xi_l^{(n+1)}\right) \right|$$

$$= \left| \prod_{\substack{l=1 \\ l \neq k_0}}^{n+1} \left(\xi_{k_0}^{(n+1)} - \xi_l^{(n+1)}\right) \right|^2 \left| v\left(\xi_1^{(n+1)}, \ldots, \xi_{k_0-1}^{(n+1)}, \xi_{k_0+1}^{(n+1)}, \ldots, \xi_{n+1}^{(n+1)}\right) \right|^2$$

$$\leqslant \left| \prod_{\substack{l=1 \\ l \neq k_0}}^{n+1} \left(\xi_{k_0}^{(n+1)} - \xi_l^{(n+1)}\right) \right|^2 v_n^2$$

Multiplying these inequalities for $k_0 = 1, 2, \ldots, n+1$, gives

$$v_{n+1}^{2(n+1)} \leqslant v_{n+1}^2 v_n^{2(n+1)}, \quad v_{n+1}^{2(n-2)} \leqslant v_n^{2(n+1)}$$

Hence

$$\sqrt[n+1]{d_{n+1}} = v_{n+1}^{\frac{2}{(n+1)n}} \leqslant v_n^{\frac{2}{n(n-1)}} = \sqrt[n]{d_n}$$

Thus the sequence of numbers $\sqrt[n]{d_n}$, $n = 2, 3, \ldots$, is non-increasing and is bounded from below. Consequently there exists the limit (2). The quantity d was introduced by Fekete [1, 3] and is called the transfinite diameter of the set $\overline{B}$.

We now write

$$\tau_n^*(z) = \prod_{k=1}^{n} \left(z - \xi_k^{(n)}\right), \quad \mu_n^* = \sup_{z \in \overline{B}} |\tau_n^*(z)|$$

where $\xi_k^{(n)}$ are the Van der Monde points.

Theorem There exist

$$\lim_{n\to\infty} \sqrt[n]{\mu_n^*} = \mu^* \text{ and } \mu = \tilde{\mu} = \mu^* = d$$

PROOF If a set $\overline{B}$ contains a finite number of points only, the theorem is clearly valid. Let us assume that $\overline{B}$ contains an infinite number of points. Let us first establish the inequality

$$\tilde{\mu}_n \leqslant \mu_n^* \leqslant \frac{v_{n+1}}{v_n} \leqslant (n+1)\mu_n, \quad n = 2, 3, \ldots \tag{3}$$

For $z \in \overline{B}$ we have

$$v(z, \xi_1^{(n)}, \ldots, \xi_n^{(n)}) = \left[\prod_{k=1}^{n}(z - \xi_k^{(n)})\right] \prod_{1 \leqslant k < l \leqslant n} (\xi_k^{(n)} - \xi_l^{(n)})$$
$$= \left[\prod_{k=1}^{n}(z - \xi_k^{(n)})\right] v(\xi_1^{(n)}, \ldots, \xi_n^{(n)})$$

and consequently

$$v_{n+1} \geqslant v_n \sup_{z \in \overline{B}} \left|\prod_{k=1}^{n}(z - \xi_k^{(n)})\right| = v_n \mu_n^* \geqslant v_n \mu_n \tag{4}$$

From this the two left-hand inequalities in (3) follow. Further

$$v(\xi_1^{(n+1)}, \ldots, \xi_{n+1}^{(n+1)}) = \begin{vmatrix} 1 & , 1 & , \ldots, & 1 \\ \xi_1^{(n+1)} & , \xi_2^{(n+1)} & , \ldots, & \xi_{n+1}^{(n+1)} \\ \cdot & \cdot & \cdot & \cdot \\ (\xi_1^{(n+1)})^{n-1}, & (\xi_2^{(n+1)})^{n-1}, & \ldots, & (\xi_{n+1}^{(n+1)})^{n-1} \\ \tau_n(\xi_1^{(n+1)}), & \tau_n(\xi_2^{(n+1)}), & \ldots, & \tau_n(\xi_{n+1}^{(n+1)}) \end{vmatrix}$$

and expanding the determinant for the elements of the last row we have

$$v(\xi_1^{(n+1)}, \ldots, \xi_{n+1}^{(n+1)}) = \sum_{k=1}^{n+1} (-1)^{n+k+1} \tau_n(\xi_k^{(n+1)}) v(\xi_1^{(n+1)},$$
$$\ldots, \xi_{k-1}^{(n+1)}, \xi_{k+1}^{(n+1)}, \ldots, \xi_{n+1}^{(n+1)})$$

Hence

$$v_{n+1} \leqslant \sum_{k=1}^{n+1} |\tau_n(\xi_k^{(n+1)})| \, |v(\xi_1^{(n+1)}, \ldots, \xi_{k-1}^{(n+1)}, \xi_{k+1}^{(n+1)}, \ldots, \xi_{n+1}^{(n+1)})|$$
$$\leqslant (n+1)\mu_n v_n$$

and consequently the right-hand part of the inequality (3) is valid. From the inequality (3) it follows that

$$\sqrt[n]{\tilde{\mu}_n} \leqslant \sqrt[n]{\mu_n^*} \leqslant \sqrt[n]{\frac{v_{n+1}}{v_n}} \leqslant \sqrt[n]{(n+1)}\,\sqrt[n]{\mu_n} \tag{5}$$

Letting n tend to ∞ and using (1), we have $\tilde{\mu} \leqslant \mu$ and since $\mu \leqslant \tilde{\mu}$ $\mu = \tilde{\mu}$. From (5) it follows that the limit μ^* exists and

$$\lim_{n \to \infty} \sqrt[n]{\frac{v_{n+1}}{v_n}} = \mu = \tilde{\mu} = \mu^*$$

Hence, and from (5), we find that for $d \neq 0$

$$\lim_{n \to \infty} \frac{d_{n+1}}{d_n} = \lim_{n \to \infty} \left[\sqrt[n]{\left(\frac{v_{n+1}}{v_n}\right)^2}\, v_n^{-\frac{2}{n(n-1)}} \right] = \frac{\mu^2}{d} \tag{6}$$

In this case as we know

$$\lim_{n \to \infty} \sqrt[n]{d_n} = \lim_{n \to \infty} \frac{d_{n+1}}{d_n}$$

and consequently $d = \frac{\mu^2}{d}$, e.g. $\mu = d$.

Let $d = 0$ and assume $\tilde{\mu} > 0$, so for any ε, $0 < \varepsilon < \tilde{\mu}$, and sufficiently large n from (5) (see also (6)) we have

$$\frac{(\tilde{\mu} - \varepsilon)^2}{\varepsilon} \leqslant \frac{d_{n+k}}{d_{n+k-1}}, \quad k = 1, 2, \ldots$$

Multiplying these inequalities for $k = 1, 2, \ldots, m$ we find

$$\frac{(\tilde{\mu} - \varepsilon)^{\frac{2m}{n+m}}}{\varepsilon^{\frac{m}{n+m}}}\, d_n^{\frac{1}{n+m}} \leqslant \sqrt[n+m]{d_{n+m}}$$

Letting m tend to ∞ in this inequality, in the limit we have

$$(\tilde{\mu} - \varepsilon)^2 \leqslant 0$$

This contradiction proves that $\tilde{\mu} = 0$ but $\mu = \tilde{\mu} = \mu^*$, and the theorem is proved.

NOTE 1 From (6) it also follows that for $d \neq 0$

$$\lim_{n \to \infty} \frac{d_{n+1}}{d_n} = d$$

NOTE 2 If $\bar{B}$ is a closed set of type $\mathfrak{M}$ then, as was shown in 1.3.2 (see Theorem 1), $\mu^* = c$ and consequently in this case

$$\mu = \tilde{\mu} = \mu^* = d = c$$

NOTE 3 The Chebyshev constant and the transfinite diameter of the set $\bar{B}$ and its boundary L coincide since the points $\xi_k^{(n)}$ for $\bar{B}$ and for L lie on L and are one and the same.

EXAMPLE Calculate the Chebyshev constant for the circle $|z-a| \leqslant R$ (or the circumference $|z-a| = R$). Let us find the capacity c of the set $|z-a| \leqslant R$ (see Section 1.2). The function $w = \varphi(z)$ mapping the domain $|z-a| > R$ on to $|w| > 1$, will be

$$w = \frac{1}{R}(z-a)$$

and consequently $c = R$ and $\mu = \tilde{\mu} = \mu^* = d = R$.

For more details about the Chebyshev constant and transfinite diameter see Goluzin [1], pp. 319-358.

1.3.5 *Certain other formulations of the problem of interpolation*

Let F, B and G be three non-empty sets in the complex z plane such that $F \subset G$ and $B \subset G$. We shall say that the condition $\{F, B, G\}$ is fulfilled if for every function $f(z)$ regular on G for any choice of the sequence of interpolating points $z_k^{(n)}$, $k = 1, 2, \ldots, n+1$, $n = 1, 2, \ldots$, from any bounded sub-set $F^* \subset F$ the sequence of interpolating polynomials $p_n(z)$ for $f(z)$ uniformly converge to $f(z)$ as $n \to \infty$ on every bounded sub-set $B^* \subset B$.

Let us consider the following three problems:

1. For a given set B and $G (B \subset G)$ to find the biggest set F such that the condition $\{F, B, G\}$ is fulfilled.
2. For a given set F and $G (F \subset G)$ to find the biggest set B such that the condition $\{F, B, G\}$ is fulfilled.
3. For a given set F and B to find the smallest set G such that the condition $\{F, B, G\}$ is fulfilled.

The last of these problems was formulated and solved by Krylov [1]. Let us solve these problems. The solution which we give below to Krylov's problem is slightly different in method from his own. We shall first state the concept underlying our method of solving these problems and formulating corresponding theorems.

Let the point $\xi_0 \in F$. If all the interpolating points coincide with the point ξ_0, then the sequence of interpolating polynomials for the function $f(z)$ regular on the set G is the sequence of partial sums of the Taylor series for the function $f(z)$ about the points ξ_0, and such an interpolating process for the function $f(z)$ will uniformly converge to the function $f(z)$ inside the greatest open circle K_{ξ_0} with its centre at the point ξ_0 and belonging to the domain of regularity of the function $f(z)$. For the interpolating process to converge uniformly on the set $\overline{B}$ it is then necessary and sufficient that $\overline{B} \subset K_{\xi_0}$.

We shall now consider a set on the finite complex plane (i.e. the plane from which the point ∞ is removed). A set M of the finite plane will be said to be closed if every finite boundary point of M belongs to M.

Theorem 1 Let B and G be two closed sets in the finite z plane, B being a bounded, non-empty set and G being different from the finite plane and $B \subset G$. Let us denote by F the set of the centres of all closed discs K such that $K \subset G$ and $B \subset K$. The set F is the greatest set for which the condition $\{F, B, G\}$ is fulfilled. F is then a closed, convex set in the finite z plane.

PROOF We assume that the set F is not empty.

Let us prove that F is a closed set. Let us assume that F is not a closed set, e.g. there is a sequence of points ξ_j, $j = 1, 2, \ldots$, belonging to the set F, which converges to some point $\xi_0 \notin F$. Let K be the largest closed circle with its centre at the point ξ_0 and belonging to G. Since the point $\xi_0 \notin F$, there is a point $z \in B$ not belonging to K. Besides this there is a point ξ which belongs simultaneously to the boundary of K and the boundary of G. Consequently

$$|\xi - \xi_0| < |z - \xi_0| \tag{1}$$

Let K_{ξ_j}, $j = 1, 2, \ldots$, be a closed circle with its centre at ξ_j and radius $r_j = |\xi - \xi_j|$. Clearly every closed circle K'_{ξ_j} with its centre at ξ_j such that $K'_{\xi_j} \in G$ and $B \subset K'_{\xi_j}$ has a radius

$r'_j \leqslant r_j$, for ξ is a boundary point of G. Moreover if $B \subset K'_{\xi_j}$ then $B \subset K_{\xi_j}$. That is $B \subset K_{\xi_j}$ and, in particular, $z \in K_{\xi_j}$. Consequently

$$|z - \xi_j| \leqslant |\xi - \xi_j|$$

Letting j tend to ∞ in this last inequality, in the limit we obtain the inequality

$$|z - \xi_0| \leqslant |\xi - \xi_0|$$

But this inequality contradicts inequality (1). F is therefore a closed set.

Let us next prove that F is a convex set. Let ξ_0 and ξ_1 be two arbitrary points from the set F and let K_{ξ_0} and K_{ξ_1} be the two, smallest, closed discs - with centres respectively at the points ξ_0 and ξ_1 - containing the set B. Then the intersection $K = K_{\xi_0} \cap K_{\xi_1}$ also contains B and the sum $K' = K_{\xi_0} \cup K_{\xi_1}$ is contained in G. The smallest, closed disc K_{ξ_λ} with its centre at an arbitrary point ξ_λ lying on the segment $\overline{\xi_0\xi_1}$ of the straight line joining the points ξ_0 and ξ_1, and containing K belongs to K'. Consequently $B \subset K_{\xi_\lambda} \subset G$ and $\xi_\lambda \in F$. Hence we conclude that the whole segment $\xi_0\xi_1 \subset F$ and due to the arbitrary nature of the points ξ_0 and ξ_1 from the set F, this set F is convex.

Let us show that no point $\xi_0 \in F$ $(\xi_0 \in G)$ can be added to F such that the condition $\{F \cup \xi_0, B, G\}$ is fulfilled. Let K_{ξ_0} be the smallest closed disc with its centre at ξ_0 containing B. The disc K_{ξ_0} does not belong to G since $\xi_0 \bar{\in} F$, and consequently there is a point z_0 lying inside K_{ξ_0} and not belonging to G. The function $\frac{1}{z - z_0}$ is regular on G and the radius of convergence $|z_0 - \xi_0|$ of its Taylor series about the point ξ_0 is less than the radius of the circle K_{ξ_0}. That is, the sequence of partial sums of the indicated Taylor series does not converge for all points of the set B. This in turn means that the condition $\{F \cup \xi_0, B, G\}$ is not fulfilled for any point $\xi_0 \bar{\in} F$.

Let us now prove that the condition $\{F, B, G\}$ is fulfilled. Let $z_k^{(n)}$, $k = 1, 2, \ldots, n$, $n = 1, 2, \ldots$, be an arbitrary, bounded sequence of interpolating points from the set F. It can be shown that the closure F^* of the set of points $z_k^{(n)}$, $k = 1, 2, \ldots, n, n = 1, 2, \ldots$, is a bounded set and $F^* \subset F$. Let K_ξ be the smallest closed disc containing the set B and having its centre at the point ξ. Let us denote $G^* = \bigcup_{\xi \in F^*} K_\xi$. It is not difficult to see that $G^* \subset G$ since G contains each of the circles K_ξ, $\xi \in F^*$, and that G^* is a bounded set. Let us prove that G^* is a closed set. Let z_j, $j = 1, 2, \ldots$, be a sequence of points of the set G^* which converges to some point z_0. For each point z_j there is a point $\xi_j \in F^*$ such that $z_j \in K_{\xi_j}$. Let us find a point $z'_j \in B$ also lying on the boundary of the disc K_{ξ_j} and let us denote the radius of the circle K_{ξ_j} by r_j. Clearly we have

$$|z'_j - \xi_j| = r_j, \quad |z_j - \xi_j| \leqslant r_j \tag{2}$$

Without reducing the generality of the calculation we can assume that the sequences $\{z'_j\}$, $\{\xi_j\}$, $\{r_j\}$ converge, for example, to z'_0, ξ_0, and r_0 respectively. Letting $j \to \infty$ in (2) we obtain

$$|z'_0 - \xi_0| = r_0, \quad |z_0 - \xi_0| \leqslant r_0$$

e.g. the closed disc K_0 with its centre at ξ_0 and radius r_0 has a point $z'_0 \in B$ on the boundary of B and the point $z_0 \in K_0$. That is the disc $K_{\xi_0} \supset K_0$ and $z_0 \subset G^*$ whence it also follows that the set G^* is closed.

Let us prove that G^* is a star-like set with respect to every point of the set B. Let us take the point $z_1 \in G^*$ furthest from the point z_0 on an arbitrary ray radiating from the arbitrary point $z_0 \in B$. There is a point $\xi \in F^*$ such that $z_1 \in K_\xi$. But $B \subset K_\xi$ which means that $z_0 \in K_\xi$. Consequently the segment $\overline{z_0 z_1} \subset K_\xi$ and the segment $z_0 z_1 \subset G^*$. Hence it follows that G^* is star-like with respect to the point z_0. From this and the definition for G^* it follows that G^* is a closed, simply-connected set.

Let $f(z)$ be an arbitrary function regular on the set G, e.g. regular on some open set $G' \supset G$. Let $L_\rho, \rho > 1$ be the contour line of the domain CG^* belonging to the set G' and such that the function $f(z)$ is regular in the closed domain bounded by L_ρ. Let us denote by δ the distance from the line L_ρ to G^*. Since L_ρ and G^* are bounded, closed sets with no common points, $\delta > 0$. Let z be an arbitrary point of the set B and ζ be an arbitrary point on L_ρ. The disc $K_{z_k^{(n)}} \subset G^*$ and $z \in K_{z_k^{(n)}}$, whence

$$|z - z_k^{(n)}| + \delta \leqslant |\zeta - z_k^{(n)}| \tag{3}$$

Let us show that the remainder term

$$r_n(z) = \frac{1}{2\pi i} \int_{L_\rho} \frac{f(\zeta)}{\zeta - z} \frac{\omega_n(z)}{\omega_n(\zeta)} d\zeta, \quad \omega_n(z) = \prod_{k=1}^{n} \left(z - z_k^{(n)}\right)$$

in the Hermite formula uniformly tends to zero on the set B as $n \to \infty$. Denoting

$$R = \sup_{\substack{z \in B \\ \xi \in F}} |z - \xi|, \quad M = \sup_{\zeta \in L_\rho} |f(\zeta)|$$

for $z \in B$, we have (using inequality (3))

$$|r_n(z)| \leqslant \frac{1}{2\pi} \frac{M}{\delta} \int_{L_\rho} \prod_{k=1}^{n} \left| \frac{z - z_k^{(n)}}{\zeta - z_k^{(n)}} \right| |d\zeta| \leqslant \frac{1}{2\pi} \frac{M}{\delta} \left(\frac{R}{R+\delta} \right)^n l_\rho$$

where l_ρ is the length of L_ρ. Hence it follows that $r_n(z) \rightrightarrows 0$ on the set B as $n \to \infty$, and the theorem is proved.

NOTE 1 The proof can also easily be carried out for the case when G is the finite plane. In this case the theorem means that for any closed, bounded set B and any integral function $f(z)$ for any choice of points from a bounded part of G the interpolating process converges to $f(z)$ uniformly on B.

Theorem 2 Let F and G be two closed sets in the z plane and $F \subset G$. Let K_ξ be the greatest, closed disc contained in G and having its centre at the point $\xi \in F$.† Let us denote $B = \bigcap_{\xi \in F} K_\xi$. If it is not empty the set B is the largest set for which the condition $\{F, B, G\}$ is fulfilled. B is then a closed, convex set of points of the finite z plane.

PROOF In the case where G is the whole finite z plane, B is also the whole finite z plane. Let B^* and F^* be two arbitrary, closed, bounded sets contained respectively in B and F. As in the proof of Theorem 1, we construct from F^* and B^* a set G^* and prove that the condition $\{F^*, B^*, G^*\}$ is fulfilled. The conclusion of the theorem follows immediately from this for the given case.

In what follows we assume that the set G differs from the

†If the set F is unbounded the term closed disc will also cover any closed, half plane which can be obtained by amalgamating discs $K \subset G$ when the centres of the discs tend to infinity for the set F. The whole plane is also called a closed disc.

whole finite z plane. In this case if the set B is not empty it is bounded, closed and convex.

Let us prove that the set B cannot be supplemented with any point $z_0 \notin B$ $(z_0 \in G)$ such that the condition $\{F, B \cup z_0, G\}$ is fulfilled. If $z_0 \notin B$ $(z_0 \in G)$, then there is a point $\xi_0 \in F$ such that the largest, closed disc K_{ξ_0} with its centre at the point ξ_0, and belonging to G does not contain the point z_0. Let ζ_0 be an arbitrary point not belonging to G, and such that $|\xi_0 - \zeta_0| < |z_0 - \xi_0|$. Such a point ζ_0, clearly exists. The function $\frac{1}{z-\zeta_0}$ is regular on G and its Taylor series about the point ξ_0 only converges in the disc $|z - \xi_0| < |\xi_0 - \zeta_0| < |z_0 - \xi_0|$. That is the sequence of partial sums of the indicated Taylor series diverges at the point z_0. This means that the condition $\{F, B \cup z_0, G\}$ is not fulfilled for any point $z_0 \notin B$.

Let us now prove that the condition $\{F, B, G\}$ is fulfilled. Let F^* be the set of centres of all closed circles K such that $K \subset G$ and $B \subset K$. From Theorem 1 it follows that the condition $\{F^*, B, G\}$, is fulfilled and this means that the condition $\{F, B, G\}$ is also fulfilled since $F \subset F^*$. The theorem is proved.

Theorem 3 Let F and B be two closed, bounded sets of the z plane. Let K_ξ be the least closed disc containing the set B and having its centre at the point $\xi \in F$. Let us then denote $G = \bigcup_{\xi \in F} K_\xi$. The set G is the smallest set for which the condition $\{F, B, G\}$ is fulfilled. G is then a closed set of the finite z plane which is star-like with respect to any point $z_0 \in B$.

PROOF That the set is closed and star-like is proved in the same manner as it was proved in Theorem 1. As in the proof of Theorem 1 we can also show that the condition $\{F, B, G\}$ is fulfilled.

Let us now prove that the condition $\{F, B, G^*\}$ is not fulfilled for any closed set G^* smaller than G, (e.g. such that $G^* \subset G$, but $G^* \neq G$). Let the point $z_0 \in G$ such that $z_0 \notin G^*$. From the definition of G it follows that there is a point $\xi_0 \in F$ such that the smallest closed disc K_{ξ_0} with its centre at the point ξ_0 and containing the set B contains the point z_0. The function $\frac{1}{z-z_0}$ is regular on the set G^* since the distance from the point z_0 to the set G^* is positive. The Taylor series of the function $\frac{1}{z-z_0}$ about the point ξ_0 converges only in the circle $|z-\xi_0|<|z_0-\xi_0|$ with a radius $|z_0-\xi_0|$ not greater than the radius of the circle K_{ξ_0}. That is, the sequence of partial sums of the indicated Taylor series diverges on the boundary of K_{ξ_0} and consequently does not converge for all points of the set B. This means that the condition $\{F, B, G^*\}$ is not fulfilled. The theorem is proved.

NOTE 2 If B is an unbounded set, then clearly G is the whole finite plane and the theorem is obviously true.

NOTE 3 The theorems stated above can also be proved with fewer assumptions concerning F, B and G.

EXAMPLES: (1) Let G be the disc $|z|<R$, B the disc $|z-a|\leqslant r, 0<a<R,\ 0<r<R-a$. Find the greatest set F such that the condition $\{F, B, G\}$ is fulfilled. Let us draw through $z=0$ an arbitrary ray and on it find the centre ξ of the largest circle K belonging to G and containing B. We have

$$R-|\xi|=|\xi-a|+r$$

Assuming that $\xi=x+yi$, we have the ellipse:

$$\frac{\left(x-\frac{a}{2}\right)^2}{\left(\frac{R-r}{2}\right)^2}+\frac{y^2}{\frac{(R-r)^2-a^2}{4}}=1$$

This curve is also the boundary of the desired closed set F.

(2) Let G be the disc $|z|<R$, F the segment $x=a$, $0<a<R$, $0\leqslant y\leqslant b$, $a^2+b^2<R^2$. Find the largest set B such that the condition $\{F, B, G\}$ is fulfilled. Let us introduce the two largest, closed discs K_1 and K_2, belonging to G and with centres at the points $z_1=a$ and $z_2=a+bi$ respectively. The common part $K_1\cap K_2$ of these circles is also B, since the largest circle belonging to G, and with its centre at any point $z_3\in F$, contains the points of intersection ξ and ξ' of the circumferences of the discs K_1 and K_2 and consequently contains $K_1\cap K_2$. Clearly if the inequality

$$(R-a)+(R-\sqrt{a^2+b^2})\geqslant b$$

is fulfilled, the set B is not empty.

(3) Let F be the segment $-a \leqslant x \leqslant a$ of the real axis, B the segment $y=b$, $c \leqslant x \leqslant d$. Find the smallest set G such that the condition $\{F, B, G\}$ is fulfilled. Let us construct the two smallest, closed discs with centres at the points $z=-a$ and $z=a$, and containing the set B. The union of these discs will be the desired set G. If we denote the intersection of these discs by B^*, the condition $\{F, B^*, G\}$ will also be fulfilled.

1.4 INTERPOLATING PROCESSES FOR FUNCTIONS WHICH ARE REGULAR ON SETS OF TYPE $\mathfrak{M}_s$

Up to now we have investigated the approximation by interpolating polynomials of functions regular on bounded sets of type $\mathfrak{M}$. We shall denote by $\mathfrak{M}_s$, $s=1, 2, \ldots$, any closed set $\overline{B}$ of the extended z plane, the complement $C\overline{B}$ which consists of s simply-connected domains B^j, $j=1, 2, \ldots, s$, which are mutually disjoint. The boundary of a domain B^j is denoted by L^j, and $L=\bigcup_{j=1}^{s} L^j$ is the boundary of the set $\overline{B}$. The class of sets of type $\mathfrak{M}$ is the same as a subclass of bounded sets from $\mathfrak{M}_1$.

A set $\overline{B}$ of type $\mathfrak{M}_s$ is called a set of type $\widetilde{\mathfrak{M}}_s$, if all points of the boundary L of the set $\overline{B}$ are accessible from the complement $C\overline{B}$. A set $\overline{B}$ of type $\mathfrak{M}_s$ is called a set of type $\overline{\mathfrak{M}}_s$, if every boundary L^j of a domain B^j is a closed Jordan curve. If $\overline{B}$ is a set of type $\overline{\mathfrak{M}}_s$, then $\overline{B}$ is a set of type $\widetilde{\mathfrak{M}}_s$ but not always conversely.

1.4.1 *Formulating the problem*

Let $\overline{B}$ be an arbitrary set of type $\mathfrak{M}_s$. In the domain B^j, $j=1, 2, \ldots, s$, let us choose an arbitrary point a_j (the point ∞ may be among the points a_j) and on $\overline{B}$ let us take a sequence of interpolating points $z_k^{(n)}$, $k=1, 2, \ldots, n+1$ $n=0, 1, \ldots$, some of which may coincide for a fixed n (see Note 2, Section 1.1.1). For each $j=1, 2, \ldots, s$ we set up a sequence of non-negative integers $m_j^{(n)}$, $n=0, 1, \ldots$, such that $\sum_{j=1} m_j^{(n)}=n$. Let there be a function $f(z)$ regular on the set $\overline{B}$. For every $n=0, 1, \ldots$, let us construct the function $f(z)$ with the points $z_k^{(n)}$, $k=1, 2, \ldots, n+1$ the generalised interpolating polynomial given by

$$p_n(z)=\sum_{j=1}^{s}\sum_{l=1}^{m_j^{(n)}}\frac{b_{l,j}^{(n)}}{(z-a_j)^l}+b_0^{(n)} \qquad (1)$$

of order n. If some $a_j=\infty$, instead of the factor $\frac{1}{(z-a_j)^l}$ $l=1, 2, \ldots, m_j^{(n)}$, we must write z^l. Let us find the necessary and sufficient conditions which must be applied to the interpolating points $z_k^{(n)}$ and the sequences $m_j^{(n)}$, such that for every function $f(z)$, which is regular on the set $\bar{B}$, the shown sequence $p_n(z)$, $n=0, 1, \ldots$, of generalised interpolating polynomials (1) for $f(z)$ uniformly converges on the set $\bar{B}$ to the function $f(z)$.

In all the discussions (but not in formulating the theorems) we assume that the set $\bar{B}$ is bounded. In the opposite case we use $\zeta=\frac{1}{z-a}$ with suitably chosen a.

Let us introduce the following notations: $w=\varphi_j(z)$ is the function mapping conformally, and one-to-one the domain B^j on to the domain $|w|>1$, the expansion of which in the neighbourhood of the point a_j is given by

$$c_j\varphi_j(z)=\frac{1}{z-a_j}+c_0^{(j)}+c_1^{(j)}(z-a_j)+\ldots+c_k^{(j)}(z-a_j)^k+\ldots \qquad c_j>0$$

If $a_j=\infty$, the expansion will be

$$c_j\varphi_j(z)=z+c_0^{(j)}+\frac{c_1^{(j)}}{z}+\ldots+\frac{c_k^{(j)}}{z^k}+\ldots, \qquad c_j>0$$

$z=\psi_j(w)$ is the inverse function for the function $w=\varphi_j(z)$, L_ρ^j is the image of the circle $|w|=\rho$, $\rho>1$ under the mapping $z=\psi_j(w)$; $L_\rho=\bigcup_{j=1}^{s}L_\rho^j$ is an internal contour line of the set $\bar{B}$; B_ρ^j is the domain bounded by L_ρ^j and contained in B^j.

By analogy with the case of a bounded set of type $\mathfrak{M}$ (see (4) and (5), Section 1.1.1 and Section 1.2) we would expect that the uniform convergence of the generalised interpolating polynomials $P_n(z)$ to the function $f(z)$ on $\bar{B}$ depends on the behaviour of the sequence of generalised polynomials

$$\omega_n(z)=\frac{\prod_{k=1}^{n+1}(z-z_k^{(n)})}{\prod_{j=1}^{s}(z-a_j)^{m_j^{(n)}}} \qquad (2)$$

It will be recalled (see Section 1.1.1) that if $z_k^{(n)} = \infty$, the factor $z - z_k^{(n)}$, disappears in this formula, and if $a_j = \infty$, the factor $(z - a_j)^{m_j^{(n)}}$ disappears. We must investigate the behaviour of the sequence $\omega_n(z)$ in the domains B^j, $j = 1, 2, \ldots, s$. We shall consider in these domains functions slightly different from function (2):

$$\omega_{j,n}(z) = \frac{\prod_{k=1}^{n+1}\left(1 - \frac{z - a_j}{z_k^{(n)} - a_j}\right)}{(z - a_j)^{m_j^{(n)}+1}\prod_{\substack{\nu=1 \\ \nu \neq j}}^{s}\left(1 - \frac{z - a_j}{a_\nu - a_j}\right)^{m_\nu^{(n)}}} = \frac{\prod_{k=1}^{n+1}(z - z_k^{(n)})\prod_{\substack{\nu=1 \\ \nu \neq j}}^{s}(a_j - a_\nu)^{m_\nu^{(n)}}}{(z - a_j)\prod_{\nu=1}^{s}(z - a_j)^{m_\nu^{(n)}}\prod_{k=1}^{n+1}(a_j - z_k^{(n)})} \tag{3}$$

For $a_j = \infty$

$$\omega_{j,n}(z) = z^{m_j^{(n)}+1}\frac{\prod_{k=1}^{n+1}\left(1 - \frac{z_k^{(n)}}{z}\right)}{\prod_{\substack{\nu=1 \\ \nu \neq j}}^{s}\left(1 - \frac{a_\nu}{z}\right)^{m_\nu^{(n)}}} = \omega_n(z) \tag{4}$$

The function $\omega_{j,n}(z)$ differs from the function $\omega_n(z)$ by the factor $\frac{A_j^{(n)}}{z - a_j}$ where $A_j^{(n)}$ is a constant. The constant factor has no influence on the evaluation of the remainder term for the interpolation: $r_n(z) = f(z) - p_n(z)$, but the factor $\frac{1}{z - a_j}$ simplifies the discussions below.

1.4.2 Principle of convergence for sets of type $\mathfrak{M}_s$

By analogy with Section 1.2 we introduce the function

$$\theta_{j,n}(z) = \frac{\sqrt[m_j^{(n)}+1]{\omega_{j,n}(z)}}{c_j\varphi_j(z)}, \quad z \in B^j, \; j = 1, 2, \ldots, s; \; n = 0, 1, \ldots \tag{1}$$

where the continuous branch of the root chosen is that for

which $\theta_{j,n}(a_j)=1$. In the domain B^j the function (1) is single-valued, regular and does not become zero. Let us assume

$$M_{j,n}(\rho)=\sup_{z\in L_\rho^j}|\omega_{j,n}(z)|,\quad \rho>1,\quad M_{j,n}=\sup_{z\in L^j}|\omega_{j,n}(z)| \tag{2}$$

Then

$$|\theta_{j,n}(z)|<\frac{\sqrt[m_j^{(n)}+1]{M_{j,n}}}{c_j},\quad z\in B^j,\quad j=1,2,\ldots,s;\quad n=0,1,\ldots \tag{3}$$

Let

$$m^{(n)}=\min_{j=1,2,\ldots,s} m_j^{(n)}$$

Theorem 1 If a sequence of generalised interpolating polynomials $p_n(z)$, $n=0,1,\ldots$, for any function $f(z)$, regular on a set $\overline{B}$ of type $\mathfrak{M}_s$, $s=1,2,\ldots,p$ is to converge uniformly to the function $f(z)$ on $\overline{B}$, it is necessary and sufficient that $\lim_{n\to\infty} m^{(n)}=\infty$ and that as $n\to\infty$

$$|\theta_{j,n}(z)|\rightrightarrows 1,\quad j=1,2,\ldots,s \tag{4}$$

in B^j or

$$\lim_{n\to\infty}\frac{\sqrt[m_j^{(n)}+1]{M_{j,n}}}{c_j}=1,\quad j=1,2,\ldots,s \tag{5}$$

COROLLARY If the function $f(z)$ is regular in the domain B_{ρ_0}, $\rho_0>1$ and condition (4) or (5) is fulfilled, then for any ρ_1, $1<\rho_1<\rho_0$

$$|f(z)-p_n(z)|\leqslant\frac{M_{\rho_1}}{\rho_1^{m^{(n)}}},\quad z\in\overline{B} \tag{6}$$

where M_{ρ_1} is a constant independent of n.

PROOF: NECESSITY That the condition $\lim_{n\to\infty} m^{(n)}=\infty$ is necessary was proved in the introduction to this chapter. The proof that conditions (4) and (5) are necessary is analogous to the proof that conditions (7) and (8) are necessary in Theorem 1, Section 1.2.

SUFFICIENCY Let (4) be fulfilled and let the function $f(z)$

be regular on $\bar{B}$ and also regular in some domain B_{ρ_0}, $\rho_0 > 1$. Let us arbitrarily fix ρ_1 such that $1 < \rho_1 < \rho_0$, and let us choose ρ_2 and ρ_3 such that $1 < \rho_3 < \rho_1 < \rho_2 < \rho_0$ and $\rho_1\rho_3 < \rho_2$ and also such that the contour L does not pass through the point ∞. From (4) it follows that in a closed domain $\bar{D}^j$, $j = 1, 2, \ldots, s$, bounded by the contour lines $L^j_{\rho_3}$ and $L^j_{\rho_2}$, we have

$$|\omega_n(z)| = \frac{|z-a_j|}{|A_j^{(n)}|}\,[c_j|\varphi_j(z)| + \varepsilon_{j,n}(z)]^{m_j^{(n)}+1}, \quad z \in \bar{D}^j$$

where $A_j^{(n)}$ is a constant and $\varepsilon_{j,n}(z) \rightrightarrows 0$ as $n \to \infty$ in $\bar{D}^j$.

Hence for $z \in L^j_{\rho_3}$ and $\zeta \in L^j_{\rho_2}$ we obtain

$$\frac{|\omega_n(z)|}{|\omega_n(\zeta)|} \leqslant \frac{|z-a_j|}{|\zeta-a_j|}\left(\frac{c_j\rho_2+\varepsilon_{j,n}}{c_j\rho_3-\varepsilon_{j,n}}\right)^{m_j^{(n)}+1}, \quad \varepsilon_{j,n} = \sup_{z\in\bar{D}^j}|\varepsilon_{j,n}(z)|$$

where $\varepsilon_{j,n} \to 0$ as $n \to \infty$.

Further, assuming that $z \in L_{\rho_3}$, from the formula for the remainder term of interpolation we have (see Theorem 2, Section 1.1.2)

$$\begin{aligned} |r_n(z)| = |f(z) - p_n(z)| &= \left|\sum_{j=1}^{s} \frac{1}{2\pi i}\int_{L^j_{\rho_2}} \frac{f(\zeta)}{\zeta - z}\,\frac{\omega_n(z)}{\omega_n(\zeta)}\,d\zeta\right| \\ &\leqslant \sum_{j=1}^{s}\frac{M_j}{2\pi}\left(\frac{c_j\rho_2+\varepsilon_{j,n}}{c_j\rho_3-\varepsilon_{j,n}}\right)^{m_j^{(n)}+1}\int_{L^j_{\rho_2}}\frac{|z-a_j|}{|\zeta-z|\,|\zeta-a_j|}\,|d\zeta| \end{aligned} \tag{7}$$

where $M_j = \sup_{z\in L^j_{\rho_2}}|f(z)|$. Let us map the z plane to the w plane for $a_j \neq \infty$ by means of the function $w = \frac{1}{z-a_j}$. The contour lines $L^j_{\rho_2}$ and $L^j_{\rho_3}$ are correspondingly transformed into some contour lines $\tilde{L}^j_{\rho_2}$ and $\tilde{L}^j_{\rho_3}$ and, assuming that $\zeta = a_j + \frac{1}{\xi}$, we have

$$\int_{L^j_{\rho_2}}\frac{|z-a_j|}{|\zeta-z|\,|\zeta-a_j|}\,|d\zeta| = \int_{\tilde{L}^j_{\rho_2}}\frac{|d\xi|}{|\xi-w|} \leqslant \frac{l^j_{\rho_2}}{\delta^j_{\rho_2\rho_3}}$$

where $l^j_{\rho_2}$ gives the length of $\tilde{L}^j_{\rho_2}$ and $\delta^j_{\rho_2\rho_3}$ is the distance between $\tilde{L}^j_{\rho_2}$ and $\tilde{L}^j_{\rho_3}$. Continuing as in the proof of sufficiency in Theorem 1, Section 1.2, we conclude that conditions (4) and (5) are sufficient on condition that $\lim\limits_{n\to\infty} m^{(n)}=\infty$. The theorem is proved.

The same remarks apply to this theorem as to Theorem 1, Section 1.2.

If in Theorem 1 we put $n=ms$, $m=1, 2, \ldots$, and $m_j^{(n)}=m$, $j=1, 2, \ldots, s$ then inequality (6) becomes

$$|f(z)-p_{ms}(z)|\leqslant\frac{M_{\rho_1}}{\rho_1^m},\quad z\in\bar{B}$$

EXAMPLE Let $\bar{B}$ be the ring $r\leqslant|z|\leqslant R$, $0<r\leqslant R$ (for $r=R$ the circle $|z|=r$). The set $\bar{B}$ is a set of type $\mathfrak{M}_2$ with complementary domains B^1: $|z|<r$ and B^2: $|z|>R$. Let $z_k^{(2n)}$, $k=1, 2, \ldots, 2n+1$, be the interpolating points uniformly distributed on the circumference $|z|=r_0$, $r\leqslant r_0\leqslant R$, $a_1=0$, $a_2=\infty$. Putting $m_j^{(2n)}=n$, $j=1, 2$, we have

$$\omega_{2n}(z)=\frac{1}{z^n}\prod_{k=1}^{2n+1}\left(z-z_k^{(2n)}\right)$$

$$\omega_{1,2n}(z)=\frac{1}{z^{n+1}}\prod_{k=1}^{2n+1}\left(1-\frac{z}{z_k^{(2n)}}\right),\quad \omega_{2,2n}(z)=z^{n+1}\prod_{k=1}^{2n+1}\left(1-\frac{z_k^{(2n)}}{z}\right)$$

$$\theta_{1,2n}(z)=\sqrt[n+1]{\prod_{k=1}^{2n+1}\left(1-\frac{z}{z_k^{(2n)}}\right)},\quad \theta_{2,2n}(z)=\sqrt[n+1]{\prod_{k=1}^{2n+1}\left(1-\frac{z_k^{(2n)}}{z}\right)}$$

We can show that $|\theta_{j,2n}(z)|\rightrightarrows 1$ inside B^j, $j=1, 2$, as $n\to\infty$ e.g. (4) is fulfilled. Let $z_k^{(2n)}=r_0e^{i\theta_k^{(2n)}}$. Then the points $\theta_k^{(n)}$ are uniformly distributed in the interval $[0, 2\pi]$, and, from the corollary to Lemma 2 in Section 1.3.1, on every closed set $E\subset B^1$ as $n\to\infty$ we have

$$\begin{aligned}\ln|\theta_{1,2n}(z)| &= \operatorname{Re}\left\{\frac{2n+1}{n+1}\,\frac{1}{2n+1}\sum_{k=1}^{2n+1}\ln\left(1-\frac{z}{r_0}e^{-i\theta_k^{(2n)}}\right)\right\}\\ &\rightrightarrows \operatorname{Re}\left\{2\cdot\frac{1}{2\pi}\int_0^{2\pi}\ln\left(1-\frac{z}{r_0}e^{-i\theta}\right)d\theta\right\}\\ &= \operatorname{Re}\left\{\frac{2}{2\pi i}\int_{|\zeta|=1}\ln\left(1-\frac{z}{r_0\zeta}\right)\frac{d\zeta}{\zeta}\right\}=0\end{aligned}$$

and consequently $|\theta_{1,2n}(z)| \rightrightarrows 1$ in B^1 as $n \to \infty$. Analogously we can show that $|\theta_{2,2n}(z)| \rightrightarrows 1$ in B^2 as $n \to \infty$.

Hence we conclude that: if the function $f(z)$ is regular in the ring $\frac{r}{\rho} < |z| < R\rho$, $\rho > 1$, and $p_{2n}(z)$, $n = 1, 2, \ldots$, are the generalised interpolating polynomials constructed with the points $z_k^{(2n)}$ for the function $f(z)$, then

$$|f(z) - p_{2n}(z)| \leqslant \frac{M_{\rho_1}}{\rho_1^n}, \quad r \leqslant |z| \leqslant R$$

where $1 < \rho_1 < \rho$ and the constant M_{ρ_1} is independent of n.

Theorem 2 If for a given function $f(z)$, on a set B of type $\mathfrak{M}_s$, for every natural m there is a generalised polynomial $p_{ms}(z)$ with poles at the points $a_j \in B^j$ and of an order not higher than ms such that

$$|f(z) - p_{ms}(z)| \leqslant \frac{M}{\rho^m}, \quad z \in \bar{B}, \ \rho > 1$$

where M is a constant independent of m, then the function $f(z)$ is regular in the domain B_ρ.*

The proof of this theorem is analogous to the proof of Theorem 2, Section 1.2.

From this theorem it follows that in the inequality (6) if the condition $m^{(n)} = n/s$ is fulfilled it is impossible that $\rho_1 > \rho_0$ if the function $f(z)$ is regular in B_{ρ_0} and has a singular point on L_{ρ_0}.

In the following section we shall indicate the selection principles of interpolating points for which conditions (4) or (5) are fulfilled.

1.4.3 *Principles of choice of interpolating points*

Let us first consider the analogue of the Fejer points (see Section 1.3.1). Let $\bar{B}$ be a set of type $\mathfrak{M}_s$. We shall say that the points $z_{k,j}^{(n)}$, $j = 1, 2, \ldots, s$; $k = 1, 2, \ldots, r_j^{(n)}$; $\sum_{j=1}^{s} r_j^{(n)} = n + 1$, $n = 0, 1, \ldots$, lying on the boundary L of the set $\bar{B}$, are

*More exactly, there is a function $f_*(z)$ regular on ρ such that $f_*(z) = f(z)$ for $z \in \bar{B}$.

uniformly distributed on L, if the points

$$z_{k,j}^{(n)},\quad k=1,\ 2,\ \ldots,r_j^{(n)}$$
$$n=0,1,\ \ldots$$

are uniformly distributed on L^j, $j=1,\ 2,\ \ldots,\ s$ (e.g. points $w_{k,j}^{(n)}=\varphi_j(z_{k,j}^{(n)})$ are equidistant on the circle $|w|=1$) and

$$\lim_{n\to\infty}\frac{r_j^{(n)}}{n}=r_j>0,\quad j=1,\ 2,\ \ldots,\ s \tag{1}$$

Let $m_j^{(n)}$, $j=1,\ 2,\ \ldots,\ s$; $n=0,\ 1,\ \ldots,$ be non-negative integers and $\sum_{j=1}^{s}m_j^{(n)}=n$. We apply the following condition to the numbers $m_j^{(n)}$

$$\lim_{n\to\infty}\frac{m_j^{(n)}}{r_j^{(n)}}=1,\quad j=1,\ 2,\ \ldots,\ s \tag{2}$$

We assume

$$\omega_n(z)=\frac{\prod_{j=1}^{s}\prod_{k=1}^{r_j^{(n)}}(z-z_{k,j}^{(n)})}{\prod_{j=1}^{s}(z-a_j)^{m_j^{(n)}}} \tag{3}$$

and introduce the functions $\omega_{j,n}(z)$ in accordance with Formula (3), Section 1.4.1 and functions $\theta_{j,n}(z)$ in accordance with Formula (1), Section 1.4.2.

Theorem 1 Let $\bar{B}$ be a set of type $\mathfrak{M}_s$, $s\geqslant 1$, and let the points $z_{k,j}^{(n)}$, $j=1,\ 2,\ \ldots,\ s$; $k=1,\ 2,\ \ldots,\ r_j^{(n)}$; $n=0,\ 1,\ \ldots,$ lie on the boundary L of the set $\bar{B}$, where $z_{k,j}^{(n)}\in L^j$ and conditions (1) and (2) are fulfilled. For the points $z_{k,j}^{(n)}$ to be uniformly distributed on L it is necessary and sufficient that as $n\to\infty$

$$|\theta_{j,n}(z)|\rightrightarrows 1,\ j=1,\ 2,\ \ldots,\ s \tag{4}$$

in B^j (see Formulae (1), Section, 1.4.2, (3), Section 1.4.1, and (3) and (2) above).

PROOF: NECESSITY Let $w_{k,j}^{(n)}=\varphi_j(z_{k,j}^{(n)})=e^{i\theta_{k,j}^{(n)}}$. Let us fix $j=j_0$, $j_0=1,\ 2,\ \ldots,\ s$. To be definite, let $a_{j_0}\neq\infty$. Let us introduce the functions

$$f_j(z, \theta)=\ln\left|\frac{z-\psi_j(e^{i\theta})}{(z-a_{j_0})(\psi(e^{i\theta})-a_{j_0})}\right|, \quad 0\leqslant\theta\leqslant 2\pi, \quad z\in L_\rho^{j_0}, \quad \rho>1$$

(If $a_{j_0}=\infty$ then we would introduce functions $f_j(z, \theta)=\ln|z-\psi_j(e^{i\theta})|$.) As the function $f_j(z, \theta)$, $j=1, 2, \ldots, s$ is equicontinuous for θ on $[0, 2\pi]$ for $z\in L_\rho^j$, $\rho>1$, and the points $\theta_{k,j}^{(n)}$ are uniformly distributed on $[0, 2\pi]$ on the basis of the corollary to Lemma 2, Section 1.3.1 we conclude that as $n\to\infty$

$$I_j^{(n)}=\frac{1}{r_j^{(n)}}\sum_{k=1}^{r_j^{(n)}}\ln\left|\frac{z-\psi_j\left(e^{i\theta_{k,j}^{(n)}}\right)}{(z-a_{j_0})\left(\psi_j\left(e^{i\theta_{k,j}^{(n)}}\right)-a_{j_0}\right)}\right| \rightrightarrows \frac{1}{2\pi}\int_0^{2\pi}\ln\left|\frac{z-\psi_j(e^{i\theta})}{(z-a_{j_0})(\psi_j(e^{i\theta})-a_{j_0})}\right|d\theta=I_j \tag{5}$$

on $L_\rho^{j_0}$, whereby

$$I_j^{(n)}=\frac{1}{r_j^{(n)}}\ln\left|\frac{\prod_{k=1}^{r_j^{(n)}}\left(z-z_{k,j}^{(n)}\right)}{(z-a_{j_0})^{r_j^{(n)}}\prod_{k=1}^{r_j^{(n)}}\left(z_{kj}^{(n)}-a_{j_0}\right)}\right|,$$

$$I_j=\operatorname{Re}\left\{\frac{1}{2\pi i}\int_{|\zeta|=1}\ln\frac{\psi_{j_0}(w)-\psi_j(\zeta)}{(\psi_j(\zeta)-a_{j_0})(\psi_{j_0}(w)-a_{j_0})}\frac{d\zeta}{\zeta}\right\}$$

Calculating these integrals from the residue theorem (proof of Theorem 1, Section 1.3.1), we find

$$I_j=\begin{cases}\ln\left|\dfrac{z-a_j}{(z-a_{j_0})(a_j-a_{j_0})}\right| & \text{for } j\neq j_0\\ \ln\left[c_{j_0}|\varphi_{j_0}(z)|\right] & \text{for } j=j_0\end{cases}$$

Thus Formulae (5) are equivalent to formulae:

$$\frac{1}{r_{j_0}^{(n)}}\ln\left|\left(\frac{a_{j_0}-a_j}{z-a_j}\right)^{r_j^{(n)}}\prod_{k=1}^{r_j^{(n)}}\frac{z-z_{k,j}^{(n)}}{z_{k,j}^{(n)}-a_{j_0}}\right|\rightrightarrows 0, \quad j\neq j_0, \quad z\in L_\rho^{j_0}$$

$$\frac{1}{r_{j_0}^{(n)}}\ln\left|\frac{1}{(z-a_{j_0})^{r_{j_0}^{(n)}}\left(c_{j_0}\varphi_{j_0}(z)\right)^{r_{j_0}^{(n)}}}\prod_{k=1}^{r_{j_0}^{(n)}}\frac{z-z_{k,j_0}^{(n)}}{z_{k,j_0}^{(n)}-a_{j_0}}\right|\rightrightarrows 0, \quad z\in L_\rho^{j_0}$$

Adding these relations and multiplying by $\frac{r_{j_0}^{(n)}}{m_{j_0}^{(n)}+1}$ we find

$$\ln \frac{\sqrt[m_{j_0}^{(n)}+1]{|\omega_{j_0,n}(z)|}}{c_{j_0}|\varphi_{j_0}(z)|} + \ln \left| \frac{(z-a_{j_0})^{\frac{1}{m_{j_0}^{(n)}+1}} \prod_{\substack{j=1 \\ j\neq j_0}}^{s} (a_{j_0}-a_j)^{\left(\frac{r_j^{(n)}}{m_j^{(n)}}-1\right)\frac{m_j^{(n)}}{m_{j_0}^{(n)}+1}}}{(c_{j_0}\varphi_{j_0}(z))^{\frac{r_{j_0}^{(n)}}{m_{j_0}^{(n)}+1}-1} \prod_{j=1}^{s} (z-a_j)^{\left(\frac{r_j^{(n)}}{m_j^{(n)}}-1\right)\frac{m_j^{(n)}}{m_{j_0}^{(n)}+1}}} \right| \rightrightarrows 0$$

for $z \in L_\rho^{j_0}$. As the second term uniformly tends to zero on $L_\rho^{j_0}$

$$f_j(a_\nu) = f_j(\psi_\nu(\infty)) = \frac{1}{2\pi i} \int_{|w|=1} f_j\left(\psi_\nu\left(\frac{1}{w}\right)\right)\frac{dw}{w} = \frac{1}{2\pi}\int_0^{2\pi} f_j(\psi_\nu(e^{i\theta}))\,d\theta \tag{6}$$

Hence (4) follows. The necessity is proved.

SUFFICIENCY According to the condition, the points $z_{k,j}^{(n)} \in L$ and conditions (1), (2) and (4) are fulfilled. From (4) it follows that as $n \to \infty$

$$\theta_{j,n}(z) \rightrightarrows 1, \quad i = 1, 2, \ldots, s \tag{7}$$

in B^j. The functions $\theta_{j,n}(z)$ do not become zero in the domain B^j and hence the functions $\ln\theta_{j,n}(z)$, normalised by the condition $\ln\theta_{j,n}(a_j)=0$, are single-valued and regular in B^j, and on the basis of (7) above and (1), Section 1.4.2, as $n \to \infty$

$$\ln\theta_{j,n}(z) = \ln \frac{\sqrt[m_j^{(n)}+1]{\omega_{j,n}(z)}}{c_j\varphi_j(z)} \rightrightarrows 0$$

in B^j. Differentiating with respect to z, we obtain, as $n \to \infty$

$$\frac{1}{m_j^{(n)}+1}\sum_{\nu=1}^{s}\sum_{k=1}^{r_\nu^{(n)}} \frac{1}{z-z_{k,\nu}^{(n)}} - \frac{1}{m_j^{(n)}+1}\sum_{\nu=1}^{s}\frac{m_\nu^{(n)}}{z-a_\nu} - \frac{1}{z-a_j} \rightrightarrows \frac{\varphi_j'(z)}{\varphi_j(z)} \tag{8}$$

on L_ρ^j, $\rho > 1$. Let us denote by $\overline{D}^j$ the set which is complementary to B^j, and by $\overline{D}^j$ the domain bounded by L_ρ^j, $\rho > 1$, and containing D_ρ^j. Let the function $f_j(z)$ be regular in $\bar{D}^j_{\rho_0}$, $\rho_0 > 1$. Fixing ρ, $1 < \rho < \rho_0$, such that the contour line L_ρ^j does not pass through the point ∞, we multiply (8) by $f_j(z)$ and integrate the relation obtained along L_ρ^j. As $n \to \infty$ we find

$$\frac{1}{m_j^{(n)}+1} \sum_{\nu=1}^{s} \sum_{k=1}^{r_\nu^{(n)}} f_j(z_{k,\nu}^{(n)}) - \frac{1}{m_j^{(n)}+1} \sum_{\substack{\nu=1 \\ \nu \neq j}}^{s} m_\nu^{(n)} f_j(a_\nu) \tag{9}$$

$$\to \frac{1}{2\pi i} \int_{L_\rho^j} f_j(z) \frac{\varphi_j'(z)}{\varphi_j(z)}\, dz$$

Clearly this formula is valid for the case where the point $z=\infty$ belongs to $\bar{D}^j$. Taking into account that the function $f_j\left(\psi_\nu\left(\frac{1}{w}\right)\right)$ $\nu \neq j$, is regular in the disc $|w| < 1$ and continuous in $|w| \leqslant 1$ and consequently

$$f_j(a_\nu) = f_j(\psi_\nu(\infty)) = \frac{1}{2\pi i} \int_{|w|=1} f_j\left(\psi_\nu\left(\frac{1}{w}\right)\right) \frac{dw}{w} = \frac{1}{2\pi} \int_0^{2\pi} f_j(\psi_\nu(e^{i\theta}))\, d\theta$$

and assuming that in the integral of Formula (9) $z = \psi_j(\rho e^{i\theta})$, and then making ρ unity, we conclude from (9) that as $n \to \infty$

$$\frac{1}{m_j^{(n)}+1} \sum_{\nu=1}^{s} \sum_{k=1}^{r_\nu^{(n)}} f_j\left(\psi_\nu\left(e^{i\theta_{k,\nu}^{(n)}}\right)\right) \to \sum_{\nu=1}^{s} \frac{m_\nu^{(n)}}{m_j^{(n)}+1} \frac{1}{2\pi} \int_0^{2\pi} f_j(\psi_\nu(e^{i\theta}))\, d\theta \tag{10}$$

for every function $f_j(z)$ regular on $\bar{D}^j$.

Now let $f(z)$ be an arbitrary function regular on $\bar{B}$ and let $\rho > 1$ such that $f(z)$ is regular on $\bar{B}_\rho$ and L_ρ does not pass through the point ∞. Then

$$f(z) = \sum_{j=1}^{s} f_j(z), \cdot f_j(z) = \frac{1}{2\pi i} \int_{L_\rho^j} \frac{f(\zeta)}{\zeta - z}\, d\zeta$$

(if the point $\infty \in B$, $f(\infty)$ must be added to one of the functions $f_j(z)$). The function $f_j(z)$, $j = 1, 2, \ldots, s$, is regular on $\bar{D}^j$. Assuming that these are the functions occurring in Formula (10), let us fix j_0, $j_0 = 1, 2, \ldots s$, and multiply the j-th relation in (10) by $\frac{m_j^{(n)}+1}{r_{j_0}^{(n)}}$ and then sum all the relations obtained. We then find that as $n\to\infty$

$$\frac{1}{r_{j_0}^{(n)}}\sum_{\nu=1}^{s}\sum_{k=1}^{r_\nu^{(n)}} f\left(\psi_\nu\left(e^{i\theta_{k,\nu}^{(n)}}\right)\right)\to\sum_{\nu=1}^{s}\frac{m_\nu^{(n)}}{r_{j_0}^{(n)}}\frac{1}{2\pi}\int_0^{2\pi} f(\psi_\nu(e^{i\theta}))\,d\theta \qquad (11)$$

for every function $f(z)$ regular on $\bar{B}$. Noting that a set $\bar{B}$ of type $\mathfrak{M}_s$ satisfies the conditions of Lemma 2, Section 1.7.3, it is easy to show that Formula (11) holds for every function $f(z)$ regular on B and continuous on $\bar{B}$.

Let us introduce a sub-class $\mathscr{K}$ of such functions $f(z)$. Let the function $\xi=\lambda(z)$ map conformally and one-to-one the domain $B=\bar{B}\setminus L$ on to some finite domain B_* bounded by analytic Jordan curves. We note that $\xi=\lambda(z)$ is regular in B and continuous on $\bar{B}$ and its inverse function $z=\mu(\xi)$ is regular on B_* and continuous on $\bar{B}_*$. Let us denote by L_*^j, $j=1, 2, \ldots, s$ the image of L^j when mapped by this function. According to a well-known theorem (see Gakhov [1]† pp. 374-378) there is a function $g(\xi)=u(\xi)+iv(\xi)$, regular in B_* and continuous on $\bar{B}_*$, such that its real part $u(\xi)$ is constant on each of the contours L_*^j except the contour $L_*^{j_0}$ on which it is a previously given analytic function. Hence it follows that the function $f(z)=g(\lambda(z))=u(\lambda(z))+iv(\lambda(z))$ is regular in B, continuous on $\bar{B}$ and $u(\lambda(z))$ is constant on every contour L^j: $u(\lambda(z))=\lambda_j$, $z\in L^j$, $j\neq j_0$. Functions $f(z)$ of this type make up the class $\mathscr{K}$. Now let $f(z)=u(\lambda(z))+iv(\lambda(z))\in\mathscr{K}$ and $\chi(\theta)=u(\lambda(\psi_{j_0}(e^{i\theta})))$, $0\leqslant\theta\leqslant 2\pi$. Considering Formula (11) for this function $f(z)$ and taking the real parts of each side, we find

$$\frac{1}{r_{j_0}^{(n)}}\sum_{k=1}^{r_{j_0}^{(n)}}\chi(\theta_{k,j_0}^{(n)})+\frac{1}{r_{j_0}^{(n)}}\sum_{\substack{\nu=1\\ \nu\neq j_0}}^{s} r_\nu^{(n)}\lambda_\nu\to\frac{m_{j_0}^{(n)}}{r_{j_0}^{(n)}}\frac{1}{2\pi}\int_0^{2\pi}\chi(\theta)\,d\theta+\frac{1}{r_{j_0}^{(n)}}\sum_{\substack{\nu=1\\ \nu\neq j_0}}^{s} m_\nu^{(n)}\lambda_\nu$$

Hence and from (1) and (2) as $n\to\infty$ we have

$$\frac{1}{r_{j_0}^{(n)}}\sum_{k=1}^{r_{j_0}^{(n)}}\chi(\theta_{k,j_0}^{(n)})\to\frac{1}{2\pi}\int_0^{2\pi}\chi(\theta)\,d\theta$$

†In the monograph mentioned here only the continuity of $u(\xi)$ is asserted, but from the proof it follows directly that, under the given circumstances, $v(\xi)$ is also continuous on $\bar{B}$.

Proceeding as in the proof of the sufficiency of Theorem 1, Section 1.3.1, we conclude that the points $\theta^{(n)}_{k,j_0}$ are uniformly distributed on $[0, 2\pi]$ and the points $z^{(n)}_{k,j_0} = \psi_j\left(e^{i\theta^{(n)}_{k,j}}\right)$ are uniformly distributed on L^j for any $j_0 = 1, 2, \ldots, s$. The theorem is proved.

Let us generalise Theorem 2, Section 1.3.1.

Theorem 2 If the function $f(z)$ is regular on the set $\overline{B}$ of type $\mathfrak{M}_s$, in particular regular in the domain $B_{\rho_0}, \rho_0 > 1$, then there is a sequence of generalised polynomials $p_n(z)$ of an order not higher than n such that

$$|f(z) - p_n(z)| \leqslant \frac{M_{\rho_1}}{\rho_1^{n/s}}, \quad z \in \overline{B}$$

where ρ_1 is an arbitrary number such that $1 < \rho_1 < \rho_0$ and M_{ρ_1} is a constant independent of n.

Examples of such generalised polynomials $p_n(z)$ are the interpolating polynomials with points uniformly distributed on the boundary L of a set $\overline{B}$ with the condition that in Equation (1)

$$r_j = \frac{1}{s}$$

It is possible to prove theorems analogous to Theorems 2, 3 and 4 of Section 1.3.1, but we shall not do this.

We can consider the analogy of the Van der Monde points for the set $\overline{B}$ of type $\mathfrak{M}_s$. The discussion does not differ in principle from that in Section 1.3.2, but instead of the Van der Monde determinant we have to investigate the function

$$v_n\left(\xi_1^{(n)}, \xi_2^{(n)}, \ldots, \xi_{n+1}^{(n)}\right) = \frac{\prod\limits_{1 \leqslant k < j \leqslant n+1} \left(\xi_k^{(n)} - \xi_j^{(n)}\right)}{\prod\limits_{k=1}^{n+1} \prod\limits_{j=1}^{s} \left(\xi_k^{(n)} - a_j\right)^{m_j^{(n)}}}$$

where $\sum\limits_{j=1}^{s} m_j^{(n)} = n$. It is assumed that the conditions $\lim\limits_{n \to \infty} m_j^{(n)} = \infty$, $j = 1, 2, \ldots, s$ are fulfilled.

When investigating the analogue of the Chebyshev points for a set $\overline{B}$ of type $\mathfrak{M}_s$ instead of considering the polynomial $p(z)$ introduced in Section 1.3.3, we investigate for fixed n the generalised polynomial

$$p^*(z)=\frac{\prod_{k=1}^{n+1}\left(z-\xi_k^{(n)}\right)}{\prod_{j=1}^{s}\left(z-a_j\right)^{m_j^{(n)}}} \tag{12}$$

where $\xi_k^{(n)}$ are now arbitrary points from $\overline{B}$ and $\sum_{j=1}^{s} m_j^{(n)}=n$, and we assume that

$$\mu_{n+1}(p^*)=\max_{j=1,2,\ldots,s}\ \sup_{z\in L^j}\frac{\sqrt[m_j^{(n)}+1]{|p_j(z)|}}{c_j},\quad \tilde{\mu}_{n+1}=\inf_{\xi_k^{(n)}\in\overline{B}}\ \mu_{n+1}(p^*) \tag{13}$$

where $p_j(z)$ is defined with respect to $p^*(z)$ in the same way as $\omega_{j,n}(z)$ is for $\omega_n(z)$ (see Formulae (3) and (4), Section 1.4.1).

Proceeding basically as in Section 1.3.3 we can show that there is a generalised polynomial $\tilde{\tau}_{n+1}(z)$ of form (12) such that

$$\mu_{n+1}(\tilde{\tau}_{n+1})=\tilde{\mu}_{n+1} \tag{14}$$

The zeros $z_k^{(n)}$, $k=1, 2, \ldots, n+1$; $n=1, 2, \ldots,$ of the generalised polynomial $\tilde{\tau}_{n+1}(z)$ can be chosen as interpolating points. These are called the Chebyshev interpolating points for $\overline{B}$. We have:

$$\omega_n(z)=\frac{\prod_{k=1}^{n+1}\left(z-z_k^{(n)}\right)}{\prod_{j=1}^{s}\left(z-a_j\right)^{m_j^{(n)}}}=\tilde{\tau}_{n+1}(z)$$

and

$$M_{j,n}=\sup_{z\in L^j}|\omega_{j,n}(z)|\leqslant c_j\tilde{\mu}_{n+1}$$

Theorem 3 If $\overline{B}$ is a set of type $\mathfrak{M}_s$, $z_k^{(n)}$ are the Chebyshev interpolating points for $\overline{B}$ and the conditions $\lim_{n\to\infty} m_j^{(n)}=\infty$, $j=1, 2, \ldots, s$ are fulfilled, then $|\theta_{j,n}(z)|\rightrightarrows 1$ in B^j, $j=1, 2, \ldots, s$, as $n\to\infty$ and $\lim_{n\to\infty}\sqrt[m_j^{(n)}+1]{M_{j,n}}=c_j$.

PROOF Let $z_k^{*(n)}$ be the Van der Monde points for $\overline{B}$ and

$$\omega_n^*(z)=\frac{\prod_{k=1}^{n+1}\left(z-z_k^{*(n)}\right)}{\prod_{j=1}^{s}\left(z-a_j\right)^{m_j^{(n)}}}, \quad M_{j,n}^*=\sup_{z\in L^j}|\omega_{j,n}^*(z)|$$

We have

$$1\leqslant\frac{\sqrt[m_j^{(n)}+1]{M_{j,n}}}{c_j}\leqslant\max_{j=1,2,\ldots,s}\frac{\sqrt[m_j^{(n)}+1]{M_{j,n}}}{c_j}\leqslant\max_{j=1,2,\ldots,s}\frac{\sqrt[m_j^{(n)}+1]{M_{j,n}^*}}{c_j}$$

But for the Van der Monde points

$$\lim_{n\to\infty}\frac{\sqrt[m_j^{(n)}+1]{M_{j,n}^*}}{c_j}=1, \quad j=1,2,\ldots,s$$

whence

$$\lim_{n\to\infty}\frac{\sqrt[m_j^{(n)}+1]{M_{j,n}}}{c_j}=1, \quad j=1,2,\ldots,s$$

and $|\theta_{j,n}(z)|\rightrightarrows 1$ in B^j as $n\to\infty$. The theorem is proved.

EXAMPLE Let $\overline{B}$ be the ring $r\leqslant|z|\leqslant R$, $0<r\leqslant R<+\infty$ (for $r=R$ the circle), B^1 be the circle $|z|<R$, and B^2 the domain $|z|>R$, $a_1=0$, $a_2=\infty$. Let us assume that $m_1^{(n)}=m$, $0<m<n$, and $m_2^{(n)}=n-m$. Let us find $\tilde{\tau}_{n+1}(z)$.

We must investigate the generalised polynomials

$$p^*(z)=\frac{1}{z^m}\prod_{k=1}^{n+1}(z-\xi_k)=z^{n+1-m}+\frac{a}{z^m}+\sum_{k=-m+1}^{n-m}a_kz^k, \quad \xi_k\in\overline{B} \tag{15}$$

Note that $r^{n+1}\leqslant|a|\leqslant R^{n+1}$. In our case

$$p_1(z)=\frac{1}{z^{m+1}}\prod_{k=1}^{n+1}\left(1-\frac{z}{\xi_k^{(n)}}\right)=(-1)^{n+1}\frac{p^*(z)}{az}, \quad p_2(z)=p^*(z) \tag{16}$$

and we must find the generalised polynomial $\tilde{\tau}_{n+1}(z)$ of the form (15) such that $\mu(\tilde{\tau}_{n+1})=\inf_{p^*}\mu(p^*)=\tilde{\mu}_{n+1}$ (see (13) and (14)). If we choose as the generalised polynomial $p^*(z)$

$$q(z)=z^{n+1-m}+\frac{a}{z^m}$$

which is possible where $r^{n+1}\leqslant|a|\leqslant R^{n+1}$, we obtain

$$\mu_{n+1}(q) = \max\left\{ \sqrt[m+1]{1+\frac{r^{n+1}}{|a|}},\quad \sqrt[n+1-m]{1+\frac{|a|}{R^{n+1}}} \right\}$$

and the quantity $\mu_{n+1}(q)$ will have its minimum value if $|a|$ is the positive root of the equation

$$\sqrt[m+1]{1+\frac{r^{n+1}}{|a|}} = \sqrt[n+1-m]{1+\frac{|a|}{R^{n+1}}} \tag{17}$$

We can show that every polynomial $q(z) = z^{n+1-m} + \frac{a}{z^m}$ where $|a|$ is the root of Equation (17), is a polynomial $\tilde{\tau}_{n+1}(z)$ and that there are no other generalised polynomials $\tau_{n+1}(z)$ of the form (15), and also that

$$\tilde{\mu}_{n+1} = \sqrt[m+1]{1+\frac{r^{n+1}}{|a|}} = \sqrt[n+1-m]{1-\frac{|a|}{R^{n+1}}} \tag{18}$$

where a is the positive root of the Equation (17). Now let

$$\tau^*_{n+1}(z) = q(z) + \omega(z),\ q(z) = z^{n+1-m} + \frac{xe^{i\theta}}{z^m}$$

$$x > 0,\quad 0 \leqslant \theta < 2\pi$$

$$\omega(z) = \sum_{k=-m+1}^{n-m} a_k z^k$$

be a generalised polynomial such that $\mu_{n+1}(\tau^*_{n+1}) = \inf_{\xi_k^{(n)} \in \bar{B}} \mu(p^*)$

Let us compare two numbers

$$\mu_1' = \sqrt[m+1]{1+\frac{r^{n+1}}{x}} \text{ and } \mu_2' = \sqrt[n+1-m]{1+\frac{x}{R^{n+1}}}$$

For example, let $\mu_1' \geqslant \mu_2'$. Clearly $\mu_1' \geqslant \tilde{\mu}_{n+1} \geqslant \mu_2'$ (see (18)). If $\varepsilon_l = e^{i\theta_l}$, $\theta_l = \frac{2\pi}{n+1} l + \frac{\theta}{n+1}$, $l = 0, 1, \ldots, n$, then

$$\frac{\sqrt[m+1]{|\tau^*_{1,\,n+1}(z\varepsilon_l)|}}{c_1} \leqslant \tilde{\mu}_{n+1},\quad \tau^*_{1,\,n+1}(z) = (-1)^{n+1}\frac{\tau^*_{n+1}(z)}{zxe^{i\theta}}$$

and from this we find $|q(r\varepsilon_l) + \omega(r\varepsilon_l)|^2 \leqslant \frac{x^2}{r^{2m}}\tilde{\mu}_{n+1}^{2(m+1)}$ or

$$\left(r^{n+1-m}+\frac{x}{r^m}\right)^2+2\mathrm{Re}\left\{e^{-i\theta}\left(r^{n+1-m}+\frac{x}{r^m}\right)\varepsilon_l^m\omega\left(r\varepsilon_l\right)\right\}+$$
$$+|\omega(r\varepsilon_l)|^2\leqslant\frac{x^2}{r^{2m}}\tilde{\mu}_{n+1}^{2(m+1)}$$

Summing this inequality for $l=0,1,\ldots,n$, we have

$$2\,\mathrm{Re}\left\{e^{-i\theta}\frac{x}{r^m}\mu_1'^{m+1}\sum_{l=0}^{n}\varepsilon_l^m\omega\left(r\varepsilon_l\right)\right\}$$
$$+\sum_{l=0}^{n}|\omega(r\varepsilon_l)|^2\leqslant\frac{x^2}{r^{2m}}\left(\tilde{\mu}_{n+1}^{2(m+1)}-\mu_1'^{2(m+1)}\right)$$

The right-hand side is not positive and the sum under the sign for the real part is equal to zero since $\sum\limits_{l=0}^{n}\varepsilon_l^\nu=0$ for $\nu=1,2,\ldots,n$. Consequently the left-hand side is not negative. Hence we conclude firstly that $\omega(r\varepsilon_l)=0$, $l=0,1,\ldots,n$, e.g. the polynomial $z^{m-1}\omega(z)$ of a degree not greater than $n-1$ has $n+1$ roots and consequently $\omega(z)\equiv 0$. Secondly we conclude that $\mu_{n+1}=\mu_1'$, e.g. x is a root of the Equation (17) and the proof is completed.

We shall not stop to consider the analogy to the fourth principle of choice of interpolating points (see Section 1.3.5). However, let us take one theorem relating to this group of problems.

Theorem 4 For any function $f(z)$ regular in the ring $1/R<|z|<R$, $R>1$, for any choice of interpolating points $z_k^{(n)}$ from the ring $1/\rho\leqslant|z|\leqslant\rho$, $1<\rho<R$, if the sequence of generalised polynomials $p_n(z)$ of the form

$$p_n(z)=\sum_{k=-n}^{n}a_k z^k$$

for the function $f(z)$ is to converge uniformly to $f(z)$ in the ring $1/r\leqslant|z|\leqslant r$, $1\leqslant r<R$, it is necessary and sufficient that the inequality $R>R_0$, be fulfilled, where R_0 is the greater root of $R_0(1+\rho r)^2=r(R_0-r)^2$.

PROOF: NECESSITY Noting that $R_0>r$, let us investigate the function

$$f(z)=\frac{z}{z+R_0}$$

regular in the ring $1/R_0 < |z| < R_0$ and let us assume $z_k^{(n)} = -\rho$. Then

$$\omega_{2n}(z) = \frac{1}{z^n}(z+\rho)^{2n+1}$$

and the remainder term of interpolation is

$$r_{2n}(z) = \frac{1}{2\pi i} \int\limits_{|\zeta|=R_1} \frac{f(\zeta)}{\zeta - z} \frac{\omega_{2n}(z)}{\omega_{2n}(\zeta)} d\zeta - \frac{1}{2\pi i} \int\limits_{|\zeta|=\frac{1}{R_1}} \frac{f(\zeta)}{\zeta - z} \frac{\omega_{2n}(z)}{\omega_{2n}(\zeta)} d\zeta$$

where $r < R_1 < R_0$, $R_1 > \rho$. At the point $z = \frac{1}{r}$

$$r_{2n}\left(\frac{1}{r}\right) = \int\limits_{|\zeta|=R_1} \frac{\zeta^{n+1} r^n \left(\frac{1}{r}+\rho\right)^{2n+1}}{(\zeta+R_0)\left(\zeta-\frac{1}{r}\right)(\zeta+\rho)^{2n+1}} d\zeta$$

$$- \frac{1}{2\pi i} \int\limits_{|\zeta|=\frac{1}{R_1}} \frac{\zeta^{n+1} r^n \left(\frac{1}{r}+\rho\right)^{2n+1}}{(\zeta+R_0)\left(\zeta-\frac{1}{r}\right)(\zeta+\rho)^{2n+1}} d\zeta$$

$$= -\frac{(-R_0)^{n+1} r^n \left(\frac{1}{r}+\rho\right)^{2n+1}}{\left(-R_0-\frac{1}{r}\right)(-R_0+\rho)^{2n+1}}$$

$$= (-1)^n \frac{R_0(1+\rho r)}{(1+rR_0)(R_0-r)} \left[\frac{R_0(1+\rho r)^2}{r(R_0-r)^2}\right]^n$$

The expression in the square brackets is equal to one and consequently $r_{2n}\left(\frac{1}{r}\right)$ tends to zero. From this it follows that for a concrete function regular in the ring $\frac{1}{R_0} < |z| < R_0$, there is no convergence, and so necessity is proved.

SUFFICIENCY If the function $f(z)$ is regular in the ring $\frac{1}{R} < |z| < R$, $R > R_0$, then, assuming $R_0 < R_1 < R$, for $\omega_{2n}(z) = \frac{1}{z^n} \prod_{k=1}^{2n+1} (z - z_k^{(n)})$, for any choice of points $z_k^{(n)}$ from the ring $\frac{1}{\rho} \leqslant |z| \leqslant \rho$ and any z from the ring $\frac{1}{r} \leqslant |z| \leqslant r$, we have for $|\zeta| = R_1$ and sufficiently large n that

$$\left|\frac{\omega_{2n}(z)}{\omega_{2n}(\zeta)}\right| \leqslant \frac{R_1^n}{|z|^n} \left(\frac{|z|+\rho}{R_1-\rho}\right)^{2n+1} \leqslant R_1^n r^n \left(\frac{\frac{1}{r}+\rho}{R_1-\rho}\right)^{2n+1}$$

$$= \frac{1+r\rho}{r(R_1-\rho)} \left[\frac{R_1(1+r\rho)^2}{r(R_1-\rho)^2}\right]^n$$

and for $|\zeta| = \frac{1}{R_1}$ and n sufficiently large

$$\left|\frac{\omega_{2n}(z)}{\omega_{2n}(\zeta)}\right| \leqslant \frac{R_1(1+r\rho)}{R_1-\rho}\left[\frac{R_1(1+r\rho)^2}{r(R_1-\rho)^2}\right]^n$$

Noting that $\frac{R_1(1+r\rho)^2}{r(R_1-\rho)^2} < 1$, from the formula for the remainder term of interpolation we easily find that $|r_{2n}(z)| \rightrightarrows 0$ in the ring $1/r \leqslant |z| \leqslant r$. The theorem is proved.

1.4.4 *Application to trigonometrical interpolation*

We recall that: the set G of points ζ of a plane ζ is called 2π-strip if the set G', obtained by displacing the set G in a parallel direction on the vector 2π, coincides with G.

Let $\overline{G}$ be a closed, 2π-strip set of points of the plane lying in some band $a < \operatorname{Im}\xi < b$ and such that its complement with respect to the ζ plane consists of two simply-connected domains, and let the function $g(\zeta)$ be regular on set $\overline{G}$ in the sense that $g(\zeta)$ is regular in some 2π-strip domain $G' \supset G$ and has a period of 2π. In some band $\beta \leqslant \operatorname{Re}\zeta < \beta + 2\pi$ let us take a sequence of interpolating points $\zeta_k^{(n)}$, $k = 1, 2, \ldots, 2_{n+1}, n = 1, 2, \ldots$, belonging to $\overline{G}$. As before if some value $\zeta_{k_0}^{(n)}$ among the points $\zeta_k^{(n)}$, $k = 1, 2, \ldots, 2n+1$, is encountered α times, this value must be an interpolating point of multiplicity α. For every $n = 1, 2, \ldots$ let us construct the trigonometric interpolating polynomial $S_n(\zeta) = \sum_{k=-n}^{n} A_k e^{i\zeta k}$ of order n.

Let us introduce the function $z = e^{i\zeta}$. This function maps the set $\overline{G}$ on to some closed set with an infinite number of sheets, the projection of this set on to the z plane being denoted by $\overline{B}$. This is a set of the type $\mathfrak{M}_2$ the complement of which with respect to an extended plane consists of the domains B^1 and B^2 respectively containing the points $z = 0$ and $z = \infty$. The points $\zeta_k^{(n)}$ become the points $z_k^{(n)} = e^{i\zeta_k^{(n)}}$; the trigonometrical polynomial $S_n(\zeta)$ is transformed into the generalised polynomial $p_n(z) = \sum_{k=-n}^{n} A_k z^k$, and the function $g(\zeta)$ is transformed into the function $f(z) = g\left(\frac{1}{i}\ln z\right)$.

Since the rotation around $z = 0$ means that $\frac{1}{i}\ln z$ is increased by 2π and since the function $g(z)$ has a period of

2π it can be shown that the function $f(z)$ is single-valued and regular on $\overline{B}$, and in view of the rule for the differentiation of complex functions, the generalised polynomial $p_n(z)$ will be the interpolating polynomial for $f(z)$ with points $z_k^{(n)}$ $k=1, 2, \ldots, 2n+1$. Besides this

$$\max_{\zeta\in\overline{G}} |g(\zeta)-S_n(\zeta)| = \max_{z\in\overline{B}} |f(z)-p_n(z)|.$$

The example shown above establishes the connection between the problems of interpolation with generalised polynomials and the problems of interpolation with trigonometrical polynomials.

1.5 INTERPOLATING PROCESSES FOR FUNCTIONS WHICH ARE REGULAR ON SETS OF TYPE $\mathfrak{M}_s^*$

Up to now we have considered approximation by generalised interpolating polynomials of functions which are regular on sets of the type $\mathfrak{M}_s$. We shall denote as a set of the type $\mathfrak{M}_s^*$, $s=1, 2, \ldots,$ any closed set $\overline{B}$ of the extended z plane not containing isolated points, the complement $C\overline{B}$ of which with respect to the extended plane consists of s finitely connected domains B^j $j=1, 2, \ldots, s$, pairs of which do not contain common points. The boundary of a domain B^j is denoted by L^j, and $L=\bigcup_{j=1}^{s} L^j$ denotes the boundary of the set $\overline{B}$. A set $\overline{B}$ of type $\mathfrak{M}_s^*$ is called a set of type $\widetilde{\mathfrak{M}}_s^*$ if all points of the boundary L of the set $\overline{B}$ are accessible from the complement $C\overline{B}$. A set $\overline{B}$ of type $\mathfrak{M}_s^*$ will be called a set of type $\overline{\mathfrak{M}}_s^*$ if the boundary L of the set $\overline{B}$ consists of a finite number of closed Jordan curves. If $\overline{B}$ is a set of type $\overline{\mathfrak{M}}_s^*$ then $\overline{B}$ is a set of type $\widetilde{\mathfrak{M}}_s^*$ but not always vice-versa.

1.5.1 *Statement of problems*

Let $\overline{B}$ be an arbitrary set of type $\mathfrak{M}_s^*$. Just as in Section 1.4.1 and retaining the same notation, let us formulate the problem of the approximation of functions $f(z)$ regular on $\overline{B}$ by means of the generalised interpolating polynomials

$$p_n(z) = \sum_{j=1}^{s} \sum_{l=1}^{m_j^{(n)}} \frac{b_{l,j}^{(n)}}{(z-a_j)^l} + b_0^{(n)} \tag{1}$$

of order n and let us show that the required conditions for the uniform convergence of the generalised interpolating polynomials $p_n(z)$ to $f(z)$ on $\overline{B}$ (for any function $f(z)$ regular on $\overline{B}$), are the conditions

$$\lim_{n \to \infty} m_j^{(n)} = \infty, \quad j = 1, 2, \ldots, s. \tag{2}$$

In Section 1.4 we made use of the functions $w = \varphi_j(z)$. In this section we shall use Green's function instead of the functions $\varphi_j(z)$. Let G be a finitely connected domain, the boundary L of which consists of l closed Jordan curves L_k, $k = 1, 2, \ldots, l$ $(l = 1, 2, \ldots)$. There is a single function $g(z, \zeta)$, $z \in G$, $\zeta \in G$, which for any $\zeta \in G$ satisfies the conditions (see Goluzin [1] pp. 287-291):

1. $g(z, \zeta)$ is a harmonic function of z in G except for the point $z = \zeta$;
2. when $z \to \zeta$ the difference $g(z, \zeta) - \ln \frac{1}{|z - \zeta|}$ if $\zeta \neq \infty$, or the difference $g(z, \infty) - \ln|z|$ remains finite;
3. when z tends arbitrarily to any point of the boundary L, the function $g(z, \zeta)$ tends to zero. Such a function $g(z, \zeta)$ is called Green's function for the domain G.

Let us note various properties of Green's function:

1. $g(z, \zeta) > 0$ for $z \in G$, $\zeta \in G$, $z \neq \zeta$.
2. If the function $z^* = \varphi(z)$ maps G one-to-one conformally on to the domain G^* bounded by a finite number of closed Jordan curves, and $z = \varphi^*(z^*)$ is the inverse function for $z^* = \varphi(z)$, then $g(\varphi^*(z^*), \varphi^*(\zeta^*))$ is Green's function for the domain G^*.
3. If G is a simply-connected domain $(l = 1)$ and $w = \varphi(z, \zeta)$ is a function of z conformally mapping one-to-one the domain G on to the domain $|w| > 1$ such that $\varphi(\zeta, \zeta) = \infty$, then the function $g(z, \zeta) = \ln|\varphi(z, \zeta)|$ is Green's function for the domain G.
4. Let $h(z, \zeta)$ be the conjugate function for $g(z, \zeta)$. If the domain G is not simply connected $(l > 1)$, then the function $h(z, \zeta)$ is not single-valued in the domain G. On each component L_k of the boundary L let us choose an arbitrary point z_k and construct in G the sections connecting in turn the points z_k such that we obtain a simply-connected domain G'. Fixing an arbitrary element of the function $h(z, \zeta)$ and continuing it analytically into G', we obtain a single-valued branch $h_*(z, \zeta)$ of the function $h(z, \zeta)$. We denote by L_k' the directed contour L_k going from the point z_k to the same point in a positive direction, and we also assume that

$L'=\bigcup_{k=1}^{l} L'_k$. The function $h_*(z, \zeta)$ is continuous on $G' \cup L'$. Moreover for any function $u(z)$ which is harmonic in G and continuous in $\bar{G}$ we have the formula

$$u(z)=\frac{1}{2\pi}\int_{L'} u(\zeta)\, dh_*(\zeta, z), \quad z \in G \tag{3}$$

We note that for a fixed $z \in G$ the function $h_*(z, \zeta)$ increases monotonically on every component L'_k when L'_k is described in the positive direction. We are thus able to introduce the parameter $\sigma=h_*(\zeta, z)-A_k$ on the k-th component of L'_k so that we can then choose the numbers A_k such that when all the components of L' are described in some way the parameter σ changes continuously in some interval (a, b).

Then $\zeta(\zeta \in L')$ will be some function of σ. We denote this by $\zeta(\sigma)$. Formula (3) can then be written:

$$u(z)=\frac{1}{2\pi}\int_a^b u(\zeta(\sigma))\, d\sigma$$

If we assume that $u(z)\equiv 1$ then we can show that $b-a=2\pi$ and can assume that $a=0$ and $b=2\pi$.

Let us now introduce the Green's function for an l simply-connected domain G, the complement of which on the extended plane does not contain isolated points. There is always a function $z^*=\varphi(z)$ which maps one-to-one conformally the domain G on to the domain G^* with the same degree of connectedness and bounded by closed analytic Jordan curves (see, for example, Goluzin [1] p. 258). If $g^*(z^*, \zeta^*)$ is Green's function for the domain G^* then the function $g^*(\varphi(z), \varphi(\zeta))$ $z \in G$ is called the Green's function for the domain G. From the properties of conformal mapping it follows that conditions 1 to 3 given above are fulfilled for the Green's function $g(z, \zeta)$ of the domain G, and it will have properties analogous to properties 1 to 4. It is also easy to show that there is a unique function $g(z, \zeta)$ for the domain G satisfying the conditions 1 to 3 indicated.

Let $g(z, \zeta)$ be the Green's function for a finitely-connected domain G. We assume that

$$u(z, \zeta)=\begin{cases} g(z, \zeta)-\ln\dfrac{1}{|z-\zeta|} & \text{for } \zeta\neq\infty \\ g(z, \infty)-\ln|z| & \text{for } \zeta=\infty \end{cases}$$

and call the quantity $\gamma(\zeta) = u(\zeta, \zeta)$, $\zeta \in G$ the Robin's constant of the domain G with respect to the point ζ, and the quantity $c = c(\zeta) = e^{-\gamma(\zeta)}$ the capacity of the set E, complementary to G, with respect to the point ζ. We note that if G is a simply-connected domain, the definitions for the capacity of the set E given here and given earlier are equivalent.

Returning to the problem posed at the beginning of the section we denote by $g_j(z, \zeta)$, $j = 1, 2, \ldots, s$ Green's function for the domain B^j (complementary to $\overline{B}$) and by B^j_ρ, $\rho > 1$ the domain defined by the inequality $g_j(z, a_j) > \ln \rho$. Every domain B^j_ρ, $\rho > 1$ contains a point a_j and there is $\rho^j_* > 1$ such that all domains B^j_ρ, $1 < \rho < \rho^j_*$, will have the same degree of connectedness as the domain B^j. Let $\rho_* = \min \rho^j_*$. The domain B^j_ρ, $1 < \rho < \rho_*$ has a boundary denoted by L^j_ρ and $L_\rho = \bigcup_{j=1}^{s} L^j_\rho$ is called the internal contour line of the set $\overline{B}$. Note that L_ρ consists of a finite number of closed analytic Jordan curves. We denote by B_ρ the domain bounded by L_ρ and containing $\overline{B}$.

As in Section 1.4.1 we introduce the functions

$$\omega_n(z) = \frac{\prod_{k=1}^{n+1} \left(z - z_k^{(n)}\right)}{\prod_{k=1}^{s} (z - a_j)^{m_j^{(n)}}}$$

and the functions $\omega_{j,n}(z)$ as in the Formulae (3), Section 1.4.1.

1.5.2 Convergence principle for sets of type $\mathfrak{M}_s^*$

Let $\gamma_j(\zeta)$ be the Robin's constant of B^j. Let us introduce the functions

$$\vartheta_{j,n}(z) = \frac{1}{m_j^{(n)} + 1} \ln|\omega_{j,n}(z)| - g_j(z, a_j) + \gamma_j(a_j), \quad z \in B^j \tag{1}$$

respectively harmonic in the domains B^j, and assume

$$m^{(n)} = \min_{j=1,2,\ldots,s} m_j^{(n)}, \quad M_{j,n} = \sup_{z \in L^j} |\omega_{j,n}(z)|, \quad c_j = e^{-\gamma_j(a_j)}$$

Theorem 1 For the sequence of generalised polynomials $P_n(z)$, $n = 1, 2, \ldots$, associated with any function $f(z)$ regular on a set $\overline{B}$ of type $\mathfrak{M}_s^*$, $s \geqslant 1$ to converge uniformly to $f(z)$

on $\bar{B}$ it is necessary and sufficient that $\lim_{n\to\infty} m^{(n)}=\infty$ and that either

$$\vartheta_{j,n}(z)\rightrightarrows 0, \quad j=1, 2, \ldots, s \tag{2}$$

in B^j, when $n\to\infty$ or

$$\lim \frac{\sqrt[m_j^{(n)}+1]{M_{j,n}}}{c_j}=1, \quad j=1, 2, \ldots, s \tag{3}$$

COROLLARY If the function $f(z)$ is regular in the domain B_{ρ_0}, $1<\rho_0<\rho_*$ and conditions (2) or (3) are fulfilled, then for any ρ_1, $1<\rho_1<\rho_0$

$$|f(z)-p_n(z)|\leqslant\frac{M_{\rho_1}}{\rho_1^{m^{(n)}}}, \quad z\in\bar{B} \tag{4}$$

where M_{ρ_1} is a constant independent of n.

The proof is analogous to the proof of Theorem 1, Section 1.4.2, but is based on the corresponding theorems for harmonic functions.

For a set of type $\mathfrak{M}_s^*$ there is a theorem which is formulated analogously to Theorem 1, Section 1.4.2 (only $\mathfrak{M}_s$ has to be replaced by $\mathfrak{M}_s^*$). The proof is analogous to the proof of the theorem mentioned above but is based on the following lemma.

Lemma 1 Let $P(z)$ be a function which is regular in the domain B^j complementary to a set $\bar{B}$ of type $\mathfrak{M}_s^*$ apart from a pole of order m_j at the points $a_j\in B^j$. If

$$\lim_{\rho\to 1+0}\sup_{z\in L_\rho^j}|P(z)|=M$$

then

$$|P(z)|\leqslant M\rho^{m_j}, \quad z\in L_\rho^j, \quad \rho>1$$

This is easily proved by considering the function

$$v(z)=\ln|P(z)|-m_j g_j(z, a_j)$$

which is harmonic in B^j apart from the points at which the function $P(z)$ becomes zero (at these $v(z)$ is equal to $-\infty$), and by applying to this function the maximum modulus principle.

1.5.3 Principles of choice of interpolating points

Let $\bar{B}$ be a set of type $\mathfrak{M}_s^*$, $s \geqslant 1$ and B^j be the complementary domains to $\bar{B}$. As in Section 1.5.1 we introduce a parameter σ and a function $\zeta_j(\sigma)$, $0 \leqslant \sigma < 2\pi$ with respect to the function $-h_j(z, \zeta)$ conjugate to Green's function $g_j(z, \zeta)$ Then (see (4), Section 1.5.1) for any function $u(z)$ harmonic in B^j and continuous on $\bar{B}^j$ we have

$$u(a_j) = \frac{1}{2\pi} \int_0^{2\pi} u(\zeta_j(\sigma))\, d\sigma \tag{1}$$

Let us consider the analogy to Fejer's points (see Sections 1.3.1 and 1.4.3). We shall say that the points $z_{k,j}^{(n)}$,

$$j = 1, 2, \ldots, s; \quad k = 1, 2, \ldots, r_j^{(n)};$$

$$n = 1, 2, \ldots; \ \sum_{j=1}^{s} r_j^{(n)} = n + 1$$

lying on the boundary L of the set $\bar{B}$ are equidistant on L if the points $z_{k,j}^{(n)}$, $k = 1, 2, \ldots, r_j^{(n)}$, $n = 1, 2, \ldots$, are uniformly distributed on L^j in the sense that the points $\sigma_{kj}^{(n)}$ defined by the condition $z_{k,j}^{(n)} = \zeta_j(\sigma_{k,j}^{(n)})$ are uniformly distributed in the interval $[0, 2\pi]$ and

$$\lim \frac{r_j^{(n)}}{n} = r_j > 0, \quad j = 1, 2, \ldots, s \tag{2}$$

Let $m_j^{(n)}$, $j = 1, 2, \ldots, s$; $n = 1, 2$, be non-negative integers and $\sum_{j=1}^{s} m_j^{(n)} = n$. Let us apply to the numbers $m_j^{(n)}$ the condition

$$\lim_{n\to\infty} \frac{m_j^{(n)}}{r_j^{(n)}} = 1, \quad j = 1, 2, \ldots, s \tag{3}$$

Let us introduce the functions $\omega_n(z)$ according to Formulae (3), Section 1.4.1 and the functions $\vartheta_{j,n}(z)$ as in Formula (1), Section 1.5.2.

Theorem 1 Let $\bar{B}$ be a set of type $\mathfrak{M}_s^*$, $s \geqslant 1$ and let the points

$$z_{k,j}^{(n)}, \quad j = 1, 2, \ldots, s; \quad k = 1, 2, \ldots, r_j^{(n)}; \quad n = 1, 2, \ldots$$

lie on the boundary L of the set $\bar{B}$, whereby $z_{k,j}^{(n)} \in L^j$ and conditions (2) and (3) are fulfilled. In order that the points $z_{k,j}^{(n)} \in L^j$ shall be uniformly distributed on L it is necessary and sufficient that when $n \to \infty$

$$\vartheta_{j,n}(z) \rightrightarrows 0, \quad j = 1, 2, \ldots, s \tag{4}$$

in B^j.

We shall not stop to prove this theorem but merely point out that it is based on Formula (1).

Analogies to the Van der Monde and Chebyshev points can be investigated just as in Section 1.4.3.

NOTE We can consider interpolation by means of generalised polynomials with several poles

$$a_{\nu,j}, \quad \nu = 1, 2, \ldots, l_j; \ a_{\nu,j} \in B^j, \ j = 1, 2, \ldots, s$$

In this case we can consider these poles as dependent on the order n of the generalised polynomial. A set $\bar{B}$ of a more general nature can be considered similarly.

1.6 INTERPOLATING PROCESSES FOR FUNCTIONS WHICH ARE CONTINUOUS ON CLOSED SETS AND REGULAR AT INTERNAL POINTS

Up till now we have considered uniform approximation of functions which are regular on closed sets by means of polynomials (and rational functions). In this way we were able, by making use of the regularity of the function in the domain containing the set under consideration, to introduce the external contour lines to evaluate the remainder term of interpolation.

In the present section we shall investigate the question of the approximation of functions which are continuous on a closed set and regular at its internal points also by means of interpolating polynomials.

This problem is considerably more complex than the previous one and we shall limit ourselves to investigating the case where the function is regular in the disc $|z|<1$ and continuous in the closed disc $|z|\leqslant 1$.

1.6.1 *Interpolating processes for functions regular in a disc and continuous in a closed disc*

In this section we shall consider functions $f(z)$ regular in the disc $|z|<1$ and continuous in the closed disc $|z|\leqslant 1$. For each such function $f(z)$ let us introduce a function

$$\omega(f,\delta)=\sup_{|z'-z''|\leqslant\delta}|f(z')-f(z'')|,\quad \delta>0;\quad |z'|\leqslant 1;\quad |z''|\leqslant 1$$

which is usually called the modulus of continuity for the function $f(z)$* in $|z|\leqslant 1$ and let us construct the sequence $\{L_n(f,z)\}_1^\infty$ of Lagrange interpolating polynomials such that $L_n(f,z)$ is a polynomial of degree not greater than $n-1$ with interpolating points at the n-th roots of -1:

$$L_n\left(f,z_k^{(n)}\right)=f\left(z_k^{(n)}\right),\quad k=1,2,\ldots,n \tag{1}$$

where $z_k^{(n)}=e^{i\theta_k^{(n)}}$, $\theta_k^{(n)}=\frac{2k-1}{n}\pi$, $k=1,2,\ldots,n$. In our case

$$\omega_{n-1}(z)=\prod_{k=1}^{n}\left(z-z_k^{(n)}\right)=z^n+1$$

and

$$L_n(f,z)=\sum_{k=1}^{n}\frac{\omega_{n-1}(z)}{\left(z-z_k^{(n)}\right)\omega_{n-1}'\left(z_k^{(n)}\right)}f\left(z_k^{(n)}\right)=\sum_{k=1}^{n}\frac{z^n+1}{z_k^{(n)}-z}\,\frac{z_k^{(n)}}{n}\,f\left(z_k^{(n)}\right) \tag{2}$$

The interpolating points $z_k^{(n)}$ are equidistant on $|z|=1$.

Let us show that the interpolating process converges uniformly on the compact sub-sets of $|z|<1$.

Theorem 1 (Fejer [2] *)* Let the function $f(z)$ be regular in the disc $|z|<1$ and continuous in the closed disc $|z|\leqslant 1$. Then the sequence $L_n(f,z)$ of Lagrange interpolating polynomials with interpolating points equidistant on $|z|=1$

*The properties of the modulus of continuity can be found in the book by I. P. Natanson [1] pages 106–111 where they are stated for real functions. They can be transferred in an obvious manner to the case of complex functions of a complex variable.

uniformly converges on the compact sub-sets of the disc $|z| < 1$ to the function $f(z)$.

PROOF Let us fix a point z from the circle $|z| < 1$. We have

$$L_n(f, z) = \sum_{k=1}^{n} \frac{z^n+1}{z_k^{(n)}-z} \frac{z_k^{(n)}}{n} f(z_k^{(n)}) = (1+z^n)\frac{1}{2\pi}\sum_{k=1}^{n} \frac{f\left(e^{i\theta_k^{(n)}}\right) e^{i\theta_k^{(n)}}}{e^{i\theta_k^{(n)}}-z} \frac{2\pi}{n}$$

When $n \to \infty$ the factor $1+z^n$ tends to one and the function $\frac{f(e^{i\theta})e^{i\theta}}{e^{i\theta}-z}$ is continuous in the interval $0 \leqslant \theta \leqslant 2\pi$. Consequently when $n \to \infty$ for every z of $|z| < 1$

$$L_n(f, z) \to \frac{1}{2\pi}\int_0^{2\pi} \frac{f(e^{i\theta})}{e^{i\theta}-z} e^{i\theta}\, d\theta = \frac{1}{2\pi i}\int_{|z|=1} \frac{f(\zeta)}{\zeta - z}\, d\zeta = f(z)$$

The uniform convergence of the sequence $L_n(f, z)$ on the compact sub-sets of the disc $|z| < 1$ follows from Vitali's theorem since the polynomials $L_n(f, z)$ are uniformly bounded on the compact sub-sets of the disc $|z| < 1$. In fact if $|f(z)| \leqslant M$ in $|z| \leqslant 1$ then

$$|L_n(f, z)| \leqslant 2\frac{1}{n}\frac{M}{1-|z|} n = \frac{2M}{1-|z|}$$

We note that the proof is entirely analogous for any equidistant system of points. The theorem is proved.

Let us now consider the behaviour of the interpolation process on the boundary of the disc. First we shall prove three lemmas.

Lemma 1 (Gronwall [1]*)* Let $z_k = e^{i\theta_k}$, $\theta_k = \frac{2k-1}{n}\pi$, $k = 1, 2, \ldots, n$ and $\mathscr{P}_{n-1}$ be the class of all polynomials $P_{n-1}(z)$ satisfying the conditions

$$|P_{n-1}(z_k)| \leqslant M, \quad M > 0, \quad k = 1, 2, \ldots, n \tag{3}$$

Then for quantities

$$l_n = \sup_{P_{n-1} \in \mathscr{P}_{n-1}} \sup_{|z| \leqslant 1} |P_{n-1}(z)|$$

we have

$$\frac{M}{\pi}\ln n < l_n = \frac{M}{n}\sum_{k=1}^{n} \frac{1}{\sin\frac{2j-1}{2n}\pi} < \frac{M}{\pi}(\ln n + 2\pi) \tag{4}$$

PROOF A polynomial $P_{n-1}(z)$ of the class in question can be written in the Lagrange form

$$P_{n-1}(z) = \sum_{k=1}^{n} \frac{1+z^n}{z_k - z} \frac{z_k}{n} w_k$$

where w_k are arbitrary points satisfying the condition $|w_k| \leqslant M$, $k=1, 2, \ldots, n$. We note that $\sup\limits_{|z|\leqslant 1} |P_{n-1}(z)|$ is reached on the circle $|z|=1$ and in what follows we assume $|z|=1$. Let $\sup\limits_{|z|\leqslant 1} |P_{n-1}(z)|$ be reached at the point $z=z_0 = e^{i\theta_0}$ ($|z_0|=1$). We have

$$|P_{n-1}(z_0)| \leqslant \sum_{k=1}^{n} \left| \frac{z_0^n+1}{z_k - z_0} \right| \frac{M}{n}$$

For the equal sign to hold in this inequality it is necessary and sufficient that $|w_k| = M$, $k=1, 2, \ldots, n$ and that

$$\arg \left\{ \frac{z_0^n+1}{z_k - z_0} z_k w_k \right\}$$

does not depend on k, e.g.

$$\arg \frac{z_k}{z_k - z_0} + \arg w_k = \alpha = \text{const}$$

or

$$\arg w_k = \alpha - \arg \frac{e^{i\theta_k}}{e^{i\theta_k} - e^{i\theta_0}} = \alpha - \arg \frac{e^{i\theta_k}}{e^{i\frac{\theta_k+\theta_0}{2}} \left(e^{i\frac{\theta_k-\theta_0}{2}} - e^{-i\frac{\theta_k-\theta_0}{2}} \right)}$$

$$= \alpha - \frac{\theta_k - \theta_0}{2} - \arg \frac{1}{2i \sin \frac{\theta_k - \theta_0}{2}}$$

$$= \alpha - \frac{\theta_k - \theta_0}{2} + \frac{\pi}{2} + \frac{\pi}{2} \left(1 - \operatorname{sign} \sin \frac{\theta_k - \theta_0}{2} \right) = \mu_k$$

Hence it follows that to solve the problem set, all we have to consider is the class of polynomials of the form

$$P_{n-1}(z) = \sum_{k=1}^{n} \frac{z^n+1}{z_k - z} \frac{z_k}{n} M e^{i\mu_k}$$

Let us transform the expression on the right-hand side

$$P_{n-1}(z) = -\sum_{k=1}^{n} \frac{z^n - z_k^n}{z - z_k} \frac{z_k}{n} M e^{i\mu_k} = -\frac{M}{n} \sum_{k=1}^{n} \sum_{j=0}^{n-1} z^{n-1-j} z_k^j z_k e^{i\mu_k}$$

$$= -\frac{M}{n} \sum_{j=0}^{n-1} \left(\sum_{k=1}^{n} z_k^{j+1} e^{i\mu_k} \right) z^{n-1-j}$$

$$= -\frac{M}{n} \sum_{j=0}^{n-1} \left(\sum_{k=1}^{n} e^{i\frac{2k-1}{n}\pi(j+1)} e^{i\alpha} e^{-i\frac{2k-1}{2n}\pi} e^{i\frac{\theta_0}{2}} ie^{i\frac{\pi}{2}\left(1-\operatorname{sign}\sin\frac{\theta_k-\theta_0}{2}\right)} \right) z^{n-1-j}$$

$$= -i\frac{M}{n} e^{i\left(\alpha+\frac{\theta_0}{2}\right)} \sum_{j=0}^{n-1} \left(\sum_{k=1}^{n} e^{i\frac{(2k-1)\left(j+\frac{1}{2}\right)\pi}{n}} \operatorname{sign}\sin\frac{\theta_k - \theta_0}{2} \right) z^{n-1-j}$$

If θ_0 lies between θ_{k_0} and θ_{k_0+1} then the sum in the round brackets has the form

$$-\sum_{k=1}^{k_0} e^{i\frac{2k-1}{2n}\pi(2j+1)} + \sum_{k=k_0+1}^{n} e^{i\frac{2k-1}{2n}\pi(2j+1)}$$

$$= -\frac{e^{i\frac{2j+1}{2n}\pi} - e^{i\frac{2j+1}{2n}\pi(2k_0+1)}}{1 - e^{i\frac{2j+1}{n}\pi}} + \frac{e^{i\frac{2j+1}{2n}\pi(2k_0+1)} - e^{i\frac{2j+1}{2n}\pi(2n+1)}}{1 - e^{i\frac{2j+1}{n}\pi}}$$

$$= \frac{2e^{i\frac{2j+1}{2n}\pi(2k_0+1)}}{1 - e^{i\frac{2j+1}{n}\pi}} = -\frac{e^{i\frac{2j+1}{n}k_0\pi}}{i\sin\frac{2j+1}{2n}\pi}$$

and

$$P_{n-1}(z) = \frac{M}{n} e^{i\left(\alpha+\frac{\theta_0}{2}\right)} \sum_{j=0}^{n-1} \frac{e^{i\frac{2j+1}{n}k_0\pi}}{\sin\frac{2j+1}{2n}\pi} z^{n-1-j}$$

$$= \frac{M}{n} e^{i\left(\alpha+\frac{\theta_0}{2}+\frac{2n-1}{n}k_0\pi\right)} \sum_{j=0}^{n-1} \frac{1}{\sin\frac{2j+1}{2n}\pi} \left(e^{-i\frac{2\pi}{n}k_0} z\right)^{n-1-j}$$

or

$$P_{n-1}(z) = \frac{M}{n} e^{i\left(\alpha+\frac{\theta_0}{2}-\frac{\pi k_0}{n}\right)} \sum_{j=1}^{n} \frac{1}{\sin\frac{2j-1}{2n}\pi} \left(e^{-i\frac{2\pi}{n}k_0} z\right)^{n-j}$$

The modulus of this polynomial reaches its greatest value on the circle $|z| = 1$ for $z = e^{i\frac{2\pi}{n}k_0}$ and this modulus is

$$I_n = \frac{M}{n} \sum_{j=1}^{n} \frac{1}{\sin \frac{2j-1}{2n} \pi}$$

If θ_0 coincides with θ_{k_0} then, proceeding in an analogous manner, we obtain the same result. Let us estimate this quantity from below. To do this we write it in the form

$$I_n = \frac{M}{2\pi} \left\{ \left(\frac{1}{2 \sin \frac{\pi}{2n}} + \frac{1}{2 \sin \frac{2n-1}{2n} \pi} \right) \frac{2\pi}{n} \right.$$
$$\left. + \left[\frac{1}{2 \sin \frac{\pi}{2n}} + \frac{1}{\sin \frac{3\pi}{2n}} + \dots + \frac{1}{\sin \frac{2(n-1)-1}{2n} \pi} + \frac{1}{2 \sin \frac{2n-1}{2n} \pi} \right] \frac{2\pi}{n} \right\}$$

and note that if the integral

$$\int_{\frac{\pi}{2n}}^{\frac{2n-1}{2n}\pi} \frac{1}{\sin \theta} d\theta$$

is calculated by means of the trapezium formula, then we obtain the expression in square brackets multiplied by $2\pi/n$. Since the function $1/\sin\theta$ is convex from above, we obtain from the integral

$$I_n > \frac{M}{2\pi} \left(\frac{2\pi}{n \sin \frac{\pi}{2n}} + \int_{\frac{\pi}{2n}}^{\frac{2n-1}{2n}\pi} \frac{d\theta}{\sin \theta} \right)$$
$$= \frac{M}{2\pi} \left(\frac{2\pi}{n \sin \frac{\pi}{2n}} + 2 \ln \cot \frac{\pi}{4n} \right) \geqslant \frac{M}{2\pi} \left(\frac{2\pi}{n \frac{\pi}{2n}} + 2 \ln \frac{4n}{\pi} \right)$$
$$= \frac{M}{\pi} \left(2 + \ln \frac{4}{\pi} + \ln n \right) > \frac{M}{\pi} \ln n$$

Let us now estimate I_n from above

$$I_n = \frac{M}{2\pi} \left\{ \left(\frac{1}{\sin \frac{\pi}{2n}} + \frac{1}{\sin \frac{2n-1}{2n} \pi} \right) \frac{2\pi}{n} + \sum_{j=2}^{n-1} \frac{1}{\sin \frac{2j-1}{2n} \pi} \frac{2\pi}{n} \right\}$$

The last sum is the integral for $\int_{\frac{\pi}{n}}^{\frac{n-1}{n}\pi} \frac{d\theta}{\sin \theta}$ and since the function $y=\frac{1}{\sin\theta}$ is convex from above, we have

$$I_n<\frac{M}{2\pi}\left\{\frac{4\pi}{n\sin\frac{\pi}{2n}}+\int_{\frac{\pi}{n}}^{\frac{n-1}{n}\pi}\frac{d\theta}{\sin\theta}\right\}=\frac{M}{2\pi}\left\{\frac{4\pi}{n\sin\frac{\pi}{2n}}+2\ln\operatorname{ctg}\frac{\pi}{2n}\right\}$$
$$\leqslant\frac{M}{2\pi}\left\{\frac{4\pi}{n\cdot\frac{2}{\pi}\cdot\frac{\pi}{2n}}+2\ln\frac{1}{\frac{\pi}{2n}}\right\}<\frac{M}{\pi}\{2\pi+\ln n\}$$

The lemma is proved.

Lemma 2 Let the function $u_k(z)$ be regular in the disc $|z|<1$ and continuous in the disc $|z|\leqslant 1$ and let the series

$$f(z)=\sum_{k=1}^{\infty}u_k(z)$$

uniformly converge in the circle $|z|\leqslant 1$ to the function $f(z)$. Then

$$L_n(f,\ z)=\sum_{k=1}^{\infty}L_n(u_k,\ z)$$

and the series on the right-hand side uniformly converges in the circle $|z|\leqslant 1$.

PROOF Note firstly that the function $f(z)$ is regular in the circle $|z|<1$ and continuous in $|z|\leqslant 1$. We denote

$$r_m(z)=\sum_{k=m+1}^{\infty}u_k(z),\quad \rho_m=\sup_{|z|\leqslant 1}|r_m(z)|$$

We have

$$L_n(f,\ z)=\sum_{k=1}^{m}L_n(u_k,\ z)+L_n(r_m,\ z)$$

and

$$|L_n(r_m,\ z)|\leqslant\frac{1}{\pi}\rho_m(\ln n+2\pi)$$

Hence the lemma is proved.

Note that this lemma is valid for any choice of interpolating points.

Lemma 3 For every real θ and every natural number n the inequality

$$\left| \sum_{k=1}^{n} \frac{\sin k\theta}{k} \right| \leqslant \lambda = \int_{0}^{\pi} \frac{\sin x}{x}\, dx \tag{5}$$

holds.

PROOF Let us assume $\lambda_n(\theta) = \sum_{k=1}^{n} \frac{\sin k\theta}{k}$.

Since the function $\lambda_n(\theta)$ is odd and periodic, it is sufficient to consider the behaviour in the interval $[0, \pi]$. It is easy to show that the points

$$\theta'_k = \frac{2k}{n}\pi, \qquad k = 1, 2, \ldots, \left[\frac{n-1}{2}\right]$$
$$\theta''_k = \frac{2k-1}{n+1}\pi, \quad k = 1, 2, \ldots, \left[\frac{n}{2}\right],$$

are respectively the points of minima and maxima of the function $\lambda_n(\theta)$ and there are no other extrema. Let us show that the maxima and minima of this function diminish as n increases. For example for the maxima we have

$$\lambda_n(\theta''_{k+1}) - \lambda_n(\theta''_k)$$
$$= \frac{1}{2} \int_{\frac{2k-1}{n+1}\pi}^{\frac{2k+1}{n+1}\pi} \left(\frac{\sin \frac{2n+1}{2}\theta}{\sin \frac{\theta}{2}} - 1 \right) d\theta < \frac{1}{2} \int_{\frac{2k-1}{n+1}\pi}^{\frac{2k+1}{n+1}\pi} \frac{\sin \frac{2n+1}{2}\theta}{\sin \frac{\theta}{2}}\, d\theta$$

Since the factor under the integral sign is negative for $\frac{2k-1}{n+1}\pi < \theta < \frac{4k}{2n+1}\pi$ and positive for $\frac{4k}{2n+1}\pi < \theta < \frac{2k+1}{n+1}\pi$, writing the last integral in the form of the sum of two integrals with respect to the indicated intervals and evaluating these integrals, we find:

$$\lambda_n(\theta''_{k+1}) - \lambda_n(\theta''_k) < \frac{1}{2 \sin \frac{2k}{2n+1}\pi} \int_{\frac{2k-1}{n+1}\pi}^{\frac{2k+1}{n+1}\pi} \sin \frac{2n+1}{2}\theta\, d\theta$$
$$= \frac{\cos \frac{2k+1}{2(n+1)}\pi - \cos \frac{2k-1}{2(n+1)}\pi}{(2n+1) \sin \frac{2k}{2n+1}\pi} < 0$$

Moreover, the last of the minima of the function $\lambda_n(\theta)$ is positive. In fact, for example, if n is even $(n=2m)$ we have

$$\lambda_n(\theta'_{m-1})=\lambda_n(\theta'_{m-1})-\lambda_n(\pi)=\frac{1}{2}\int\limits_{\frac{m-1}{m}\pi}^{\pi}\left[1-\frac{\sin\frac{4m+1}{2}\theta}{\sin\frac{\theta}{2}}\right]d\theta$$

$$\geqslant\frac{1}{2}\int\limits_{\frac{m-1}{m}\pi}^{\pi}\left[1-\frac{\sin\frac{4m+1}{2}\theta}{\sin\frac{2m}{4m+1}\pi}\right]d\theta=\frac{\pi}{2m}-\frac{\sin\frac{\pi}{2m}}{(4m+1)\sin\frac{2m}{4m+1}\pi}>0$$

From the above it follows that $\lambda_n(\theta)>0$ for $0<\theta<\pi$ and

$$\sup_{\theta}|\lambda_n(\theta)|=\lambda_n(\theta''_1)=\sum_{k=1}^{n}\frac{\sin\frac{k}{n+1}\pi}{k}=\sum_{k=1}^{n}\frac{\sin\frac{k}{n+1}\pi}{\frac{k}{n+1}\pi}\,\frac{\pi}{n+1}$$

The last sum is an approximate sum for the integral

$$\int\limits_{0}^{2\pi}\frac{\sin x}{x}\,dx$$

Since the function under the integral sign diminishes in the interval $[0, 2\pi]$ and in the sum the value of the function under the integral sign is evaluated at the right-hand ends of the sub-intervals, the sum is smaller than the integral, e.g. the estimate (5) holds. The lemma is proved.

We shall also require the following lemma.

Lemma 4 The inequality

$$\ln(n+1)<\sum_{k=1}^{n}\frac{1}{k}<\ln n+1$$

is valid.

PROOF By using the mean-value theorem it is easy to show that the inequalities

$$\frac{1}{k+1}<\ln(k+1)-\ln k<\frac{1}{k},\qquad k=1,\ 2,\ \ldots$$

are true. Summing the inequality on the left-hand side for

$\ldots, n-1$ $k=1, 2,$ and on the right-hand for $k=1, 2, \ldots, n$ we obtain the desired inequality and the lemma is proved.

Theorem 2 (Fejer [1] *)* There is a function $f(z)$ regular in the circle $|z|<1$ and continuous in the closed circle $|z|\leqslant 1$ such that the sequence $\{L_n(f, z)\}_1^\infty$ of Lagrange interpolating polynomials for equidistant interpolating points on $|z|=1$ diverges at the point $z=1$ more exactly

$$\overline{\lim_{n\to\infty}}\, L_n(f, 1)=\infty$$

PROOF Let us introduce the polynomials

$$P_{2n}(z)=\frac{1}{n}+\frac{z}{n-1}+\frac{z^2}{n-2}+\ldots+\frac{z^{n-1}}{1}-\frac{z^{n+1}}{1}-\frac{z^{n+2}}{2}-\ldots-\frac{z^{2n}}{n},\quad n=1, 2, \ldots,$$

and evaluate their moduli in the circle $|z|<1$. Assuming $z=e^{i\theta}$, $0\leqslant\theta\leqslant 2\pi$ we have

$$|P_{2n}(e^{i\theta})|=|e^{in\theta}|\left|\frac{e^{-n\theta i}}{n}+\frac{e^{-(n-1)\theta i}}{n-1}+\ldots+\frac{e^{-\theta i}}{1}-\frac{e^{\theta i}}{1}-\ldots-\frac{e^{(n-1)\theta i}}{n-1}-\frac{e^{n\theta i}}{n}\right|$$

$$=2\left|\frac{\sin n\theta}{n}+\frac{\sin (n-1)\theta}{n-1}+\ldots+\frac{\sin\theta}{1}\right|\leqslant 2\lambda=2\int_0^\pi\frac{\sin x}{x}\,dx$$

Thus all polynomials $P_{2n}(z)$, $n=1, 2, \ldots,$ are bounded on the circle $|z|=1$, that is, also inside the circle $|z|<1$, by the number 2λ. For

$$z_k^{(n)}=e^{i\theta_k^{(n)}},\quad \theta_k^{(n)}=\frac{2k-1}{n}\pi,\quad k=1, 2, \ldots,$$

we find:

$$P_{2n}(z_k^{(n)})=\frac{1}{n}+\frac{z_k^{(n)}}{n-1}+\frac{\left(z_k^{(n)}\right)^2}{n-2}+\ldots+\frac{\left(z_k^{(n)}\right)^{n-1}}{1}+\frac{z_k^{(n)}}{1}+\ldots+\frac{\left(z_k^{(n)}\right)^{n-1}}{n-1}-\frac{1}{n}$$

$$=\left(1+\frac{1}{n-1}\right)z_k^{(n)}+\left(\frac{1}{2}+\frac{1}{n-2}\right)(z_k^{(n)})^2+\ldots+\left(\frac{1}{n-1}+1\right)(z_k^{(n)})^{n-1}$$

Because of the uniqueness of the interpolating polynomial

$$L_n(P_{2n}, z) = \left(1 + \frac{1}{n-1}\right)z + \left(\frac{1}{2} + \frac{1}{n-2}\right)z^2 + \dots + \left(1 + \frac{1}{n-1}\right)z^{n-1}$$

whence

$$L_n(P_{2n}, 1) = 2\left(1 + \frac{1}{2} + \dots + \frac{1}{n-1}\right) > 2 \ln n$$

Analogously it can be shown that if m is an odd number, then

$$L_n(P_{2nm}, 1) = 2\sum_{\nu=1}^{n-1}\left(\frac{1}{\nu} - \frac{1}{n+\nu} + \frac{1}{2n+\nu} - \dots + \frac{1}{(m-1)n+\nu}\right) > 0$$

Besides this if $2m < n$ then

$$L_n(P_{2m}, z) = P_{2m}(z), \quad L_n(P_{2m}, 1) = 0$$

Let us now introduce the function

$$f(z) = \sum_{k=1}^{\infty} \frac{P_{2\cdot 3^{k^3}}(z)}{k^2}$$

The series on the right-hand side uniformly converges in the disc $|z| \leqslant 1$ since it is majorised by the convergent series $\sum_{k=1}^{\infty} \frac{2\lambda}{k^2}$ whence it follows that the function $f(z)$ is regular in $|z| < 1$ and continuous in $|z| \leqslant 1$. For this function on the basis of Lemma 2 and the above

$$L_{3^{n^3}}(f, 1) = \sum_{k=1}^{\infty} \frac{L_{3^{n^3}}\left(P_{2\cdot 3^{k^3}}, 1\right)}{k^2} > \frac{L_{3^{n^3}}\left(P_{2\cdot 3^{n^3}}, 1\right)}{n^2} \geqslant \frac{2 \ln 3^{n^3}}{n^2} = 2n \ln 3$$

whence the theorem is proved.

NOTE The theorem remains valid if $L_n(f, z)$ is an interpolating polynomial with points $z_k^{(n)}$ arbitrarily distributed on the circumference $|z| = 1$ (Alper [2]).

We recall that the equidistant points on the circle correspond in the interval $-1 \leqslant x \leqslant 1$ of the real axis to points: (see end of Section 1.3.1)

$$x_k^{(n)} = \cos \frac{2k+1}{2n} \pi, \quad k = 0, 1, \dots, n-1$$

In the constructive theory of functions of a real variable there is an analogous theorem to the Fejer theorem we have proved, e.g. there is a function continuous in the interval $-1 \leqslant x \leqslant 1$ for which the sequence of interpolating polynomials with the given interpolating points diverges. However, it can be shown (see Natanson [1] p. 549) that if we take all the given points with multiplicity two and assume that the derivative of the polynomial is equal to zero for each of the points, then for any function $f(x)$ continuous on $-1 \leqslant x \leqslant 1$ the sequence of such interpolating polynomials for $f(z)$ uniformly converges to $f(x)$ on $-1 \leqslant x \leqslant 1$. Let us show that in the case of a circle there is no analogous theorem to this.

For the function $f(z)$ regular in the circle $|z| < 1$ and continuous in $|z| \leqslant 1$ let us introduce interpolating polynomials $L_{n,m}(f, z)$, $m \geqslant 2$ of degree not greater than $nm-1$ satisfying the conditions

$$\begin{gathered} L_{n,m}(f, z_k^{(n)}) = f(z_k^{(n)}), \quad k = 1, 2, \ldots, n; \quad n = 1, 2, \ldots; \\ z_k^{(n)} = e^{i\frac{2k-1}{n}\pi} \\ \left[\frac{d^\nu}{dz^\nu} L_{n,m}(f, z)\right]_{z=z_k^{(n)}} = 0, \quad \nu = 1, 2, \ldots, m-1. \end{gathered} \tag{6}$$

We shall limit ourselves to considering the polynomials $L_{n,2}(f, z)$.

By using the results of Section 1.1.1, we find

$$\begin{aligned} L_{n,2}(f, z) = (z^n + 1)^2 \Bigg[& \frac{n-1}{n} \frac{1}{2\pi i} \sum_{k=1}^{n} \frac{f\left(e^{i\theta_k^{(n)}}\right) e^{i\theta_k^{(n)}}}{e^{i\theta_k^{(n)}} - z} i \frac{2\pi}{n} \\ & + \frac{1}{n} \frac{1}{2\pi i} \sum_{k=1}^{n} \frac{f\left(e^{i\theta_k^{(n)}}\right) e^{i\theta_k^{(n)}}}{\left(e^{i\theta_k^{(n)}} - z\right)^2} i \frac{2\pi}{n} \Bigg] \end{aligned} \tag{7}$$

In the disc $|z| < 1$ the first term in the square brackets tends to the integral

$$\frac{1}{2\pi i} \int_0^{2\pi} \frac{f(e^{i\theta})}{e^{i\theta} - z} e^{i\theta} i \, d\theta = \frac{1}{2\pi i} \int_{|\zeta|=1} \frac{f(\zeta)}{\zeta - z} d\zeta = f(z)$$

when $n \to \infty$ and the second, disregarding the multiplier $\frac{1}{n}$, tends to the integral

$$\frac{1}{2\pi i} \int_{|\zeta|=1} \frac{\zeta f(\zeta)}{(\zeta - z)^2} d\zeta = \frac{d}{dz}(z f(z))$$

and since when $n \to \infty$ and $|z| < 1$ the factor $(z^n + 1)^2 \to 1$ when $n \to \infty$

$$L_{n,2}(f, z) \to f(z)$$

in the circle $|z| < 1$ the convergence being uniform on compact sub-sets of the disc $|z| < 1$ this following, as in Theorem 1, from Vitali's theorem.

Let us now investigate the behaviour of $L_{n,2}(f, z)$ at the point $z = 1$.

If $|f(z)| \leqslant M$ in the circle $|z| \leqslant 1$ the second term in the square bracket in Formula (7) has the value

$$I_2 \leqslant \frac{M}{n^2} \sum_{k=1}^{n} \frac{1}{\left| e^{i\theta_k^{(n)}} - 1 \right|^2} = \frac{M}{2\pi n} \sum_{k=1}^{n} \frac{1}{\sin^2 \frac{\theta_k^{(n)}}{2}} \frac{\pi}{n}$$

$$= \frac{M}{2\pi n} \left[\left(\frac{1}{\sin^2 \frac{\pi}{2n}} + \frac{1}{\sin^2 \frac{2n-1}{2n}\pi} \right) \frac{\pi}{n} + \sum_{k=2}^{n-1} \frac{1}{\sin^2 \frac{2k-1}{2n}\pi} \frac{\pi}{n} \right]$$

$$\leqslant \frac{M}{2\pi n} \left[\frac{2\pi}{n \sin^2 \frac{\pi}{2n}} + \int_{\frac{\pi}{n}}^{\frac{n-1}{n}\pi} \frac{d\theta}{\sin^2 \theta} \right] = \frac{M}{2\pi} \left[\frac{2\pi}{n^2 \sin^2 \frac{\pi}{2n}} + \frac{2}{n \operatorname{tg} \frac{\pi}{n}} \right]$$

When $n \to \infty$ the last expression tends to $\frac{5M}{\pi^2}$ and consequently the sum under consideration is bounded for all n. The first term only differs from $L_n(f, 1)$ by the factor $\frac{n-1}{2n}$ and so for the function $f(z)$ considered in Theorem 2 we have

$$\overline{\lim_{n \to \infty}} L_{n,2}(f, 1) = \infty$$

Thus we have proved the theorem below.

Theorem 3 (Lozinsky [1]*)* There is a function $f(z)$, regular in the disc $|z| < 1$ and continuous in $|z| \leqslant 1$ such that the sequence of interpolating polynomials $L_{n,2}(f, z)$ does not converge in the closed disc $|z| \leqslant 1$ to $f(z)$ more exactly such that

$$\overline{\lim_{n \to \infty}} L_{n,2}(f, 1) = \infty$$

Lozinsky proved the validity of this theorem for interpolating polynomials $L_{n,m}(z)$ for any natural $m \geqslant 2$.

However, for any function $f(z)$ regular in the disc $|z|<1$ and continuous in $|z|\leqslant 1$ we can construct a sequence of polynomials uniformly converging to $f(z)$ on $|z|\leqslant 1$. Let us mention some methods of obtaining such a sequence. To do this let us consider 'approximate' interpolation.

Let us substitute for the given function $f(z)$ regular in the disc $|z|<1$ and continuous in $|z|\leqslant 1$ the 'approximate' function $f_r(z)=f(rz)$, $0<r<1$ regular in the disc $|z|<\frac{1}{r}$ and continuous in $|z|\leqslant\frac{1}{r}$. For the function $f_r(z)$ let us construct an interpolating polynomial $p_{n-1}(z)=L_n(f_r, z)$ with points equidistant on the circumference $|z|=1$. By using Hermite's formula for the remainder term of interpolation and assuming that $M=\sup_{|z|=1}|f(z)|$ we have on the circumference $|z|=1$

$$|f(z)-p_{n-1}(z)|\leqslant|f(z)-f_r(z)|+|f_r(z)-p_{n-1}(z)|$$
$$\leqslant\omega(f, 1-r)+\frac{1}{2\pi}\int\limits_{|\zeta|=\frac{1}{r}}\frac{|f_r(\zeta)|\cdot|z^n+1|}{|\zeta-z|\cdot|\zeta^n+1|}|d\zeta|\leqslant\omega(f, 1-r)$$
$$+\frac{1}{2\pi}\int\limits_0^{2\pi}\frac{2M}{\left|\frac{1}{r}e^{i\theta}-z\right|\left(\frac{1}{r^n}-1\right)r}d\theta$$
$$=\omega(f, 1-r)+\frac{Mr^n}{\pi(1-r^n)}\int\limits_0^{2\pi}\frac{d\theta}{|e^{i\theta}-z|}$$

Let us evaluate the last integral. By using the well-known inequality $\sin x\geqslant\frac{2}{\pi}x$, $0\leqslant x\leqslant\frac{\pi}{2}$ we have

$$\int\limits_0^{2\pi}\frac{d\theta}{|e^{i\theta}-rz|}=2\int\limits_0^{\pi}\frac{d\theta}{\sqrt{(1-r)^2+4r\sin^2\frac{\theta}{2}}}$$
$$\leqslant 2\int\limits_0^{\pi}\frac{d\theta}{\sqrt{(1-r)^2+4r\frac{\theta^2}{\pi^2}}}=\frac{\pi}{\sqrt{r}}\ln\frac{1+\sqrt{r}}{1-\sqrt{r}}$$

Hence, and from the above, it follows that

$$|f(z)-p_{n-1}(z)|\leqslant\omega(f, 1-r)+\frac{Mr^{n-\frac{1}{2}}}{1-r^n}\ln\frac{1+\sqrt{r}}{1-\sqrt{r}}, \quad |z|\leqslant 1 \tag{8}$$

Assuming $n \geqslant 2$, $r = 1 - \frac{\ln(n+1)}{n}$ and noting that

$$\left(1 - \frac{\ln(n+1)}{n}\right)^{n-\frac{1}{2}} = \frac{1}{\left(1 + \frac{\ln(n+1)}{n - \ln(n+1)}\right)^{\left(\frac{n - \ln(n+1)}{\ln(n+1)} + \frac{1}{2}\right)\frac{\ln(n+1)}{n - \frac{1}{2}\ln(n+1)}\left(n - \frac{1}{2}\right)}} \leqslant \frac{1}{n+1}$$

we find

$$|f(z) - p_{n-1}(z)| \leqslant \omega\left(f, \frac{\ln(n+1)}{n}\right) + \frac{M}{n} \ln \frac{4n}{\ln(n+1)}, \quad |z| \leqslant 1 \tag{9}$$

Obviously we can assume that $f(0) = 0$. Then using the properties of the modulus of continuity, we have

$$M \leqslant \omega(f, 1) = \omega\left(f, \frac{n}{\ln(n+1)} \frac{\ln(n+1)}{n}\right) \leqslant \left(\frac{n}{\ln(n+1)} + 1\right) \omega\left(f, \frac{\ln(n+1)}{n}\right)$$

and inequality (9) can be altered as follows:

$$|f(z) - p_{n-1}(z)| \leqslant \left(1 + \left(\frac{1}{\ln(n+1)} + \frac{1}{n}\right) \ln \frac{4n}{\ln(n+1)}\right) \omega\left(f, \frac{\ln(n+1)}{n}\right), \quad |z| \leqslant 1$$

and consequently

$$|f(z) - p_{n-1}(z)| \leqslant C_1 \omega\left(f, \frac{\ln(n+1)}{n}\right), \quad |z| \leqslant 1, \quad n = 2, 3, \ldots,$$

where C_1 is an absolute constant (we can put $C_1 = 1 + \left(\frac{1}{\ln 3} + \frac{1}{2}\right) \ln \frac{8}{\ln 3} < 3.76$).

Let us consider one more method of approximation. Let us introduce the partial sum of the Maclaurin series of the function $f(rz)$

$$p^*_{n-1}(z) = \frac{1}{2\pi i} \int\limits_{|\zeta| = \frac{1}{r}} \frac{f(r\zeta)(\zeta^n - z)}{(\zeta - z)\zeta^n} d\zeta$$

of degree not higher than $n-1$. Proceeding as above, we find

$$|f(z) - p^*_{n-1}(z)| \leqslant \omega(f, 1 - r) + \frac{M}{2} r^{n-\frac{1}{2}} \ln \frac{1 + \sqrt{r}}{1 - \sqrt{r}}, \quad |z| \leqslant 1$$

or, assuming $n \geqslant 3$ and $r = 1 - \frac{\ln n}{n}$

$$|f(z) - p^*_{n-1}(z)| \leqslant \omega\left(f, \frac{\ln n}{n}\right) + \frac{M}{2n} \ln \frac{4n}{\ln n}, \quad |z| \leqslant 1$$

and consequently

$$|f(z) - p^*_{n-1}(z)| \leqslant C_2 \omega\left(f, \frac{\ln n}{n}\right), \quad |z| \leqslant 1, \quad n = 3, 4, \ldots$$

where C_2 is an absolute constant (we can put $C_2 = 1 + \left(\frac{1}{\ln 3} + \frac{1}{3}\right) \ln \frac{12}{\ln 3} < 4\right)$.

Thus we have proved the following theorem.

Theorem 4 For every function $f(z)$ regular in the disc $|z| < 1$ and continuous in $|z| \leqslant 1$ there is a sequence of polynomials $p_{n-1}(z)$, $n = 3, 4, \ldots$, such that

$$|f(z) - p_{n-1}(z)| \leqslant C^{(0)} \omega\left(f, \frac{\ln n}{n}\right), \quad |z| \leqslant 1, \quad n = 2, 3, 4, \ldots$$

where $C^{(0)}$ is an absolute constant.

NOTE The theorem remains valid when $\ln n / n$ is altered to $1/n$ and $C^{(0)} = 12$. The same result also holds for a ring.

Let $\alpha_k^{(n)}$ and $\beta_k^{(n)}$ be given numbers,

$$z_k^{(n)} = e^{i \frac{2k-1}{n} \pi}, \; k = 1, 2, \ldots, n; \; n = 1, 2, \ldots; \; R_n(z)$$

be a polynomial of degree not greater than $2n - 1$ satisfying the conditions

$$R_n(z_k^{(n)}) = \alpha_k^{(n)}, \quad R''_n(z_k^{(n)}) = \beta_k^{(n)}, \quad k = 1, 2, \ldots, n$$

Theorem 5 (Kish [1] *)* Let $f(z)$ be a function regular in the disc $|z| < 1$ and continuous in $|z| \leqslant 1$ for which

$$\lim_{\delta \to 0} [\omega(f, \delta) \ln \delta] = 0 \tag{10}$$

Then the sequence of polynomials $R_n(z)$, $n = 1, 2, \ldots$, of a degree not higher than $2n - 1$ satisfying the conditions

$$R_n(z_k^{(n)}) = f(z_k^{(n)}), \quad R''_n(z_k^{(n)}) = \beta_k^{(n)}, \quad k = 1, 2, \ldots, n$$

in which $z_k^{(n)} = e^{i \frac{2k-1}{n}}$, $k = 1, 2, \ldots, n$ and $\beta_k^{(n)}$ are arbitrary numbers such that

$$\lim_{n\to\infty} \frac{\beta_k^{(n)} \ln n}{n^2} = 0$$

uniformly converges to $f(z)$ in the circle $|z| \leqslant 1$.

We shall not stop to prove this theorem.

Theorem 4 can be generalised.

Theorem 6 For every function $f(z)$ regular in the disc $|z| < 1$ and having in $|z| \leqslant 1$ a continuous derivative of order m, the modulus of continuity of which is $\omega_m(f, \delta)$ there is a sequence of polynomials $p_n(z)$, $n=m+2, m+3, \ldots,$ such that

$$|f(z) - p_n(z)| \leqslant C^{(m)} \left(\frac{\ln n}{n}\right)^m \omega_m\left(f, \frac{\ln n}{n}\right) \tag{11}$$

where $C^{(m)}$ is an absolute constant.

This is easily proved by applying Theorem 4 to $f^{(m)}(z)$ and successively repeating the procedure (see Mergelian [2]).

Theorems 4 and 6 can be extended to domains B bounded by a closed, smooth Jordan curve and also, with certain alterations to the right-hand sides of the inequalities, to domains bounded by a piecewise smooth Jordan curve (see Mergelian [1, 2], and Alper [1]).

There are reverse theorems to Theorems 4 and 6 and their analogues, these reverse theorems relating to simply-connected, bounded domains so that it is possible to determine the properties of a function $f(z)$ which is regular in the simply-connected domain B from the known rate of decrease of the quantity

$$\rho_n(f, \bar{B}) = \inf_{p_n(z)} \sup_{z \in \bar{B}} |f(z) - p_n(z)|$$

where $p_n(z)$ is a polynomial of degree not greater than n. In the case of the real axis such theorems were first obtained by Bernstein [1, 2]. For regular functions of a complex variable there are corresponding results, for example in the work of Mergelian [1, 2], and Dziadyk [1, 2].

In the following section we shall give one entirely general method proposed by Mergelian for changing the function $f(z)$ continuous in a closed set $\bar{B}$ and regular at its internal points, to an 'approximate' function regular on $\bar{B}$.

1.7 UNIFORM APPROXIMATION OF FUNCTIONS ON CLOSED, BOUNDED SETS BY MEANS OF POLYNOMIALS AND RATIONAL FUNCTIONS

To solve the problems indicated we shall have to use the extension of functions which are continuous on a closed, bounded set to the whole plane. The possibility of such an extension with the continuity and maximum modulus retained is well known. However, simple examples show that extension retaining the modulus of continuity is not always possible. In connection with this we shall introduce a somewhat unusual definition for the modulus of continuity. In the new definition of the modulus of continuity we in fact retain its monotonicity with respect to δ and add the property of convexity.

Let $f(z)$ be a function which is continuous on a closed, bounded set $\bar{B}$ and

$$\omega(f, \delta) = \sup_{|z'-z''| \leqslant \delta} |f(z') - f(z'')|,$$

$$z' \in \bar{B}, \quad z'' \in \bar{B}, \quad \delta \geqslant 0$$

If d is the diameter of the set $\bar{B}$, then $\omega(f, \delta) = \omega(f, d)$ for $\delta \geqslant d$.

In this section we shall denote as the modulus of continuity of the function $f(z)$ the function

$$\omega_f(\delta) = \sup_{\substack{y \geqslant 0 \\ 0 \leqslant x \leqslant \delta}} \left[\frac{x}{x+y}\,\omega(f, \delta+y) + \frac{y}{x+y}\,\omega(f, \delta-x)\right], \qquad \delta \geqslant 0$$

where we assume that for $x=0$, $y=0$ the expression in the square brackets is equal to $\omega(f, \delta)$.

Let us note certain properties of the function $\omega_f(\delta)$:

1. $\omega_f(0) = 0$;
2. $\omega_f(\delta) \geqslant \omega(f, \delta), \quad \delta \geqslant 0$
3. $\omega_f(\delta) \leqslant \omega(f, d), \quad \delta \geqslant 0$
4. the function $\omega_f(\delta)$ is continuous and does not diminish for $\delta \geqslant 0$
5. $\omega_f\left(\frac{1}{2}(\delta_1+\delta_2)\right) \geqslant \frac{1}{2}(\omega_f(\delta_1) + \omega_f(\delta_2)), \quad \delta_1 \geqslant 0, \quad \delta_2 \geqslant 0$;
6. $\omega_f(\delta_1+\delta_2) \leqslant \omega_f(\delta_1) + \omega_f(\delta_2), \quad \delta_1 \geqslant 0, \quad \delta_2 \geqslant 0$;
7. $\omega_f(\lambda\delta) \leqslant \lambda\omega_f(\delta), \quad \delta \geqslant 0, \quad \lambda \geqslant 1$.

Theorem (see Lebedev [2]) Let $\bar{B}$ be a closed, bounded set of points of the complex plane z; $f(z)$ be a function, continuous on $\bar{B}$. There is a function $f_*(z)$ defined on the whole complex z plane such that

(a) $f_*(z)=f(z)$ for $z\in\bar{B}$;
(b) $\omega_{f_*}(\delta)=\omega_f(\delta)$, $\delta>0$;
(c) $|f_*(z)|\leqslant \sup\limits_{z\in\bar{B}}|f(z)|$ for every z.

If we do not introduce this concept for the modulus of continuity and this theorem, then, when extending the continuous function whilst retaining the property of continuity, as is done in the following, we should have to use the ordinary modulus of continuity of the extended function. The actual fact that a continuous function can be extended is easily proved (see Alexandrov [1] pp. 284-287).

1.7.1 *Generalised Cauchy formula for functions with continuous derivatives of the first order with respect to x and y*

Let us prove the following theorem.

Theorem 1 Let the bounded domain D have a boundary L consisting of a finite number of closed, partially smooth Jordan curves, and let $G(z)$, $z=x+yi$ be a function with continuous partial derivatives of the first order with respect to x and y in the closed domain $\bar{D}$ (e.g. partial derivatives exist and are continuous in D and the values of these partial derivatives can be extended to the whole closed domain $\bar{D}$ whilst retaining the continuity). Then

$$G(z)=\frac{1}{2\pi i}\int_L \frac{G(\zeta)}{\zeta-z}\,d\zeta+\frac{1}{2\pi}\int\int_D \frac{g(\zeta)}{\zeta-z}\,d\sigma_\zeta, \quad z\in D \qquad (1)$$

where $d\sigma_\zeta$ is the area element for the integration variable $\zeta=\xi+i\eta$ and

$$g(z)=\frac{\partial G(z)}{\partial x}+i\,\frac{\partial G(z)}{\partial y}, \quad z=x+yi$$

If the point z is outside $\bar{D}$, then the right-hand part of (1) is equal to zero.

PROOF Let $f(z)$ be a function regular on $\bar{D}$ so that

$$\int_L f(z)G(z)\,dz = \int_L P\,dx - Q\,dy + i\int_L Q\,dx + P\,dy$$

where $P = \operatorname{Re}\{f(z)G(z)\}$ and $Q = \operatorname{Im}\{f(z)G(z)\}$. By the use of Green's formula and the Cauchy-Riemann conditions for the function $f(z)$ we find

$$\int_L f(z)G(z)\,dz = \int_D\int\left[\frac{\partial P}{\partial y} + \frac{\partial Q}{\partial x} + i\left(\frac{\partial Q}{\partial y} - \frac{\partial P}{\partial x}\right)\right]d\sigma_z$$

$$= -\int_D\int\left[\frac{\partial}{\partial x}(f(z)G(z)) + i\frac{\partial}{\partial y}(f(z)G(z))\right]d\sigma_z$$

$$= -i\int_D\int\left[G(z)\left(\frac{\partial f(z)}{\partial x} + i\frac{\partial f(z)}{\partial y}\right) + f(z)\left(\frac{\partial G(z)}{\partial x} + i\frac{\partial G(z)}{\partial y}\right)\right]d\sigma_z$$

$$= -i\int_D\int f(z)g(z)\,d\sigma_z.$$

Let us apply this formula to the function $f(z) = \frac{1}{z - z_0}$, $z_0 \in D$ and to the domain D_ε obtained from D by excluding a sufficiently small disc $|z - z_0| < \varepsilon$ with boundary C_ε:

$$\int_L \frac{G(z)}{z - z_0}\,dz - \int_{C_\varepsilon} \frac{G(z_0)}{z - z_0}\,dz - \int_{C_\varepsilon} \frac{G(z) - G(z_0)}{z - z_0}\,dz = -i\int_{D_\varepsilon}\int \frac{g(z)}{z - z_0}\,d\sigma_z. \tag{2}$$

But

$$\int_{C_\varepsilon} \frac{G(z_0)}{z - z_0}\,dz = 2\pi i G(z_0),$$

$$\left|\int_{C_\varepsilon} \frac{G(z) - G(z_0)}{z - z_0}\,dz\right| \leqslant 2\pi \sup_{z \in C_\varepsilon} |G(z) - G(z_0)|$$

and consequently if ε tends to zero in (2), in the limit we have

$$\int_L \frac{G(z)}{z - z_0}\,dz - 2\pi i G(z_0) = -i\int_{D_\varepsilon}\int \frac{g(z)}{z - z_0}\,d\sigma_z$$

which is equivalent to (1). If the point z_0 is outside $\overline{D}$ then we do not have to exclude the small disc and the last formula does not contain the term $2\pi i G(z_0)$. The theorem is proved.

NOTE 1 If the function $G(z)=U(z)+iV(z)$ is regular on some open set, then on this set $g(z)\equiv 0$ on the basis of the Cauchy-Riemann conditions.

NOTE 2 Clearly Formula (1) remains valid when the conditions applied to the function $g(z)$ are such that Green's formula is applicable. In particular, this formula is valid if $g(z)$ is bounded and has discontinuities on a finite number of piecewise smooth Jordan curves belonging to $\bar{D}$. The latter can be proved in the usual manner, in particular by subdividing the domain D into parts and applying to the subdomains the formula proved in the theorem.

1.7.2 *Analytic measure of sets*

Let $\bar{B}$ be a bounded, closed set of points of the z plane and B^1 be the domain complementary to $\bar{B}$ which contains the point ∞. Let us denote by $\mathfrak{R}(\bar{B})$ the class of functions $w=\mu(z)$, $\mu(\infty)=0$ regular on B^1 and satisfying in B^1 the condition $|\mu(z)|\leqslant 1$. In the neighbourhood of the point ∞ every function $\mu(z)\in\mathfrak{R}(\bar{B})$ can be expanded for any finite ζ in a series with respect to the negative powers of $z-\zeta$

$$\mu(z)=\frac{a(\mu)}{z-\zeta}+\frac{a_2}{(z-\zeta)^2}+\dots+\frac{a_n}{(z-\zeta)^n}+\dots,$$
$$a(\mu)=\lim_{z\to\infty} z\mu(z),$$

and the coefficient $a(\mu)$ does not depend on the choice of the point ζ. The quantity

$$\gamma(\bar{B})=\sup_{\mu(z)\in\mathfrak{R}(\bar{B})}|a(\mu)|\geqslant 0 \tag{1}$$

is said to be the analytic measure of the set $\bar{B}$.* The following are certain properties of $\gamma(\bar{B})$.

1. If $\bar{B}_1\subset\bar{B}_2$ are two closed, bounded sets, then $\gamma(\bar{B}_1)\leqslant\gamma(\bar{B}_2)$. This follows from the fact that the class $\mathfrak{R}(\bar{B})$ is 'not narrowed' when the set $\bar{B}$ is 'extended'.
2. The analytic measures of the set $\bar{B}$, of the boundary of the domain B^1, and of the complement CB^1 to the domain B^1 are all equal. This property is obvious.
3. There is a unique function $\mu_0(z)\in\mathfrak{R}(\bar{B})$ such that $a(\mu_0)=\gamma(\bar{B})$.

PROOF By taking a sequence of functions $\mu_n(z)\in\mathfrak{R}(\bar{B})$,

*In some works the term 'analytic capacity of the set $\bar{B}$' is used.

$n=1, 2, \ldots$ such that $a(\mu_n) \geqslant 0$ and $\lim\limits_{n\to\infty} a(\mu_n) = \gamma(\bar{B})$ and choosing a sub-sequence uniformly converging on the compact sub-sets of B^1 to some function $\mu_0(z)$ we can easily show that $\mu_0(z) \in \mathfrak{R}(\bar{B})$ and $a(\mu_0) = \gamma(\bar{B})$. The uniqueness of the function $\mu_0(z)$ (for proof see Khavinson [1]) will not be utilised.

4. If a set B lies in the disc $|z-\zeta| \leqslant R$, $R>0$ then for every function

$$\mu(z) = \frac{a(\mu)}{z-\zeta} + \frac{a_2}{(z-\zeta)^2} + \ldots \in \mathfrak{R}(\bar{B}) \tag{2}$$

there is

$$|\mu(z)| \leqslant \frac{1}{x-R}\gamma(\bar{B}), \quad |z-\zeta| \geqslant x > R \tag{3}$$

$$|a_n| \leqslant e\gamma(\bar{B}) R^{n-1} n, \quad n=2, 3, \ldots, \tag{4}$$

and if $\gamma(\bar{B})=0$ then the class $\mathfrak{R}(\bar{B})$ consists of the function $\mu(z) \equiv 0$ only.

PROOF Fixing z_0, $|z_0-\zeta| \geqslant x > R$ let us investigate the function regular in B^1

$$\mu_*(z) = \frac{(x-R)(\mu(z)-\mu(z_0))}{(z-z_0)(1-\overline{\mu(z_0)}\mu(z))} = \frac{(x-R)\mu(z_0)}{z-\zeta} + \ldots$$

Since $|\mu_*(z)| < 1$ for every z belonging to the intersection of B^1 and the disc $|z-\zeta| < R$, $\mu_*(z) \in \mathfrak{R}(\bar{B})$ whence it follows that $(x-R)|\mu(z_0)| \leqslant \gamma(\bar{B})$, i.e. the inequality (3) holds. To prove the inequality (4) we make use of the formula

$$a_n = \frac{1}{2\pi i} \int\limits_{|z-\zeta|=x>R} \mu(z)(z-\zeta)^{n-1}\,dz, \quad n=2, 3, \ldots,$$

and of inequality (3):

$$|a_n| \leqslant \gamma(\bar{B}) \frac{x^n}{x-R}$$

assuming $x = \frac{n}{n-1}R$ we have inequality (4).

For $n=2$ the inequality (4) can be slightly strengthened on condition that $|a(\mu)| = \gamma(\bar{B})$.

5. If the set $\bar{B}$ lies in the disc $|z-\zeta| \leqslant R$, $R>0$ and in the expression (2) $|a(\mu)| = \gamma(\bar{B})$ then

$$|a_2| \leqslant 2\gamma(\bar{B}) R \tag{5}$$

PROOF We can assume that $\gamma(\bar{B})>0$ and $a(\mu)=\gamma(\bar{B})$. Taking some $x>R$ let us consider the function

$$\varphi(w)=\frac{\mu\left(\zeta+\frac{x}{w}\right)\frac{x-R}{\gamma(\bar{B})\,w}-\frac{x-R}{x}}{1-\frac{x-R}{x}\mu\left(\zeta+\frac{x}{w}\right)\frac{x-R}{\gamma(\bar{B})\,w}}$$

Bearing in mind (3), the inequality $x>R$ and Schwartz's lemma, we find that the denominator does not become zero in the disc $|w|<1$. The function $\varphi(w)$ is regular in this disc and the following expansion holds:

$$\varphi(w)=\frac{a_2(x-R)}{\gamma(\bar{B})(2xR-R^2)}w+\alpha_2 w^2$$

From the expression for $\varphi(w)$ it follows that $|\varphi(w)|<1$ for $|w|<1$ and consequently

$$\left|\frac{a_2(x-R)}{\gamma(\bar{B})(2xR-R^2)}\right|<1$$

If x tends to infinity, we have inequality (5).

Let D^1 be a simply-connected domain containing the point ∞, the boundary of which contains not less than two points. Let us denote by $w=\Phi(z)$, $\Phi(\infty)=\infty$, $\Phi'(\infty)=1$ the function mapping one-to-one conformally the domain D^1 on to the domain $|w|>\rho$. The quantity $\rho=\rho(D^1)$ is called the conformal radius of the domain D^1 with respect to the point ∞. In the neighbourhood of $z=\infty$ the following expansion holds:

$$\Phi(z)=z+\alpha_0+\frac{\alpha_1}{z}+\ldots,$$

and $\Phi(z)\equiv z+\alpha_0$ in the case, and only in the case, where D^1 is the domain $|z+\alpha_0|>\rho$, $\rho>0$.

6. Let D^1 be a simply-connected domain containing the point ∞, the boundary of which contains not less than two points, and let $\bar{D}$ be the complement to D^1. If $\bar{B}\subset\bar{D}$ then $\gamma(\bar{B})\leqslant\rho(D^1)$ and the equal sign holds in the case, and only in the case, where $D^1=B^1$ where B^1 is the component of $C\bar{B}$ which contains the point ∞. If B^1 is a simply-connected domain then $\gamma(B)=\rho(B^1)$.

PROOF Let $\mu_0(z)\in\mathfrak{R}(\bar{B})$ and $a(\mu_0)=\gamma(\bar{B})$. The function $\mu_0(z)\Phi(z)$ is regular in D^1 and $|\mu_0(z)\Phi(z)|\leqslant\rho$ for $z\in D^1$, which when $z=\infty$ gives $\gamma(\bar{B})\leqslant\rho=\rho(D^1)$. The equals sign holds in the case, and only in the case, where $\mu_0(z) = \frac{\rho}{\Phi(z)}$, $z \in D^1$. Since $D^1 \subset B^1$ and $|\mu_0(z)| \to 1$ when z tends to the boundary of D^1, $\mu_0(z) = \frac{\rho}{\Phi(z)} \in \mathfrak{R}(\bar{B})$ in the case, and only in the case, where $D^1 = B^1$. The latter follows directly if we take B^1 for D^1.

The following can be proved very easily.

7. If $\bar{B}$ lies in the disc $|z-\zeta| \leqslant R$ then $\gamma(\bar{B}) \leqslant R$ and the equal sign holds in the case, and only in the case, where CB^1 is the closed circle $|z-\zeta| \leqslant R$.

We recall that the diameter of a bounded, closed set $\bar{B}$ is, by definition, the quantity

$$d = d(\bar{B}) = \sup_{z_1, z_2 \in \bar{B}} |z_1 - z_2|$$

8. If B^1 is a simply-connected domain then $\gamma(\bar{B}) \geqslant \frac{1}{4} d$ and $\bar{B}$ lies in some closed disc with radius $R = \frac{4}{\sqrt{3}} \gamma(\bar{B})$.

PROOF Let $w = \Phi(z)$, $\Phi(\infty) = 0$, $\Phi'(\infty) = 1$ be the above-mentioned function for the domain B^1 and $z = \psi(w)$ be the inverse function. Then if $z_1 \in \bar{B}$ the function

$$f(\zeta) = \frac{\rho}{\psi\left(\frac{\rho}{\zeta}\right) - z_1}$$

is regular and schlicht in the circle $|\zeta| < 1$ where $f(0) = 0$ and $f'(0) = 1$. On the basis of Koebe's theorem (see Goluzin [1] p. 56) every point not belonging to the image of the disc $|\zeta| < 1$ by the function $f(\zeta)$ is separated from the origin of coordinates by not less than $1/4$, e.g. $\frac{\rho}{|z_2 - z_1|} \geqslant \frac{1}{4}$ for z_1, $z_2 \in \bar{B}$ and consequently $\gamma(\bar{B}) = \rho \geqslant \frac{1}{4} d(\bar{B})$. The inequality $R \leqslant \frac{4}{\sqrt{3}} \gamma(\bar{B})$ follows directly from Young's theorem relating to the possibility of covering $\bar{B}$ with a closed disc of radius $\frac{1}{\sqrt{3}} d(\bar{B})$.

Let D be an arbitrary bounded set of points of the z plane and $\bar{B}$, $\bar{B} \subset D$, be an arbitrary closed set. The quantity

$$\gamma(D) = \sup_{\bar{B} \subset D} \gamma(\bar{B})$$

is called the analytic measure of the set D.

In addition we state the following properties of the analytic measure of the arbitrary set D.

9. The analytic measure of an open set is positive. This follows from properties (3) and (4).
10. When an open set is extended its analytic measure increases, e.g. if D_1 and D_2 are open sets and $\bar{D}_1 \subset D_2$ then $\gamma(D_1) < \gamma(D_2)$. This is easily proved by the maximum modulus principle.
11. If D is an open, bounded set and λ, $0 < \lambda < 1$ is a given number, then there is a closed set $\bar{B} \subset D$ consisting of a finite number of polygons, each of which consists of a finite number of links such that $\gamma(\bar{B}) \geqslant \lambda\gamma(D)$.

PROOF There is a closed set $\bar{B}_1 \in D$ such that $\gamma(\bar{B}_1) \geqslant \lambda\gamma(D)$. Since the distance from $\bar{B}_1$ to the boundary of the set D is greater than zero, only a finite number of the domains D_ν, $\nu = 1, 2, \ldots, k$ making up the set D contain points of the set $\bar{B}_1$. In each of the domains D_ν, $\nu = 1, 2, \ldots, k$ let us trace a closed polygon consisting of a finite number of links and containing all points of the set $\bar{B}_1$ lying in D_ν and let us denote by $\bar{B}$ the set of all points of these polygons. From properties (1) and (2) it follows that the set $\bar{B}$ is the desired set.

In connection with this see Section 1.3.4.

1.7.3 *Approximation of functions continuous on a closed set and regular at its internal points by means of functions regular on a closed set*

Let the function $f(z)$ be continuous on the bounded, closed set $\bar{B}$ of the z plane and regular at its internal points. We shall investigate the conditions applied to the set $\bar{B}$ and the function $f(z)$ of the above type such that when they are fulfilled for every $\varepsilon > 0$ there is a function $f_*(z)$ regular on $\bar{B}$ such that

$$|f(z) - f_*(z)| \leqslant \varepsilon, \quad z \in \bar{B}$$

Let L be the boundary of the set $\bar{B}$. For a given $\delta > 0$ we denote by $\bar{K}^*$ the closed set of those points $z \in \bar{B}$ which are separated from $C\bar{B}$, (L) by not more than $\delta/2$, and by $K_\zeta^0 = K_\zeta$, $\zeta \in L$ the disc $|z - \zeta| < \delta$. The union of all circles K_ζ. contains the set $\bar{K}^*$ and, by the Heine-Borel theorem, there is a finite number of points $\zeta_1, \zeta_2, \ldots, \zeta_m$ such that the circles K_{ζ_ν}, $\nu = 1, 2, \ldots, m$ cover $\bar{K}^*$.

It is not difficult to see that for any $\delta_0 > 0$ there is such a $\delta, 0 < \delta \leqslant \delta_0$, that the boundary of the set $\bigcup_{\nu=1}^{m} K_{\zeta_\nu}$ consists of a finite number of disjoint closed Jordan curves. We shall assume this in the following.

Let us form a non-empty, open set $D_\nu = K_{\zeta_\nu} \cap C\bar{B}$ and let $\gamma = \min_{\nu=1,\ldots,m} \gamma(D_\nu)$. From property (9), Section 1.7.2, we have $\gamma > 0$. If λ, $0 < \lambda < \frac{\gamma}{\delta}$ is an arbitrary number (note that $\gamma \leqslant \delta$), then $\gamma(D_\nu) > \lambda\delta$. Let F_ν^*, $\nu = 1, 2, \ldots, m$ be the set of points of a finite number of polygons (closed Jordan curves) of a finite number of links lying in D and such that $\gamma(F_\nu^*) \geqslant \lambda\delta$ (property (11), Section 1.7.2). Different polygons can have common points, but in every case the set of these polygons is the aggregate of the boundaries of a finite number of bounded domains. We call those of these domains that lie entirely in at least one of the discs K_{ζ_ν} domains of type A. The polygons referred to do not contain points of the set $\bar{B}$, but points from the set $\bar{B}$ may be found in domains of type A.

Let $f(z)$ be continuous on B and regular at internal points of this set and let $\omega_f(\delta)$ be its modulus of continuity. Our problem is to define a function regular on $\bar{B}$, e.g. regular on some open set containing $\bar{B}$ and close to $f(z)$ on $\bar{B}$. The process of constructing $f_\delta(z)$ is connected with the choice of δ, the circles K_{ζ_ν} and the polygons mentioned above.

The function $f_\delta(z)$ can be constructed very easily for a set $\bar{B}$ belonging to domains of the type A. Each such domain G belongs to at least one disc K_{ζ_ν} but can also belong to several such discs. Let us take one of these discs K_{ζ_ν} and put $f_\delta(z) = f(\zeta_\nu)$ for $z \in G$. Let us do this for all domains of type A containing points from B. Thus for the closed set $\bar{B}_0$ in $\bar{B}$ contained in domains of type A we approximate to function $f(z)$ with the function $f_\delta(z)$ which is equal to the respective constant in domains G of type A containing points of $\bar{B}$, whereby

$$|f(z) - f_\delta(z)| \leqslant \omega_f(\delta) \leqslant 2\omega_f\left(\frac{\delta}{2}\right), \qquad z \in \bar{B}_0 \tag{1}$$

As the distance from $\bar{B}_0$ to $\bar{B} - \bar{B}_0$ is positive, the set $\bar{B}_* = \bar{B} - \bar{B}_0$ is closed and its boundary L_* is contained in L. We must cover $\bar{B}_*$ with some open set D not containing points from $\bar{B}_0$ and then construct in D a function $f_\delta(z)$ approximating to $f(z)$. Let us now pass on to this rather complicated construction.

Let us introduce an open set $K=\bigcup_{\nu=1}^{m} K_{\zeta_\nu}$ and let F_0^* be the boundary of the set $\bar{B}\cup K$. From the fact that $\bar{B}$ is closed and the construction of K, it follows directly that every point $z\in F_0^*$ cannot be an internal and boundary point of $\bar{B}$ and cannot be an internal point of K. Hence it follows that the set $\bar{B}\cup K$ is open and F_0^* consists of a finite number of arcs of the circles $|z-\zeta|=\delta$ or complete circles, whereby F_0^* is one or a finite number of closed curves of the type mentioned, since in the opposite case there is a point from the set $\bar{B}$ which can unite the Jordan curve not intersecting F_0^* with some point of the complement of the set $\bar{B}\cup K$. The number of complementary domains to this closed set $F^*=\bigcup_{j=0}^{m} F_j^*$ is finite, one of them contains the point ∞ and this last domain does not contain points of $\bar{B}$. Let us unite with F^* all domains of type A and also those of the domains complementary to F^* which do not contain points of the set $\bar{B}$, and let F be the closed set obtained in this way. Let us also introduce the closed sets $F_\nu=F\cap K_{\zeta_\nu}$, $\nu=1, 2, \ldots, m$ whereby clearly $\gamma(F_\nu)\geqslant\lambda\delta$. Every set CF_ν is a connected, open set (domain) containing the point ∞, since every domain complementary to F_ν not containing the point ∞ must belong to a disc K_{ζ_ν} since its boundary consists of points of the set F_ν and must be one of the domains which are united to F. The complement D of the set F will be the bounded, open set mentioned above. The set D consists of a finite number of mutually disjoint bounded, finitely-connected domains the boundary of each of which is a single or several piecewise smooth Jordan curves. The boundary of D is denoted by Γ. It is not difficult to see that $D\subset\bar{B}_*\cup K=\bar{B}\cup K$. In fact let us assume that $z'\in D'$ where D' is one of the domains in D but $z'\notin\bar{B}_*\cup K$ and deduce from this a contradiction. On account of the construction of D the domain D' must contain the points $\bar{B}_*$, and the set $\bar{B}'$ of points $\bar{B}_*$ in D' is a closed set, the distance of which from the boundary of D' is positive. Hence it follows that at least one of the points z'' belonging to L_* and therefore to L belongs to the domain D'. Thus $z'\notin\bar{B}\cup K$ and $z''\in\bar{B}\cup K$. Both these points are in the domain D' and there is a Jordan arc l joining z' and z'' and belonging to D'. From the above it follows that the arc l belonging to D' must have common points with the boundary F_0^* of the set $\bar{B}\cup K$. This contradicts the definition for D and thus we have proved that $D\subset\bar{B}_*\cup K$.

Let $\bar{E}_\delta$ be the set of those points $\bar{B}_*$ separated from $C\bar{B}_*$ by not less than $\delta/2$. Each point $\bar{E}_\delta$ is obviously an internal point of $\bar{B}_*$, and every closed disc with its centre at $\bar{E}_\delta$ and radius $\delta/2$ belongs to $\bar{B}_*$. From $D\subset\bar{B}_*\cup K$ and the definition of K it follows directly that the open set $D^*=D-\bar{E}_\delta$ consisting of the points of D not belonging to $\bar{B}_*$ and those points of $\bar{B}_*$ separated from $C\bar{B}_*$ by less than $\delta/2$ belongs to the open set K. To construct the function $f_\delta(z)$ on D we need to construct D^* in the form of a union of disjoint sets D^*_ν, $\nu=1, 2, \ldots, m$, such that $D^*_\nu\subset K_{\zeta_\nu}$. Let D^*_1 be the set of all points $z\in D^*\cap K_{\zeta_1}$ and D^*_ν, $\nu=2, 3, \ldots, m$ be the set of all points $z\in D^*\cap K_{\zeta_\nu}$ not belonging to $\bigcup_{j=1}^{\nu-1} D^*_j$. Clearly we obtain the representation mentioned above in the form $D^*=\bigcup_{j=1}^{m} D^*_j$ where some of the terms may be empty sets. Note that the distance of any point $z\in D^*_\nu$ from ζ_ν is less than δ.

In what follows we shall require the lemma below.

Lemma 1 There is a function $g(z, \zeta)$, $z\in D, \zeta\in\bar{D^*}$ regular with respect to z on D for every $\zeta\in\bar{D^*}$ and continuous with respect to ζ on $\bar{D^*_\nu}$, $\nu=1, 2, \ldots, m$ for every $z\in D$ such that

$$\left.\begin{aligned} |g(z, \zeta)| &\leqslant \left(\frac{1}{\lambda}+\frac{3}{\lambda^2}\right)\frac{1}{\delta}; \\ \left|\frac{1}{\zeta-z}-g(z, \zeta)\right| &\leqslant \left(4+\frac{8}{\lambda}+\frac{24}{\lambda^2}\right)\frac{\delta^2}{|z-\zeta|^3} \end{aligned}\right\} \tag{2}$$

for $z\in D, \zeta\in\bar{D^*}$.

PROOF Let the function $\mu(z)\in\mathfrak{R}(F_\nu)$ and $a^{(\nu)}=\lim_{z\to\infty}(z\mu(z))=\gamma(F_\nu)$. In the neighbourhood of the point $z=\infty$ we have the expansion

$$\mu(z)=\frac{a^{(\nu)}}{z-\zeta_\nu}+\frac{a_2^{(\nu)}}{(z-\zeta_\nu)^2}+\ldots$$

Here $|a_2^{(\nu)}|\leqslant 2a^{(\nu)}\delta$ (property (5), Section 1.7.2). We can write $\mu(z)$ for any ζ in the form

$$\mu(z)=\frac{a^{(\nu)}}{z-\zeta}+\frac{\alpha_2^{(\nu)}}{(z-\zeta)^2}+\ldots,$$

where $\alpha_2^{(\nu)}=a_2^{(\nu)}-a^{(\nu)}(\zeta-\zeta_\nu)$, and then if $\zeta\in D^*_\nu$, and $|\alpha_2^{(\nu)}| \leqslant |a_2^{(\nu)}| + a^{(\nu)}|\zeta - \zeta_\nu| \leqslant 3a^{(\nu)}\delta$. For $\zeta \in D_\nu^*$ let us put

$$g_\nu(z, \zeta) = -\frac{\mu(z)}{a^{(\nu)}} + \frac{\alpha_2^{(\nu)}}{a^{(\nu)}}\left[\frac{\mu(z)}{a^{(\nu)}}\right]^2 = -\frac{1}{z-\zeta} + \frac{\alpha_3^{(\nu)}}{(z-\zeta)^3} + \dots$$

This function, as a function of z, is defined in the complementary domain to $F_\nu F_\nu^1$ which contains the point ∞, and

$$|g_\nu(z, \zeta)| \leqslant \frac{1}{a^{(\nu)}} + \frac{3a^{(\nu)}\delta}{a^{(\nu)}}\left(\frac{1}{a^{(\nu)}}\right)^2 \leqslant \left(\frac{1}{\lambda} + \frac{3}{\lambda^2}\right)\frac{1}{\delta}$$

Clearly the function

$$\varphi_\nu(z) = (z-\zeta)^2 + (z-\zeta)^3 g_\nu(z, \zeta)$$

is regular with respect to z in F_ν^1 for $\zeta \in D_\nu^*$ and in accordance with the maximum modulus principle for the given values for z and ζ

$$|\varphi_\nu(z)| \leqslant |z-\zeta|^2 + |z-\zeta|^3 |g_\nu(z, \zeta)| \leqslant (2\delta)^2 + (2\delta)^3\left(\frac{1}{\lambda} + \frac{3}{\lambda^2}\right)\frac{1}{\delta} = \left(4 + \frac{8}{\lambda} + \frac{24}{\lambda^2}\right)\delta^2$$

whence

$$\left|\frac{\varphi_\nu(z)}{(z-\zeta)^3}\right| = \left|\frac{1}{\zeta - z} - g_\nu(z, \zeta)\right| \leqslant \left(4 + \frac{8}{\lambda} + \frac{24}{\lambda^2}\right)\frac{\delta^2}{|z-\zeta|^3}$$

Assuming $g(z, \zeta) = g_\nu(z, \zeta)$ for $\zeta \in \overline{D}_\nu^*$, $z \in D$, the lemma is proved.

Let us now construct with respect to the function $f(z)$ a function $G_\delta(z)$ defined for the whole finite plane $z = x + yi$ which is continuous and has continuous derivatives with respect to x and y. To do this let us extend $f(z)$, given on $\overline{B}$, to the whole plane whilst retaining its modulus of continuity $\omega_f(\delta)$ and maximum modulus and let us construct the average function

$$G_\delta(z) = \int\int f(\zeta) K_\delta(|z-\zeta|)\, d\sigma_\zeta \tag{3}$$

where the averaging kernel $K_\delta(r)$ is defined in the following manner:

$$K_\delta(r) = \begin{cases} \frac{12}{\pi\delta^2}\left(1 - \frac{2r}{\delta}\right), & 0 \leqslant r \leqslant \frac{\delta}{2} \\ 0, & r > \frac{\delta}{2} \end{cases}$$

so that in Formula (3) the integration is carried out over the disc $|\zeta - z| \leqslant \frac{\delta}{2}$. Note that

$$\iint\limits_{|\zeta - z| \leqslant \frac{\delta}{2}} K_\delta(|\zeta - z|)\, d\sigma_\zeta = \int_0^{2\pi}\int_0^{\frac{\delta}{2}} K_\delta(r)\, r\, dr\, d\varphi = 1 \tag{4}$$

Formula (3) can be written in the form

$$G_\delta(z) = \frac{12}{\pi\delta^2}\int_0^{2\pi}\int_0^{\frac{\delta}{2}} f(x+\xi,\ y+\eta)\left(1 - \frac{2r}{\delta}\right) r\, dr\, d\varphi$$

where

$$f(\alpha,\ \beta) = f(\alpha + \beta i), \quad z = x + yi, \quad \xi = r\cos\varphi, \quad \eta = r\sin\varphi$$

That $G_\delta(z)$ is continuous on the whole z plane is obvious. Assuming $f(x+\xi,\ y+\eta) = \frac{\partial}{\partial\xi} f_1(x+\xi,\ y+\eta)$, where $f_1(\alpha,\ \beta)$ is the indefinite integral of $f(\alpha,\ \beta)$ with respect to α, and integrating by parts we can easily show that $G_\delta(z)$ has a derivative with respect to x which is continuous over the whole plane and which is obtained from Formula (3) by differentiating under the integral sign. An analogous argument can be applied to the derivative with respect to y.

Let us estimate on the whole plane the difference

$$|f(z) - G_\delta(z)| \leqslant \int\limits_{|\zeta - z| \leqslant \frac{\delta}{2}} |f(z) - f(\zeta)|\, K_\delta(|\zeta - z|)\, d\sigma_\zeta$$

$$\leqslant \int_0^{2\pi}\int_0^{\frac{\delta}{2}} \omega_f\left(\frac{\delta}{2}\right) K_\delta(r)\, r\, dr\, d\varphi = \omega\left(\frac{\delta}{2}\right)$$

Bearing in mind that $f(z)$ is regular at internal points of the set $\overline{B}$ and continuous on $\overline{B}$ for $z \in \overline{E}_\delta$ we find

$$f(z) = \frac{1}{2\pi}\int_0^{2\pi} f(z_{\text{,}} + re^{i\varphi})\, d\varphi, \qquad 0 \leqslant r \leqslant \frac{\delta}{2}$$

Multiplying both parts of this equation by $2\pi K_\delta(r)\, r\, dr$ and integrating with respect to r from $r = 0$ to $r = \frac{\delta}{2}$ on the basis of (4) we obtain $f(z)=G_\delta(z)$. Thus on the plane z

$$|f(z)-G_\delta(z)|\leqslant\omega_f\left(\frac{\delta}{2}\right)$$
$$f(z)=G_\delta(z),\qquad z\in\overline{E}_\delta. \tag{5}$$

Further we shall use the theorem from Section 1.7.1 and we need an estimate for the function

$$g_\delta(z)=\frac{\partial G_\delta(z)}{\partial x}+i\frac{\partial G_\delta(z)}{\partial y}$$

It is not difficult to see that the integrals of $\frac{\partial}{\partial x}K_\delta(|\zeta-z|)$ and $\frac{\partial}{\partial y}K_\delta(|\zeta-z|)$ over the disc $|\zeta-z|\leqslant\frac{\delta}{2}$ are zero, whence it follows that

$$g_\delta(z)=\iint\limits_{|\zeta-z|\leqslant\frac{\delta}{2}}[f(\zeta)-f(z)]\left[\frac{\partial}{\partial x}K_\delta(|\zeta-z|)+i\frac{\partial}{\partial y}K_\delta(|\zeta-z|)\right]d\sigma_\zeta$$

and

$$|g_\delta(z)|\leqslant\int_0^{2\pi}\int_0^{\frac{\delta}{2}}\omega_f\left(\frac{\delta}{2}\right)\left|\frac{\partial}{\partial r}K_\delta(r)\cos\varphi\right.$$
$$\left.+i\frac{\partial}{\partial r}K_\delta(r)\sin\varphi\right|r\,dr\,d\varphi=\frac{6\omega_f\left(\frac{\delta}{2}\right)}{\delta}$$

If $z\in\overline{E}_\delta$ then from $G_\delta(z)=f(z)$ and from the Cauchy-Riemann equation it follows that $g_\delta(z)=0$. Thus

$$|g_\delta(z)|\leqslant\frac{6}{\delta}\omega_f\left(\frac{\delta}{2}\right)\quad\text{on the plane } z;$$
$$g_\delta(z)=0,\qquad z\in\overline{E}_\delta \tag{6}$$

Applying the theorem from Section 1.7.1 to the function $G_\delta(z)$ and the domains forming the open set D, on the basis of the last equation for $g_\delta(z)$ we have

$$G_\delta(z)=\frac{1}{2\pi i}\int\limits_\Gamma\frac{G_\delta(\zeta)}{\zeta-z}d\zeta+\frac{1}{2\pi}\iint\limits_{D^*}\frac{g_\delta(\zeta)}{\zeta-z}d\sigma_\zeta,\quad z\in D$$

Let us now put

$$f_\delta(z)=\frac{1}{2\pi i}\int\limits_\Gamma\frac{G_\delta(\zeta)}{\zeta-z}d\zeta+\frac{1}{2\pi}\iint\limits_{D^*}g_\delta(\zeta)g(z,\zeta)\,d\sigma_\zeta,\qquad z\in D$$

Bearing in mind the properties of the function $g(z, \zeta)$ indicated in Lemma 1, formulating the difference quotient with respect to z and using the possibility of passing to the limit (see Smirnov [2] p. 162) it is not difficult to show that the last integral is a function which is regular with respect to z in D, and that therefore $f_\delta(z)$ is regular in D. For $z \in D$ we have

$$
\begin{aligned}
|O_\delta(z) - f_\delta(z)| &\leqslant \frac{1}{2\pi} \iint\limits_{D^*} |g_\delta(\zeta)| \left| \frac{1}{\zeta - z} - g(z, \zeta) \right| d\sigma_\zeta \\
&\leqslant \frac{3}{\pi\delta} \omega_f\left(\frac{\delta}{2}\right) \left\{ \iint\limits_{\substack{|\zeta - z| \geqslant 2\delta \\ \zeta \in D^*}} \left| \frac{1}{\zeta - z} - g(z, \zeta) \right| d\sigma_\zeta \right. \\
&\left. + \iint\limits_{\substack{|\zeta - z| \leqslant 2\delta \\ \zeta \in D^*}} \frac{d\sigma_\zeta}{|\zeta - z|} + \iint\limits_{\substack{|\zeta - z| \leqslant 2\delta \\ \zeta \in D^*}} |g(z, \zeta)| \, d\sigma_\zeta \right\} \\
&\leqslant \frac{3}{\pi\delta} \omega_f\left(\frac{\delta}{2}\right) \left\{ \left(4 + \frac{8}{\lambda} + \frac{24}{\lambda^2}\right) \delta^2 \iint\limits_{|\zeta - z| \geqslant 2\delta} \frac{d\sigma_\zeta}{|\zeta - z|^3} \right. \\
&\left. + \iint\limits_{|\zeta - z| \leqslant 2\delta} \frac{d\sigma_\zeta}{|\zeta - z|} + \left(\frac{1}{\lambda} + \frac{3}{\lambda^2}\right) \frac{1}{\delta} \iint\limits_{|\zeta - z| \leqslant 2\delta} d\sigma_\zeta \right\} \\
&\leqslant 6\omega_f\left(\frac{\delta}{2}\right) \left\{ \left(4 + \frac{8}{\lambda} + \frac{24}{\lambda^2}\right) \cdot \frac{1}{2} + 2 + \left(\frac{1}{\lambda} + \frac{3}{\lambda^2}\right) \cdot 2 \right\} \\
&= 12\left(2 + \frac{3}{\lambda} + \frac{9}{\lambda^2}\right) \omega_f\left(\frac{\delta}{2}\right)
\end{aligned}
\tag{7}
$$

Hence and from (5) we obtain the basic inequality:

$$|f(z) - f_\delta(z)| \leqslant \left(25 + \frac{36}{\lambda} + \frac{108}{\lambda^2}\right) \omega_f\left(\frac{\delta}{2}\right), \qquad z \in \bar{B}_* \tag{8}$$

To this should be added inequality (1) which holds for points $z \in \bar{B}_0$. The positive constant δ appearing in (8) and (1) can be chosen arbitrarily but λ depends not only on δ but also on the choice of circles K_{ζ_ν}, $\nu = 1, 2, \ldots, m$. We recall that this quantity λ satisfied $\gamma(K_{\zeta_\nu} \cap C\bar{B}) > \lambda\delta$, $\nu = 1, 2, \ldots, m$ and we used only this inequality in proving inequality (8). In this connection let us introduce a new notation:

$$\gamma_\zeta^\delta(\bar{B}) = \gamma(K_\zeta^\delta \cap C\bar{B}), \quad z \in L, \qquad \gamma^\delta(\bar{B}) = \inf_{\zeta \in L} \gamma_\zeta^\delta(\bar{B})$$

Besides this let us denote by D_δ the open set which is the union of D and the domains G of type A containing points of $\bar{B}$. From the above the following lemma may be derived.

Lemma 2 (Mergelian) Let $\bar{B}$ be a closed, bounded set of points of the z plane for given $\delta > 0$ and $\gamma^{\delta}(\bar{B}) > 0$ Then for any λ, $0 < \lambda < (1/\delta)\,\gamma^{\delta}(\bar{B})$ there is an open set D_{δ}, containing $\bar{B}$ and possessing the property that for every function $f(z)$ continuous on $\bar{B}$ and regular at the internal points of $\bar{B}$, there is a function $f_{\delta}(z)$ regular on D_{δ}, such that for points $z \in \bar{B}$ inequality (8) holds.

NOTE 1 The condition $\gamma^{\delta}(\bar{B}) > 0$ is fulfilled if there is a number $d > 0$ such that in any disc K_{ζ}^{δ} there is a Jordan arc belonging to $C\bar{B}$ with diameter not less than d since then (see property (8), Section 1.7.2) $\gamma_{\zeta}^{\delta}(\bar{B}) \geqslant \frac{d}{4}$ and $\gamma^{\delta}(\bar{B}) \geqslant \frac{d}{4}$. Such a number $d > 0$ exists if the complement $C\bar{B}$ to the set $\bar{B}$ consists of a finite number of finitely-connected domains. Let us now select a point z_j in each of the domains B^j complementary to $\bar{B}$ and then denote by δ_0 the distance of this set of points from $\bar{B}$. Let ζ be an arbitrary point belonging to L and let K_{ζ}^{0} be the disc $|z - \zeta| < \delta$, $\delta > 0$. Let us assume $\zeta \in L^1$. In the disc $|z - \zeta| < \varepsilon < \min(\delta, \delta_0)$ let us take an arbitrary point $z_1' \in B^1$ and let us join it to z_1 by a Jordan curve lying in B^1. Clearly K_{ζ}^{δ} has an arc of this curve with a diameter not less than $d = \min(\delta - \varepsilon, \delta_0 - \varepsilon)$. If $\delta < \delta_0$ then (see 1.7.2, property (8)), $\gamma(K_{\zeta}^{\delta} \cap C\bar{B}) \geqslant \frac{\delta - \varepsilon}{4}$ and since ε and the point $\zeta \in L$ are arbitrary, we have $\gamma^{\delta}(\bar{B}) \geqslant \frac{\delta}{4}$.

1.7.4 *Theorems of Mergelian and Vitushkin*

Let us show that the regular function $f_{\delta}(z)$ can be aproximated as well as is desired on $\bar{B}$ by means of rational functions. Let ε be the distance of $\bar{B}$ from the boundary of D_{δ}. In accordance with the Heine-Borel theorem the set $\bar{B}$ can be covered with a finite number of open discs of radius $\varepsilon/2$ with centres at the points of $\bar{B}$. Let B_{ε} be the union of all these discs. The complement $C\bar{B}_{\varepsilon}$ clearly consists of a finite number of finitely connected domains the boundary of each of which is one or more piecewise smooth Jordan arc, whereby $\bar{B} \subset B_{\varepsilon}$ and $\bar{B}_{\varepsilon} \subset D_{\delta}$. On the basis of the statements in Section 1.5, the function $f_{\delta}(z)$ can be uniformly approximated as well as is desired on $\bar{B}_{\varepsilon}$ by means of rational functions with poles outside B_{ε}. From the above follows (see inequalities (1) and (7), Section 1.7.3) Theorem 1.

Theorem 1 (Mergelian [2] *)* Let $\bar{B}$ be a closed, bounded set of the z plane. If the function $f(z)$ is continuous on $\bar{B}$, regular at all internal points of $\bar{B}$ and for some sequence of numbers $\delta_n > 0$, $\delta_n \to 0$ when $n \to \infty$, we have

$$\lim_{n\to\infty} \frac{\delta_n^2 \omega_f\left(\frac{\delta_n}{2}\right)}{\gamma^{\delta_n}(\bar{B})^2} = 0 \tag{1}$$

then the function $f(z)$ can be uniformly approximated on $\bar{B}$ as well as is desired by means of rational functions with poles outside $\bar{B}$.*

NOTE 1 Condition (1) will obviously be fulfilled for any function $f(z)$ continuous on $\bar{B}$ and regular at all internal points of $\bar{B}$, if the following condition is fulfilled.

(a) There is $\lambda > 0$ such that for any sufficiently small $\delta > 0$ the inequality $\gamma^{\delta}(\bar{B}) > \lambda\delta$ is fulfilled.

Theorem 2 (Mergelian [2] *)* Let $\bar{B}$ be a closed, bounded set of the z plane and let condition (a) above be fulfilled. For the function $f(z)$ given on $\bar{B}$ to be approximated as well as is desired on $\bar{B}$ by means of rational functions with poles outside $\bar{B}$, it is necessary and sufficient that the function $f(z)$ be continuous on $\bar{B}$ and regular at all internal points of the set $\bar{B}$.

That it is sufficient follows from Theorem 1 (see also Note 1) and that it is necessary, from the investigations at the beginning of the chapter.

We made the formulation of Theorems 1 and 2 somewhat more rigorous than as proved by Mergelian. In his formulation instead of $\gamma_{\zeta}^{\delta}(\bar{B})$ he used the diameter of the Jordan arc mentioned in Note 1, Section 1.7.3. The theorem was presented in the form given by Vitushkin. In all the discussions in this section we have followed Mergelian's basic presentation but have substituted throughout the analytic measure $\gamma_{\zeta}^{\delta}(\bar{B})$ for the Jordan arc mentioned above.

NOTE 2 In Theorems 2 and 3 it can be assumed that in each of the complementary domains to $\bar{B}$ the rational function can have only one pole at a point given in advance.

NOTE 3 Condition (a) for a set $\bar{B}$ is fulfilled if the complementary $C\bar{B}$ of the set B consists of a finite number of

* See also M. M. Dzhrbashian [4,6], G. Tumarkin [1,3], and V.Erokhin [1].

finitely-connected domains (see Note 1, Section 1.7.3).

From Notes 2 and 3 and Theorem 2 and taking into account the statements at the beginning of the chapter follows the theorem below.

Theorem 3 (Mergelian [2]*)* For the function $f(z) \neq$ const, given and bounded on the set B of points of the extended z plane to be uniformly approximated as well as is desired by means of polynomials on B it is necessary and sufficient that the set B be bounded and that the function $f(z)$ can be extended to the closed set $\bar{B}_*$, complementary to that domain amongst those complementary to $\bar{B}$ which contains the point $z = \infty$ in such a way that the function $f(z)$ will be continuous on $\bar{B}_*$ and regular at every internal point of the set $\bar{B}_*$.

Let us prove another theorem directly connected with Theorem 3.

Theorem 4 (Mergelian [2]*)* For any function $f(z)$ defined and continuous on a closed set $\bar{B}$ of the extended z plane and regular at each internal point of $\bar{B}$ to be uniformly approximated as well as is desired on $\bar{B}$ by means of polynomials, it is necessary and sufficient that the complement $C\bar{B}$ of the set $\bar{B}$ be connected and contain the point ∞.

Theorem 3 gives the necessary and sufficient conditions both for the function $f(z) \neq$ const and for the set B in order that the function should be approximated as well as is desired on $\bar{B}$ by means of polynomials. Theorem 2 gives only the sufficient condition for the set $\bar{B}$ and the necessary and sufficient conditions for the function in order that the function, given on $\bar{B}$, can be approximated as well as is desired on $\bar{B}$ by means of rational functions. Note that up till now we have not solved the problem of the necessary and sufficient conditions which must be satisfied for a closed set $\bar{B}$ of the extended z plane in order that every function $f(z)$ continuous on $\bar{B}$ and regular at all its internal points can be approximated as well as is desired on $\bar{B}$ by means of rational functions with poles outside $\bar{B}$. However, the analogous problem has been solved for all functions $f(z)$ continuous on $\bar{B}$. From the statements at the beginning of the chapter it follows that the set $\bar{B}$ should not have internal points and consequently the boundary of $C\bar{B}$ should coincide with $\bar{B}$.

Theorem 5 (Vitushkin [2]*)* Let $\bar{B}$ be a closed bounded set of the z plane without internal points. For every function continuous on $\bar{B}$ to be approximated as well as is desired by means of rational functions with poles outside B, it is necessary and sufficient that for any $\delta > 0$ the condition $\gamma_z^\delta(\bar{B}) = \delta$ be fulfilled for every $z \in \bar{B}$.

PROOF: SUFFICIENCY This follows directly from Theorem 2.

NECESSITY Let K be a disc of radius $\delta > 0$ with its centre at the point $z_0 \in \bar{B}$ and $D = K \cap C\bar{B}$ *. We can now show that for any $\varepsilon > 0$ there is a closed set $F \subset D$ and a function $\mu(z) \in \mathfrak{R}(F)$ such that $|a(\mu)| > \delta - \varepsilon$, where $\varepsilon > 0$ is arbitrary. Hence on the basis of the inequality $|a(\mu)| \leqslant \gamma_{z_0}^\delta(\bar{B}) \leqslant \delta$ and since ε is arbitrary, it follows that $\gamma_z^\delta(\bar{B}) = \delta$ and the necessity will be proved.

Let us introduce the function $f(z) = \frac{\delta}{z - z_0}$ for $|z - z_0| \geqslant \delta$ and equals $\frac{\overline{z - z_0}}{\delta}$ for $|z - z_0| < \delta$. The modulus of this function never exceeds one and the function is continuous on $\bar{B}$ and consequently for any integer $s > 1$ there is a rational function $R_s(z)$ such that

$$|f(z) - R_s(z)| < \frac{1}{s}, \quad z \in \bar{B}.$$

The set F^* of points z, $|z - z_0| \leqslant \delta$ such that

$$|f(z) - R_s(z)| \geqslant \frac{2}{s}$$

is closed, and since it is separated from $\bar{B}$ by a positive distance, there are only a finite number of domains forming $C\bar{B}$ which contain points of F^*. Extending F^* we can obtain a closed set $F \subset C\bar{B}$ which consists of a finite number of closed domains, the external boundary of each of which is a piecewise smooth Jordan curve, and on F the inequality

$$|f(z) - R_s(z)| \geqslant \frac{3}{2s}$$

is fulfilled.

Let us denote by G the domain complementary to F which contains the point ∞, and by Γ let us denote its boundary. In the closed domain $\bar{G}$ let us define the function

* The discussion below remains valid if the point $z_0 \bar{\in} \bar{B}$.

$$f_s(z)=\begin{cases}\dfrac{\delta}{z-z_0}, & |z-z_0|\geqslant\delta\\ R_s(z)+r_s(z), & z\in\overline{K}\cap\overline{G}\end{cases}$$

where $r_s(z)=\left(\frac{\overline{z-z_0}}{\delta}-R_s(z)\right)\left(\frac{|z-z_0|}{\delta}\right)^s$. This function is continuous on $\overline{G}$ and has continuous derivatives with respect to x and to y $(z=x+yi)$ for $|z-z_0|>\delta$ and for $z\in\overline{K}\cap\overline{G}$. Note that

$$|f_s(z)|\leqslant 1+\frac{4}{s}, \quad z\in\overline{G}$$

In fact, for $|z-z_0|>\delta$ this inequality is obvious and for $z\in\overline{K}\cap\overline{G}$ we have

$$|f_s(z)|\leqslant|R_s(z)-f(z)|+|f(z)|$$
$$+|f(z)-R_s(z)|\left(\frac{|z-z_0|}{\delta}\right)^s\leqslant\frac{2}{s}+1+\frac{2}{s}=1+\frac{4}{s}$$

Besides this for the function $g_s(z)=\frac{\partial}{\partial x}f_s(z)+i\frac{\partial}{\partial y}f_s(z)$, $z=x+yi$ we have $g_s(z)=0$ for $|z-z_0|>\delta$ and for $z\in\overline{K}\cap\overline{G}$

$$|g_s(z)|=\left|\frac{2}{\delta}\left(\frac{|z-z_0|}{\delta}\right)^s+s(f(z)-R_s(z))\frac{|z-z_0|^{s-1}}{\delta^s}(z-z_0)\right|$$
$$\leqslant\frac{4}{\delta^s}|z-z_0|^{s-1}$$

By using Theorem 1, Section 1.7.1, we have

$$f_s(z)=\frac{1}{2\pi i}\int_\Gamma\frac{f_s(\zeta)}{\zeta-z}d\zeta+\frac{1}{2\pi}\iint_{\overline{K}\cap\overline{G}}\frac{g_s(\zeta)}{\zeta-z}d\sigma_\zeta, \quad z\in G$$

and introducing another function regular in G

$$\varphi_s(z)=\frac{1}{2\pi i}\int_\Gamma\frac{f_s(\zeta)}{\zeta-z}d\zeta, \quad z\in G$$

we find

$$|f_s(z)-\varphi_s(z)|\leqslant\frac{1}{2\pi}\iint_{\overline{K}\cap\overline{G}}\frac{|g_s(\zeta)|}{|\zeta-z|}d\sigma_\zeta\leqslant\frac{2}{\pi\delta^s}\iint_{0\leqslant|\zeta-z_0|\leqslant\delta}\frac{|\zeta-z_0|^{s-1}}{|\zeta-z|}d\sigma_\zeta \tag{2}$$

Let us find an upper estimate for the integral on the right for $z\in\overline{G}$. If $z=z_0$ then we can find easily that it is $4/s$. Further, assuming that $|z-z_0|=\rho\neq 0$ and introducing a system of polar coordinates with a pole at the point z_0, we have

$$|f_s(z)-\varphi_s(z)| \leqslant \frac{2}{\pi\delta^s}\int_0^\delta \left[\int_0^{2\pi} \frac{r^s}{\sqrt{r^2-2r\rho\cos\theta+\rho^2}}\,d\theta\right] dr$$

Proceeding as in the proof of inequality (11), Section 1.6.1, we find

$$|f_s(z)-\varphi_s(z)| \leqslant \frac{2}{\delta^s}\int_0^\delta r^{s-\frac{1}{2}}\frac{1}{\sqrt{\rho}}\ln\frac{\sqrt{r}+\sqrt{\rho}}{|\sqrt{r}-\sqrt{\rho}\,|}\,dr$$

Assuming now that $0<\rho<\delta$ (the case of $\rho\geqslant\delta$ is simpler) we have

$$|f_s(z)-\varphi_s(z)| \leqslant \frac{2}{\delta^s}\int_0^\rho r^{s-\frac{1}{2}}\frac{1}{\sqrt{\rho}}\ln\frac{1+\sqrt{\frac{r}{\rho}}}{1-\sqrt{\frac{r}{\rho}}}\,dr$$

$$+\frac{2}{\delta^s}\int_\rho^\delta r^{s-\frac{1}{2}}\frac{1}{\sqrt{\rho}}\ln\frac{1+\sqrt{\frac{\rho}{r}}}{1-\sqrt{\frac{\rho}{r}}}\,dr = \frac{4}{\delta^s}\sum_{k=0}^{\infty}\frac{1}{(2k+1)\rho^{k+1}}\int_0^\rho r^{k+s}\,dr$$

$$+\frac{4}{\delta^s}\sum_{k=0}^{\infty}\frac{\rho^k}{2k+1}\int_\rho^\delta r^{s-k-1}\,dr = 4\left(\frac{\rho}{\delta}\right)^s\sum_{k=0}^{\infty}\frac{1}{(2k+1)(k+s+1)}$$

$$+\frac{4}{\delta^s}\sum_{k=0}^{s-1}\frac{\rho^k}{(2k+1)(s-k)}\left(\delta^{s-k}-\rho^{s-k}\right)+\frac{4}{\delta^s}\frac{\rho^s}{2s+1}\ln\frac{\delta}{\rho}$$

$$+\frac{4}{\delta^s}\sum_{k=s+1}^{\infty}\frac{\rho^k}{(2k+1)(k-s)}\left(\rho^{s-k}-\delta^{s-k}\right)\leqslant 4\sum_{k=0}^{\infty}\frac{1}{(k+1)(k+s+1)}$$

$$+4\sum_{k=0}^{s-1}\frac{1}{(2k+1)(s-k)}+\frac{4}{es(2s+1)}+4\sum_{k=s+1}^{\infty}\frac{1}{(k+1)(k-s)}$$

$$=\frac{4}{s}\sum_{k=1}^{s}\frac{1}{k}+\frac{4}{s+\frac{1}{2}}\sum_{k=1}^{2s}\frac{1}{k}+\frac{4}{es(2s+1)}+\frac{4}{s+1}\sum_{k=1}^{s+1}\frac{1}{k}\leqslant\frac{4}{s}(\ln s+1)$$

$$+\frac{4}{s}(\ln 2s+1)+\frac{4}{s}\frac{1}{e(2s+1)}+\frac{4}{s+1}(\ln(s+1)+1)$$

$$\leqslant\frac{4}{s}\left[3\ln s+3+\ln 2+\frac{1}{5e}\right]\leqslant\frac{\ln s}{s}\left[12+4\,\frac{3+\ln 2+\frac{1}{5e}}{\ln 2}\right]\leqslant 40\,\frac{\ln s}{s}$$

In evaluating the sums of the form $\sum_{k=1}^{n}\frac{1}{k}$ we made use of Lemma 4, Section 1.6.1. Thus

$$|f_s(z)-\varphi_s(z)|\leqslant C\frac{\ln s}{s}, \quad z\in G$$

where C is some constant. We now have

$$|\varphi_s(z)|\leqslant|f_s(z)|+|\varphi_s(z)-f_s(z)|\leqslant 1+\frac{4}{s}+C\frac{\ln s}{s}, \quad z\in G$$

and for sufficiently large

$$|\varphi_s(z)|\leqslant 1+\varepsilon_1, \quad \varepsilon_1=\frac{\varepsilon}{1+\delta-\varepsilon} \tag{3}$$

Let us now multiply (2) by $|z|$ and let z tend to ∞. In the limit we have

$$\left|\lim_{z\to\infty}(z\varphi_s(z))-\delta\right|\leqslant\frac{4\delta}{s+1}$$

which for sufficiently large s gives

$$\left|\lim_{z\to\infty}(z\varphi_s(z))\right|\geqslant\delta-\varepsilon_1 \tag{4}$$

Let us fix s such that inequalities (3) and (4) are fulfilled. Then the function $\mu(z)=\frac{\varphi_s(z)}{1+\varepsilon_1}\in\Re(F)$ and from (4) it follows that

$$|a(\mu)|\geqslant\frac{\delta-\varepsilon_1}{1+\varepsilon_1}=\delta-\varepsilon$$

The theorem is proved.

NOTE 4 One can give an example of a closed set without internal points on which not every continuous function can be uniformly approximated as well as is desired by means of rational functions (see Mergelian [2]).

Theorem 6 (Mergelian [2]*)* Let $\bar{B}$ be a bounded, closed set of points of the z plane; for some $\delta>0$ there is $\lambda>0$ such that $\gamma^{2\delta}(\bar{B})>2\lambda\delta$; let $f(z)$ be a function continuous on $\bar{B}$ and regular on the open set B_1 of the points $z\in\bar{B}$ separated from $C\bar{B}$ by more than $\delta/2$. Then $f_\delta(z)$, regular on B, is such that

$$|f(z)-f_\delta(z)|\leqslant\left(49+\frac{72}{\lambda}+\frac{216}{\lambda^2}\right)\omega_f\left(\frac{\delta}{2}\right), \quad z\in\bar{B} \tag{5}$$

PROOF As in Section 1.7.3, let us extend the function $f(z)$ to the whole z plane and let us define a function $G_\delta(z)$ by means of Formula (3) of the section. Let us denote by E_δ the set of those points $z \in B_1$ which are separated from $C\bar{B}_1$ by more than $\delta/2$. It is obvious that the distance of any point $z \in \bar{B} - E_\delta$ to $C\bar{B}$ is not greater than δ. It is easy to see that conditions (5) and (6) of Section 1.7.3 will be fulfilled for functions $f(z)$ and $G_\delta(z)$. As in the section mentioned we introduce a closed set $\bar{K}^*$ of those points $z \in \bar{B}$ separated from $C\bar{B}$ by not more than δ (and not $\delta/2$), and instead of the discs K_ζ^δ, $\zeta \in L$ we shall consider the discs $K_\zeta^{2\delta}$ of radius 2δ with the centre at $\zeta \in L$. For the rest we shall retain the notation and procedure of Section 1.7.3, altering δ to 2δ throughout. Inequality (1) of Section 1.7.3 then becomes the inequality

$$|f(z) - f_\delta(z)| \leqslant \omega_f(2\delta) \leqslant 4\omega_f\left(\frac{\delta}{2}\right), \quad z \in \bar{B}_0$$

where $\bar{B}_0$ is the analogue of $\bar{B}_0$ in the section: in the inequalities (2) of the section, δ becomes 2δ; in inequality (7) of the section, the right-hand side is doubled and inequality (8) of the section is identical with inequality (5). The theorem is proved.

NOTE 4 If $\bar{B}$ is a bounded, closed set of the z plane the complement to which is connected (domain) then on the strength of Note 3 we can choose $\lambda < 1/4$ and as close to 1/4 as desired. Then inequality (5) has the form

$$|f(z) - f_\delta(z)| \leqslant C'\omega_f\left(\frac{\delta}{2}\right), \quad z \in \bar{B}$$

where C' is some constant greater than $C_0 = 3{,}793$ which can be chosen as close to C_0 as desired.

Let us give an application of Theorem 6. Let $\bar{B}$ be a closed, bounded set of the z plane the complement of which is connected (domain); L be the boundary of $\bar{B}$; $\bar{B}_{(x)}$, $x \geqslant 0$ be a family of closed, bounded sets embedded in one another; $\bar{B}_0 = \bar{B}$, $B_{(x')} \subset B_{(x'')}$, $x'' \geqslant x'$ such that every point $z \neq \infty$ belongs to all $B_{(x)}$ for sufficiently large x; $d(\zeta, x)$ be the distance from the point $\zeta \in L$ to $C\bar{B}_{(x)}$, $d(x) = \sup_{\zeta \in L} d(\zeta, x)$; $\rho(n, x, \bar{\bar{B}})$ be the exact upper bound of the quantity $\rho_n(f, \bar{B})$ (see introduction to the chapter) with respect to all possible functions $f(z)$ regular in $\bar{B}_{(x)}$ and with modulus limited to unity. Note that $d(x)$ is a non-diminishing and $\rho(n, x, \bar{B})$ a non-increasing function of x; $d(0)=0$ and $\rho(n, x, \bar{B}) \leqslant 1$.

Theorem 7 (Mergelian [2]*)* Let $\bar{B}$ be a closed, bounded set of points of the z plane the complement to which is connected; R be the radius of the smallest disc K such that $\bar{B} \subset \bar{K}$; $f(z)$ a function, continuous on $\bar{B}$ and regular at all internal points of the set $\bar{B}$. Then

$$\rho_n(f, \bar{B}) \leqslant \left(C + \frac{mR}{\gamma(\bar{B})}\right) \omega_f(d(x_n)) \tag{6}$$

where $m>0$ is an arbitrary number, x_n is the root of the equation $\rho(n, x, \bar{B}) = \frac{m}{\gamma(\bar{B})} d(x)$ and C is an absolute constant (we can put $C=7{,}588$).

PROOF Let us extend $f(z)$ to the whole plane whilst preserving $\omega_f(\delta)$ and the maximum modulus. Let $x>0$. On the basis of Theorem 6 and Note 4 there is a function $f_x(z)$ regular on $\bar{B}_{(x)}$ such that

$$|f(z) - f_x(z)| \leqslant C' \omega_f(d(x)), \quad z \in \bar{B}_{(x)}$$

where C' is the same as in Note 4 and consequently

$$|f_x(z)| \leqslant \sup_{z \in \bar{B}_{(x)}} |f(z)| + C' \omega_f(d(x)), \quad z \in \bar{B}_{(x)}$$

In proving inequality (6) we can assume that $f(z)$ is equal to zero at the centre of the circle indicated in formulating the theorem, and therefore

$$\sup_{z \in \bar{B}_{(x)}} |f(z)| \leqslant \omega_f(R) + \omega_f(d(x)) = \omega_f\left(\frac{R}{d(x)} d(x)\right) + \omega_f(d(x))$$
$$\leqslant \left(\frac{R}{d(x)} + 1\right) \omega_f(d(x))$$

and consequently

$$|f_x(z)| \leqslant \left(\frac{R}{d(x)} + C' + 1\right) \omega_f(d(x))$$

For the function $f_x(z)$ there is a polynomial $\pi_n(z)$ of degree not greater than n such that

$$|f_x(z) - \pi_n(z)| \leqslant \left(\frac{R}{d(x)} + C' + 1\right) \omega_f(d(x)) \rho(n, x, \bar{B}), \quad z \in \bar{B}$$

and now for $z \in \bar{B}$ we have

$$|f(z) - \pi_n(z)| \leqslant |f(z) - f_x(z)| + |f_x(z) - \pi_n(z)|$$
$$\leqslant \left[C' + \left(\frac{R}{d(x)} + C' + 1\right)\rho(n, x, \bar{B})\right]\omega_f(d(x))$$

By using the inequality $\rho(n, x, \bar{B}) \leqslant 1$ and assuming $x = x_n$ we find

$$|f(z) - \pi_n(z)| \leqslant \left(2C' + 1 + \frac{mR}{\gamma(\bar{B})}\right)\omega_f(d(x_n)), \quad z \in \bar{B}$$

Hence and from Note 4 the theorem is proved.

NOTE 5 From property (7), Section 1.7.2, of the analytic measure it follows that if $\bar{B}$ is a set of type $\mathfrak{M}$ then

$$\frac{R}{\gamma(\bar{B})} \leqslant \frac{4}{\sqrt{3}}$$

NOTE 6 From the properties of the functions $\rho(n, x, \bar{B})$ and $d(x)$ it follows that in inequality (6) instead of x_n we can choose any number x'_n such that

$$\rho(n, x'_n, \bar{B}) \leqslant \frac{md(x'_n)}{\gamma(\bar{B})} \tag{7}$$

In the following we shall assume that $\bar{B}$ is a set of type $\mathfrak{M}$ and $\bar{B}_x$ is a closed, simply-connected domain $\bar{B}_{1+x}$ bounded by the contour line L_{1+x} when $\bar{C}B$ is conformally mapped on to the domain $|w| > 1$ such that B is B_1. Then $d(x)$ and $\rho(n, x, \bar{B})$ will be denoted by $d_0(x)$ and $\rho_0(n, x, B)$. In the following we shall need a lower estimate for $d_0(x)$ to use Note 5.

Lemma 1 If $\bar{B}$ is a set of type $\mathfrak{M}$ then $d_0(x) \geqslant \gamma(\bar{B})\, x$, $x \geqslant 0$.

PROOF Let $s(r)$ be the area of B_r, $r \geqslant 1$ and $l(r)$ be the length of L_r, $r > 1$. From geometrical considerations it is easy to show that there is the following inequality:

$$s(r) - s(1) \leqslant l(r)\rho_0 - \pi\rho_0^2, \quad \rho_0 = d_0(r - 1), \quad r > 1$$

and from this we find that

$$\rho_0 \geqslant \frac{1}{2\pi}\left[l(r) - \sqrt{[l(r)]^2 - 4\pi(s(r) - s(1))}\right] \tag{8}$$

Note that: (a) $s(r) \leqslant \pi\left(\frac{l(r)}{2\pi}\right)^2 = \frac{(l(r))^2}{4\pi}$ and that therefore the expression under the root is not negative; (b) when $l(r)$ increases the right-hand side of inequality (8) decreases; (c) when $s(r)-s(1)$ diminishes the right-hand side of inequality (8) decreases.

Let

$$\psi(w)=\gamma\left(w+\alpha_0+\frac{\alpha_1}{w}+\frac{\alpha_2}{w^2}+\ldots\right),\quad \gamma=\gamma(\bar{B})$$

From the (external) area theorem (see Goluzin [1] p. 54) it follows that

$$s(r)=\pi\gamma^2\left(r^2-\sum_{\nu=1}^{\infty}\nu\,|\alpha_\nu|^2\frac{1}{r^{2\nu}}\right)$$

and therefore

$$s(r)-s(1)\geqslant\pi\gamma(r^2-1)\left(1+\sum_{\nu=1}^{\infty}\nu^2\,|\alpha_\nu|^2\frac{1}{r^{2\nu}}\right)$$

For $l(r)$ we have

$$l(r)=\int_{|w|=r}|\psi'(w)|\,|dw|=r\int_0^{2\pi}|\psi'(re^{i\theta})|\,d\theta$$

$$\leqslant r\sqrt{2\pi\int_0^{2\pi}|\psi'(re^{i\theta})|^2\,d\theta}=2\pi\gamma r\sqrt{1+\sum_{\nu=1}^{\infty}\nu^2\,|\alpha_\nu|^2\frac{1}{r^{2\nu+2}}}$$

Assuming $y=\sum_{\nu=1}^{\infty}\nu^2\,|\alpha_\nu|^2\,r^{-2\nu}$ and taking into account the observation made above, inequality (8) can be changed to

$$\rho_0\geqslant\gamma\left(\sqrt{r^2+y}-\sqrt{(2-r^2)\,y+1}\right) \qquad (9)$$

where the expression under the second square root is non-negative for possible values of y. The right-hand side of inequality (9) as a function of y increases as $y\geqslant 0$ and for $y=0$ we obtain the conclusion of the lemma.

NOTE 7 If the set $\bar{B}$ in Theorem 7 is a set of type $\mathfrak{M}$ then in inequality (6), on the basis of Note 6, we can choose for x_n any number which satisfies the following condition

$$\rho(n,\,x_n,\,\bar{B})\leqslant mx_n \qquad (10)$$

Theorem 8 (Mergelian [2] *)* Let $\bar{B}$ be a set of type $\mathfrak{M}$, $f(z)$ a function continuous on B and regular at all internal points of the set $\bar{B}$. Then

$$\rho_n(f, \bar{B}) \leqslant C\omega_f\left(d_0\left(\frac{2\ln(n+2)}{n+1}\right)\right), \quad n=2, 3, \ldots, \tag{11}$$

where C is an absolute constant (we can put $C=7{,}592$).

PROOF We shall make use of Theorem 7 by putting $R=\frac{4}{\sqrt{3}}\gamma(\bar{B})$ (see Note 5), $m=\sqrt{3}$ and $x_n=\frac{2\ln(n+2)}{n+1}$. Then in view of Note 6, all we have to prove is that inequality (10) is fulfilled. From Theorem 2, Section 2.1.3 (statement 2) it follows (see also Note 1 in Section 2.1):

$$\rho(n, r-1, \bar{B}) \leqslant \left[\frac{2\sqrt{e}}{r-1}+\sqrt{\left(\frac{n+1}{r^2-1}+\frac{r^2}{(r^2-1)^2}\right)\ln(n+2)}\,\right]\frac{1}{r^n}$$
$$n+1 \geqslant \frac{1}{\sqrt{r}-1},$$

whence

$$\rho(n, x, \bar{B}) \leqslant \left[\frac{2\sqrt{e}}{x}+\sqrt{\left(\frac{n+1}{2x}+\frac{1}{x^2}\right)\ln(n+2)}\,\right]\frac{1}{(1+x)^n}, \quad \frac{2n+3}{(n+1)^2}\leqslant x \tag{12}$$

It is sufficient to show that

$$\rho(n, x_n, \bar{B}) \leqslant \left[\frac{\sqrt{e}}{\ln 4}+\frac{1}{2}\sqrt{1+\frac{1}{\ln 4}}\,\right]\frac{n+1}{(n+2)^{\frac{2n}{n+1+\ln(n+2)}}} \tag{13}$$
$$\leqslant \sqrt{3}\,\frac{2\ln(n+2)}{n+1}=\sqrt{3}\,x_n$$

The first inequality follows directly from (12) and from $\left(1+\frac{1}{y}\right)^{y+\frac{1}{2}} > e\,(y>0)$. To prove the second inequality let us consider the function

$$u(n)=\ln\frac{(n+1)^2}{(n+2)^{\frac{2n}{n+1+\ln(n+2)}}\ln(n+2)}$$
$$=2\ln(n+1)-\frac{2n\ln(n+2)}{n+1+\ln(n+2)}-\ln\ln(n+2)$$

Its derivative

$$u'(n)=\frac{-2n(n+2)(\ln(n+2))^3+(2n+3)(n+1)(\ln(n+2))^2+2(n+1)^2\ln(n+2)-(n+1)^3}{(n+1)(n+2)[n+1+\ln(n+2)]^2\ln(n+2)}$$

is clearly negative for n satisfying the condition

$$-(\ln(n+2))^3+(\ln(n+2))^2+\ln(n+2)-1<0$$

in particular for $n \geqslant 2$. Therefore it is sufficient to verify inequality (13) for $n=2$. The theorem is proved.

If we apply certain conditions to the boundary L of a set $\bar{B}$ of type $\mathfrak{M}$, then we can obtain good upper bounds for $d_0(x)$ and Formula (11) can be used to evaluate the upper bound of $\rho_n(f, \bar{B})$. Such upper and lower bounds can be found in the works of Mergelian [1, 2].

In the last few years the American mathematicians Bishop, Vermer, Rudin and others have discovered a new approach to the problem of uniform approximation of functions of one or more complex variables by means of polynomials and rational functions, based on the application of methods of functional analysis.

SUPPLEMENTARY LITERATURE

In addition to the works given in the text we should like to mention the following monographs and review articles dealing with material related to the contents of this chapter: Goncharov [1] Chapter 5, Walsch [1] with review article by Mergelian; review articles by Mergelian [2, 5, 6] and Evgrafov [1]. We should also like to point out the following rather specialised works: Akhiezer [1], Vitushkin [1], Gelfond [1], Gonchar [1, 2], Go Chuo-Joo [1], Dzhrbashian [1], Levin [1], Chapter 4, Pommerenke [1], and Ronian [1].

CHAPTER TWO

Faber polynomials and problems of representing a regular function by a polynomial series

In this chapter we shall introduce a certain class of polynomials defined for any set $\bar{B}$ of type $\mathfrak{M}$, e.g. a bounded, closed set $\bar{B}$ of points of the z plane the complement B^1 of which is a simply-connected domain. This class of polynomials was introduced by Faber [1] and was successfully applied by him in later works [2]. For the disc $|z-z_0| \leqslant R$, $R > 0$ it is the class of polynomials $(z-z_0)^k$, $k=0, 1, \ldots$ As for Taylor series, every function regular on $\bar{B}$ can be expanded into a series of Faber polynomials for $\bar{B}$, uniformly converging on $\bar{B}$. The Faber polynomials and their various generalisations can be applied to a whole series of problems in the theory of regular functions.

2.1 FABER POLYNOMIALS AND THEIR APPLICATIONS

Let $\bar{B}$ be a set of type $\mathfrak{M}$ and let the function

$$z = \Psi(w) = w + \beta_0 + \frac{\beta_1}{w} + \ldots = w + M\left(\frac{1}{w}\right)$$

map one-to-one conformally the domain $|w| > \rho$ on to the domain B^1 such that $\Psi(\infty)=\infty$ and let the Maclaurin series $M(t)$ converge in the disc $|t|<\frac{1}{\rho}$. Here, and in the following, Maclaurin series will be denoted by $M(t)$ (sometimes with indices). Let $w=\Phi(z)$ be the inverse function to $z=\Psi(w)$. In the neighbourhood of the point $z=\infty$ its expansion is of the form

$$\Phi(z)=z+\alpha_0+\frac{\alpha_1}{z}+\cdots$$

and is connected with the function $w=\varphi(z)$ introduced in Section 1.2 by the formula $w=\Phi(z)=\frac{\varphi(z)}{\varphi'(\infty)}=c\varphi(z)$ where $\varphi'(\infty)=\frac{1}{\rho}$, $\rho=c$ and $\psi(w)=\Psi(\rho w)$. We shall denote the image of the circle $|w|=r$, $r>\rho$, when mapped by the function $z=\Psi(w)$, by L_r, the bounded domain with boundary L_r as B_r and the unbounded domain as B_r^1. The boundary of the domain B^1 will be denoted by L or L_ρ and the boundary of the set $\bar{B}$ will be denoted in the same way. $\bar{B}$ and B^1 will be denoted respectively by $\bar{B}_\rho$ and B_ρ^1.

NOTE 1 If we maintain the normalisation of the function $z=\psi(w)$ from Chapter 1, e.g. as indicated above, alter w to ρw, then instead of L_r, B_r, B_r^1, $r>\rho$, we should have respectively $L_{r/\rho}$, $B_{r/\rho}$ and $B_{r/\rho}^1$. In this way in all the following formula if desired we can convert to the notation for contour lines and domains used in Chapter 1.

NOTE 2 The capacity of the set $\bar{B}_r$, $r\geqslant\rho$ is clearly equal to r (see Section 1.2).

2.1.1 Definition of the Faber polynomials

Let us investigate the expression $\frac{d\zeta}{\zeta-z}$ under the integral sign in Cauchy's formula where ζ is the integration variable on the z plane. Substituting for the variable $\zeta=\Psi(t)$

$$\frac{d\zeta}{\zeta-z}=\frac{\Psi'(t)\,dt}{\Psi(t)-z}$$

we find the following function of t:

$$K(z,t)=\frac{t\Psi'(t)}{\Psi(t)-z} \qquad (1)$$

where t is added for convenience in later calculations.

As a function of the complex variable t the function $K(z, t)$ is defined for $|t| > \rho$. From the formula

$$K(z, t) = \frac{t\left[1 - \frac{1}{t^2} M'\left(\frac{1}{t}\right)\right]}{-z + \left[t + M\left(\frac{1}{t}\right)\right]} = \frac{1 - \frac{1}{t^2} M'\left(\frac{1}{t}\right)}{1 - \frac{z}{t} + \frac{1}{t} M\left(\frac{1}{t}\right)} \tag{2}$$

and (1) it follows that the function $K(z, t)$ is regular at the point $t = \infty$ for any z, $K(z, \infty) = 1$ and the function $K(z, t)$ can have a singularity only at points t satisfying the equation $\Psi(t) = z$. If $z \in \overline{B}$ there are no such values of t, and $K(z, t)$ is regular in the domain $|t| > \rho$. If $z \in B^1$ because the function $\Psi(t)$ is schlicht for $|t| > \rho$ there is a single value $t = \tau$ such that $(\tau) = z$ $(\tau = \Phi(z))$ and the function $K(z, t)$ is regular in the domain $|t| > r$ where $r = |\tau| > \rho$ for every $z \in L_r$. At the point $t = \tau$ this function has a pole of the first order. By using Formula (2) for sufficiently large $|t|$ we have

$$K(z, t) = \left[1 - \frac{1}{t^2} M'\left(\frac{1}{t}\right)\right] \sum_{k=0}^{\infty} \left[\frac{1}{t} z - \frac{1}{t} M\left(\frac{1}{t}\right)\right]^k$$

or

$$K(z, t) = \frac{t\Psi'(t)}{\Psi(t) - z} = \sum_{k=0}^{\infty} P_k(z) \frac{1}{t^k} \tag{3}$$

where $P_k(z)$ is a polynomial of degree k with leading coefficient one k. For every fixed $z \in \overline{B}$ the series (3) converges in the domain $|t| > \rho$ for every fixed $z \in \overline{B}_r$, $r > \rho$, converges in the domain $|t| > r$ and for every fixed $z \in B_r$ uniformly converges in the closed domain $|t| \geqslant r$.

The polynomials $P_k(z)$, $k = 0, 1, \ldots,$ are called Faber polynomials for the set $\overline{B}$ and the function $\frac{t\Psi'(t)}{\Psi(t) - z}$ is called the generating function of Faber polynomials.

From (3) we have:

$$P_k(z) = \frac{1}{2\pi i} \int\limits_{|t| = r_1} \frac{t\Psi'(t)}{\Psi(t) - z} t^{k-1} \, dt, \quad z \in \overline{B}_r, \ \rho \leqslant r < r_1 < \infty$$

Altering the integration variable in accordance with the formula $t = \Phi(\zeta)$ we have

$$P_k(z) = \frac{1}{2\pi i} \int\limits_{L_{r_1}} \frac{\Phi(\zeta)^k}{\zeta - z} \, d\zeta, \quad z \in \overline{B}_r, \ \rho \leqslant r < r_1 < \infty \tag{4}$$

The function under the integral sign has no singular points in B_r^1 except for a pole of order k at the point $z=\infty$. Thus it is possible to alter the contour of integration L_r to the circumference C_R with centre at the point $z=0$ and greater radius R such that the domain B_r lies inside C_R, and means that the integral can be calculated by using the expansion as a Laurent series in the region of the point $\zeta=\infty$:

$$\Phi(\zeta)^k=\zeta^k+C_1^{(k)}\zeta^{k-1}+\ldots+C_k^{(k)}+\frac{\alpha_1^{(k)}}{\zeta}+\frac{\alpha_2^{(k)}}{\zeta^2}+\ldots$$

$$\frac{1}{\zeta-z}=\frac{1}{\zeta}+\frac{z}{\zeta^2}+\ldots+\frac{z^k}{\zeta^{k+1}}+\ldots$$

Then integrating we have

$$P_k(z)=z^k+C_1^{(k)}z^{k-1}+\ldots+C_k^{(k)}$$

Thus the Faber polynomial $P_k(z)$ is the principal part of the expansion of the function $\Phi(z)^k$ as a Laurent series in the neighbourhood of the point $z=\infty$ and consequently

$$P_k(z)=\Phi(z)^n+\omega_k(z) \tag{5}$$

where $\omega_k(z)$ is a function which is regular in the domain B^1 and $\omega(\infty)=0$.

This relation between the polynomials $P_k(z)$ and the functions $\Phi(z)^k$ can be taken as the definition for Faber polynomials.

Let us assume that in the following $z\in B^1$. The function

$$K(z, t)=\frac{t\Psi'(t)}{\Psi(t)-z}=\frac{t\Psi'(t)}{\Psi(t)-\Psi(\tau)}, \quad \tau=\Phi(z)$$

can be considered as a function of the variables t and τ, defined for $|t|>\rho$ and $|\tau|>\rho$. It has a singularity only at $t=\tau$. Let us determine its residue at the simple pole $t=\tau$ for fixed τ:

$$r_\tau=\lim_{t\to\tau}\frac{t\Psi'(t)(t-\tau)}{\Psi(t)-\Psi(\tau)}=\frac{\tau\Psi'(\tau)}{\Psi'(\tau)}=\tau$$

Analogously the residue as a function τ at the simple pole for fixed t is $-t$. Hence it follows directly that

$$\frac{t\Psi'(t)}{\Psi(t)-\Psi(\tau)}=-\frac{t}{\tau-t}+M(t, \tau)$$

where $M(t, \tau)$ is a function which is regular for $|t|>\rho$ and $|\tau|>\rho$. It is useful to note that

$$\frac{\tau}{t-\tau}=\frac{-t}{\tau-t}-1$$

Collecting in the double-power series for $M(t, \tau)$ the terms with the same degree in t we find

$$\frac{t\Psi'(t)}{\Psi(t)-\Psi(\tau)}=\frac{t}{t-\tau}+a_0(\tau)+a_1(\tau)\frac{1}{t}+a_2(\tau)\frac{1}{t^2}+\dots$$
$$|\tau|>\rho, \; |t|>\rho$$

From (2) it follows that at $t=\infty$ and $\tau\neq\infty$ the left-hand side of the equation is equal to one and since then the right-hand side of the equation is equal to $1+a_0(\tau)$ $a_0(\tau)\equiv 0$. For $\tau=\infty$ and $t\neq\infty$ the left-hand side and first term of the right-hand side are equal to zero and therefore $a_k(\infty)=0$, $k=1, 2, \dots$, e.g.

$$\frac{t\Psi'(t)}{\Psi(t)-\Psi(\tau)}=\frac{t}{t-\tau}+\sum_{k=1}^{\infty}\frac{1}{\tau}M_k\left(\frac{1}{\tau}\right)t^{-k} \tag{6}$$

where $M_k(1/\tau)$ is a Laurent series with negative powers of τ, converging for $|\tau|>\rho$. Expanding the fraction $\frac{t}{t-\tau}$ for $|t|>|\tau|$ into a series of negative powers of t, we have

$$\frac{t\Psi'(t)}{\Psi(t)-\Psi(\tau)}=1+\sum_{k=1}^{\infty}\left(\tau^k+\frac{1}{\tau}M_k\left(\frac{1}{\tau}\right)\right)\frac{1}{t^k}$$

and comparing this with (3), we find the following expression for the Faber polynomials:

$$P_k(z)=\tau^k+\frac{1}{\tau}M_k\left(\frac{1}{\tau}\right), \quad k=1, 2, \dots, \quad z\in B^1, \quad \tau=\Phi(z) \tag{7}$$

or comparing with (5)

$$P_k(z)=\Phi(z)^k+\omega_k(z), \quad \omega_k(z)=\frac{1}{\tau}M_k\left(\frac{1}{\tau}\right), \quad \tau=\Phi(z) \tag{8}$$

By using this formula we can obtain the Faber polynomials from the following considerations. Let

$$P_k(z)=z^k+c_1z^{k-1}+\dots+c_k$$

be an arbitrary polynomial. Let us put $z=\Psi(\tau)$ and expand the right-hand side into a power series of τ:

$$P_k(z)=P_k(\Psi(\tau))=\tau^k+\gamma_1\tau^{k-1}+\ \dots\ +\gamma_k+\frac{\gamma_{k+1}}{\tau}+\frac{\gamma_{k+2}}{\tau^2}+\dots$$

Let us define the coefficients c_s of the polynomial $P_k(z)$ such that $\gamma_1=\gamma_2=\dots=\gamma_k=0$. It is easy to see that this can be done and in one way only, e.g. the given conditions define a unique polynomial $P_k(z)$. From (7) it now follows that the polynomial $P_k(z)$ obtained in this way is the Faber polynomial for the set $\bar{B}$.

This method of obtaining Faber polynomials can be chosen to define them.

From the definition for Faber polynomials we obtain directly the following results.

NOTE 1 Faber polynomials $P_k(z)$ for a set $\bar{B}$ of type $\mathfrak{M}$ are at the same time Faber polynomials for the domain $\bar{B}_r$, $\rho<r<\infty$.

NOTE 2 If a set $\bar{B}$ of type $\mathfrak{M}$ is a closed domain, the boundary L of which is an analytic curve, e.g. the function $z=\Psi(w)$ is regular and schlicht in the domain $\rho_1<|w|<\infty$, where $\rho_1<\rho$, then on the z plane there will be circular images L_r for $r>\rho_1$, where, for $\rho_1<r<\rho$, L_r lie inside the domain B and the polynomials $P_k(z)$ are the Faber polynomials for the closed domain $\bar{B}_r$ with the boundary L_r, $\rho_1<r<+\infty$.

EXAMPLES OF FABER POLYNOMIALS: (1) Let $\bar{B}$ be the circle $|z-z_0|<\rho$. Then

$$w=\Phi(z)=z-z_0 \text{ and } \Phi(z)^k=(z-z_0)^k, \quad k=0,\ 1,\ \dots$$

so that in this case the Faber polynomials will be the polynomials

$$P_k(z)=(z-z_0)^k, \quad k=0,\ 1,\ \dots$$

(2) Let $\bar{B}$ be the ellipse with foci at the points ± 1. The function $z=\Psi_*(w)=\frac{1}{2}\left(\frac{1}{w}+w\right)$ maps some domain $|w|>\rho'$ on to the outside of the ellipse $\bar{B}$ but is not normalised in the way we need. If we alter w to $2w$, then we have $z=\Psi(w)=w+\frac{1}{4}w^{-1}$. Hence $w=\Phi(z)=\frac{1}{2}(z+\sqrt{z^2-1})$ (for a suitable choice of branch of the root). The Faber polynomials will then have the form

$$P_k(z) = \frac{1}{2^k}[(z+\sqrt{z^2-1})^k + (z-\sqrt{z^2-1})^k]$$

since

$$P_k(\Psi(\tau)) = \frac{1}{2^k}\left[(z+\sqrt{z^2-1})^k + \frac{1}{(z+\sqrt{z^2-1})^k}\right]\Bigg|_{z=\Psi(\tau)}$$

$$= \tau^k + 2^{-2k}\frac{1}{\tau^k}$$

The same Faber polynomials are also obtained for the segment $-1 \leqslant z \leqslant 1$ (see Note 2). The polynomials $P_k(z)$ thus obtained are also Chebyshev polynomials for the segment and ellipses (see Section 1.3.4, the introduction to Chapter 5 and Section 1.1).

2.1.2 Some properties of the Faber polynomials

Let us consider Formula (6) of Section 2.1.1

$$M(t, \tau) = \frac{t\Psi'(t)}{\Psi(t) - \Psi(\tau)} - \frac{t}{t-\tau} = \sum_{k=1}^{\infty} \frac{1}{\tau} M_k\left(\frac{1}{\tau}\right)\frac{1}{t^k} \tag{1}$$

where the function $M(t, \tau)$ is regular with respect to τ and to t for $|\tau| > \rho$, $|t| > \rho$. Let us denote

$$G(r) = \max_{|t|=r,\, |\tau|=r} |M(t, \tau)|, \quad \rho < r < \infty$$

Here and in the following we shall denote by $G(r)$ (sometimes with indices) non-negative constants determined by the choice of r. By using the well-known estimates for the coefficients of a power series we have

$$\left|\frac{1}{\tau} M_k\left(\frac{1}{\tau}\right)\right| \leqslant G(r) r^k, \quad |\tau| = r \tag{2}$$

Since $M_k(1/\tau)$ is regular in the domain $|\tau| \geqslant r$, this estimate also holds for $|\tau| > r$. Hence and from Formula (7), Section 2.1.1

$$P_k(z) = \tau^k + \frac{1}{\tau} M_k\left(\frac{1}{\tau}\right) = \tau^k\left(1 + \theta_k(z) G(r)\left(\frac{r}{|\tau|}\right)^k\right), \quad |\tau| \geqslant r \tag{3}$$

where

$$|\theta_k(z)| \leqslant 1, \quad r > \rho, \quad \tau = \Phi(z)$$

For $|\tau| > r$ Formula (3) gives the asymptotic representation of the Faber polynomials for large k. It can be written in the form

$$P_k(z) = \Phi(z)^k + \omega_k(z)$$
$$\omega_k(z) = \theta_k^*(z) G(r) r^k, \quad z \in \bar{B}_r^1, \quad |\theta_k^*(z)| \leqslant 1 \tag{4}$$

From (3) for arbitrary r and r_0, $r_0 > r > \rho$, $|\tau| \geqslant r_0$ we have

$$\left| \frac{P_{k+1}(z)}{P_k(z)} - \tau \right| = \left| \tau \frac{1 + \theta_{k+1}(z) G(r) \left(\frac{r}{|\tau|}\right)^{k+1}}{1 + \theta_k(z) G(r) \left(\frac{r}{|\tau|}\right)^k} - \tau \right|$$
$$\leqslant |\tau| \frac{\left[|\theta_{k+1}(z)| \left(\frac{r}{|\tau|}\right)^{k+1} + |\theta_k(z)| \left(\frac{r}{|\tau|}\right)^k \right] G(r)}{1 - G(r) \left(\frac{r}{|\tau|}\right)^k} \leqslant r_0 \frac{2G(r) \left(\frac{r}{r_0}\right)^k}{1 - G(r) \left(\frac{r}{r_0}\right)^k}$$

Hence it follows that as $k \to \infty$

$$\frac{P_{k+1}(z)}{P_k(z)} \rightrightarrows \tau, \quad \tau = \Phi(z) \tag{5}$$

inside the domain B^1.

Analogously as $k \to \infty$ we have

$$\sqrt[k]{|P_k(z)|} \rightrightarrows |\tau|, \quad \tau = \Phi(z) \tag{6}$$

inside the domain B^1. This last property can also be obtained directly from (5).

Let us now estimate $G(r)$. Noting that Formula (1) can be written in the form

$$\frac{t\Psi'(t)}{\Psi(t) - \Psi(\tau)} - \frac{t}{t - \tau} = \sum_{k=1}^{\infty} \omega_k(z) \frac{1}{t^k}, \quad z = \Psi(\tau)$$

and bearing in mind the relation $\psi(w) = \Psi(\rho w)$ between $\psi(w)$ and the function $\Psi(w)$ introduced in Chapter 1, inequalities (13) for $m = 1$ and (14) from Section 1.3.1, can be written in the form

$$\left| \frac{t\Psi'(t)}{\Psi(t) - \Psi(\tau)} - \frac{t}{t - \tau} \right| \leqslant \frac{|t| \rho}{|t|^2 - \rho^2} \sqrt{\ln \frac{|\tau|^2}{|\tau|^2 - \rho^2}}, \quad |t| > \rho, \quad |\tau| > \rho \tag{7}$$

$$\sum_{k=1}^{\infty} \frac{|\omega_k(z)|^2}{k\rho^{2k}} \leqslant \ln \frac{|\tau|^2}{|\tau|^2-\rho^2}, \quad \tau=\Phi(z), \quad z \in B^1 \tag{8}$$

From (7) we have

$$G(r) \leqslant \frac{r\rho}{r^2-\rho^2} \sqrt{\ln \frac{r^2}{r^2-\rho^2}}, \quad r>\rho$$

Inequality (8) can be used to prove the following lemma.

Lemma Let $\bar{B}$ be a set of type $\mathfrak{M}$ and $P_n(z)=\Phi(z)^n+\omega_n(z)$ be the Faber polynomials for $\bar{B}$. Then

$$\omega_n(z)=\begin{cases} \sqrt{\ln \frac{r^2}{r^2-\rho^2}}\sqrt{n}\,\rho^n & \qquad (9) \\ & z \in \bar{B}_r^1, \quad r>\rho \\ \left(1+\sqrt{e}+\sqrt{n\ln(n+1)}\right)\frac{\rho}{r}\rho^n & \qquad (10) \end{cases}$$

$$P_n(z)=\begin{cases} r^n+\sqrt{\ln \frac{r^2}{r^2-\rho^2}}\sqrt{n}\,\rho^n, \quad z \in \bar{B}_r, \quad r>\rho & (11) \\ \left(\sqrt{e}+\sqrt{n\ln(n+1)}\right)\rho^n, \quad z \in \bar{B} & (12) \end{cases}$$

If the boundary L of the set $\bar{B}$ is an analytic curve and the function $z=\Psi(w)$ is regular and schlicht in the domain $\rho_1<|w|<\infty$, $\rho_1<\rho$ (see Goluzin [1], p. 51), then

$$|\omega_n(z)| \leqslant \sqrt{\ln \frac{r^2}{r^2-\rho_1^2}}\sqrt{n}\,\rho_1^n, \quad z \in \bar{B}_r^1, \quad r>\rho_1 \tag{13}$$

$$|P_n(z)| \leqslant r^n+\sqrt{\ln \frac{r^2}{r^2-\rho_1^2}}\sqrt{n}\,\rho_1^n, \quad z \in \bar{B}_r, \quad r>\rho_1 \tag{14}$$

PROOF Equation (9) follows directly from (8). Equation (11) follows from (9) and (4). Assuming in (11) that $\left(\frac{r}{\rho}\right)^2 = 1+\frac{1}{n}$ we obtain (12). From (4) and (12) and by using the maximum modulus principle we find (10). Inequalities (13) and (14) follow respectively from (9) and (11) on the strength of Note 2, Section 2.1.1. The lemma is proved.

NOTE 1 Let $\bar{B}$ be a convex set* of type $\mathfrak{M}$. We can show

*A set $\bar{B}$ is said to be convex if the segment $\overline{z_1 z_2}$ connecting two points z_1 and z_2 belonging to $\bar{B}$ also belongs to $\bar{B}$.

that then

$$|P_n(z)| \leqslant 2\rho^n, \quad z \in \bar{B}, \quad n = 0, 1, \ldots$$

and consequently

$$|\omega_n(z)| \leqslant 3\rho^n, \quad z \in \bar{B}^1, \quad n = 0, 1, \ldots$$

We can find other conditions for the set $\bar{B}$ which are fulfilled if the inequality

$$|P_n(z)| < a\rho^n, \quad z \in \bar{B} \tag{15}$$

holds, where a is some constant.

NOTE 2 Proceeding in a slightly different manner, we can obtain sufficiently good estimates for the derivatives of Faber polynomials. We shall not dwell on this.

Note two more properties of Faber polynomials.

1. Since the function

$$\frac{1}{t^n} P_n(\Psi(t)) = 1 + \frac{1}{t^{n+1}} M_n\left(\frac{1}{t}\right)$$

is regular in the domain $|t| \geqslant r$, $r > \rho$, then

$$\sup_{z \in L_r} |P_n(z)| \geqslant r^n, \quad \inf_{z \in L_r} |P_n(z)| \leqslant r^n, \quad r > \rho \tag{16}$$

Hence it follows, in particular, that the curve, defined by the equation $|P_n(z)| = r^n$ (lemniscate), has common points with the contour line L_r, $r > \rho$.

2. From (6) and the formula $\Phi(z) = \frac{\varphi(z)}{\varphi'(\infty)}$ we see that $\frac{\sqrt[n]{|P_n(z)|}}{c|\varphi(z)|} \rightrightarrows 1$ as $n \to \infty$ in B^1, $c = \frac{1}{\varphi'(\infty)}$. Besides from (3)

$$|P_n(z)| \geqslant |\tau|^n \left(1 - O(r)\left(\frac{r}{|\tau|}\right)^n\right), \quad |\tau| > r, \ \rho < r < +\infty, \ \tau = \Phi(z)$$

whence it follows that for each r, $\rho < r < +\infty$, however close to ρ, there is a value of N such that for $n > N$ all zeros of $P_n(z)$ belong to B_r, and consequently for every function $f(z)$, regular on $\bar{B}$, for sufficiently large n, an interpolating polynomial $p_n(z)$ can be constructed, the interpolating points of which are the zeros of $P_{n+1}(z)$. As $n \to \infty$ the sequence $p_n(z)$ will converge on B to $f(z)$ (see Section 1.2). See also Suetin [6].

2.1.3 Representation of functions regular in a closed domain by a series of Faber polynomials

Theorem 1 Let $P_k(z)$, $k=0, 1, 2, \ldots$, be a sequence of Faber polynomials for the set $\bar{B}$ of type $\mathfrak{M}$. If the function $f(z)$ is regular in the domain B_R, $\rho<R<\infty$, and there is a singular point on the contour line L_R, then

1. The function $f(z)$ can be expanded into a series of Faber polynomials

$$f(z)=\sum_{k=0}^{\infty} a_k P_k(z) \tag{1}$$

where

$$a_k=\frac{1}{2\pi i}\int_{|t|=r_1} f(\Psi(t))\frac{1}{t^{k+1}}\,dt, \quad \rho<r_1<R \tag{2}$$

2. In this case

$$\overline{\lim_{k\to\infty}}\sqrt[k]{|a_k|}=\frac{1}{R} \tag{3}$$

and series (1) uniformly converges on the compact sub-sets of B_R and diverges outside $\bar{B}_R$.

3. The expansion of the function $f(z)$ into a series of Faber polynomials uniformly converging in some domain $D\supset\bar{B}$ is unique.

4. Conversely: if (3) holds, then the series (1) uniformly converges on the compact sub-sets of B_R and diverges outside $\bar{B}_R$ and the function $f(z)$ defined by the series (1) is regular in the domain B_R and has a singular point on the line L_R.

PROOF (1) Let $z\in B_R$, in particular, $z\in L_{r_2}$, $\rho<r_2<R$. Let us choose r_1 such that $r_2<r_1<r$. Then

$$f(z)=\frac{1}{2\pi i}\int_{L_{r_1}}\frac{f(\zeta)}{\zeta-z}\,d\zeta=\frac{1}{2\pi i}\int_{|t|=r_1} f(\Psi(t))\frac{\Psi'(t)}{\Psi(t)-z}\,dt$$

$$=\frac{1}{2\pi i}\int_{|t|=r_1} f(\Psi(t))\sum_{k=0}^{\infty}P_k(z)\frac{1}{t^{k+1}}\,dt$$

$$=\frac{1}{2\pi i}\int_{|t|=r_1}\sum_{k=0}^{\infty} f(\Psi(t))\frac{1}{t^{k+1}}P_k(z)\,dt$$

The series under the integral sign uniformly converges on

the circle $|t| = r_1$. Integrating term by term we find

$$f(z) = \sum_{k=0}^{\infty} a_k P_k(z), \quad a_k = \frac{1}{2\pi i} \int\limits_{|t|=r_1} f(\Psi(t)) \frac{1}{t^{k+1}} dt$$

We have obtained Formulae (1) and (2) whereby in the formulae for a_k we can choose r_1 arbitrarily in the interval $\rho < r_1 < R$ such that the function under the integral sign is regular in the ring $\rho < |t| < R$.

2. From the formula for the coefficients we have

$$|a_k| \leqslant \frac{1}{2\pi} M(r_1) \frac{1}{r_1^{k+1}} \cdot 2\pi r_1 = \frac{M(r_1)}{r_1^k}, \quad M(r_1) = \sup_{|t|=r_1} |f(\Psi(t))|$$

and

$$\sqrt[k]{|a_k|} \leqslant \sqrt[k]{M(r_1)} \frac{1}{r_1}$$

whence

$$\overline{\lim_{k\to\infty}} \sqrt[k]{|a_k|} \leqslant \frac{1}{r_1}$$

As r_1 is arbitrary, $\rho < r_1 < R$, we have

$$\overline{\lim_{k\to\infty}} \sqrt[k]{|a_k|} \leqslant \frac{1}{R}$$

Let us assume that it is the inequality sign which holds. Let us put

$$\overline{\lim_{k\to\infty}} \sqrt[k]{|a_k|} = \frac{1}{r_0} < \frac{1}{R}$$

Then if $z \in L_{r_1}$, $\rho < r_1 < r_0$, from (3) and (5), Section 2.1.2, we obtain for the terms of the series (1)

$$\overline{\lim_{k\to\infty}} \sqrt[k]{|a_k P_k(z)|} = \frac{r_1}{r_0} < 1$$

Let the number q be such that $\frac{r_1}{r_0} < q < 1$. Then for sufficiently large k

$$\sqrt[k]{|a_k P_k(z)|} < q$$

or

$$|a_k P_k(z)| < q^k$$

K

Hence it follows that the series (1) uniformly converges on L_{r_1}, that is in the domain B_{r_1}. As r_1 is arbitrary, $\rho < r_1 < r_0$, this means that the series (1) uniformly converges on the compact sub-sets of B_{r_0} to a regular function. This function coincides with $f(z)$ on $B_R \subset B_{r_0}$. Because of the uniqueness of analytic continuation the function $f(z)$ is regular on B_{r_0}. But this contradicts the assumption of the theorem that the function $f(z)$ has a singular point on $L_R (R < r_0)$. That is, our assumption is wrong, and

$$\overline{\lim_{k\to\infty}} \sqrt[k]{|a_k|} = \frac{1}{R}$$

and putting $r_0 = R$ in the above we find that the series (1) converges uniformly on the compact sub-sets of B_R.

Let us now show that the series (1) diverges outside $\overline{B}_R$. Let $z \in L_{r'}$, $R < r' < \infty$. Then

$$\overline{\lim_{k\to\infty}} \sqrt[k]{|a_k P_k(z)|} = \frac{r'}{r} > 1$$

and there is an infinite set of values k such that

$$\sqrt[k]{|a_k P_k(z)|} > 1$$

for such k

$$|a_k P_k(z)| > 1$$

This proves that the series (1) diverges outside $\overline{B}_R$.

3. Let the function $f(z)$ be expanded into a series of Faber polynomials

$$f(z) = \sum_{k=0}^{\infty} a_k P_k(z)$$

and let the series converge uniformly in some domain $D \supset \overline{B}$ and also converge uniformly on some contour line $L_{r''}$, $r'' > \rho$. Assuming $z = \Psi(t)$, $|t| = r''$, we have

$$f(\Psi(t)) = \sum_{k=0}^{\infty} a_k P_k(\Psi(t)) = \sum_{k=0}^{\infty} a_k \left(t^k + \frac{1}{t} M_k\left(\frac{1}{t}\right)\right)$$

Multiplying both halves of this equation by $\frac{1}{2\pi i} \frac{1}{t^{n+1}}$ and integrating term by term along the circle $|t| = r''$ we have

$$\frac{1}{2\pi i}\int\limits_{|t|=r''} f(\Psi(t))\,\frac{1}{t^{n+1}}\,dt$$

$$=\sum_{k=0}^{\infty} a_k\,\frac{1}{2\pi i}\int\limits_{|t|=r''}\left(t^k+\frac{1}{t}\,M_k\left(\frac{1}{t}\right)\right)\frac{1}{t^{n+1}}\,dt=a_n$$

i.e.

$$a_n=\frac{1}{2\pi i}\int\limits_{|t|=r''} f(\Psi(t))\,\frac{1}{t^{n+1}}\,dt,\quad n=0,\ 1,\ \ldots$$

for any expansion of the function $f(z)$ into a series of Faber polynomials in any domain $D\supset\overline{B}$. Hence it follows that the expansion is unique.

4. Let (3) hold. Then, as in the second part of the proof, let us show that the series (1) converges uniformly on the compact sub-sets of B_R and that the function $f(z)$, defined by the series (1), is regular in B_R. If the function $f(z)$ did not have a singular point on L_R then it would be regular in some domain B_{r_0} wider than $B_R, \rho<R<r_0<\infty$. But then as in the second part of the proof we could show that

$$\overline{\lim_{k\to\infty}}\sqrt[k]{|a_k|}=\frac{1}{r_0}<\frac{1}{R}$$

But this contradicts condition (3). That is the function $f(z)$ has a singular point on L_R and the theorem is completely proved.

NOTE 1 The theorem about the expansion by Faber polynomials is proved under the condition that the function is regular on $\overline{B}$. If $\overline{B}$ is a closed, bounded domain the boundary of which is an analytic curve, it is not difficult to show that the expansion theorem also holds for any function $f(z)$ which is regular in some closed domain $\overline{B}_{\rho_1}$, $\rho_1<\rho$ (see Note 2, Section 2.1.1), where the coefficients of the expansion are defined by the former Formula (2) with r_1, $\rho_1<r_1$, such that L_{r_1} lies in the domain of regularity of the function $f(z)$.

NOTE 2 Part (4) of the theorem is analogous to the Cauchy-Hadamard theorem relating to the radius of convergence of a power series.

Let us now introduce an estimate for the modulus of the difference $f(z)-S_n(z)$ where $f(z)$ is a function regular on B and $S_n(z)$ is the partial sum of the expansion of $f(z)$ by Faber polynomials:

$$S_n(z) = \sum_{k=0}^{n} a_k P_k(z)$$

Theorem 2 Let $\overline{B}$ be a set of type $\mathfrak{M}$, $f(z)$ be a regular function in B_R, $\rho < R < \infty$, and

$$\frac{1}{2\pi}\int_0^{2\pi} |f(\Psi(re^{i\theta}))|\,d\theta \leqslant M$$

for all r, $\rho < r < R$, where M is some constant. Then

1. $|f(z) - S_n(z)|$

$$\leqslant M\left[\frac{R}{R-r} + \left(\frac{\rho}{r}\right)^{n+1}\sqrt{\left[\frac{(n+1)R^2}{R^2-\rho^2} + \frac{\rho^2 R^2}{(R^2-\rho^2)^2}\right]\ln\frac{r^2}{r^2-\rho^2}}\right]\left(\frac{r}{R}\right)^{n+1}$$

$$z \in \overline{B}_r, \quad \rho < r < R$$

2. $|f(z) - S_n(z)|$

$$\leqslant M\left[\frac{2\sqrt{e}\,R}{R-\rho} + \sqrt{\left[\frac{(n+1)R^2}{R^2-\rho^2} + \frac{\rho^2 R^2}{(R^2-\rho^2)^2}\right]\ln(n+2)}\right]\left(\frac{\rho}{R}\right)^{n+1}$$

$$z \in \overline{B}, \quad n+1 > \frac{\sqrt{\rho}}{\sqrt{R}-\sqrt{\rho}}$$

3. If the boundary L of the set $\overline{B}$ is an analytic curve and the function $z = \Psi(w)$ is regular and schlicht in the domain $\rho_1 < |w| < \infty$, $\rho_1 < \rho$, then

$|f(z) - S_n(z)|$

$$\leqslant M\left[\frac{R}{R-r} + \left(\frac{\rho_1}{r}\right)^{n+1}\sqrt{\left[\frac{(n+1)R^2}{R^2-\rho_1^2} + \frac{\rho_1^2 R^2}{(R^2-\rho_1^2)^2}\right]\ln\frac{r^2}{r^2-\rho_1^2}}\right]\left(\frac{r}{R}\right)^{n+1}$$

$$z \in \overline{B}_r, \quad \rho \leqslant r < R$$

4. If the set $\overline{B}$ has the property $|P_n(z)| \leqslant a\rho^n$, $z \in \overline{B}$, $n = 0, 1, \ldots$ where a is a constant, then

$$|f(z) - S_n(z)| \leqslant Ma\frac{R}{R-r}\left(\frac{r}{R}\right)^{n+1}, \quad z \in \overline{B}_r, \ \rho \leqslant r < R$$

PROOF (1) The coefficients a_k can be represented by the formulae

$$a_k = \frac{1}{2\pi i}\int_{|t|=r_1} f(\Psi(t))\frac{1}{t^{k+1}}\,dt, \quad \rho < r < r_1 < R$$

and for $z \in L_r$ we have

$$|f(z)-S_n(z)| \leqslant \sum_{k=n+1}^{\infty} |a_k||P_k(z)| \leqslant M \sum_{k=n+1}^{\infty} (|\Phi(z)|^k + |\omega_k(z)|) \frac{1}{r_1^k}$$

$$= M\left[\sum_{k=n+1}^{\infty}\left(\frac{r}{r_1}\right)^k + \sum_{k=n+1}^{\infty} \sqrt{k}\left(\frac{\rho}{r_1}\right)^k \frac{|\omega_k(z)|}{\sqrt{k}\rho^k}\right]$$

$$\leqslant M\left[\frac{r^{n+1}}{r_1^n(r_1-r)} + \sqrt{\sum_{k=n+1}^{\infty} k\left(\frac{\rho}{r_1}\right)^{2k} \sum_{k=n+1}^{\infty} \frac{|\omega_k(z)|^2}{k\rho^{2k}}}\right]$$

By using the inequality (8), Section 2.1.2, and letting r_1 tend to the limit R, we prove statement (1) of the theorem. Assuming that $\left(\frac{r}{\rho}\right)^2 = 1 + \frac{1}{n+1}$, we prove statement (2) of the theorem. Statement (3) follows directly from (1) (see also Note 1). Statement (4) is easily proved. The theorem is proved.

NOTE 3 In formulating the theorem the constant M can be defined by the inequality $|f(z)| \leqslant M$, $z \in B_R$, for example by assuming

$$M = \sup_{z \in B_R} |f(z)|$$

Various other formulae defining M can be given. These depend on the different methods of estimating the coefficients

$$a_k = \frac{1}{2\pi r_1^k} \int_0^{2\pi} f(\Psi(r_1 e^{i\theta}))\, e^{-ik\theta}\, d\theta$$

$$\rho < r_1 < R$$

For example, changing θ to $\theta + \frac{\pi}{k}$, we have

$$a_k = -\frac{1}{2\pi r_1^k} \int_0^{2\pi} f\left(\Psi\left(r_1 e^{i\left(\theta+\frac{\pi}{k}\right)}\right)\right) e^{-ik\theta}\, d\theta$$

Summing we have

$$a_k = \frac{1}{4\pi r_1^k} \int_0^{2\pi} \left[f(\Psi(r_1 e^{i\theta})) - f\left(\Psi\left(r_1 e^{i\left(\theta+\frac{\pi}{k}\right)}\right)\right)\right] e^{-ik\theta}\, d\theta$$

Let us assume

$$\omega_1(\delta, f(\Psi(r_1e^{i\theta}))) = \sup_{0<h\leqslant\delta} \frac{1}{2\pi}\int_0^{2\pi} |f(\Psi(r_1e^{i\theta})) - f(\Psi(r_1e^{i(\theta+h)}))|\,d\theta$$

$$\omega_1(\delta) = \sup_{\rho<r_1<R} \omega_1(\delta, f(\Psi(r_1e^{i\theta})))$$

and if $\omega_1\left(\frac{\pi}{n+1}\right) < +\infty$, we have

$$|a_k| \leqslant \frac{1}{2r_1^k}\,\omega_1\left(\frac{\pi}{n+1}\right), \quad k = n+1,\ n+2, \ldots$$

We can then put $M = \frac{1}{2}\omega_1\left(\frac{\pi}{n+1}\right)$. Note that $\omega_1(\delta)$ can be estimated by the modulus of continuity $\omega(\delta, f)$ of the function $f(z)$ on $\overline{B}_R$, if the function $f(z)$ is continuous on $\overline{B}_R$: $\omega_1(\delta) \leqslant c\omega(\delta, f)$, where c depends only on R/ρ.

Proceeding in an analogous manner to that applied before, we find, for example, the inequalities

$$|a_k| \leqslant \frac{1}{2\pi k r_1^k} \bigvee_0^{2\pi} f(\Psi(r_1e^{i\theta}))$$

$$|a_k| \leqslant \frac{1}{2kr_1^k}\,\frac{1}{2\pi}\int_0^{2\pi} |f'(\Psi(r_1e^{i\theta}))\,\Psi'(r_1e^{i\theta})|\,d\theta$$

where $\bigvee_0^{2\pi} f(\Psi(r_1e^{i\theta}))$ is the variation of the function $f(\Psi(r_1e^{i\theta}))$ with respect to θ in the interval $(0, 2\pi)$.

NOTE 4 We can prove a theorem analogous to Theorem 2 for the case where the function $f(z)$ is regular on $\overline{B}_R$ apart from a finite number of poles and logarithmic singularities lying on L_r. To evaluate the coefficients we then follow the method of Mergelian [2] pp. 104-106. It is also possible to obtain an upper estimate of $|f'(z) - S'_n(z)|$ by using the estimates for the derivatives of the Faber polynomials.

We shall further consider the approximation of the function $f(z)$, regular in the domain B_R, $\rho < R < +\infty$, by means of polynomials $p_n(z)$ with the special form

$$p_n(z) = \frac{1}{2\pi i}\int_{L_{r_1}} \frac{f(\zeta)\,(P_{n+1}(\zeta) - P_{n+1}(z))}{(\zeta - z)\,\Phi(\zeta)^{n+1}}\,d\zeta, \quad \rho < r_1 < R, \quad z \in \overline{B} \qquad (4)$$

They were introduced by Mergelian. For these it is not difficult to prove the following theorem.

Theorem 3 Let $\overline{B}$ be a set of type $\mathfrak{M}$; $f(z)$ a function regular in the domain

$$B_{R},\ \rho < R < \infty;\ \frac{1}{2\pi}\int_{0}^{2\pi} |f(\Psi(re^{i\theta}))|\,d\theta \leqslant M,\ \rho < r < R$$

and $p_n(z)$ the polynomial defined by Formula (4). Then

$$1.\ |f(z) - p_n(z)| \leqslant M\frac{R+\rho}{R-\rho}\left[\sqrt{n+1}\left(\sqrt{\ln(n+2)} + \sqrt{\ln\frac{R^2}{R^2-\rho^2}}\right) + \sqrt{e}\right]\left(\frac{\rho}{R}\right)^{n+1} \quad z \in \overline{B}$$

$$2.\ |f(z) - p_n(z)| \leqslant M\frac{R+r}{R-r}\left[1 + \left(\sqrt{\ln\frac{r^2}{r^2-\rho^2}} + \sqrt{\ln\frac{R^2}{R^2-\rho^2}}\right)\sqrt{n+1}\left(\frac{\rho}{r}\right)^{n+1}\right]\left(\frac{r}{R}\right)^{n+1} \quad z \in \overline{B}_r,\ \rho < r < R$$

Parallel with (4) we can consider the polynomials

$$p_n^*(z) = \frac{1}{2\pi i}\int_{L_{r_1}} \frac{f(\zeta)\left(P_{n+1}(z_0) - P_{n+1}\left(\frac{z_0}{\zeta}z\right)\right)}{(\zeta - z)P_{n+1}(z_0)}\,d\zeta,\quad \rho < r_1 < R,\quad z \in \overline{B}$$

where z_0 is a point on L_{r_1}, such that $|P_{n+1}(z_0)| \geqslant r_1^{n+1}$ (see Formula (16), Section 2.1.2). By using this formula and applying the lemma in Section 2.1.2 we can easily prove a theorem analogous to Theorem 3 above.

There is a wide literature dealing with problems of the approximation of functions by means of Faber polynomials. See, for example, Alper and Ivanov [1], Ignateva and Vostretsov [1], Seleznev [1] and Suetin [1].

2.1.4 An analogy to Faber polynomials for unbounded and multiply-connected domains

Let us carry out an analogous investigation to that above, but for unbounded and multiply-connected domains without going into details.

Let $\overline{D}$ be a closed set of the extended z plane the complement D^1 of which is the simply-connected domain containing some point $z_0 \neq \infty$. Let the function

$$z=\Psi(w,\ z_0)=z_0+\frac{\alpha_1}{w}+\frac{\alpha_2}{w^2}+\ \ldots,\ \alpha_1=1$$

map one-to-one conformally some domain $|w|>\rho$ on to the domain D^1 and $w=\Phi(z,\ z_0)$ be the inverse function. Let us denote by l_r, $\rho<r<+\infty$, the image of the circle $|w|=r$ when mapped by the function $z=\Psi(w,\ z_0)$, by D_r the domain bounded by l_r and containing $\bar{D}$, and by D_r^1 the complement of $\bar{D}_r$ with respect to the extended plane.

Note that if we map the plane z on to the plane ζ by means of the function $\zeta=\frac{1}{z-z_0}$, the set $\bar{D}$ is mapped on to some set $\bar{B}$ of type $\mathfrak{M}$ and the functions $w=\Phi(\zeta)$ and $\zeta=\Psi(w)$, introduced in Section 2.1 for $\bar{B}$ will be

$$w=\Phi(\zeta)=\Phi\left(\frac{1}{\zeta}+z_0,\ z_0\right),\quad \zeta=\Psi(w)=\frac{1}{\Psi(w,\ z_0)-z_0}$$

the lines l_r being mapped on to the lines L_r for $\bar{B}$, the domains D_r and D_r^1 respectively being mapped on to domains B_r and B_r^1 corresponding to the set $\bar{B}$. Then if

$$P_k(\zeta)=b_k^{(k)}\zeta^k+b_{k-1}^{(k)}\zeta^{k-1}+\ \ldots\ +b_0^{(k)},\quad b_k^{(k)}=1,\quad k=0,\ 1,\ \ldots$$

are the Faber polynomials for $\bar{B}$, then the generalised polynomials (rational functions) $Q_k(z)=P_k\left(\frac{1}{z-z_0}\right)$ are analogous to the Faber polynomials for the set $\bar{D}$. Clearly

$$Q_k(\Psi(\tau,\ z_0))=\tau^k+\frac{1}{\tau}M_k\left(\frac{1}{\tau}\right),\quad \left|\frac{1}{\tau}M_k\left(\frac{1}{\tau}\right)\right|\leqslant G(r)r^k,\quad |\tau|\geqslant r>\rho$$

An analogous theorem to Theorem 1, Section 2.1.3, can obviously also be formulated for the set $\bar{D}$. Note that the theorem indicated is obtained by applying Theorem 1, Section 2.1.3, to the function $f_1(\zeta)=f\left(\frac{1}{\zeta}+z_0\right)$ and then reconverting to the variable z. Then Formula (2), Section 2.1.3, has the form

$$a_k=\frac{1}{2\pi i}\int\limits_{|t|=r_1}f(\Psi(t,\ z_0))\frac{1}{t^{k+1}}\,dt,\quad \rho<r_1$$

Without stopping to consider arbitrary multiply-connected domains, let us now consider the case of a doubly-connected closed domain $\bar{B}$. Let the complement of $\bar{B}$ comprise two simply-connected domains: the domain B^1 containing the point ∞ and the domain B^2 containing some point z_0. As before let us introduce the function

$$z=\Psi_1(w)=w+\alpha_0+\frac{\alpha_1}{w}+\frac{\alpha_2}{w^2}+\ldots$$

mapping one-to-one conformally some domain $|w|>\rho_1$ on to B^1, and the function

$$z=\Psi_2(w)=z_0+\frac{\alpha_1}{w}+\frac{\alpha_2}{w^2}+\ldots$$

mapping one-to-one conformally some domain $|w|>\rho_2$ on to B^2 and let the inverse functions of these be denoted respectively by $w=\Phi_1(z)$ and $w=\Phi_2(z, z_0)$. We also introduce the contour lines L_r for the domain B^1 and the contour lines l_r for B^2.

Let us denote the doubly-connected domain bounded by L_{r_1}, $\rho_1<r_1<\infty$, and l_{r_2}, $\rho_2<r_2<\infty$ by B_{r_1,r_2}. For the set $\bar{B}_1$ complementary to B^1 with respect to the extended plane, we introduce, as in Section 2.1, Faber polynomials $P_k^{(1)}(z)$, $k=0, 1, \ldots$, and for the image $\bar{B}_2$ of the set $\bar{D}_2$ by the function $\zeta=\frac{1}{z-z_0}$, complementary to B^2 with respect to the extended plane, let us introduce the Faber polynomials $P_k^{(2)}(\zeta), k=0, 1, \ldots$ (the generalised polynomials $P_k^{(2)}\left(\frac{1}{z-z_0}\right)$ will then be analogous to the Faber polynomials for the set $\bar{D}_2$).

Let us now consider the analogy to Theorem 1, Section 2.1.3, for the case in question. Let the function $f(z)$ be regular in the domain B_{r_1,r_2}, $\rho_1<r_1<\infty$, $\rho_2<r_2<\infty$, and let there be singular points on L_{r_1} and l_{r_2}. Assuming $\rho_1<r_1'<r_1$, $\rho_2<r_2'<r_2$, according to Cauchy's formula for $z\in B_{r_1', r_2'}$ we have

$$f(z)=\frac{1}{2\pi i}\int\limits_{L_{r_1'}}\frac{f(\xi)}{\xi-z}\,d\xi+\frac{1}{2\pi i}\int\limits_{l_{r_2'}}\frac{f(\xi)}{\xi-z}\,d\xi$$

$$=\frac{1}{2\pi i}\int\limits_{L_{r_1'}}\frac{f(\xi)}{\xi-z}\,d\xi+\frac{1}{2\pi i}\int\limits_{l_{r_2'}}\frac{f(\xi)(z-z_0)}{(\xi-z)(\xi-z_0)}\,d\xi+\frac{1}{2\pi i}\int\limits_{l_{r_2'}}\frac{f(\xi)}{\xi-z_0}\,d\xi$$

Altering the variable in the first integral on the right-hand side of the equation according to the formula $\xi=\Psi_1(t)$ and in the second according to the formula $\xi = \Psi_2(t, z_0)$, we find

$$f(z) = \frac{1}{2\pi i} \int_{|t|=r_1'} f(\Psi_1(t)) \frac{\Psi_1'(t)}{\Psi_1(t) - z} dt$$

$$+ \frac{1}{2\pi i} \int_{|t|=r_2'} f(\Psi_2(t, z_0)) \frac{\Psi_2'(t)}{\Psi_2(t) - \frac{1}{z - z_0}} dt + \frac{1}{2\pi i} \int_{l_{r_2'}} \frac{f(\xi)}{\xi - z_0} d\xi$$

where $\zeta = \Psi_2(t) = \frac{1}{\Psi_2(t, z_0) - z_0}$ is the function mapping the domain $|t| > \rho_2$ on to the set B_2.

Proceeding in a manner not differing in principle from that used in proving Theorem 1, Section 2.1.3 we arrive at the following theorem.

Theorem 1 If the function $f(z)$ is regular in the domain B_{r_1, r_2}, $\rho_1 < r_1 < \infty, \rho_2 < r_2 < \infty$, and there are singular points on L_{r_1} and l_{r_2}, then

1. The function $f(z)$ can be expanded into the series

$$f(z) = \sum_{k=0}^{\infty} a_k P_k^{(1)}(z) + \sum_{k=0}^{\infty} b_k P_k^{(2)}\left(\frac{1}{z - z_0}\right) + A \tag{1}$$

where

$$a_k = \frac{1}{2\pi i} \int_{|t|=r_1'} f(\Psi_1(t)) \frac{1}{t^{k+1}} dt, \quad b_k = \frac{1}{2\pi i} \int_{|t|=r_2'} f(\Psi_2(t, z_0)) \frac{1}{t^{k+1}} dt$$

$$A = \frac{1}{2\pi i} \int_{l_{r_2'}} \frac{f(\xi)}{\xi - z_0} d\xi \tag{2}$$

and

$$\rho_1 < r_1' < r_1, \quad \rho_2 < r_2' < r_2$$

2. In this case

$$\overline{\lim_{k \to \infty}} \sqrt[k]{|a_k|} = r_1, \quad \overline{\lim_{k \to \infty}} \sqrt[k]{|b_k|} = r_2 \tag{3}$$

the series $\sum_{k=1}^{\infty} a_k P_k^{(1)}(z)$ uniformly converges on the compact sub-sets of the domain bounded by L_{r_1} and containing

$\overline{B}$, and the series $\sum_{k=1}^{\infty} b_k P_k^{(2)}\left(\frac{1}{z-z_0}\right)$ uniformly converges on the compact sub-sets of the domain bounded by l_{r_2} and containing $\overline{B}$.

3. The expansion of $f(z)$ into a series of the form (1), uniformly converging in some domain $B' \supset \overline{B}$, is unique;
4. Conversely: if (3) holds, then (2) is valid for series

$\sum_{k=1}^{\infty} a_k P_k^{(1)}(z)$ and $\sum_{k=1}^{\infty} b_k P_k^{(2)}\left(\frac{1}{z-z_0}\right)$ and the function defined by the series (1) is regular in the domain B_{r_1, r_2} and there are singular points on L_{r_1} and l_{r_2}.

We can also prove theorems analogous to Theorems 2 and 3, Section 2.1.3.

2.1.5 Asymptotic form of best approximating polynomials and Chebyshev polynomials

Theorem 1 Let $\overline{B}$ be a set of type $\mathfrak{M}$; the function $f(z)$ be regular in the domain B_R, $\rho < R < +\infty$, and let there be a singular point on L_R; r, $\rho \leqslant r < R$, be a fixed number; $\rho_n(f, \overline{B}_r)$ be the measure of best approximation of the function $f(z)$ by means of polynomials of a degree not greater than n (see introduction to Chapter 1). Then

$$\varlimsup_{n\to\infty} \sqrt[n+1]{\rho_n(f, \overline{B}_r)} = \frac{r}{R}, \quad \rho \leqslant r < R \tag{1}$$

PROOF Denoting

$$M(R_1) = \frac{1}{2\pi}\int_0^{2\pi} |f(\Psi(R_1 e^{i\theta}))|\, d\theta, \quad r < R_1 < R$$

and altering the constant M to $M(R_1)$ in Theorem 2, Section 2.1.3, we find an upper bound for $\rho_n(f, \overline{B}_r)$, from which follows

$$\varlimsup_{n\to\infty} \sqrt[n+1]{\rho_n(f, \overline{B}_r)} \leqslant \frac{r}{R_1}$$

and since R_1 is arbitrary, $r < R_1 < R$, we have

$$\overline{\lim_{n\to\infty}} \sqrt[n+1]{\rho_n(f, \bar{B}_r)} \leqslant \frac{r}{R} \tag{2}$$

Let us now find a lower bound for $\rho_n(f, \bar{B}_r)$. Let $\pi_n(z)$ be the best approximating polynomial of degree not greater than n for $f(z)$ on $\bar{B}_r$ (see introduction to Chapter 1) and $D_n(z) = f(z) - \pi_n(z)$. Let us expand the function $D_n(z)$ by means of Faber polynomials:

$$f(z) = \pi_n(z) + \sum_{k=0}^{\infty} b_k^{(n)} P_k(z) = \sum_{k=0}^{\infty} a_k P_k(z) \tag{3}$$

Since the expansion of the function $f(z)$ into a series of Faber polynomials is unique, we can conclude that $a_k = b_k^{(n)}$, $k = n+1, n+2, \ldots$, where

$$b_k^{(n)} = \frac{1}{2\pi i} \int_{|t|=r} D_n(\Psi(t)) \frac{1}{t^{k+1}} dt, \quad k = 0, 1, \ldots, \rho < r < R$$

Hence we have

$$|b_k^{(n)}| \leqslant \rho_n(f, \bar{B}_r) \frac{1}{r^k}, \quad k = 0, 1, \ldots$$

and consequently

$$\rho_n(f, \bar{B}_r) \geqslant |a_{n+1}| r^{n+1}, \quad \rho < r < R \tag{4}$$

Considering the converse, it is easy to show that $\rho_n(f, B_r)$ depends continuously on r, $r \geqslant \rho$. Hence it follows that (4) also holds for $r = \rho$, and on the basis of Theorem 1, Section 2.1.3,

$$\overline{\lim_{n\to\infty}} \sqrt[n+1]{\rho_n(f, \bar{B}_r)} \geqslant \frac{r}{R}, \quad \rho \leqslant r < R$$

Hence and from (2), (1) follows, and the theorem is proved.

Let us now find the asymptotic form of the polynomials $\pi_n(z)$ of best approximation for $f(z)$ on $\bar{B}_r$ for large n. To do this let us take the value of the difference of the n-th partial sum of the series of Faber polynomials for $f(z)$ and the polynomial $\pi_n(z)$.

From (3) we have

$$\left| \sum_{k=0}^{n} a_k P_k(z) - \pi_n(z) \right| \leqslant \sum_{k=0}^{n} |b_k^{(n)}| |P_k(z)|, \quad z \in \bar{B}_r \tag{5}$$

From (3) it also follows that

$$D_n(\Psi(t)) = \sum_{k=0}^{\infty} b_k^{(n)} t^k + \sum_{k=1}^{\infty} \frac{a_k^{(n)}}{t^k}, \quad \rho < |t| < R$$

whence

$$I_n = \frac{1}{2\pi} \int_0^{2\pi} |D_n(\Psi(re^{i\theta}))|^2 \, d\theta$$

$$= \sum_{k=0}^{\infty} |b_k^{(n)}|^2 r^{2k} + \sum_{k=1}^{\infty} \frac{|a_k^{(n)}|^2}{r^{2k}}, \quad \rho < r < R$$

But $I_n \leqslant \sup_{|t|=r} |D_n(\Psi(t))|^2$, and therefore

$$\sum_{k=0}^{\infty} |b_k^{(n)}|^2 r^{2k} \leqslant [\rho_n(f, \bar{B}_r)]^2, \quad \rho \leqslant r < R. \tag{6}$$

By using Cauchy's inequality, we have

$$\sum_{k=0}^{n} |b_k^{(n)}| |P_k(z)| \leqslant \sqrt{\sum_{k=0}^{n} |b_k^{(n)}|^2 r^{2k} \sum_{k=0}^{n} |P_k(z)|^2 \frac{1}{r^{2k}}}, \quad z \in \bar{B}_r \tag{7}$$

To estimate the first sum under the root sign we can make use of inequality (6). To estimate the second sum we can apply the lemma in Section 2.1.2 and the inequality ($\alpha_k \geqslant 0$, $a > 0$)

$$\sum_{k=0}^{n} (a + \sqrt{k}\alpha_k)^2 = (n+1) a^2 + 2a \sum_{k=1}^{n} \sqrt{k}\alpha_k + \sum_{k=1}^{n} k\alpha_k^2$$

$$\leqslant (n+1) a^2 + 2a \sqrt{\sum_{k=1}^{n} k \sum_{k=1}^{n} \alpha_k^2} + \sum_{k=1}^{n} k\alpha_k^2$$

$$\leqslant (n+1)\left(a + \sqrt{\sum_{k=1}^{n} \alpha_k^2}\right)^2$$

We then find

$$\sum_{k=0}^{n} |P_k(z)|^2 \frac{1}{r^{2k}}$$

$$\leqslant \begin{cases} \left(1 + \sqrt{\frac{\rho^2}{r^2 - \rho^2} \ln \frac{r^2}{r^2 - \rho^2}}\right)^2 (n+1), & z \in \bar{B}_r, \ r > \rho \\ (\sqrt{e} + \sqrt{\ln[(n+1)!]})^2 (n+1), & z \in \bar{B}, \ r = \rho \end{cases} \tag{8}$$

Comparing (5), (6), (7) and (8), we have

$$\left|\sum_{k=0}^{n} a_k P_k(z) - \pi_n(z)\right|$$

$$\leqslant \begin{cases} \sqrt{[\rho_n(f, \bar{B}_r)]^2 - \sum\limits_{k=n+1}^{\infty} |a_k|^2 r^{2k}} \\ \qquad \times\left(1+\sqrt{\frac{\rho^2}{r^2-\rho^2}\ln\frac{r^2}{r^2-\rho^2}}\right)\sqrt{n+1}, \quad z \in \bar{B}_r,\ \rho < r < R \\ \sqrt{[\rho_n(f, \bar{B}_r)]^2 - \sum\limits_{k=n+1}^{\infty} |a_k|^2 r^{2k}} \\ \qquad \times\left(\sqrt{e}+\sqrt{\ln[(n+1)!]}\right)\sqrt{n+1}, \quad z \in \bar{B} \end{cases} \tag{9}$$

Whence and on the strength of Stirling's formula, we find

$$\overline{\lim_{n\to\infty}} \sqrt[n+1]{\left|\sum_{k=0}^{n} a_k P_k(z) - \pi_n(z)\right|}$$

$$\leqslant \overline{\lim_{n\to\infty}} \sqrt[n+1]{\rho_n(f, \bar{B}_r)} = \frac{r}{R}, \quad z \in \bar{B}_r,\ \rho < r < R$$

and

$$\pi_n(z) = \sum_{k=0}^{n} a_k P_k(z)$$

$$+\begin{cases} \theta_n(z)\rho_n(f, \bar{B}_r)\left(1+\sqrt{\frac{\rho^2}{r^2-\rho^2}\ln\frac{r^2}{r^2-\rho^2}}\right)\sqrt{n+1} \\ \qquad\qquad z \in \bar{B}_r,\ \rho < r < R \\ \theta_n(z)\rho_n(f, \bar{B}_r)\left(\sqrt{e}+\sqrt{\ln[(n+1)!]}\right)\sqrt{n+1}, \quad z \in \bar{B} \end{cases}$$

where $\theta_n(z)$ is a polynomial of degree not greater than n, whereby $|\theta_n(z)| \leqslant 1$, $z \in \bar{B}_r, \rho \leqslant r < R$. The last formula gives the asymptotic form for the polynomials $\pi_n(z)$ of best approximation for $f(z)$ on $\bar{B}_r$ for large n.

By using Formula (9), let us find the asymptotic form of the Chebyshev polynomials $T_n(z) = z^n + c_1^{(n)} z^{n+1} + \ldots + c_n^{(n)}$ for the domain $\bar{B}_r$, $\rho < r < \infty$ (see Section 1.3.4) for large n. The polynomial $T_n(z)$ can be written in the form

$$T_n(z) = z^n - \pi^*_{n-1}(z)$$

where $\pi^*_{n-1}(z)$ is a polynomial of degree not greater than $n-1$ which is the best approximation for z^n on $\bar{B}_r$. Denoting

$$m_n = \sup_{z \in \bar{B}_r} |T_n(z)|$$

we find

$$m_n \leqslant \sup_{z \in \bar{B}_r} |P_n(z)| \leqslant r^n + \sqrt{\ln \frac{r^2}{r^2 - \rho^2}} \sqrt{n} \rho^n \tag{10}$$

Expanding z^n by means of Faber polynomials:

$$z^n = \sum_{k=1}^{n} a_k P_k(z), \quad a_n = 1$$

we have

$$T_n(z) - P_n(z) = \sum_{k=0}^{n-1} a_k P_k(z) - \pi^*_{n-1}(z)$$

and Formula (9) gives

$$|T_n(z) - P_n(z)| \leqslant \sqrt{m_n^2 - r^{2n}} \left(1 + \sqrt{\frac{\rho^2}{r^2 - \rho^2} \ln \frac{r^2}{r^2 - \rho^2}}\right) \sqrt{n}, \quad z \in \bar{B}_r$$

By using Formula (10) we obtain

$$|T_n(z) - P_n(z)| \leqslant r^n G(r, n) \left(\frac{\rho}{r}\right)^{n/2}, \quad z \in \bar{B}_r \tag{11}$$

where

$$G(r, n) = \sqrt{2 \sqrt{\ln \frac{r^2}{r^2 - \rho^2}} + \ln \frac{r^2}{r^2 - \rho^2} \sqrt{n} \left(\frac{\rho}{r}\right)^n} \times \left(1 + \sqrt{\frac{\rho^2}{r^2 - \rho^2} \ln \frac{r^2}{r^2 - \rho^2}}\right) n^{3/4}$$

whence

$$T_n(z) = P_n(z) + \theta^*_n(z) r^n G(r, n) \left(\frac{\rho}{r}\right)^{n/2}, \quad z \in \bar{B}_r, \quad \rho < r < +\infty \tag{12}$$

where $\theta^*_n(z)$ is a polynomial of degree not greater than $n-1$, whereby $|\theta^*_n(z)| \leqslant 1$, $z \in \bar{B}_r$, and $G(r, n) \left(\frac{\rho}{r}\right)^{n/2} \to 0$ as $n \to \infty$. We obtain from (12) the asymptotic behaviour of the Chebyshev polynomials for the domain $\bar{B}_r$, $\rho < r < +\infty$, for large n.

2.1.6 *A method of analytic continuation of power series*

Let B^1 be a simply-connected domain of the extended complex z plane, the boundary L of which contains more than one point, and let the point $a \in B^1$. Let us denote by $w = \Phi(z)$ the function mapping one-to-one conformally the domain B^1 on to the domain $|w| > \rho$ and having in the neighbourhood of the point a the expansion

$$\Phi(z) = \frac{1}{z-a} + \alpha_0 + \alpha_1(z-a) + \alpha_2(z-a)^2 + \dots \quad \text{for} \quad a \neq \infty$$

$$\Phi(z) = z + \alpha_0 + \frac{\alpha_1}{z} + \frac{\alpha_2}{z^2} + \dots \quad \text{for} \quad a = \infty$$

and let $z = \Psi(w)$ be the function inverse to $w = \Phi(z)$; L_r, $\rho < r < \infty$ be the image of the circumference $|w| = r$ when mapped by the function $z = \Psi(w)$; B^1_r be the domain bounded by the curve L_r and containing the point a. For $k = 1, 2, \dots$ we have in the neighbourhood of the point $z = a$

$$\Phi(z)^k = \frac{a_k^{(k)}}{(z-a)^k} + \frac{a_{k-1}^{(k)}}{(z-a)^{k-1}} + \dots + a_0^{(k)} + \sum_{\nu=1}^{\infty} \alpha_\nu^{(k)} (z-a)^\nu \quad \text{for} \quad a \neq \infty$$

$$\Phi(z)^k = a_k^{(k)} z^k + a_{k-1}^{(k)} z^{k-1} + \dots + a_0^{(k)} + \sum_{\nu=1}^{\infty} \frac{\alpha_\nu^{(k)}}{z^\nu} \quad \text{for} \quad a = \infty$$

Here $a_k^{(k)} = 1$. When $a = \infty$ the polynomial

$$P_k(z) = a_k^{(k)} z^k + a_{k-1}^{(k)} z^{k-1} + \dots + a_0^{(k)}$$

is the Faber polynomial for the closed set $\bar{B}$ which is the complement to B^1 with respect to the extended plane, and when $a \neq \infty$ the generalised polynomial $P_k\left(\frac{1}{z-a}\right)$ is analogous to the Faber polynomial for the set $\bar{B}$, complementary to B^1.

Theorem 1 Let there be given the series

$$f(z) = \sum_{k=1}^{\infty} \frac{a_k}{z^{k+1}} \tag{1}$$

which converges in the neighbourhood of the point $z = \infty$.

1. For the sum of the series (1) to be a function $f(z)$ regular in the domain B^1, it is necessary and sufficient that

$$\overline{\lim_{k\to\infty}} \sqrt[k]{|\delta_k|} \leqslant \rho \tag{2}$$

where

$$\delta_k = a_1^{(k)} a_1 + a_2^{(k)} a_2 + \dots + a_k^{(k)} a_k$$

2. For $f(z)$ to be a function regular in the domain B_r^1, $\rho < r < \infty$, and for there to be a singular point on L_r, it is necessary and sufficient that

$$\overline{\lim_{k\to\infty}} \sqrt[k]{|\delta_k|} = r \tag{3}$$

3. For all $z \in B^1$ if (2) holds or for all $z \in B_r^1$ if (3) holds and $r > \rho$ the series

$$\sum_{k=1}^{\infty} \frac{-\delta_k \Phi'(z)}{[\Phi(z)]^{k+1}}$$

converges uniformly on the compact sub-sets of B^1 or B_r^1 respectively and its sum is equal to $f(z)$.

PROOF The function $f(\Psi(w))\Psi'(w)$ is regular in some neighbourhood $|w| > r_0$ of the point $w = \infty$ and has the expansion

$$F(w) = f(\Psi(w))\Psi'(w) = \sum_{k=0}^{\infty} \frac{\gamma_k}{w^{k+1}}, \quad |w| > r_0 \tag{4}$$

If r_1, $r_1 > r_0$, is sufficiently large, then

$$\gamma_k = \frac{1}{2\pi i} \int_{|w|=r_1} f(\Psi(w))\Psi'(w) w^k \, dw = \frac{1}{2\pi i} \int_{L_{r_1}} f(z) \Phi(z)^k \, dz$$

The last integral is equal to the coefficient of $1/z$ with the minus sign in the Laurent expansion of the function $f(z)\Phi(z)^k$ in the neighbourhood of the point ∞. Consequently $\gamma_k = -\delta_k$. For the function $f(z)$ satisfying the conditions of the theorem to be regular in the domain B_r^1, $\rho < r < \infty$, and for there to be a singular point on L_r (since it is regular in B^1), it is necessary and sufficient that the function $F(w)$ be regular in the domain $|w| > r$ and that there be a singular point on

L

$|w|=r$ (correspondingly, that it be regular in $|w|>\rho$). Hence, and from the Cauchy-Hadamard theorem relating to the radius of convergence of a power series, applied to series (4), we prove statements (1) and (2) of the theorem. To prove (3) we must put $w=\Phi(z)$ in Formula (4). The theorem is proved.

Analogously we can prove, when $a \neq \infty$, the theorem below.

Theorem 2 Let the series

$$f(z)=\sum_{k=0}^{\infty} a_{k+1}(z-a)^k \tag{5}$$

converging in the neighbourhood of the point a be given.

1. For the sum of the series (1) to be the function $f(z)$, regular in the domain B^1 and to have at the point $z=\infty$, if $\infty \in B^1$, a zero of not lower than the second order, it is necessary and sufficient that

$$\overline{\lim_{k\to\infty}} \sqrt[k]{|\delta_k|} \leqslant \rho \tag{6}$$

where

$$\delta_k = a_1^{(k)}a_1 + a_2^{(k)}a_2 + \ldots + a_k^{(k)}a_k$$

2. For $f(z)$ to be a function regular in the domain B_r^1, $\rho<r<\infty$, and to have a singular point on L_r and a zero of not lower than the second order at the point $z=\infty$, if $\infty \in B_r^1$, it is necessary and sufficient that

$$\overline{\lim_{k\to\infty}} \sqrt[k]{|\delta_k|} = r \tag{7}$$

3. Also for all $z \in B^1$ if (6) holds or for all $z \in B_r^1$ if (7) holds and $r>\rho$ the series

$$\sum_{k=0}^{\infty} \frac{\delta_k \Phi'(z)}{\Phi(z)^{k+1}} \tag{8}$$

converges uniformly on the compact sub-sets of B^1 or B_r^2 respectively and its sum is equal to $f(z)$.

In proving this we have to take into account that if the point $\infty \in B^1$, then there is a point w_0, $|w_0|>\rho$, which is a simple pole of the function $\Psi(w)$.

NOTE 1 In Theorem 2 the requirement that the function $f(z)$ has at the point $z=\infty$ a zero of not lower than the second order can be eliminated by using a slightly more complex formula for δ_k and the series (8). For example, we can consider the function

$$F(w)=f(\Psi(w))\Psi'(w)(w-w_0)^2, \quad \text{for} \quad w_0=\Phi(\infty)$$

Then, in formulating the theorem, δ_k must be altered to $\delta_k'=w_0^2\delta_k-2w_0\delta_{k-1}+\delta_{k-2}$, $k=1, 2, \ldots,$ where we assume that δ_0 and δ_1 are equal to zero. The series (8) has to be altered to the series

$$\sum_{k=0}^{\infty}\delta_k^1\frac{\Phi'(z)}{(\Phi(z)-\Phi(\infty))^2}\frac{1}{\Phi(z)^{k+1}}$$

Part (1) of Theorem 1 was proved by Khavin [1, 2] by the method of functional analysis. In article [2] a statement of type (1) is proved not only for Faber polynomials but also for a whole series of other polynomials.

The proof given here is due to Lebedev [1].

2.2 GENERALISED FABER POLYNOMIALS

2.2.1 Definition of the generalised Faber polynomials

As in Section 2.1 let us denote a set of type $\mathfrak{M}$ by $\bar{B}$ and retain the remaining notation from Section 2.1. Let

$$R(t)=\sum_{k=0}^{\infty}\frac{c_k}{t^k}, \quad c_0\neq 0$$

be a function regular in the domain $|t|>\rho$. Let us multiply the generator function $\frac{t\Psi'(t)}{\Psi(t)-z}$ of Faber polynomials by the function $R(t)$ and, assuming that $z\in L_r$, $\rho<r<\infty$, let us expand the product obtained by powers of t:

$$K(z, t)=\frac{t\Psi'(t)}{\Psi(t)-z}R(t)=\sum_{k=0}^{\infty}P_k(z)\frac{1}{t^k}\sum_{k=0}^{\infty}\frac{c_k}{t^k}=\sum_{n=0}^{\infty}\Pi_n(z)\frac{1}{t^n} \tag{1}$$

where $\Pi_n(z)=c_0P_n(z)+c_1P_{n-1}(z)+\ldots+c_nP_0(z)$ and $P_k(z)$ are the Faber polynomials. For every fixed $z \in B_r$, $\rho < r < \infty$ the series converges in the domain $|t| > r$ and for every $z \in \overline{B}$ converges in the domain $|t| > \rho$; consequently for every fixed $z \in B_r$ it converges uniformly in the closed domain $|t| \geqslant r$.

The polynomials $\Pi_n(z)$, $n = 0, 1, \ldots$, are called the generalised Faber polynomials and the function $K(z, t)$ is the generating function of the polynomials $\Pi_n(z)$.

From the expansion (1) we have

$$\Pi_n(z) = \frac{1}{2\pi i} \int\limits_{|t|=r_1} \frac{t\Psi'(t)}{\Psi(t) - z} R(t)\, t^{n-1}\, dt, \quad z \in \overline{B}_r, \quad \rho \leqslant r < r_1 < \infty$$

or

$$\Pi_n(z) = \frac{1}{2\pi i} \int\limits_{L_{r_1}} \frac{R(\Phi(\zeta))\, \Phi(\zeta)^n}{\zeta - z}\, d\zeta, \quad z \in \overline{B}_r, \quad \rho \leqslant r < r_1 < \infty \tag{2}$$

Hence by a method analogous to that introduced to find (4), Section 2.1.1, we find that, if in the neighbourhood of the point $\zeta = \infty$ we have an expansion

$$R(\Phi(\zeta))\, \Phi(\zeta)^n = c_0 \zeta^n + c_1^{(n)} \zeta^{n-1} + \ldots + c_n^{(n)} + \frac{\alpha_1^{(n)}}{\zeta} + \frac{\alpha_2^{(n)}}{\zeta^2} + \ldots$$

then

$$\Pi_n(z) = c_0 z^n + c_1^{(n)} z^{n-1} + \ldots + c_n^{(n)}$$

and consequently

$$\Pi_n(z) = R(\Phi(z))\, \Phi(z)^n + \omega_n(z) \tag{3}$$

where $\omega_n(z)$ is a function which is regular in the domain B^1 and $\omega_n(\infty) = 0$.

This relation between the polynomials $\Pi_n(z)$ and the functions $R(\Phi(z))\, \Phi(z)^n$ can be used to define the generalised Faber polynomials.

Let $z \in B^1$. Assuming $z \in \Psi(\tau)$ in the formula for the generating function and proceeding as in Section 2.1.1 for $|t| > \rho$, $|\tau| > \rho$ we have

$$M(t, \tau) = \frac{t\Psi'(t)}{\Psi(t) - \Psi(\tau)} R(t) - \frac{\tau R(\tau)}{t - \tau} = \sum_{n=0}^{\infty} a_n(\tau) \frac{1}{t^n}$$

where $a_0(\tau)=c_0$ and $a_k(\tau)=\frac{1}{\tau}M_k\left(\frac{1}{\tau}\right)$, $k>0$. Thus for $|\tau|<|t|$ we have

$$\frac{t\Psi'(t)}{\Psi(t)-\Psi(\tau)}R(t)=c_0+\frac{\tau R(\tau)}{t-\tau}+\sum_{n=1}^{\infty}\frac{1}{\tau}M_n\left(\frac{1}{\tau}\right)\frac{1}{t^n}$$

$$=c_0+\sum_{n=1}^{\infty}\left(R(\tau)\tau^n+\frac{1}{\tau}M_n\left(\frac{1}{\tau}\right)\right)\frac{1}{t^n}$$

Comparing this expansion with (1), we find

$$\Pi_n(z)=R(\tau)\tau^n+\frac{1}{\tau}M_n\left(\frac{1}{\tau}\right), \quad \tau=\Phi(z). \tag{4}$$

Here $\frac{1}{\tau}M_n\left(\frac{1}{\tau}\right)$ is the coefficient of $\frac{1}{t^n}$ in the expansion of the function $M(t, \tau)$ as a series of powers of $1/t$ and consequently if

$$G(r)=\max_{|t|=r,\ |\tau|=r}|M(t, \tau)|, \quad \rho<r<\infty$$

then (see discussion, Section 2.1.2)

$$\left|\frac{1}{\tau}M_n\left(\frac{1}{\tau}\right)\right|\leqslant G(r)r^n, \quad |\tau|\geqslant r$$

and

$$\Pi_n(z)=\tau^n\left(R(\tau)+\theta_n(z)G(r)\left(\frac{r}{|\tau|}\right)^n\right), \quad |\tau|\geqslant r \tag{5}$$

where $|\theta_n(z)|\leqslant 1$, $\rho<r<\infty$, $\tau=\Phi(z)$. For $|\tau|>r$ Formula (5) gives the asymptotic form for the generalised Faber polynomials for large n. It can be written:

$$\begin{gathered}\Pi_n(z)=R(\Phi(z))\Phi(z)^n+\omega_n(z)\\ \omega_n(z)=\theta_n^*(z)G(r)r^n, \quad z\in B_r^1\end{gathered} \tag{6}$$

where $|\theta_n^*(z)|\leqslant 1$. After (5) we have the following lemma.

Lemma 1 On every closed set $E\subset B^1$, on which the function $R(\Phi(z))$ does not become zero,

$$\frac{\Pi_{n+1}(z)}{\Pi_n(z)}\to\tau, \quad \sqrt[n]{|\Pi_n(z)|}\to|\tau|, \quad \tau=\Phi(z)$$

uniformly for $n \to \infty$ and

$$\overline{\lim_{n\to\infty}} \sqrt[n]{|\Pi_n(z)|} \leqslant r$$

uniformly for для $z \in \bar{B}_r$, $\rho < r < \infty$.

Uniformly here means that for any $\varepsilon > 0$ there is N such that for $n > N$ the inequality

$$\sqrt[n]{|\Pi_n(z)|} \leqslant r + \varepsilon$$

is fulfilled for all $z \in \bar{B}_r$.

EXAMPLE Let us consider the particular case of generalised Faber polynomials when $\bar{B}$ is the circle $|z| \leqslant \rho$, $\rho > 0$. In this case $\Psi(t) = t$ and the generating function for the generalised Faber polynomials, which are here denoted by $q_n(z)$, will be

$$\frac{t}{t-z} R(t) = \sum_{k=0}^{\infty} z^k \frac{1}{t^k} \sum_{k=0}^{\infty} \frac{c_k}{t^k} = \sum_{n=0}^{\infty} q_n(z) \frac{1}{t^n}$$

and

$$q_n(z) = c_0 z^n + c_1 z^{n+1} + \ldots + c_n = z^n S_n\left(\frac{1}{z}\right)$$

where $S_n(1/t)$ is the n-th partial sum of the series $R(t)$. Consequently as $n \to \infty$

$$\frac{q_n(z)}{z^n} \rightrightarrows R(z)$$

inside the domain $|z| > \rho$ or

$$q_n(z) = z^n [R(z) + \varepsilon_n(z)]$$

where $\varepsilon_n(z) \rightrightarrows 0$ as $n \to \infty$ inside the domain $|z| > \rho$. Hence it follows that

$$\lim_{n\to\infty} \sqrt[n]{|q_n(z)|} = |z|$$

and the convergence is uniform on every compact set lying in the domain $|z| > \rho$ and not containing the points z in which $R(z)$ becomes zero.

Let us consider some particular cases of the function $R(t)$ and find the generalised Faber polynomials.

1. For $R(t)=[\Psi'(t)]^{-1}$

$$K(z,\ t)=\frac{t}{\Psi(t)-z}$$

If $z\in\bar{B}_r,\ \rho\leqslant r<+\infty$, the function

$$f(\zeta)=\frac{\frac{r}{\zeta}\zeta}{\Psi\left(\frac{r}{\zeta}\right)-z}=\sum_{n=0}^{\infty}(\Pi_n(z)\,r^{-n})\,\zeta^{n+1},\quad |\zeta|<1$$

is regular and schlicht in the disc $|\zeta|<1$ and $f(0)=0$, $f'(0)=1$, and therefore (see Goluzin [1] p. 201)

$$|\Pi_n(z)\,r^{-n}|\leqslant e(n+1),\quad n=0,\ 1,\ \ldots$$

or

$$|\Pi_n(z)|\leqslant e(n+1)\,r^n,\quad n=0,\ 1,\ \ldots,\quad z\in\bar{B}_r,\quad \rho\leqslant r<\infty$$

The multiplier $e(n+1)$ can be changed to $e/2(n+1)+1.8$ (see Lebedev and Milin [1] and Bazilevich [1]), but must not be replaced by $a(n+1),\ a<1$.

2. For $R(t)=[\Psi'(t)]^{-\alpha}$, α real,

$$K(z,\ t)=\frac{t\,[\Psi'(t)]^{1-\alpha}}{\Psi(t)-z}$$

If $z\in\bar{B}_r,\ \rho\leqslant r<+\infty$, then

$$K\left(z,\ \frac{r}{\zeta}\right)=\left(\frac{f(\zeta)}{\zeta}\right)^{2\alpha-1}(f'(\zeta))^{1-\alpha}=\sum_{n=0}^{\infty}(\Pi_n(z)\,r^{-n})\,\zeta^n,\quad |\zeta|<1$$

where $f(\zeta)$ is the function introduced above.

We easily obtain (see Goluzin [1] p. 146):

(a) for $\alpha\geqslant 0$

$$\left|K\left(z,\ \frac{r}{\zeta}\right)\right|=\left|\frac{f(\zeta)^2}{\zeta^2 f'(\zeta)}\right|^{\alpha}\left|\frac{\zeta f'(\zeta)}{f(\zeta)}\right|\leqslant\frac{1}{(1-|\zeta|^2)^{\alpha}}\,\frac{1+|\zeta|}{1-|\zeta|}=\frac{(1+|\zeta|)^{1-\alpha}}{(1-|\zeta|)^{1+\alpha}}$$

(b) for $\alpha\leqslant 0$, proceeding in analogous manner,

$$\left|K\left(z,\ \frac{r}{\zeta}\right)\right|\leqslant\frac{(1+|\zeta|)^{1+\alpha}}{(1-|\zeta|)^{1-\alpha}}$$

By using the estimates for the coefficients of a power series, in each case we have

$$|\Pi_n(z)\, r^{-n}| \leqslant \frac{(1+|\zeta|)^{1-|\alpha|}}{(1-|\zeta|)^{1+|\alpha|}\,|\zeta|^n}, \qquad |\zeta|<1$$

Assuming $|\zeta| = \frac{n}{n+1+|\alpha|}$, we find

$$|\Pi_n(z)| \leqslant \left(\frac{e}{1+|\alpha|}\right)^{1+|\alpha|} \frac{(n+1+|\alpha|)^{1+|\alpha|}}{(2n+1+|\alpha|)^{|\alpha|-1}}\, r^n, \qquad n=0,\ 1,\ \ldots$$

$$z \in \overline{B}_r, \quad \rho \leqslant r < +\infty$$

A rather better estimate of $|\Pi_n(z)|$ can be obtained but we shall not stop over this. As before we can find estimates for the derivatives of the generalised Faber polynomials for $R(t) = [\Psi(t)]^{-\alpha}$.

2.2.2 *Zero expansion by means of generalised Faber polynomials*

Let τ_0 be some root of the equation $R(t)=0$ from the domain $|t|>\rho$. Since $c_0 \neq 0$, $\tau_0 \neq \infty$. The series

$$\sum_{n=0}^{\infty} \frac{1}{\tau_0^n}\, \Pi_n(z)$$

uniformly converges in every domain $\overline{B}_r$, $\rho_0 < r < |\tau_0|$, since for $z \in \overline{B}_r$

$$\overline{\lim_{n\to\infty}} \sqrt[n]{\left|\frac{1}{\tau_0^n}\, \Pi_n(z)\right|} \leqslant \frac{r}{|\tau_0|} < 1$$

uniformly. From the expansion (1), Section 2.2.1, it follows that the sum of this series is equal to the value of the generating function of the polynomials $\Pi_n(z)$ at the point $t=\tau_0$, e.g.

$$\sum_{n=0}^{\infty} \frac{1}{\tau_0^n}\, \Pi_n(z) \equiv 0, \qquad z \in B_{|\tau_0|}$$

Let τ_0 ($|\tau_0| > \rho$) be some root of the equation $R(t)=0$ of multiplicity k_0. Dividing identity (1), Section 2.2.1, by t and differentiating with respect to t ν times, $\nu = 0,\ 1,\ \ldots,$

$k_0 - 1$, we find

$$\sum_{n=0}^{\infty} \frac{(n+1)(n+2)\dots(n+\nu)}{\tau_0^{n+\nu+1}} \Pi_n(z) \equiv 0, \quad z \in B_{|\tau_0|}$$

Hence it follows that if τ_j, $j=1, 2, \dots, m$, are all zeros of the function $R(t)$ of respective multiplicity k_j, lying in the domain $|t| \geqslant r > \rho$, then for any constant $c_{j,\nu}$,

$$j=1, 2, \dots, m; \quad \nu=0, 1, \dots, k_j-1$$

the following zero expansion will hold:

$$\sum_{n=0}^{\infty} a_n \Pi_n(z) \equiv 0, \quad z \in B_r$$

$$a_n = \sum_{j=1}^{m} \sum_{\nu=0}^{k_j-1} (n+1)(n+2)\dots(n+\nu) \frac{c_{j,\nu}}{\tau_j^{n+\nu+1}} \tag{1}$$

Then from

$$|a_n| \leqslant \frac{1}{r^n} \sum_{j=1}^{m} \sum_{\nu=1}^{k_j-1} (n+1)(n+2)\dots(n+\nu) |c_{j,\nu}| \frac{1}{|\tau_j|^{\nu}} \left(\frac{r}{|\tau_j|}\right)^n$$

it follows easily that

$$\overline{\lim_{n\to\infty}} \sqrt[n]{|a_n|} \leqslant \frac{1}{r}$$

The reverse is also true but we shall not stop to prove this.

Theorem 1 (Faber [3]*)* In order that there should exist a zero expansion

$$0 \equiv \sum_{n=0}^{\infty} a_n \Pi_n(z)$$

where not all $a_n = 0$, uniformly converging on the compact sub-sets of B_r, $\rho < r < \infty$, it is necessary and sufficient that the function $R(t)$ has zeros in the domain $r \leqslant |t| < \infty$. Then if

$$\overline{\lim_{n\to\infty}} \sqrt[n]{|a_n|} = \frac{1}{r}, \quad \rho < r < \infty$$

the coefficients a_n are represented by the formulae

$$a_n = \sum_{j=1}^{m} \sum_{\nu=0}^{k_j - 1} (n+1)(n+2)\dots(n+\nu)\frac{c_{j,\nu}}{\tau_j^{n+\nu+1}}$$

where τ_j, $j = 1, 2, \dots, m$, are all zeros of the function $R(t)$ lying in the domain $r \leqslant |t| < \infty$, of respective multiplicities k_j, and $c_{j,\nu}$ are some numbers.

In Section 2.2.4 we shall investigate the expansion of an arbitrary function $f(z)$, regular on $\bar{B}$, by means of polynomials $\Pi_n(z)$ assuming that the function $R(t)$ does not have zeros in the domain $\rho < |t| < \infty$ and consequently the expansion of the function $f(z)$ as a series, uniformly converging in some domain $D \supset \bar{B}$, is unique. Let us first consider the special case of the expansion of Cauchy's kernel.

2.2.3 *Expansion of the function $1/(\zeta - z)$ by means of polynomials $\Pi_n(z)$*

As a function of z for fixed $\zeta \in B_r^1, \rho < r < \infty$, the function $\frac{1}{\zeta - z}$ is regular in the domain B_r and

$$\frac{1}{\zeta - z} = \frac{1}{2\pi i} \int_{L_r} \frac{du}{(\zeta - u)(u - z)}$$

Assuming $u = \Psi(t)$ we find

$$\frac{1}{\zeta - z} = \frac{1}{2\pi i} \int_{|t|=r} \frac{1}{\zeta - \Psi(t)} \cdot \frac{t\Psi'(t)}{\Psi(t) - z} \frac{dt}{t}$$

Let $R(t)$ be a regular function not becoming zero for $|t| > \rho$, and $R_*(t) = \frac{1}{R(t)}$. Assuming

$$K(z, t) = \frac{t\Psi'(t)}{\Psi(t) - z} R(t) \text{ and } K_*(\zeta, t) = \frac{1}{\zeta - \Psi(t)} R_*(t).$$

we find

$$\frac{1}{\zeta - z} = \frac{1}{2\pi i} \int_{|t|=r} K_*(\zeta, t) K(z, t) \frac{dt}{t}, \quad z \in B_r, \quad \zeta \in B_r^1$$

Substituting series (1), Section 2.2.1, for $K(z, t)$ in this formula and integrating the result, we have

$$\frac{1}{\zeta - z} = \sum_{n=0}^{\infty} \Pi_n(z) \frac{1}{2\pi i} \int_{|t|=r} K_*(\zeta, t) \frac{1}{t^{n+1}} dt, \quad z \in B_r, \quad \zeta \in B_r^1$$

and assuming

$$Q_n(\zeta) = \frac{1}{2\pi i} \int_{|t|=r} K_*(\zeta, t) \frac{1}{t^{n+1}} dt \tag{1}$$

we obtain the expansion

$$\frac{1}{\zeta - z} = \sum_{n=0}^{\infty} Q_n(\zeta) \Pi_n(z), \quad z \in B_r \quad \zeta \in B_r^1, \quad \rho < r < \infty \tag{2}$$

The function $Q_n(\zeta)$, $n = 0, 1, \ldots$, is called the conjugate function with respect to the generalised Faber polynomial $\Pi_n(z)$. If $R(t) \equiv 1$, the functions $Q_n(\zeta)$ will be conjugate with respect to the Faber polynomials $P_n(z)$.

Changing the variable in Formula (1) by $t = \Phi(u)$:

$$Q_n(\zeta) = \frac{1}{2\pi i} \int_{L_r} \frac{R_*(\Phi(u))\, \Phi'(u)}{(\zeta - u)\, \Phi(u)^{n+1}} du, \quad \zeta \in B_r^1$$

and assuming that the function $\dfrac{R_*(\Phi(u))\, \Phi'(u)}{\Phi(u)^{n+1}}$, $n = 0, 1, \ldots$, is regular in B_r^1 and becomes zero at the point $u = \infty$ we find

$$Q_n(\zeta) = \frac{R_*(\Phi(\zeta))\, \Phi'(\zeta)}{\Phi(\zeta)^{n+1}}, \quad \zeta \in B^1 \tag{3}$$

whence it follows that the function $Q_n(\zeta)$ is regular in B^1, has a zero of the $(n+1)$-th order at the point $\zeta = \infty$ and in B^1 when

$$\sqrt[n]{|Q_n(\zeta)|} \rightrightarrows |\tau|^{-1}, \quad \tau = \Phi(\zeta) \tag{4}$$

Hence and from Lemma 1, Section 2.2.1, we conclude that for every fixed r, $\rho < r < \infty$, the series (2) for the function $\dfrac{1}{\zeta - z}$ converges absolutely and uniformly with respect to z and ζ on the compact sub-sets B_r and B_r^1 respectively.

NOTE Let the set $\bar{B}$ of type $\mathfrak{M}$ be a closed domain, the boundary of which is an analytic curve. Then the function $z = \Psi(w)$ will be schlicht in some domain $|w| > \rho_1$, $\rho_1 < \rho$. If the function $R(t)$ is regular in the domain $|t| > \rho_1$, then in the above ρ can be changed to ρ_1 throughout.

2.2.4 *Expansion of a regular function by $\pi_n(z)$ and $q_n(z)$*

In this section we shall assume that the function $R(t)$ does not become zero in the domain $|t|>\rho$.

In the following we shall require the lemma below.

Lemma 1 Let C be an arbitrary, closed, piecewise smooth Jordan curve containing the set $\overline{B}$ in its interior. Then

1. $$\frac{1}{2\pi i}\int_C \Pi_n(z)\Pi_m(z)\,dz=0,\quad n,m=0,1,\ldots,$$

2. $$\frac{1}{2\pi i}\int_C Q_n(z)Q_m(z)\,dz=0,\quad n,m=0,1,\ldots,$$

3. $$I_{n,m}=\frac{1}{2\pi i}\int_C \Pi_n(z)Q_m(z)\,dz=\begin{cases}1\text{ when } n=m,\\ 0\text{ when } n\neq m,\end{cases}$$
$$n,m=0,1,\ldots$$

PROOF Assertion (1) of the theorem is obvious. Assertions (3) for $n<m$ and (2) follow from the fact that the expression under the integral sign is regular in the unbounded domain with boundary C and at the point $z=\infty$ there is a zero of at least the second order.

To complete the proof let us expand the function $\Pi_n(z)$ as a series of $\Pi_m(z)$, $m=0,1,\ldots$ By using the expansion (2), Section 2.2.3, for the function $\frac{1}{\zeta-z}$ converging uniformly with respect to $\zeta\in C$ and for any compact set of z, lying inside C, we find

$$\Pi_n(z)=\frac{1}{2\pi i}\int_C\frac{\Pi_n(\zeta)}{\zeta-z}\,d\zeta=\sum_{m=0}^{\infty}\Pi_m(z)\frac{1}{2\pi i}\int_C\Pi_n(\zeta)Q_m(\zeta)\,d\zeta$$
$$=\sum_{m=0}^{\infty}I_{n,m}\Pi_m(z)=\sum_{m=0}^{n}I_{n,m}\Pi_m(z)$$

Comparing the coefficients of z^n, we conclude that $I_{n,n}=1$ and consequently

$$\sum_{m=0}^{n-1}I_{n,m}\Pi_m(z)\equiv 0$$

Hence, taking in turn the coefficients for powers of z, we find that $I_{n,k}=0$; $k=n-1, n-2, \dots, 1, 0$, and the lemma is proved.

Theorem 1 If the function $f(z)$ is regular in the domain $B_r, \rho<r<\infty$, and has a singular point on the contour line L_r, then the following applies:

1. The function $f(z)$ can be expanded into a series of generalised Faber polynomials:

$$f(z)=\sum_{k=0}^{\infty} a_k \Pi_k(z), \tag{1}$$

where

$$a_k=\frac{1}{2\pi i}\int_{L_{r_1}} f(\zeta) Q_n(\zeta)\, d\zeta, \qquad \rho<r_1<r; \tag{2}$$

2. In this case

$$\overline{\lim_{k\to\infty}} \sqrt[k]{|a_k|}=\frac{1}{r}, \quad \rho<r<\infty, \tag{3}$$

and the series (1) uniformly converges on the compact sub-sets of B_r and diverges outside $\overline{B}_r$.

3. The expansion of the function $f(z)$ by generalised Faber polynomials (uniformly converging in some domain $D\supset\overline{B}$) is unique.

4. Conversely: if (3) holds, then the series (1) uniformly converges on the compact sub-sets of B_r and diverges outside $\overline{B}_r$, and the function $f(z)$ defined by the series (1) is regular in the domain B_r and has a singular point on the contour line L_r.

PROOF In principle the proof does not differ from the proof for Section 2.1.3. Let us state it briefly.

(1) Let $z\in B_r$, and in particular $z\in L_{r_2}$, $\rho<r_2<r$, and let $r_2<r_1<r$. Then by using the series (2), Section 2.2.3, for the function $\frac{1}{\zeta-z}$, which uniformly converges for $\zeta\in L_{r_1}$, we have

$$f(z)=\frac{1}{2\pi i}\int_{L_{r_1}}\frac{f(\zeta)}{\zeta-z}\,d\zeta=\sum_{k=0}^{\infty}\Pi_k(z)\frac{1}{2\pi i}\int_{L_{r_1}} f(\zeta) Q_k(\zeta)\, d\zeta$$

$$=\sum_{n=0}^{\infty} a_k\Pi_k(z).$$

Hence follows expansion (1) and Formula (2).

(2) From (4), Section 2.2.3, it follows that for sufficiently large k the inequality

$$\sqrt[k]{|Q_k(\zeta)|} \leqslant \frac{1}{r_1} + \varepsilon, \quad \varepsilon > 0, \quad \zeta \in L_{r_1}$$

will be fulfilled, and consequently for such k we have

$$|a_k| = \left| \frac{1}{2\pi i} \int\limits_{L_{r_1}} f(\zeta) Q_k(\zeta) \, d\zeta \right| \leqslant \frac{1}{2\pi} M_{r_1} \left(\frac{1}{r_1} + \varepsilon \right)^n l_{r_1}$$

where $M_{r_1} = \max\limits_{\zeta \in L_{r_1}} |f(\zeta)|$ and l_{r_1} is the length of the line L_{r_1}. Hence

$$\overline{\lim_{k \to \infty}} \sqrt[k]{|a_k|} \leqslant \frac{1}{r_1} + \varepsilon$$

Since $\varepsilon > 0$ is arbitrarily small and since $r_1, r_2 < r_1 < r$, is arbitrary, we have

$$\overline{\lim_{k \to \infty}} \sqrt[k]{|a_k|} \leqslant \frac{1}{r}$$

Assuming that the inequality sign holds, and proceeding as in the proof of assertion (2) of Theorem 1, Section 2.1.3, the proof of assertion (2) is completed. In doing this we make use of the lemma in Section 2.2.1.

(3) Let there exist an expansion of the function $f(z)$ into a series of generalised Faber polynomials:

$$f(z) = \sum_{k=0}^{\infty} a_k \Pi_k(z)$$

and let the series uniformly converge in some domain $D \supset \bar{B}$. The series uniformly converges on the contour line L_{r_1}, $\rho < r_1 < \infty$, if $L_{r_1} \subset D$. Multiplying it by $\frac{1}{2\pi i} Q_m(z)$, $m = 0, 1, \ldots$, and integrating along L_{r_1}, by virtue of Lemma 1 (assertion 3), we obtain

$$\frac{1}{2\pi i} \int\limits_{L_{r_1}} f(z) Q_m(z) \, dz = a_m, \quad m = 0, 1, \ldots$$

for any expansion of the function $f(z)$ by generalised Faber polynomials. Hence it follows that the expansion is unique.

Assertion (4) of this theorem is proved in the same way as assertion (4) in Theorem 1, Section 2.1.3. The following two theorems are proved in an analogous fashion.

Theorem 2 If the function $f(z)$ is regular in the domain B_r^1, $\rho < r < \infty$, and has a singular point on the contour line L_r, then the following applies:

1. The function $f(z)$ can be expanded into a series of functions $Q_k(z)$ conjugate to the generalised Faber polynomials:

$$f(z) = \sum_{k=0}^{\infty} b_k Q_k(z) + f(\infty) \tag{4}$$

where

$$b_k = \frac{1}{2\pi i} \int_{L_{r_1}} f(\zeta)\, \Pi_k(\zeta)\, d\zeta, \quad r < r_1 < \infty \tag{5}$$

2. In this case

$$\overline{\lim_{k\to\infty}} \sqrt[k]{|b_k|} = r, \quad \rho < r < \infty \tag{6}$$

and the series (4) uniformly converges on the compact sub-sets of B_r^1 and diverges outside $\overline{B}_r^1$.

3. The expansion of the function $f(z)$ in a series of the functions $Q_k(z)$, uniformly converging in some neighbourhood of the point ∞, is unique.
4. Conversely: if (6) holds, then the series (4) uniformly converges on the compact sub-sets of B_r^1 and diverges outside $\overline{B}_r^1$ and the function $f(z)$ defined by the series (4) is regular in the domain B_r^1 and has a singular point on the contour line L_r.

NOTE 1 If the conditions of the theorem required the function $f(z)$ to be regular in $\overline{B}^1$, then the theorem would remain in force if in Formula (5) it is assumed that $\rho < r_1 < \infty$, (6) is altered to the inequality

$$\overline{\lim_{k\to\infty}} \sqrt[k]{|b_k|} \leqslant \rho$$

and assertion (4) is altered to the assertion: if (6) holds then the series (4) converges uniformly on the compact sub-sets of B^1 and its sum $f(z)$ is regular in B^1.

Theorem 3 If the function $f(z)$ is regular in the doubly-connected domain B_{r_1, r_2}, $\rho < r_1 < r_2 < \infty$, bounded by the contour lines L_{r_1} and L_{r_2}, and has singular points on the contour lines L_{r_1} and L_{r_2}, then the following applies:

1. The function $f(z)$ can be expanded as a series

$$f(z)=\sum_{k=0}^{\infty} a_k\Pi_k(z)+\sum_{k=0}^{\infty} b_k Q_k(z) \tag{7}$$

where

$$a_k=\frac{1}{2\pi i}\int_{L_{r'}} f(\zeta)\,Q_k(\zeta)\,d\zeta,\quad b_k=\frac{1}{2\pi i}\int_{L_{r'}} f(\zeta)\,\Pi_k(\zeta)\,d\zeta,\quad r_1<r'<r_2$$

2. In this case

$$\overline{\lim_{k\to\infty}}\sqrt[k]{|a_k|}=\frac{1}{r_2},\quad \overline{\lim_{k\to\infty}}\sqrt[k]{|b_k|}=r_1 \tag{8}$$

and the series $\sum_{k=0}^{\infty} a_k\Pi_k(z)$ converges uniformly on the compact sub-sets of B_{r_2} and diverges outside $\overline{B}_{r_2}$, and the series $\sum_{k=0}^{\infty} b_k Q_k(\zeta)$ converges uniformly on the compact sub-sets of $B^1_{r_1}$ and diverges outside $\overline{B}^1_{r_1}$.

3. The expansion of the function $f(z)$ as a series of the form (7), uniformly converging on some contour line L_r, $r_1<r<r_2$, is unique.

4. Conversely: if (8) holds, the assertion in (2) is true for the series $\sum_{k=0}^{\infty} a_k\Pi_k(z)$ and $\sum_{k=0}^{\infty} b_k Q_k(z)$ and the function $f(z)$ defined by the series (7) is regular in the domain $B_{r_1,\,r_2}$ and has singular points on L_{r_1} and L_{r_2}.

A remark analogous to that made for the previous theorem can also be made for this theorem (see also the remark made at the end of the previous section).

NOTE 2 The results of Sections 2.2.1 and 2.2.4, suitably modified, remain in force for the generalised Faber polynomials.

There are various applications for generalised Faber polynomials. We shall meet some of them in Section 3.3 and Section 4.3.

In the following section we shall consider other generalisations of the Faber polynomials, which include the generalisation given above as a special case.

2.2.5 *Connection between series of generalised Faber polynomials and power series*

Let g be the domain of the z plane containing a set $\overline{B}$ of type $\mathfrak{M}$, and g_1 be the intersection $g \cap B^1$. The function $w = \Phi(z)$ maps the domain g_1 on to some domain G_1 lying in the domain $|w| > \rho$ and having the circle $|w| = \rho$ as one of its boundary continuums. Let us denote by G the domain which we get if the disc $|w| \leqslant \rho$ is united with the domain G_1.

Theorem 1 (Lokhin [2])† Let

$$\overline{\lim_{n \to \infty}} \sqrt[n]{|a_n|} = \frac{1}{r}, \quad r > \rho \tag{1}$$

For the function

$$f(z) = \sum_{n=0}^{\infty} a_n \text{П}_n(z) \tag{2}$$

to be regular in the domain g containing B_r it is necessary and sufficient that the function

$$F(w) = \sum_{n=0}^{\infty} a_n w^n \tag{3}$$

be regular in the domain G.

PROOF: NECESSITY Assuming that in (2) $z = \Psi(t)$ and applying Formula (4), Section 2.2.1, we have

$$f(\Psi(t)) = \sum_{n=0}^{\infty} a_n \left(R(t) t^n + \frac{1}{t} M_n\left(\frac{1}{t}\right)\right), \quad \rho < |t| < r$$

Multiplying this series by the function $\frac{1}{2\pi i} \frac{R_*(t)}{t-w}$ and integrating the result over the circle $|t| = r', \rho < r' < r$, for $|w| < r'$, we have

$$\frac{1}{2\pi i} \int_{|t|=r'} \frac{f(\Psi(t)) R_*(t)}{t-w} dt$$
$$= \sum_{n=0}^{\infty} a_n \frac{1}{2\pi i} \int_{|t|=r'} \left(\frac{t^n}{t-w} + \frac{\frac{1}{t} M_n\left(\frac{1}{t}\right) R_*(t)}{t-w} \right) dt$$

The second term under any of the integrals on the right-hand side of the equation is a regular function of t in the

† See also M. Gagua [1].

M

closed domain $|t| \geqslant r'$ and at the point $t = \infty$ has a zero of not less than second order. Consequently its integral is zero and, by applying Cauchy's formula, we have

$$\frac{1}{2\pi i} \int\limits_{|t|=r'} \frac{f(\Psi(t))\, R_*(t)}{t-w}\, dt = \sum_{n=0}^{\infty} a_n w^n = F(w). \qquad (4)$$

By virtue of (1) this series converges uniformly on the compact sub-sets of the disc $|w| < r$, and the function $F(w)$ defined by this series is regular in $|w| < r$. We must prove that the function $F(w)$ can be analytically continued to the whole domain G (containing the disc $|w| < r$), e.g. to any point $w_0 \in G$, where it is sufficient to consider the case when $|w_0| \geqslant r$. We can easily construct a closed, piecewise smooth Jordan curve Γ such that one of the closed domains into which it divides the w plane is contained in G and contains the point w_0 and the disc $|w| \leqslant r'$. Let d be the image of the domain D, bounded by the circle $|w| = r'$ and Γ, when mapped by the function $z = \Psi(w)$. Since $\bar{d} \in g$, the function $f(z) R_*(\Phi(z))$ is regular on $\bar{d}$, and the function $f(\Psi(w)) R_*(w)$ is regular on $\bar{D}$, and consequently the function under the integral sign in (4) as a function of t, is regular on $\bar{D}$ for $|w| < r'$, and integration over the circle $|t| = r'$ can be altered to integration along Γ:

$$F(w) = \frac{1}{2\pi i} \int\limits_{\Gamma} \frac{f(\Psi(t))\, R_*(t)}{t-w}\, dt,$$

This Cauchy-type integral is a regular function not only in the disc $|w| < r'$ but in the whole domain bounded by Γ and containing this disc, and consequently the function $F(w)$ is regular in this domain and, in particular, is regular at the point w_0. Since $w_0 \in G$ is arbitrary, the function $F(w)$ is regular in G and therefore the necessity is proved.

SUFFICIENCY Assuming that $w = \Phi(\zeta)$ in (3), we obtain a series uniformly converging on every contour line $L_{r'}$, $\rho < r' < r$, and the function $F(\Phi(\zeta))$ is regular in the domain g. Multiplying this series by the function $\frac{1}{2\pi i} \frac{R(\Phi(\zeta))}{\zeta - z}$ where $z \in B_{r'}$, $\rho < r' < r$, and integrating along $L_{r'}$, in virtue of Formulae (3) and (2), Section 2.2.1, we have

$$\frac{1}{2\pi i} \int\limits_{L_{r'}} \frac{F(\Phi(\zeta))\, R(\Phi(\zeta))}{\zeta - z}\, d\zeta = \sum_{n=0}^{\infty} a_n \Pi_n(z) = f(z), \qquad z \in B_{r'}. \qquad (5)$$

This series converges uniformly on the compact sub-sets

of B_r. Proceeding in an analogous manner to the proof of the necessity, we alter the contour of integration $L_{r'}$, to another contour γ such that one of the domains into which it divides the z plane contains both $\overline{B}_{r'}$, and an arbitrary point $z_0 \in g$, and in the same way we prove that the function $f(z)$ is regular at the point z_0 and consequently in g. The theorem is proved.

NOTE 1 Let condition (1) be fulfilled for the power series (3) (and consequently the series converges in the disc $|w| < r$). If the terms of the series (3) can be combined in groups

$$P_{n_k, n_{k+1}}(w) = \sum_{n=n_k+1}^{n_{k+1}} a_n w^n, \quad n_0 = 0, \quad k = 0, 1, \ldots,$$

such that the series

$$F(w) = \sum_{k=1}^{\infty} P_{n_k, n_{k+1}}(w)$$

converges uniformly on the compact sub-sets of some domain G (containing the disc $|w| < r$) then for such a grouping of the terms of the series (2), we shall obtain a series converging uniformly on the compact sub-sets of the domain g corresponding to the domain G, and vice versa.

To prove the validity of this statement all that is necessary is to repeat part of the proof of the theorem assuming that after altering the variable and multiplying by the respective factor $\left(\frac{1}{2\pi i}\frac{R(\Phi(\zeta))}{\zeta - z}, \frac{1}{2\pi i}\frac{R_*(t)}{t - w}\right)$ the grouped series can be integrated term by term over any piecewise smooth curve lying in the domains g and G respectively.

Applying this theorem and the note following it, it is possible to translate a whole series of results which are known for power series to series of generalised Faber polynomials. Amongst these are the following: the properties of series with gaps, Polya's theorem, series with coefficients which have a finite number of values. In this respect see, for example, Krasnoshchekova [1, 2] and Varganova and Korobeinik [1]. See also Iliev [1].

2.3 GENERALISED FABER POLYNOMIALS (continuation)

Some of the results obtained in Section 2.2 can be generalised. This generalisation is carried out in this section.

2.3.1 Adjoined functions

Let the radius of convergence of the power series

$$F(z) = b_0 + b_1 z + \dots + b_n z^n + \dots$$

be equal to one and suppose that the function $F(z)$ can be analytically continued from the disc $|z| < 1$ by any path not passing through the points $z = 0, 1, \infty$. If all coefficients $b_k \neq 0$, $k = 0, 1, \dots$, then we can form the series

$$F_*(z) = \frac{1}{b_0} + \frac{1}{b_1} z + \dots + \frac{1}{b_n} z^n + \dots$$

If its radius of convergence is also equal to one and it is also possible to continue analytically the function $F_*(z)$ from the disc $|z| < 1$ by any path not passing through the points $z = 0, 1, \infty$, then the functions $F(z)$ and $F_*(z)$ are said to be adjoined. Clearly the point $z = 1$ is a singular point for both functions $F(z)$ and $F_*(z)$.

EXAMPLE Let α, β and γ be fixed numbers not equal to $0, -1, -2 \dots$ and let

$$F(\alpha, \beta, \gamma, z) = 1 + \sum_{k=1}^{\infty} \frac{\alpha(\alpha+1)\dots(\alpha+k-1)\beta(\beta+1)\dots(\beta+k-1)}{k!\,\gamma(\gamma+1)\dots(\gamma+k-1)} z^k$$

be the so-called hypergeometric series (see Smirnov [1] pp. 371-374). This series converges in the disc $|z| < 1$ and the function $F(\alpha, \beta, \gamma, z)$ can be analytically continued from the disc $|z| < 1$ on any path not passing through the points $z = 0, 1, \infty$. The functions $F(z) = F(\alpha, 1, \gamma, z)$ and $F_*(z) = F(\gamma, 1, \alpha, z)$ are allied. In particular, for $\alpha = \gamma$ we have $F(z) = F_*(z) = \frac{1}{1-z}$ and for $\alpha = 2$, $\gamma = 1$ we have

$$F(z) = \frac{1}{(1-z)^2}, \quad F_*(z) = \ln \frac{1}{1-z}$$

NOTE 1 The function $F(z)$ can be many-valued (when analytically continued). Let B be a bounded Jordan domain with boundary L, containing the point $z = 0$. We can define a single-valued function $F\left(\frac{z}{u}\right)$, $z \in B$, $u \in L$, in the following manner; for every fixed $u = u_1 \in L$ the function

$$F\left(\frac{z}{u_1}\right)=\sum_{k=0}^{\infty} b_k \frac{z}{u_1}$$

is regular in some neighbourhood of the point $z=0$. Since $\frac{z}{u_1}\neq 1$ for $z\in B$, by virtue of the monodromy theorem, $F\left(\frac{z}{u_1}\right)$ is single-valued and a regular function in the domain B. Thus the function $F\left(\frac{z}{u}\right)$ will be defined, single-valued and regular for $z\in B$ for all $u\in L$. Clearly it is continuous (and even regular) with respect to $u\in L$ for every $z\in B$. The same is true for the function $F_*\left(\frac{z}{u}\right)$.

Lemma 1 Let

$$F(z)=\sum_{k=0}^{\infty} b_k z^k \text{ and } F_*(z)=\sum_{k=0}^{\infty}\frac{1}{b_k} z^k$$

be adjoined functions and let B be a bounded Jordan domain with a piecewise smooth boundary L containing the point $z=0$ inside.

1. If the function $f(z)$ is regular on the closed domain $\overline{B}$ and in the neighbourhood of the point $z=0$ has the following expansion

$$f(z)=\sum_{k=0}^{\infty} a_k z^k$$

then the function $f_*(z)$ with the expansion

$$f_*(z)=\sum_{k=0}^{\infty} a_k b_k z^k$$

in the neighbourhood of the point $z=0$ is regular on $\overline{B}$ and the following formulae hold:

$$\begin{aligned} f_*(z)&=\frac{1}{2\pi i}\int_L f(u)\,F\left(\frac{z}{u}\right)\frac{du}{u}\\ f(z)&=\frac{1}{2\pi i}\int_L f_*(u)\,F_*\left(\frac{z}{u}\right)\frac{du}{u}\end{aligned}\tag{1}$$

2. If the function $f(\zeta)$ is regular on $\overline{B}^1$ and in the neighbourhood of the point $\zeta=\infty$ has the expansion

$$f(\zeta)=\sum_{k=0}^{\infty}\frac{a_k}{\zeta^k}$$

then the function which in the region of the point $\zeta=\infty$ has the expansion

$$f_*(\zeta)=\sum_{k=0}^{\infty}\frac{a_k b_k}{\zeta^k}$$

is regular on $\bar{B}^1$ and the following formulae hold:

$$f_*(\zeta)=\frac{1}{2\pi i}\int_L f(u)\,F\left(\frac{u}{\zeta}\right)\frac{du}{u}$$

$$f(\zeta)=\frac{1}{2\pi i}\int_L f_*(u)\,F_*\left(\frac{u}{\zeta}\right)\frac{du}{u}$$

PROOF Let us consider the right-hand side of the first of the Formulae (1). Since the function under the integral sign, as a function of u, is continuous on L for every fixed $z\in B$, the integral is a function of z defined in the domain B and since the function under the integral sign is regular with respect to z in B for every $u\in L$ the integral is a regular function of z in B.

Let C_r, $r>0$, be the circle $|z|=r$. Choosing r such that the circle C_r belongs to B and fixing z, $|z|<r$, it is easy to see that the function under the integral sign is regular with respect to u in the closed domain bounded by L and C_r, and consequently

$$\frac{1}{2\pi i}\int_L f(u)\,F\left(\frac{z}{u}\right)\frac{du}{u}=\frac{1}{2\pi i}\int_{C_r} f(u)\,F\left(\frac{z}{u}\right)\frac{du}{u}$$

$$=\sum_{k=0}^{\infty}a_k\frac{1}{2\pi i}\int_{C_r}u^{k-1}\sum_{\nu=0}^{\infty}b_\nu\frac{z^\nu}{u^\nu}\,du=\sum_{k=0}^{\infty}a_k b_k z^k=f_*(z),\quad |z|<r$$

In view of the uniqueness of analytic continuation, the integral in question is equal to $f_*(z)$ for all $z\in B$. To prove that the function $f_*(z)$ is regular on $\bar{B}$, let us alter the contour of integration L to the contour L_{ρ^*}, $\rho^*>\rho$, such that the function $f(z)$ is regular on $\bar{B}_{\rho^*}$. The integral obtained will then be a regular function in B_{ρ^*} coinciding with $f_*(z)$ on B and therefore $f_*(z)$ is a regular function in B_{ρ^*}.

Since the functions $F(z)$ and $F_*(z)$ are adjoined, from the above it follows that the integral in the second of Formulae (1) defines a function which is regular on $\bar{B}$ and has in the neighbourhood of the point $z=0$ the expansion

$$\sum_{k=0}^{\infty}(a_k b_k)\frac{1}{b_k}z^k=\sum_{k=0}^{\infty}a_k z^k$$

e.g. the function $f(z)$, and the first part of the lemma is proved. The second part is proved in an analogous manner.

COROLLARY For $z\in B$ and $\zeta\in B^1$ under the conditions for the lemma

$$\frac{1}{\zeta-z}=\frac{1}{2\pi i}\int_L F_*\left(\frac{u}{\zeta}\right)F\left(\frac{z}{u}\right)\frac{du}{\zeta u} \tag{2}$$

NOTE 2 If both the adjoined functions $F(z)$ and $F_*(z)$ are regular on the extended plane apart from the point $z=1$, then the requirement in the formulation of the corollary that the point $z=0$ should belong to the domain B is no longer necessary.

The functions $F(z)$ and $F_*(z)$ with expansions in the neighbourhood of the point $z=0$

$$F(z)=\sum_{k=0}^{\infty}b_k z^k,\quad F_*(z)=\sum_{k=0}^{\infty}\frac{1}{b_k}z^k$$

are said to be weakly adjoined if they are regular in the domain G obtained by removing from the z plane all points of the ray $1\leqslant z\leqslant+\infty$. Clearly if $F(z)$ and $F_*(z)$ are allied they are also weakly allied. The reverse is not true. For example, if $0<q<1$, then the function

$$F(z)=\frac{1}{1-z}-\frac{a}{1-qz}=\sum_{n=0}^{\infty}(1-aq^n)z^n$$

is regular in the domain G. Assuming $|a|<1$, let us form the function

$$F_*(z)=\sum_{n=0}^{\infty}\frac{1}{1-aq^n}z^n=\sum_{n=0}^{\infty}\sum_{k=0}^{\infty}a^k q^{nk}z^n$$

which is obviously regular in the disc $|z|<1$. Altering the order of summation, we have

$$F_*(z)=\sum_{k=0}^{\infty}a^k\sum_{n=0}^{\infty}(q^kz)^n=\sum_{k=0}^{\infty}\frac{a^k}{1-q^kz}$$

whence we conclude that the function $F_*(z)$ is also regular in the domain G, and consequently the functions $F(z)$ and $F_*(z)$ are weakly adjoined. Clearly they are not adjoined.

A closed set $\bar{B}$ of the z plane will be said to be star-like if on every ray λ coming from the point $z=0$ there is a point z_λ such that the closed segment $\overline{Oz_\lambda}$ of the ray λ from the point $z=0$ to the point $z=z_\lambda$ belongs to the set $\bar{B}$, and all the remaining points of the ray do not belong to $\bar{B}$. If $z_\lambda=0$, then by the segment $\overline{Oz_\lambda}$ we mean the point $z=0$, and if $z_\lambda=\infty$, then by the segment $\overline{Oz_\lambda}$ we mean the whole ray λ.

NOTE 3 If in formulating the lemma and its corollary the set $\bar{B}$ is star-like, then the statements in the lemma and corollary remain in force for the case when $F(z)$ and $F_*(z)$ are weakly adjoined functions. In fact, in the case considered, the function under the integral sign, for example in the first Formula (1), is as a function of u regular on $\bar{B}$, for every fixed $z\in B$ apart perhaps from the points u satisfying the inequality $1\leqslant\frac{z}{u}\leqslant+\infty$, e.g. the points of the closed segment connecting the points $u=0$ and $u=z$. Consequently the function under the integral sign is regular on L as a function of u. The rest of the proof is analogous to the proof of the lemma.

NOTE 4 If

$$F_j(z)=\sum_{k=0}^{\infty}b_k^{(j)}z^k,\quad F_{*j}(z)=\sum_{k=0}^{\infty}\frac{1}{b_k^{(j)}}z^k, j=1,2$$

are two pairs of adjoined functions, then the functions

$$F(z)=\frac{1}{2\pi i}\int_{|u|=r<1}F_1(u)F_2\left(\frac{z}{u}\right)\frac{du}{u},\quad |z|<r$$

$$F_*(z)=\frac{1}{2\pi i}\int_{|u|=r<1}F_{*1}(u)F_{*2}\left(\frac{z}{u}\right)\frac{du}{u},\quad |z|<r$$

are also adjoined. In fact in the neighbourhood of zero the expansions are

$$F(z)=\sum_{k=0}^{\infty} b_k^{(1)}b_k^{(2)}z^k \text{ and } F_*(z)=\sum_{k=0}^{\infty}\frac{1}{b_k^{(1)}b_k^{(2)}}z^k$$

where according to the lemma the function $F(z)$ (and $F_*(z)$) is regular in every domain of regularity of the function $F_1(z)$ (respectively $F_{*1}(z)$) and consequently $F(z)$ ($F_*(z)$) can be continued analytically on any path from the disc $|z|<1$ not passing through the points $0, 1, \infty$. Analogously if one of the pair of functions, for example the first pair $F_1(z)$ and $F_{*1}(z)$ are weakly adjoined functions and the other pair are adjoined, then $F(z)$ and $F_*(z)$ are weakly adjoined functions.

2.3.2 *Generalised Faber polynomials $\pi_n(z)$, conjugate functions $q_n(\zeta)$ and the expansion of $1/(\zeta - z)$ by means of functions $\pi_n(z)$ and $q_n(z)$*

We shall use the notation introduced in Section 2.1. Let $\bar{B}$ be a set of type $\mathfrak{M}$ containing the point $z=0$, $F(z)=\sum_{k=0}^{\infty} b_k z^k$ and $F_*(z)$ be adjoined functions and $R(t)=\sum_{k=0}^{\infty}\frac{c_k}{t^k}$, $c_0\neq 0$ be a regular function in the domain $|t|>\rho$. Let us introduce a function

$$K(z, t)=F\left(\frac{z}{\Psi(t)}\right)\frac{t\Psi'(t)}{\Psi(t)}R(t)$$

For every fixed $z\in L_r$, $\rho<r<\infty$, this function is regular in the domain $|t|>r$ and has a single singular point $\tau=\Phi(z)$ on the circle $|t|=r$ if $R(\tau)\neq 0$. Let us expand the function $K(z, t)$, $z\in L_r$, in a power series of t

$$K(z, t)=F\left(\frac{z}{\Psi(t)}\right)\frac{t\Psi'(t)}{\Psi(t)}R(t)=\sum_{n=0}^{\infty}\pi_n(z)\frac{1}{t^n}, \quad |t|>r \tag{1}$$

For every fixed $z\in B_r$ this series converges uniformly on the closed domain $|t|\geqslant r$ and

$$\pi_n(z)=\frac{1}{2\pi i}\int\limits_{|t|=r_1} F\left(\frac{z}{\Psi(t)}\right)\frac{t\Psi'(t)}{\Psi(t)}R(t)t^{n-1}\,dt, \quad z\in\bar{B}_r, \ \rho<r<r_1<\infty \tag{2}$$

or changing the integration variable by using the formula

$t=\Phi(u)$, we have

$$\pi_n(z)=\frac{1}{2\pi i}\int\limits_{L_{r_1}} F\left(\frac{z}{u}\right) R(\Phi(u))\Phi(u)^n \frac{du}{u},\ z\in\bar{B}_r,\quad \rho<r<r_1<R$$

For z with sufficiently small modulus we have from (2)

$$\pi_n(z)=\sum_{k=0}^{\infty} b_k z^k \frac{1}{2\pi i}\int\limits_{|t|=r_1} \frac{t\Psi'(t)}{\Psi(t)^{k+1}} R(t)\, t^{n-1}\, dt. \tag{3}$$

But the functions $\frac{t\Psi'(t)}{\Psi(t)^{k+1}} R(t)$ are regular in $|t|>\rho$, and for t with sufficiently large modulus we have the expansion

$$\frac{t\Psi'(t)}{\Psi(t)^{k+1}} R(t)=\frac{t\left(1-\frac{1}{t^2}M'\left(\frac{1}{t}\right)\right)}{\left[t+M\left(\frac{1}{t}\right)\right]^{k+1}}\left(c_0+M_0\left(\frac{1}{t}\right)\right)$$

$$=c_0\frac{1}{t^k}+c_{k+1}^{(k)}\frac{1}{t^{k+1}}+\ \ldots,$$

whence it follows that $\pi_n(z)$ is a polynomial of degree n. The polynomials $\pi_n(z)$, $n=0,\ 1,\ \ldots,$ are called generalised Faber polynomials and the function $K(z, t)$ is called the generating function for the polynomials $\pi_n(z)$. Note that if $F(z)=\frac{1}{1-z}$, the generating function for the polynomials $\pi_n(z)$ becomes the generating function for the polynomials $\Pi_n(z)$ (see Formula (1), Section 2.2.1).

From Formula (3) it follows that the polynomials $\pi_n(z)$ are obtained from the polynomials $\Pi_n(z)$ by multiplying the coefficient of z^k in the polynomial $\Pi_n(z)$ by b_k, $k=0,\ 1, \ldots,\ n$ and by virtue of Lemma 1, Section 2.3.1, we find

$$\begin{aligned}\pi_n(z)&=\frac{1}{2\pi i}\int\limits_{L_r}\Pi_n(u)\,F\left(\frac{z}{u}\right)\frac{du}{u},\quad z\in B_r\\ \Pi_n(z)&=\frac{1}{2\pi i}\int\limits_{L_r}\pi_n(u)\,F_*\left(\frac{z}{u}\right)\frac{du}{u},\quad z\in B_r\end{aligned}\tag{4}$$

From (2) we easily obtain

$$|\pi_n(z)|\leqslant M_{r,r_1} r_1^n,\quad z\in\bar{B}_r,\ M_{r,r_1}=\sup_{z\in\bar{B}_r,\ |t|=r_1}\left|F\left(\frac{z}{\Psi(t)}\right)\frac{t\Psi'(t)}{\Psi(t)}R(t)\right|$$

and due to the arbitrary closeness of r_1 to r, $r_1>r$,

$$\overline{\lim_{n \to \infty}} \sqrt[n]{|\pi_n(z)|} \leqslant r, \quad \rho < r < \infty \tag{5}$$

uniformly for $z \in \bar{B}_r$. By using Formula (4) it is easy to show that Theorem 1, Section 2.2.5, is valid.

Let us now consider the expansion of the function $\frac{1}{\zeta - z}$ by the polynomials $\pi_n(z)$ for a fixed ζ. From the corollary to Lemma 1, Section 2.3.1

$$\frac{1}{\zeta - z} = \frac{1}{2\pi i} \int_{L_r} F_*\left(\frac{u}{\zeta}\right) F\left(\frac{z}{u}\right) \frac{du}{\zeta u}, \quad z \in B_r, \quad \zeta \in B_r^1, \quad \rho < r < \infty$$

Assuming $u = \Psi(t)$, we have

$$\frac{1}{\zeta - z} = \frac{1}{2\pi i} \int_{|t|=r} \frac{1}{\zeta} F_*\left(\frac{\Psi(t)}{\zeta}\right) F\left(\frac{z}{\Psi(t)}\right) \frac{t\Psi'(t)}{\Psi(t)} \frac{dt}{t} \tag{6}$$

Let $R(t)$ be a function which is regular and does not become zero in the domain $|t| > \rho$ and $R_*(t) = \frac{1}{R(t)}$. Assuming

$$K(z, t) = F\left(\frac{z}{\Psi(t)}\right) \frac{t\Psi'(t)}{\Psi(t)} R(t), \quad K_*(\zeta, t) = F_*\left(\frac{\Psi(t)}{\zeta}\right) \frac{1}{\zeta} R_*(t)$$

from (6) we find

$$\frac{1}{\zeta - z} = \frac{1}{2\pi i} \int_{|t|=r} K_*(\zeta, t) K(z, t) \frac{dt}{t}, \quad z \in B_r, \ \zeta \in B_r^1, \ \rho < r < \infty$$

Substituting the series (1) into this formula and integrating the result, we have the expansion

$$\frac{1}{\zeta - z} = \sum_{n=0}^{\infty} q_n(\zeta) \pi_n(z), \quad z \in B_r, \ \zeta \in B_r^1, \quad q_n(\zeta) = \frac{1}{2\pi i} \int_{|t|=r} K_*(\zeta, t) \frac{1}{t^{n+1}} dt \tag{7}$$

The functions $q_n(\zeta)$, $n = 0, 1, \ldots$, are said to be conjugate functions to the polynomials $\pi_n(z)$.

From (7) follows

$$q_n(\zeta) = \frac{1}{2\pi i} \int_{|t|=r_1} F_*\left(\frac{\Psi(t)}{\zeta}\right) \frac{1}{\zeta} R_*(t) \frac{1}{t^{n+1}} dt \tag{8}$$

$$\zeta \in \bar{B}_r^1, \quad \rho < r_1 < r$$

where r is arbitrary, $\rho < r < \infty$, and therefore

$$\overline{\lim_{n\to\infty}} \sqrt[n]{|q_n(\zeta)|} \leqslant \frac{1}{r}, \quad \rho < r < \infty \tag{9}$$

uniformly for $\zeta \in \bar{B}_r^1$. Hence and from (5) we conclude that for every fixed r, $\rho < r < \infty$, the series (7) for the function $\frac{1}{\zeta - z}$ converges absolutely and uniformly with respect to z and ζ inside on the compact sub-sets of B_r and B_r^1 respectively.

From Formula (8) and as r, $\rho < r < +\infty$ is arbitrary, it follows that the function $q_n(\zeta)$ is regular in B^1. Expanding $F_*\left(\frac{\Psi(t)}{\zeta}\right)$ by powers of $\frac{1}{\zeta}$, we can easily prove that $q_n(\zeta)$ has a zero of the $(n+1)$-th order at the point $\zeta = \infty$ and that this function is obtained from the function $Q_n(\zeta)$, conjugate to the polynomial $\Pi_n(z)$, by multiplying the coefficient of $\frac{1}{\zeta^k}$ in the expansion of the function $Q_n(\zeta)$ by b_k^{-1}, $k = n+1$, $n+2, \ldots$ Hence and from Lemma 1, Section 2.3.1, we obtain

$$\begin{aligned} q_n(\zeta) &= \frac{1}{2\pi i} \int_{L_r} Q_n(u) F_*\left(\frac{u}{\zeta}\right) \frac{du}{u}, \quad \zeta \in B_r^1 \\ Q_n(\zeta) &= \frac{1}{2\pi i} \int_{L_r} q_n(u) F\left(\frac{u}{\zeta}\right) \frac{du}{u}, \quad \zeta \in B_r^1 \end{aligned} \tag{10}$$

NOTE 1 If $\bar{B}$ is star-like, then by $F(z)$ and $F_*(z)$ in all the above we understand weakly adjoined functions. This follows from Note 2 to Lemma 1, Section 2.3.1, and from the fact that in this case the sets $\bar{B}_r$, $\rho < r < +\infty$, will also be star-like (see Goluzin [1] p. 176).

Note that Note 1, Section 2.2.3, also holds in the case in question.

Proceeding as in Section 2.2.4, we obtain theorems for the expansion of regular functions by the polynomials $\pi_n(z)$ and $q_n(z)$, analogous to the theorems in Section 2.2.4. All we have to do is alter $\Pi_n(z)$ and $Q_n(z)$ to $\pi_n(z)$ and $q_n(z)$ throughout, remembering that the set $\bar{B}$ must contain the point $z = 0$ (see Note 2, Section 2.3.1).

2.3.3 Connection between series of polynomials $\pi_n(z)$ and power series

Let us take the series of polynomials $\pi_n(z)$:

$$f(z)=\sum_{n=0}^{\infty} a_n\pi_n(z), \quad \overline{\lim_{n\to\infty}} \sqrt[n]{|a_n|}=\frac{1}{r}, \quad r>\rho \tag{1}$$

and let the function $f(z)$ defined by this series be regular in some domain G containing B_r. The function $f_*(z)$ defined by the formula

$$f_*(z)=\frac{1}{2\pi i}\int_{L_{r'}} f(u)F_*\left(\frac{z}{u}\right)du, \quad z\in B_{r'}, \ \rho<r'<r \tag{2}$$

is also regular in the domain G on the strength of Lemma 1, Section 2.3.1, and then

$$f(z)=\frac{1}{2\pi i}\int_{L_{r'}} f_*(u)F\left(\frac{z}{u}\right)du, \quad z\in B_{r'}, \quad \rho<r'<r \tag{3}$$

Substituting series (1) in Formula (2) and integrating the result, on the basis of Formulae (4), Section 2.3.2, we obtain

$$f_*(z)=\sum_{n=0}^{\infty} a_n\Pi_n(z) \tag{4}$$

If we then substitute this series in Formula (3) and integrate the result, on the basis of the same Formulae (4), Section 2.3.2, we obtain series (1).

Assuming the functions $f(z)$ and $f_*(z)$, defined by series (1) and (4) respectively have identical domains of regularity, and assuming that the relation between the series (4) and the power series

$$F_0(w)=\sum_{n=0}^{\infty} a_n w^n \tag{5}$$

is that established in Theorem 1, Section 2.2.5, we conclude that the theorem indicated and the note on it also hold for the series (1) of polynomials $\pi_n(z)$. Hence it follows that all the properties indicated in Section 2.2.5 for series of polynomials $\Pi_n(z)$ are valid for series of polynomials $\pi_n(z)$.

2.4 BIORTHOGONAL SYSTEMS OF FUNCTIONS

A pair of sequences of functions $v_k(z)$ and $\lambda_k(z)$, $k=0, 1, \ldots,$ is said to form a biorthogonal system of functions on the curve L if

$$\frac{1}{2\pi i}\int_L v_n(z)\lambda_m(z)\,dz=\delta_{n,m}, \quad n, m=0, 1, \ldots,$$

where $\delta_{n,m}=0$, if $n\neq m$ and $\delta_{n,n}=1$. We shall also say that the system of functions $v_k(z)$ is biorthogonal on L together with the system $\lambda_k(z)$ and the system $\lambda_k(z)$ is said to be a conjugate system of functions with respect to the system $v_k(z)$, $k=0, 1 \ldots,$ on L.

In considering biorthogonal systems of functions $v_k(z)$ without making any other particular specifications, we shall assume that the curve of integration is a piecewise smooth Jordan curve and that the functions $v_k(z)$ and $\lambda_k(z)$ are continuous on the curve L. Throughout the following when Jordan curves are mentioned it will be assumed that they are piecewise smooth.

We have already encountered biorthogonal systems of functions with respect to Faber polynomials (lemmas in Sections 2.2.3 and 2.3).

From the definition of biorthogonal systems it follows directly that if the function $f(z)$, continuous on the curves L, can be expanded on L into a series with respect to the biorthogonal system of functions $v_k(z)$

$$f(z)=\sum_{k=0}^{\infty} a_k v_k(z), \quad z\in L$$

uniformly converging on L, then

$$a_k=\frac{1}{2\pi i}\int_L f(z)\lambda_k(z)\,dz, \quad k=0, 1, \ldots,$$

and at the same time the expansion is unique.

2.4.1 *Series expansion of a regular function with respect to a biorthogonal system of functions*

Let $\bar{D}$ be a closed set of the extended z plane and $\mathfrak{R}_{\bar{D}}$ the class of all functions which are regular on $\bar{D}$. The system of functions $v_k(z)$, $k=0, 1, \ldots,$ regular on $\bar{D}$, will be said to be complete in the class $\mathfrak{R}_{\bar{D}}$ if for every function $f(z) \in \mathfrak{R}_{\bar{D}}$ and for every $\varepsilon > 0$ there is a generalised polynomial

$$p(z) = \sum_{k=0}^{n} \alpha_k v_k(z)$$

such that

$$|f(z) - p(z)| < \varepsilon, \quad z \in \bar{D} \tag{1}$$

If every function $f(z) \in \mathfrak{R}_{\bar{D}}$ can be uniquely expanded in a series with respect to the system of functions $v_k(z)$:

$$f(z) = \sum_{k=0}^{\infty} a_k v_k(z)$$

uniformly converging on $\bar{D}$, then the system of functions $v_k(z)$ is called a basis in the class $\mathfrak{R}_{\bar{D}}$.

Analogously, let B be a domain and $\mathfrak{R}_B$ the class of all functions regular in B. The system of functions $v_k(z)$, $k = 0, 1, \ldots$, regular in B, is said to be complete in the class $\mathfrak{R}_B$ if for every function $f(z) \in \mathfrak{R}_B$ and every closed set $\bar{D} \subset B$ for any $\varepsilon > 0$ there is a generalised polynomial

$$p(z) = \sum_{k=0}^{n} \alpha_k v_k(z)$$

such that (1) is fulfilled. If every function $f(z) \in \mathfrak{R}_B$ can be uniquely expanded in a series with respect to the system of functions $v_k(z)$, uniformly converging on the compact subsets of B, then the system of functions $v_k(z)$ is called a basis in the class $\mathfrak{R}_B$.

Let $\bar{B}$ be a set of type $\mathfrak{M}$. Let us denote by $B_{r_0, R}$, $\rho \leqslant r_0 < R \leqslant \infty$ the doubly-connected domain bounded by the contour lines L_{r_0} and L_R (L_ρ is the boundary of B^1), and let the system of functions $v_k(z)$, $k = 0, 1, \ldots$, regular in the domain $B_{r_0, R}$, be biorthogonal on some contour line L_{r_1}, $r_0 < r_1 < R$, together with the system of functions $\lambda_k(z)$, $k = 0, 1, \ldots$, regular in $B_{r_0, R}$. By virtue of Cauchy's integral theorem the system of functions $v_k(z)$ will be biorthogonal on every closed Jordan curve lying in $B_{r_0, R}$ and not contractable to a point, with the system of functions $\lambda_k(z)$, in particular on every contour line L_r, $r_0 < r < R$. In this sense we shall say that the system of functions $v_k(z)$ is biorthogonal in $B_{r_0, R}$ with the system of functions $\lambda_k(z)$.

From the functions $\lambda_k(z)$ we can construct the functions

$$\lambda_k^*(z) = -\frac{1}{2\pi i}\int\limits_{L_r} \frac{\lambda_k(\zeta)}{\zeta - z}\,d\zeta, \quad z \in B_r^1, \quad r_0 < r < R \tag{2}$$

The functions $\lambda_k^*(z)$ do not depend on the choice of r, $r_0 < r < R$, are regular in the domain $B_{r_0}^1$ and equal to zero for $z = \infty$. Let us show that if the functions $v_k(z)$, $k = 0, 1, \ldots,$ are regular in B_R, then they are biorthogonal in $B_{r_0, R}$ together with the system of functions $\lambda_k^*(z)$. In fact if $r_0 < r < r_1 < R$, then

$$\frac{1}{2\pi i}\int\limits_{L_{r_1}} v_n(z)\lambda_m^*(z)\,dz = \frac{1}{2\pi i}\int\limits_{L_{r_1}} v_n(z)\left[-\frac{1}{2\pi i}\int\limits_{L_r}\frac{\lambda_m(\zeta)}{\zeta - z}\,d\zeta\right]dz$$

$$= \frac{1}{2\pi i}\int\limits_{L_r}\left[\frac{1}{2\pi i}\int\limits_{L_{r_1}}\frac{v_n(z)}{z-\zeta}\,dz\right]\lambda_m(\zeta)\,d\zeta = \frac{1}{2\pi i}\int\limits_{L_r} v_n(\zeta)\lambda_m(\zeta)\,d\zeta = \delta_{n,m}$$

Thus in the following it is possible to assume that the functions $\lambda_k(z)$ of the conjugate system with respect to the system of functions $v_k(z)$ with the indicated properties are regular in $B_{r_0}^1$ and become zero at the point $z = \infty$.

Analogously let us show that if the functions $\lambda_k(z)$, $k = 0, 1, \ldots,$ are regular in the domain $B_{r_0}^1$ and become zero at the point $z = \infty$, then the functions $v_k^*(z)$, defined by the formulae

$$v_k^*(z) = \frac{1}{2\pi i}\int\limits_{L_r}\frac{v_k(\zeta)}{\zeta - z}\,d\zeta, \quad z \in B_r, \quad r_0 < r < R \tag{3}$$

and regular in the domain B_R, form a biorthogonal system of functions in the domain $B_{r_0, R}$ with the system of functions $\lambda_k(z)$.

Let the system of functions $v_k(z)$, $k = 0, 1, \ldots,$ regular in the domain B_R, $\rho < R < \infty$, biorthogonal in $B_{r_0, R}$, $\rho < r_0 < R < +\infty$, together with the system of functions $\lambda_k(z)$, $k = 0, 1, \ldots,$ which are regular in the $B_{r_0}^1$ sphere and becoming zero at the point ∞, satisfy the following two conditions:

1. Uniformly for z with $z \in L_r$ for all r, $r_0 < r < R$,

$$\overline{\lim_{k\to\infty}}\sqrt[k]{|v_k(z)|} \leqslant r, \quad \overline{\lim_{k\to\infty}}\sqrt[k]{|\lambda_k(z)|} \leqslant \frac{1}{r} \tag{4}$$

2. The system of functions $v_k(z)$ is complete in some

fixed class $\mathfrak{R}_{\bar{B}_{r^*}}$, $r_0 < r^* < R$. In this case we shall say that the system of functions $v_k(z)$ and $\lambda_k(z)$ satisfy the conditions A in the domain $B_{r_0, R}$.

Lemma 1 If the systems of functions $v_k(z)$ and $\lambda_k(z)$ satisfy conditions A in the domain $B_{r_0, R}$, then for any r, $r_0 < r < R$, the following formula holds

$$\frac{1}{\zeta - z} = \sum_{k=0}^{\infty} \lambda_k(\zeta) v_k(z), \quad z \in B_r, \quad \zeta \in B_r^1 \tag{5}$$

where the series converges uniformly with respect to z and ζ on the compact sub-sets of B_r and B_r^1 respectively.

PROOF Let us fix $\zeta \in B_{r^*}^1$. For $\varepsilon > 0$ there is a generalised polynomial

$$p(z) = \sum_{k=0}^{n} \alpha_k v_k(z)$$

such that

$$\left| \frac{1}{\zeta - z} - p(z) \right| < \varepsilon, \quad z \in \bar{B}_{r^*}$$

Let us put $\beta(z) = \frac{1}{\zeta - z} - p(z)$. Multiplying this equation by $\lambda_m(z)$ and integrating with respect to L_{r^*}, we have

$$\beta_m = \frac{1}{2\pi i} \int_{L_{r^*}} \beta(z) \lambda_m(z)\, dz = \lambda_m(\zeta) - \alpha_m, \quad m = 0, 1, \ldots, n$$

Then

$$|\beta_m| \leqslant \frac{\varepsilon l_{r^*}}{2\pi} \lambda_m, \quad \lambda_m = \sup_{z \in L_{r^*}} |\lambda_m(z)|$$

where l_{r^*} is the length of L_{r^*}. For $z \in \bar{B}_r$, $r_0 < r < r^*$, we have

$$\left| \frac{1}{\zeta - z} - \sum_{k=0}^{n} \lambda_k(\zeta) v_k(z) \right| \leqslant \left| \frac{1}{\zeta - z} - p(z) \right| + \left| \sum_{k=0}^{n} \beta_k v_k(z) \right| \leqslant \varepsilon + \frac{\varepsilon l_{r^*}}{2\pi} \sum_{k=0}^{n} \lambda_k v_k \tag{6}$$

where $v_k = \sup_{z \in L_r} |v_k(z)|$. By virtue of (4),

$$\overline{\lim_{k\to\infty}} \sqrt[k]{|\lambda_k(\zeta)\, v_k(z)|} \leqslant \frac{r}{r^*} < 1$$

uniformly with respect to z and ζ, $z\in B_r$, $\zeta\in B^1_{r^*}$, and consequently the series $\sum_{k=1}^{\infty}\lambda_k v_k$ converges. Hence and from (6), since $\varepsilon>0$ is arbitrarily small, we conclude that for $z\in \bar{B}_r$ (5) holds and the series converges uniformly for $z\in\bar{B}_r$.

Now let r, $r_0<r<R$, be an arbitrary number and $r_0<r_1<r<r_2<R$. Operating on (4) it is easy to show that the series (5) converges uniformly with respect to z and ζ for $z\in\bar{B}_{r_1}$, $\zeta\in\bar{B}^1_{r_2}$, and since the choice of r_1 and r_2 is arbitrary, it converges uniformly with respect to z and ζ on the compact sub-sets of B_r and B^1_r respectively to a function which is regular with respect to z and ζ for $z\in B_r$ and $\zeta\in B^1_r$, and equation (5) holds for arbitrary $z\in B_r$ and $\zeta\in B^1_r, r_0<r<R$. The lemma is proved.

Let us prove that under conditions A there will always be an equals sign in the inequalities (4). Let us assume, for example, that in the second of the inequalities (4) for some $z_0\in L_r$, $r_0<r<R$,

$$\overline{\lim_{k\to\infty}} \sqrt[k]{|\lambda_k(z_0)|} = \frac{1}{r_2} < \frac{1}{r}$$

Assuming that $z\in\bar{B}_{r_1}$, $r<r_1<\min\{r_2, R\}$, we have

$$\overline{\lim_{k\to\infty}} \sqrt[k]{|\lambda_k(z_0)\, v_k(z)|} \leqslant \frac{r_1}{r_2} < 1$$

and consequently the series $\sum_{k=0}^{\infty}\lambda_k(z_0)\, v_k(z)$ converges uniformly on $\bar{B}_{r_1}$ to a function regular in B_{r_1}. On the other hand, the sum of this series for $z\in B_{r_1}$ is equal to $\frac{1}{z_0-z}$ and in view of the uniqueness of analytical continuation the function $\frac{1}{z_0-z}$ is regular in B_{r_1}, in particular, at the point $z_0\in B_{r_1}$. This contradiction proves our assertion.

As a corollary to this we find that for the generalised Faber polynomials $\pi_k(z)$ (see Section 2.3) and their conjugate functions $q_k(z)$

$$\overline{\lim_{k\to\infty}} \sqrt[k]{|\pi_k(z)|} = r, \quad \overline{\lim_{k\to\infty}} \sqrt[k]{|q_k(z)|} = \frac{1}{r}$$

uniformly for $z \in L_r$, $\rho < r < \infty$, since, as shown earlier, the above-mentioned functions satisfy conditions A.

It is now easy to prove that if the systems of functions $v_k(z)$ and $\lambda_k(z)$ satisfy conditions A in the domain $B_{r_0, R}$, theorems concerning the expansion of regular functions in a series, analogous to those proved in Section 2.2.4 for systems $\Pi_k(z)$ and $Q_k(z)$, are also valid for functions $v_k(z)$ and $\lambda_k(z)$, and, in particular, to assert that (1) the system of functions $v_k(z)$ forms a basis in every class $\mathfrak{R}_{\overline{B}_r}$ and in every class $\mathfrak{R}_{B_r}$, $r_0 < r < R$; (2) the system of functions $\lambda_k(z)$ forms a basis in every class $\mathfrak{R}_{\overline{B}^1_r}$ and in every class $\mathfrak{R}_{B^1_r}$, $r_0 < r < R$; (3) the combined system of functions $v_k(z)$, and $\lambda_k(z)$, $k = 0, 1, \ldots$, forms a basis in every class $\mathfrak{R}_{\overline{B}_{r_1, r_2}}$ and in every class

$$\mathfrak{R}_{B_{r_1, r_2}} \quad r_0 < r_1 < r_2 < R$$

NOTE 1 In defining conditions A, condition (1) can be altered in the following manner: uniformly with respect to z and ζ for $z \in L_r$ and $\zeta \in L_{r_1}$ for all r and r_1, $r_0 < r < R$, $r_0 < r_1 < R$,

$$\overline{\lim_{k \to \infty}} \sqrt[k]{|\lambda_k(\zeta) v_k(z)|} \leqslant \frac{r}{r_1}$$

In this case the above discussion has to be altered slightly.

NOTE 2 Let the system of functions $v_k(z)$, $k = 0, 1, \ldots$, regular in the domain B_R, $R > \rho$, form a basis in every class $\mathfrak{R}_{B_r}$, $\rho \leqslant r_0 < r < R$. Let us expand the function $\frac{1}{\zeta - z}$, $z \in B_r$, $\zeta \in \overline{B}^1_r$, in a series with respect to the system of functions $v_k(z)$:

$$\frac{1}{\zeta - z} = \sum_{k=0}^{\infty} \lambda_k(\zeta) v_k(z) \tag{7}$$

uniformly converging on the compact sub-sets of B_r. It is possible to show that the functions $\lambda_k(\zeta)$ are regular in $B^1_{r_0}$, $\lambda_k(\infty) = 0$, and that the system of functions $v_k(z)$ is bi-orthogonal on $B_{r_0, R}$ with the system $\lambda_k(z)$. In view of the uniqueness of the expansion (7) there is only one conjugate system $\lambda_k(z)$, $\lambda_k(\infty) = 0$, $k = 0, 1, \ldots$, with respect to the system $v_k(z)$.

2.4.2 Mutually adjoined functions

Let $\bar{B}$ be a set of type $\mathfrak{M}$; $F(z, \zeta)$ and $F_*(z, \zeta)$ be functions regular with respect to z and ζ for $z \in B_r$, $\zeta \in B_r^1$, for every r, $\rho \leqslant r_0 < r < R \leqslant \infty$, and such that for every r, $r_0 < r < R$, the following formulae hold

$$\left.\begin{aligned} \frac{1}{\zeta - z} &= \frac{1}{2\pi i} \int\limits_{L_r} F(z, u) F_*(u, \zeta)\, du, \\ \frac{1}{\zeta - z} &= \frac{1}{2\pi i} \int\limits_{L_r} F_*(z, u) F(u, \zeta)\, du, \end{aligned}\right\} \quad z \in B_r,\ \zeta \in B_r^1 \tag{1}$$

Such functions $F(z, \zeta)$ and $F_*(z, \zeta)$ are said to be mutually adjoined for the domain $B_{r_0, R}$ and either of them is said to be adjoined to the other.

Note that if $F(z)$ and $F_*(z)$ are mutually adjoined functions as introduced in Section 2.3, then the functions $F(z, \zeta) = F\left(\frac{z}{\zeta}\right)\frac{1}{\zeta}$ and $F_*(z, \zeta) = F_*\left(\frac{z}{\zeta}\right)\frac{1}{\zeta}$ are mutually allied in any domain $B_{r_0, R}$, $\rho < r_0 < R < \infty$, for any set $\bar{B}$ of type $\mathfrak{M}$ containing the point $z = 0$. This follows from Formula (2), Section 2.3.1.

Let us note the following properties of functions $F(z, \zeta)$ and $F_*(z, \zeta)$ mutually adjoined in the domain $B_{r_0, R}$.

1. If $f(z)$ is an arbitrary function, regular in the domain B_r, $r_0 < r < R$, then the function

$$f_*(z) = \frac{1}{2\pi i} \int\limits_{L_{r_1}} F_*(z, u) f(u)\, du \qquad z \in B_{r_1}, \quad r_0 < r_1 < r \tag{2}$$

is also regular in the domain B_r and

$$f(z) = \frac{1}{2\pi i} \int\limits_{L_{r_1}} F(z, u) f_*(u)\, du \qquad z \in B_{r_1}, \quad r_0 < r_1 < r \tag{3}$$

Conversely, if the function $f_*(z)$ is regular in the domain $B_r, r_0 < r \leqslant R$, then the function $f(z)$, defined by Formula (3) is also regular in the domain B_r and Formula (2) holds. In fact if $z \in B_{r_2}$, $r_0 < r_2 < r_1 < r$, then

$$\frac{1}{2\pi i}\int_{L_{r_2}} F(z, u) f_*(u)\, du$$

$$= \frac{1}{2\pi i}\int_{L_{r_2}} F(z, u)\left[\frac{1}{2\pi i}\int_{L_{r_1}} F_*(u, \xi) f(\xi)\, d\xi\right] du$$

$$= \frac{1}{2\pi i}\int_{L_{r_1}}\left[\frac{1}{2\pi i}\int_{L_{r_2}} F(z, u) F_*(u, \xi)\, du\right] f(\xi)\, d\xi$$

$$= \frac{1}{2\pi i}\int_{L_{r_1}} \frac{f(\xi)}{\xi - z}\, d\xi = f(z)$$

The reverse is proved in an analogous fashion.

2. If the function $f(z)$ is regular in the domain B^1_r, $r_0 \leqslant - r < R$, and $f(\infty) = 0$ then the function

$$f_*(z) = \frac{1}{2\pi i}\int_{L_{r_1}} F(u, z) f(u)\, du, \quad z \in B^1_{r_1}, \quad r < r_1 < R \qquad (4)$$

is regular in the domain B^1_r, $f_*(\infty) = 0$, and

$$f(z) = \frac{1}{2\pi i}\int_{L_{r_1}} F_*(u, z) f_*(u)\, du, \quad z \in B^1_{r_1}, \quad r < r_1 < R \qquad (5)$$

The converse is also true. The proof is based on Formula (1) and makes use of property (1).

3. Assuming that in (2) $f(z) = F(z, \zeta)$, we obtain $f_*(z) = \frac{1}{\zeta - z}$, and, substituting this function in (3), we have

$$F(z, \zeta) = \frac{1}{2\pi i}\int_{L_{r_1}} \frac{F(z, u)}{\zeta - u}\, du, \quad z \in B_{r_1}, \quad \zeta \in B^1_{r_1}$$

whence it follows that $F(z, \infty) \equiv 0$, and analogously $F_*(z, \infty) \equiv 0$.

4. The coefficient of $1/\zeta$ in the Laurent expansion of the function $F(z, \zeta)$ in the neighbourhood of the point $\zeta = \infty$

$$F(z, \zeta) = \sum_{k=1}^{\infty} \frac{\alpha_k(z)}{\zeta^k}$$

is not identically equal to zero. In fact, multiplying the second of Equations (1) by ζ and allowing ζ to tend to ∞, in the limit we have

$$1 = \frac{1}{2\pi i} \int_{L_r} F_*(z, u)\, \alpha_1(u)\, du, \quad z \in B_r$$

and consequently $\alpha_1(u) \not\equiv 0$. There is an analogous property for $F_*(z, \zeta)$.

5. Let $F_j(z, \zeta)$, $F_{*j}(z, \zeta)$, $j = 1, 2$, be mutually adjoined functions for $B_{r_0, R}$, $\rho \leqslant r_0 < R \leqslant \infty$. The functions

$$F(z, \zeta) = \frac{1}{2\pi i} \int_{L_r} F_1(z, u)\, F_2(u, \zeta)\, du$$

$$F_*(z, \zeta) = \frac{1}{2\pi i} \int_{L_r} F_{*2}(z, u)\, F_{*1}(u, \zeta)\, du$$

where $r_0 < r < R$, $z \in B_r$, $\zeta \in B_r^1$, are also mutually adjoined for the domain $B_{r_0, R}$.

6. Let $v_k(z)$, $k = 0, 1, \ldots$, be a system of functions regular in the domain B_R, $\rho < R \leqslant \infty$. Let us introduce the system of functions

$$v_k^*(z) = \frac{1}{2\pi i} \int_{L_r} F_*(z, u)\, v_k(u)\, du, \quad z \in B_r, \quad \rho \leqslant r_0 < r < R \tag{6}$$

regular in the domain B_R. Then

$$v_k(z) = \frac{1}{2\pi i} \int_{L_r} F(z, u)\, v_k^*(u)\, du, \quad z \in B_r, \quad r_0 < r < R \tag{7}$$

If the system of functions $v_k(z)$ is complete (basis) in some class $\mathfrak{R}_{B_r}$, $r_0 < r \leqslant R$, then the system of functions $v_k^*(z)$ is also complete (correspondingly a basis) in the class $\mathfrak{R}_{B_r}$.

7. Let the system of functions $v_k(z)$, $k = 0, 1, \ldots$, regular in the domain B_R, $\rho < R \leqslant \infty$, be biorthogonal in the domain $B_{r_0, R}$ with the system of functions $\lambda_k(z)$, $k = 0, 1, \ldots$, regular in $B_{r_0}^1$, $\rho \leqslant r_0 < R$, and becoming zero at the point $z = \infty$. Let us introduce the system of functions

$$\lambda_k^*(z) = \frac{1}{2\pi i} \int_{L_r} F(u, z)\, \lambda_k(u)\, du, \quad z \in B_r^1, \quad r_0 < r < R \tag{8}$$

regular in the domain $B_{r_0}^1$ and becoming zero at the point $z = \infty$. Then

$$\lambda_k(z) = \frac{1}{2\pi i} \int_{L_r} F_*(u, z)\, \lambda_k^*(u)\, du, \quad z \in B_r^1, \quad r_0 < r < R \tag{9}$$

The system of functions $v_k^*(z)$ is biorthogonal on L_r, $r_0 < r < R$, with the system of functions $\lambda_k^*(z)$.

8. Let the functions of the systems $v_k(z)$ and $\lambda_k(z)$ satisfy the conditions of (4), Section 2.4.1. Then the systems of functions $v_k^*(z)$ and $\lambda_k^*(z)$, defined respectively by (6) and (8), also satisfy the conditions of (4), Section 2.4.1, e.g.

$$\overline{\lim_{k\to\infty}} \sqrt[k]{|v_k^*(z)|} \leqslant r, \quad \overline{\lim_{k\to\infty}} \sqrt[k]{|\lambda_k^*(z)|} \leqslant \frac{1}{r} \tag{10}$$

uniformly for $z \in L_r$, $r_0 < r < R$,

We leave the proof of properties (5) to (8) to the reader.

From the above we move on to the following theorem.

Theorem 1 If the systems of functions $v_k(z)$ and $\lambda_k(z)$ satisfy conditions A in the domain $B_{r_0, R}$, $\rho \leqslant r_0 < R \leqslant \infty$, then the systems of functions $v_k^*(z)$ and $\lambda_k^*(z)$ defined respectively by Formulae (6) and (8), also satisfy conditions A in the domain $B_{r_0, R}$ Then the mutually adjoined functions $F(z, \zeta)$ and $F_*(z, \zeta)$ can be represented in the domain $B_{r_0, R}$ by the series

$$\begin{aligned} F(z, \zeta) &= \sum_{k=0}^{\infty} \lambda_k^*(\zeta)\, v_k(z), \\ F_*(z, \zeta) &= \sum_{k=0}^{\infty} \lambda_k(\zeta)\, v_k^*(z), \end{aligned} \qquad z \in B_r,\ \zeta \in B_r^1. \tag{11}$$

for every r, $r_0 < r < R$, and series (11) uniformly converge with respect to z and ζ on the compact sub-sets of B_r and B_r^1 respectively.

The following assertion which in a certain sense is the converse of the previous one is also true.

Theorem 2 Let the systems of functions $v_k(z)$ and $\lambda_k(z)$ and also the systems of functions $v_k^*(z)$ and $\lambda_k^*(z)$ satisfy conditions A in the domain $B_{r_0, R}$, $\rho \leqslant r_0 < R \leqslant \infty$. Then the functions $F(z, \zeta)$ and $F_*(z, \zeta)$, defined by series (11), are mutually adjoined in the domain $B_{r_0, R}$, and the series indicated converge uniformly with respect to z and ζ on the compact sub-sets of B_r and B_r^1 respectively. For these functions when $z \in B_r$, $\zeta \in B_r^1$, $r_0 < r < R$, we have

$$\frac{1}{2\pi i}\int_{L_r} F(z, u) F_*(u, \zeta)\, du = \sum_{k=0}^{\infty} v_k(z) \frac{1}{2\pi i}\int_{L_r} \overset{*}{\lambda}_k(u) F_*(u, \zeta)\, du$$

$$= \sum_{k=0}^{\infty} v_k(z) \sum_{\nu=0}^{\infty} \lambda_\nu(\zeta) \frac{1}{2\pi i}\int_{L_r} \overset{*}{\lambda}_k(u) \overset{*}{v}_\nu(u)\, du = \sum_{k=0}^{\infty} \lambda_k(\zeta) v_k(z) = \frac{1}{\zeta - z}$$

Analogously we can show that:

$$\frac{1}{2\pi i}\int_{L_r} F_*(z, u) F(u, \zeta)\, du = \frac{1}{\zeta - z}$$

Hence it follows that the functions $F(z, \zeta)$ and $F_*(z, \zeta)$ are mutually adjoined in the domain $B_{r_0, R}$.

Mutually adjoined functions could be defined also on the contour line L_r.

2.4.3 Approximately biorthogonal systems of functions

In this section we shall establish that the system of functions $v_k^*(z)$, $k = 0, 1, \ldots,$ which is in a certain sense close to the complete system of functions $v_k(z)$ biorthogonal with the system $\lambda_k(z)$, is itself biorthogonal with a system $\overset{*}{\lambda}_k(z)$ and is complete. The theorems in the previous section establish a close connection between biorthogonal systems of functions and mutually adjoined functions. In connection with this, bearing in mind the problem we have set ourselves, let us first prove a lemma concerning mutually adjoined functions. We shall start from the simplest pair of mutually adjoined functions

$$\tilde{F}(z, \zeta) = \tilde{F}_*(z, \zeta) = \frac{1}{\zeta - z}$$

Let $\bar{B}$ be a set of type $\mathfrak{M}$. Let us consider the function

$$F_*(z, \zeta) = \frac{1}{\zeta - z} - \lambda K_*(z, \zeta) \tag{1}$$

where λ is a parameter and $K_*(z, \zeta)$ is a function which, for every r, $\rho < r_0 \leqslant r < R \leqslant \infty$, is regular with respect to z and ζ for $z \in B_r$ and $\zeta \in B_r^1$, is continuous with respect to z and ζ for $z \in \bar{B}_{r_0}$, $\zeta \in \bar{B}_{r_0}^1$, and becomes zero for $\zeta = \infty$ and any $z \in B_r$. We ask the following question: for what values of λ is there a function

$$F(z, \zeta)=\frac{1}{\zeta-z}+\lambda K(z, \zeta), \tag{2}$$

adjoined with $F_*(z, \zeta)$ in the domain $B_{r_0, R}$ where $K(z, \zeta)$ is a function with the same properties as the function $K_*(z, \zeta)$ depending on the choice of λ?

Let us assume that for a certain value of λ there is such a function $F(z, \zeta)$ and consequently the conditions of Formula (1) Equation (3) are fulfilled:

$$\left.\begin{aligned}\frac{1}{\zeta-z}&=\frac{1}{2\pi i}\int_{L_r} F(z, u)F_*(u, \zeta)\,du\\ \frac{1}{\zeta-z}&=\frac{1}{2\pi i}\int_{L_r} F_*(z, u)F(u, \zeta)\,du\end{aligned}\right\}\quad z\in B_r,\ \zeta\in B_r^1,\ r_0<r<R \tag{3}$$

Substituting in this the expressions (1) and (2) for $F_*(z, \zeta)$ and $F(z, \zeta)$ and applying Cauchy's theorem, we have

$$\left.\begin{aligned}K(z, \zeta)&=K_*(z, \zeta)+\frac{\lambda}{2\pi i}\int_{L_r} K_*(u, \zeta)K(z, u)\,du\\ K(z, \zeta)&=K_*(z, \zeta)+\frac{\lambda}{2\pi i}\int_{L_r} K_*(z, u)K(u, \zeta)\,du\end{aligned}\right\}\quad z\in B_r,\ \zeta\in B_r^1,\ r_0<r<R$$

If $z\in\overline{B}_{r_0}$, then on the basis of the conditions applied to the functions $K(z, \zeta)$ and $K_*(z, \zeta)$, the contour of integration L_r in both integrals in these formulae can be changed to L_{r_0}, and we can then assume that $\zeta\in\overline{B}_{r_0}^1$:

$$\left.\begin{aligned}K(z, \zeta)&=K_*(z, \zeta)+\frac{\lambda}{2\pi i}\int_{L_{r_0}} K_*(u, \zeta)K(z, u)\,du\\ K(z, \zeta)&=K_*(z, \zeta)+\frac{\lambda}{2\pi i}\int_{L_{r_0}} K_*(z, u)K(u, \zeta)\,du\end{aligned}\right\}\quad z\in L_{r_0},\ \zeta\in L_{r_0} \tag{4}$$

From the theory of integral equations (see for example, Smirnov [1])†, it follows that the number $\frac{\lambda}{2\pi i}$ in the case in question is not a characteristic value of the kernel $K_*(z, \zeta)$ and the function $K(z, \zeta)$ is the resolvent of the kernel $K_*(z, \zeta)$ (for $z, \zeta\in L_{r_0}$).

† In equations (4) we can put $u=\Psi(r_0e^{i\theta})$ and $z=\Psi(r_0e^{i\phi})$ and consider the integral equation obtained as an integral equation with respect to the polar angle ϕ.

Conversely if the number $\frac{\lambda}{2\pi i}$ is not a characteristic value of the kernel $K_*(u, \zeta)$ and at the same time this kernel has the resolvent $K(z, \zeta)$, $z \in L_{r_0}$, $\zeta \in L_{r_0}$, which is continuous with respect to z and ζ for $z \in L_{r_0}$, $\zeta \in L_{r_0}$ and satisfies Equations (4). Then, since the function $K_*(u, \zeta)$ according to the conditions is regular with respect to ζ $\zeta \in B^1_{r_0}$, continuous with respect to ζ and u for $\zeta \in B^1_{r_0}$, $u \in \bar{B}_{r_0}$ and $K_*(u, \infty) = 0$, the integral in the first of Equations (4) for every fixed $z \in L_{r_0}$ is a function of ζ which is regular in $B^1_{r_0}$, continuous in $\bar{B}^1_{r_0}$ and becomes zero at the point $\zeta = \infty$. Consequently the function $K(z, \zeta)$ has the same properties. Thus the function $F(z, \zeta)$, defined by Formula (2), has the same properties as the function $F_*(z, \zeta)$, defined by Formula (1). To prove that the functions $F(z, \zeta)$ and $F_*(z, \zeta)$ are mutually adjoined it is sufficient to verify that conditions (3) are fulfilled. To do this it is sufficient to substitute expressions (1) and (2) for $K_*(z, \zeta)$ and $K(z, \zeta)$ in (4) and apply Cauchy's theorem.

In the above we required the function $K_*(z, \zeta)$ to be continuous with respect to z and ζ for $z \in \bar{B}_{r_0}$ and $\zeta \in \bar{B}^1_{r_0}$. The results remain valid, though this discussion is somewhat more complex, if the condition for $r = r_0$ is altered to the same condition for some $r = r_*$, $r_0 < r_* < R$.

Thus the following result is proved below.

Lemma 1 Let $F_*(z, \zeta) = \frac{1}{\zeta - z} - \lambda K_*(z, \zeta)$ and for every r, $\rho \leqslant r_0 < r < R \leqslant \infty$ let the function $K_*(z, \zeta)$ be regular with respect to z and ζ for $z \in B_r$, $\zeta \in B^1_r$; for some r_*, $r_0 \leqslant r_* \leqslant R$ ($r_* \neq \rho$, $r_* \neq \infty$) be continuous with respect to z and ζ for $z \in \bar{B}_{r_*}$, $\zeta \in \bar{B}^1_{r_*}$, and $K_*(z, \infty) = 0$ for $z \in B_R$. If the number $\frac{\lambda}{2\pi i}$ is not a characteristic value of the kernel

$$K_*(z, \zeta), \quad z \in L_{r_*}, \quad \zeta \in L_{r_*}$$

then there is a unique function $F(z, \zeta) = \frac{1}{\zeta - z} + \lambda K(z, \zeta)$, adjoined to $F_*(z, \zeta)$ in the domain $B_{r_0, R}$. In this case the function $K(z, \zeta)$ has the same properties as the function $K_*(z, \zeta)$ and satisfies Equations (4) where r_0 is changed to r_*.

NOTE 1 In the theory of integral equations it is proved that the number $\frac{\lambda}{2\pi i}$ is not a characteristic value of the kernel $K_*(z, \zeta)$, $z, \zeta \in L_{r_*}$, if

$$|\lambda| < \frac{1}{\sqrt{M}}, \quad M = \frac{1}{(2\pi)^2} \int_{L_{r_*}} \int_{L_{r_*}} |K_*(z, \zeta)|^2 |dz| |d\zeta|$$

and, in particular, if

$$|\lambda| < \frac{2\pi}{M_{r_*} l_{r_*}} \text{ where } M_{r_*} = \sup_{z, \zeta \in L_{r_*}} |K_*(z, \zeta)|$$

and l_{r_*} is the length of the contour line L_{r_*}.

Returning to the problem posed at the beginning of this section, let us show that the following result is true.

Theorem 1 Let $\bar{B}$ be a set of type $\mathfrak{M}$ and the systems of functions $v_k(z)$ and $\lambda_k(z)$ satisfy conditions A in the domain $B_{r_0, R}$, $\rho \leqslant r_0 < R \leqslant \infty$. Further let the functions $\omega_k(z)$, $k = 0, 1, \ldots$, be regular in the domain B_r and such that for $z \in L_r$, $r_0 < r < R$,

$$\overline{\lim_{k \to \infty}} \sqrt[k]{|\omega_k(z)|} \leqslant r$$

uniformly, and the function

$$K_*(z, \zeta) = \sum_{k=0}^{\infty} \omega_k(z) \lambda_k(\zeta)$$

for some r_*, $r_0 \leqslant r_* \leqslant R$ ($r_* \neq \rho$, $r_* \neq \infty$), is continuous with respect to z and ζ for $z \in \bar{B}_{r_*}$, $\zeta \in \bar{B}^1_{r_*}$. If $\frac{\lambda}{2\pi i}$ is not a characteristic value of the kernel $K_*(z, \zeta)$, $z \in L_{r_*}$, $\zeta \in L_{r_*}$, then for the system of functions

$$v^*_k(z, \lambda) = v_k(z) - \lambda \omega_k(z), \quad k = 0, 1, \ldots, \tag{5}$$

there is a system of functions $\lambda^*_k(z, \lambda)$, regular in the domain $B^1_{r_0}$, such that the systems of functions $v^*_k(z, \lambda)$ and $\lambda^*_k(z, \lambda)$ satisfy the conditions A in the domain $B_{r_0, R}$.

PROOF It is easy to see that for every finite λ for $z \in L_r$, $r_0 < r < R$,

$$\overline{\lim_{k \to \infty}} \sqrt[k]{|v^*_k(z, \lambda)|} \leqslant r$$

uniformly. As usual we can show that the series

$$F_*(z, \zeta) = \sum_{k=0}^{\infty} v^*_k(z, \lambda) \lambda_k(\zeta), \quad z \in B_r, \ \zeta \in B^1_r, \quad r_0 < r < R$$

converges uniformly with respect to z and ζ on the compact sub-sets of B_r and B_r^1 respectively, and consequently the function $F_*(z, \zeta)$ for every fixed r, $r_0 < r < R$, is regular with respect to z and ζ for $z \in B_r$, $\zeta \in B_r^1$. An analogous assertion holds for the function $K_*(z, \zeta)$. By using Formula (5), let us transform the series of the function

$$F_*(z, \zeta), \ z \in B_r, \ \zeta \in B_r^1, \ r_0 < r < R$$

$$F_*(z, \zeta) = \sum_{k=0}^{\infty} \overset{*}{v}_k(z, \lambda)\lambda_k(\zeta) = \sum_{k=0}^{\infty} v_k(z)\lambda_k(\zeta) - \lambda \sum_{k=0}^{\infty} \omega_k(z)\lambda_k(\zeta) = \frac{1}{\zeta - z} - \lambda K_*(z, \zeta)$$

e.g.

$$F_*(z, \zeta) = \frac{1}{\zeta - z} - \lambda K_*(z, \zeta), \quad z \in B_r, \quad \zeta \in B_r^1, \quad r_0 < r < R$$

whence it follows that the function $F_*(z, \zeta)$ satisfies the conditions of the lemma and consequently since $\frac{\lambda}{2\pi i}$ is not a characteristic value of the kernel $K_*(z, \zeta)$, there is a function $F(z, \zeta) = \frac{1}{\zeta - z} + \lambda K(z, \zeta)$, adjoined to the function $F_*(z, \zeta)$ in the domain $B_{r_0, R}$. Expanding this function into a series of the functions $v_k(z)$, we obtain

$$F(z, \zeta) = \sum_{k=0}^{\infty} \overset{*}{\lambda}_k(\zeta, \lambda) v_k(z)$$

where for $\zeta \in B_r^1$, $r_0 < r < R$,

$$\overset{*}{\lambda}_k(\zeta, \lambda) = \frac{1}{2\pi i} \int_{L_r} F(u, \zeta)\lambda_k(u)\,du = \lambda_k(\zeta) + \lambda \int_{L_r} K(u, \zeta)\lambda_k(\zeta)\,d\zeta$$

By virtue of Theorem 1, in this section, the system of functions $\overset{*}{v}_k(z, \lambda)$ and $\overset{*}{\lambda}_k(z, \lambda)$ satisfy the conditions A in the domain $B_{r_0, R}$ and the theorem is proved.

By using Note 1 we obtain the following corollary.

COROLLARY 1 If the systems of functions $v_k(z)$ and $\lambda_k(z)$ satisfy the conditions A in the domain $B_{r_0, R}$ and the system of functions $\overset{*}{v}_k(z)$ is such that

1. uniformly for $z \in L_r$ for every r, $r_0 < r < R$,

$$\overline{\lim_{k \to \infty}} \sqrt[k]{|\overset{*}{v}_k(z)|} \leqslant r$$

2. the series

$$K_*(z, \zeta) = \sum_{k=0}^{\infty} \omega_k(z) \lambda_k(\zeta), \quad \omega_k(z) = v_k(z) - \overset{*}{v}_k(z)$$

converges uniformly with respect to z and ζ for $z, \zeta \in L_{r_*}$ where r_* is some number $r_0 \leqslant r_* \leqslant R$. $(r_* \neq \rho, \ r_* \neq \infty)$

3.
$$\int_{L_{r_*}} \int_{L_{r_*}} |K_*(z, \zeta)|^2 \, |dz| \, |d\zeta| < (2\pi)^2 \qquad \textbf{(6)}$$

then there is a system of functions $\overset{*}{\lambda}_k(z)$ such that the systems of functions $\overset{*}{v}_k(z)$ and $\overset{*}{\lambda}_k(z)$ satisfy conditions A in the domain $B_{r_0, R}$.

Condition (6) defines the sufficient closeness of the system of functions $\overset{*}{v}_k(z)$ to the system of functions $v_k(z)$ for the conclusion of this corollary to be fulfilled. This closeness guarantees that $\frac{\lambda}{2\pi i}$ is not a characteristic value.

NOTE 2 If we put

$$\overset{*}{v}_k(z) = \sum_{\nu=0}^{k} \alpha_\nu^{(k)} v_\nu(z)$$

where $\alpha_\nu^{(k)}$ are certain constants with $\alpha_k^{(k)} \neq 0$, then the conclusion of the corollary is valid if only conditions (1) and (2) of the corollary are fulfilled.

To prove this let us find N such that the subsidiary system of functions $v_k^+(z)$:

$$v_k^+(z) = v_k(z), \quad k = 0, 1, \ldots, N, \quad v_k^+(z) = \overset{*}{v}_k(z), \quad k > N$$

satisfies all the conditions of the corollary. That condition (1) is satisfied for the system of functions $v_k^+(z)$ is obvious. In conditions (2) we shall have the series

$$K_*(z, \zeta) = \sum_{k=N+1}^{\infty} (v_k(z) - \overset{*}{v}_k(z)) \lambda_k(\zeta)$$

satisfying condition (2). Condition (3)

$$\int_{L_{r_*}} \int_{L_{r_*}} |K_*(z, \zeta)|^2 \, |d\zeta| \, |dz| < (2\pi)^2$$

is also fulfilled for sufficiently large N. Let us fix such an N. Consequently the conclusion of the corollary holds, e.g. there is a system of functions $\lambda_k^+(z)$ such that the systems of functions $v_k^+(z)$, $\lambda_k^+(z)$ satisfy the conditions A in the domain $B_{r_0, R}$. In particular

$$\frac{1}{2\pi i}\int\limits_{L_r} v_n^+(z)\lambda_m^+(z)\,dz = \delta_{n,m}, \quad n, m = 0, 1, \ldots, \quad r_0 < r < R$$

Hence it follows that the conditions

$$\frac{1}{2\pi i}\int\limits_{L_r} v_n^*(z)\lambda_m^+(z)\,dz = \delta_{n,m} \tag{7}$$

can only not be fulfilled for $n, m \leqslant N$. In connection with this it is natural to put $\lambda_m^*(z) = \lambda_m^+(z)$ for $m > N$ and to try to find $\lambda_m^*(z)$ for $m \leqslant N$ in the form

$$\lambda_m^*(z) = \sum_{k=0}^{N} b_k^{(m)}\lambda_k^+(z) \tag{8}$$

where the numbers $b_k^{(m)}$ are defined from conditions (7) for $m, n \leqslant N$. This condition becomes

$$\frac{1}{2\pi i}\int\limits_{L_r} \sum_{k=0}^{n} \alpha_k^{(n)} v_k(z) \sum_{k=0}^{N} b_k^{(m)}\lambda_k^+(z)\,dz = \sum_{k=0}^{n} \alpha_k^{(n)} b_k^{(m)} = \delta_{n,m}$$

Fixing $m = 0, 1, \ldots, N$, we obtain a system for finding $b_k^{(m)}, k = 0, 1, \ldots, N$, the determinant of which is not zero (by virtue of the condition $\alpha_k^{(k)} \neq 0$), and consequently uniquely defines $b_k^{(m)}$ and the function $\lambda_m^*(z)$. Then

$$\overline{\lim_{k\to\infty}} \sqrt[k]{|\lambda_k^*(z)|} = \overline{\lim_{k\to\infty}} \sqrt[k]{|\lambda_k^+(z)|} \leqslant \frac{1}{r}, \quad r_0 < r < R$$

uniformly with respect to z for $z \in L_r$. To prove the note it is sufficient to show that the system of functions $v_k^*(z)$ is complete in the class $\mathfrak{R}_{\overline{B}_r}$, $r_0 < r < R$. Let the function $f(z)$ be regular on $\overline{B}_r$. Let us expand it into a series of the functions $v_k^+(z)$:

$$f(z) = \sum_{k=0}^{N} a_k^* v_k(z) + \sum_{k=N+1}^{\infty} a_k v_k^*(z)$$

uniformly converging on B_r. The functions $v_k(z)$ are linear combinations of the functions $v_\nu^*(z)$, $\nu = 0, 1, \ldots, k$,

$$v_k(z) = \sum_{\nu=0}^{k} \beta_\nu^{(k)} v_\nu^*(z), \quad k = 0, 1, \ldots,$$

and we obtain the expansion of the function $f(z)$ by the functions $v_k^*(z)$

$$f(z) = \sum_{k=0}^{\infty} a_k v_k^*(z),$$

uniformly converging on $\bar{B}_r$ and the note is proved.

Let us clarify the form of the functions $\lambda_k^*(z)$. We have the expansion

$$\frac{1}{\zeta - z} = \sum_{k=0}^{\infty} \lambda_k(\zeta) v_k(z) = \sum_{k=0}^{\infty} \lambda_k(\zeta) \sum_{\nu=0}^{k} \beta_\nu^{(k)} v_\nu^*(z),$$

converging uniformly on $\bar{B}_r$, $r_0 < r < R$, for $\zeta \in B_r^1$. Multiplying this series $\frac{1}{2\pi i} \lambda_m^*(z)$ and integrating along L_r, we have

$$\lambda_m^*(\zeta) = \sum_{k=m}^{\infty} \beta_m^{(k)} \lambda_k(\zeta), \quad \zeta \in B_{r_0}^1, \quad m = 0, 1, \ldots$$

Analogously we have

$$\lambda_m(\zeta) = \sum_{k=m}^{\infty} \alpha_m^{(k)} \lambda_k^*(\zeta), \quad \zeta \in B_{r_0}^1, \quad m = 0, 1, \ldots$$

Theorem 2 Let $\bar{B}$ be a set of type $\mathfrak{M}$ and the systems of functions $v_k(z)$ and $\lambda_k(z)$ satisfy the conditions A for the domain $B_{r_0, R}$, $\rho \leqslant r_0 < R \leqslant \infty$. Let the functions $\delta_k(\zeta)$, $k = 0, 1, \ldots$, regular in the domain $B_{r_0}^1$ and becoming zero at the point $\zeta = \infty$, be such that

$$\overline{\lim_{k \to \infty}} \sqrt[k]{|\delta_k(\zeta)|} \leqslant \frac{1}{r}$$

uniformly for $\zeta \in L_r$, $r_0 < r < R$, and the function

$$K(z, \zeta) = \sum_{k=0}^{\infty} v_k(z) \delta_k(\zeta)$$

be continuous with respect to z and ζ for $z \in \bar{B}_{r_*}$, $\zeta \in \bar{B}_{r_*}^1$, for some r_*, $r_0 \leqslant r_* \leqslant R$. If $\frac{\lambda}{2\pi i}$ is not a characteristic value of the kernel $K(z, \zeta)$, $z \in L_{r_*}$, $\zeta \in L_{r_*}$, then for the systems of functions

$$\lambda_k^*(z, \lambda) = \lambda_k(z) + \lambda \delta_k(z), \quad k = 0, 1, \ldots,$$

there is a system of functions $v_k^*(z, \lambda)$, regular in the domain B_R, such that the systems of functions $v_k^*(z,\lambda)$ and $\lambda_k^*(z, \lambda)$ satisfy the conditions A in the domain $B_{r_0, R}$.

The proof of this theorem is analogous to the proof of Theorem 1. We can obviously formulate a corollary to this theorem analogous to the corollary to Theorem 1 and also a note relating to it.

Bearing in mind that every function $\lambda(z)$, regular in $B^1_{r_0}$, $\rho \leqslant r_0 < \infty$, and becoming zero at the point $z = \infty$, can be approximated as well as is desired inside $B^1_{r_0}$ by means of generalised polynomials in the form

$$\sum_{\nu=1}^{m} \frac{c_\nu}{(z - z_0)^\nu}$$

where z_0 is an internal point of the set $\bar{B}$, we conclude that there are biorthogonal systems of functions

$$v_k^*(z), \lambda_k^*(z) = \sum_{\nu=1}^{m_k} \frac{c_\nu^{(k)}}{(z - z_0)^\nu}$$

satisfying conditions A in the domain $B_{r_0, R}$.

Let z_0 be an internal point of the set $\bar{B}$ of type $\mathfrak{M}$. Let us denote by l_r, $\rho \leqslant r < \infty$, the image of the contour line L_r when mapped by the function $w = \frac{1}{z - z_0}$; G_r and G_r^1, $\rho \leqslant r < \infty$, be respectively the bounded and unbounded domains with boundary l_r; G and $\bar{G}^1$ be respectively the images of B^1 and $\bar{B}$ for the same mapping; G_{r_1, r_2}, $\rho < r_1 < r_2 < \infty$ be the doubly-connected domain bounded by the contour lines l_{r_1} and l_{r_2}. It is easy to show that if the systems of functions $v_k(z)$ and $\lambda_k(z)$ satisfy the conditions A in the domain $B_{r_0, R}$, $\rho \leqslant r_0 < R \leqslant \infty$, then the systems

$$v_k^*(w) = -\lambda_k\left(z_0 + \frac{1}{w}\right)\frac{1}{w}, \quad \lambda_k^*(w) = v_k\left(z_0 + \frac{1}{w}\right)\frac{1}{w}, \quad k = 0, 1, \ldots, \tag{9}$$

satisfy the conditions A in the domain $G_{r_0, R}$.

Transformation (9) of the biorthogonal systems clearly allows all the previous results of this chapter to be extended to closed sets $\bar{G}^1$ containing the point $z = \infty$ such that the complement G of a set $\bar{G}^1$ is a simply-connected domain. Hence and from the above we conclude, in particular, that for any simply-connected domain G there are systems of polynomials

$$v_k(z) = \sum_{\nu=0}^{m_k} c_\nu^{(k)} z^\nu$$

and of functions $\lambda_k(z)$, $k=0, 1, \ldots$, satisfying conditions A in the domain $G_{\rho,\infty}$. In particular every function which is regular in the domain $G_{r_0, R}$, $\rho \leqslant r_0 < R \leqslant \infty$, can be expanded into a series of functions $v_k(z)$ and $\lambda_k(z)$, converging uniformly on the compact sub-sets of $G_{r_0, R}$. Such an expansion is unique. A function regular in the domain G_r, $\rho \leqslant r < \infty$, can be expanded into a series of polynomials $v_k(z)$ converging uniformly on the compact sub-sets of G_r. With reference to this see also Goluzin [2].

2.4.4 *Transformation of biorthogonal systems of functions*

Let $\overline{B}$ be a set of type $\mathfrak{M}$ of the z plane; $\overline{D}$ be a set of type $\mathfrak{M}$ of the w plane; $w=\varphi(z)$ be a function mapping one-to-one conformally the domain B^1 on to the domain D^1, normalised by the conditions $\varphi(\infty)=\infty$, $\varphi'(\infty)>0$; $z=\psi(w)$ be the inverse function to $w=\varphi(z)$; l_r be the image of the contour line L_r, $\rho<r<\infty$ of the domain B^1; D_r and D_r^1 respectively be the bounded and unbounded domains of the w plane with boundary l_r; D_{r_1, r_2}, $\rho \leqslant r_1 < r_2 \leqslant \infty$ be the doubly-connected domain bounded by l_{r_1} and l_{r_2}.

Let $R(z)$ and $R_*(z)$ be two functions which are regular in the domain B^1 and are such that $R(z)R_*(z) \equiv 1$ in B^1. It is easy to see that if the function $f(z)$ is regular in the domain B_r, $\rho<r<\infty$, then the function $f_*(w)$ defined by

$$f_*(w) = \frac{1}{2\pi i} \int_{l_{r_1}} \frac{f(\psi(t))\,R_*(\psi(t))}{t-w}\,dt, \quad w \in D_{r_1}, \quad \rho < r_1 < r \tag{1}$$

is regular in the domain D_r and

$$f(z) = \frac{1}{2\pi i} \int_{L_{r_1}} \frac{f_*(\varphi(\xi))\,R(\xi)}{\xi - z}\,d\xi, \quad z \in B_{r_1}, \quad \rho < r_1 < r \tag{2}$$

Let the systems of functions $v_k(z)$ and $\lambda_k(z)$ satisfy the conditions A in the domain $B_{r_0, R}$, $\rho \leqslant r_0 < R < \infty$. Let us introduce the systems of functions

$$v_k^*(w)=\frac{1}{2\pi i}\int_{L_{r_1}}\frac{v_k(\psi(t))R_*(\psi(t))}{t-w}\,dt,\quad w\in D_{r_1},\quad r_0<r_1<R,\ k=0,1,\ldots,\tag{3}$$

$$\lambda_k^*(w)=\lambda_k(\psi(w))R(\psi(w))\psi'(w),\quad w\in D_{r_0}^1,\quad k=0,1,\ldots\tag{4}$$

It is easy to show that they satisfy conditions A in the domain $D_{r_0,R}$.

Note that Formulae (3), (4), and (9) in Section 2.4.3 and the converse formulae to these can be used to convert various problems relating to biorthogonal systems of functions to corresponding problems for power series.

SUPPLEMENTARY LITERATURE

In addition to the works given in the text we should like to mention some articles dealing with material close to the contents of the present chapter: Gromov [1], Dragilev [1 to 4], Evgrafov and Solovev [1], Zakharuta [1], Kazmin [1 to 4], Korobeinik [1], Kraplin [1], Lokhin [1], Markushevich [2], Mergelian [6], Khaplanov [1], and Shuvalova [1].

CHAPTER THREE

Quadratic approximation. Functions orthogonal with respect to a domain

Let B be a domain on the complex plane, $\mathscr{L}^2(B)$ be the set of all complex functions, given in B, the real and imaginary parts of which are measurable in B and the square of the modulus $|f(z)|^2$ should be Lebesque integrable on B (for concepts relating to the theory of Lebesque measure and integral see Chapter II of Smirnov [2]). The whole theory of quadratic approximation of functions is based on investigation of the class $\mathscr{L}^2(B)$. In this theory the measure of the quadratic approximation of the function $f(z) \in \mathscr{L}^2(B)$ by the function $g(z) \in \mathscr{L}^2(B)$ is expressed by the integral

$$\delta = \iint_B |f(z) - g(z)|^2 \, d\sigma$$

where $d\sigma$ is the area element.

Note that if the functions $f(z)$ and $g(z)$ belong to $\mathscr{L}^2(B)$ and their values coincide everywhere in B, apart from a set of measure zero, then $\delta = 0$. Therefore two such functions, which are said to be equivalent, are usually identified so that, strictly speaking, the elements of $\mathscr{L}^2(B)$ are not functions but equivalence classes of functions.

In this chapter we shall investigate the class $\mathscr{A}(B)$ of functions $f(z)$, regular in B and such that the integral $\iint_B |f(z)|^2 \, d\sigma$ is finite, and we shall also consider quadratic approximation in the class $\mathcal{A}(B)$. This study was started in 1922 by Bochner [1] and Carleman [1].

In this chapter we shall also consider quadratic approximation in the class $\mathcal{A}(\gamma, B)$ of functions, regular in B and such that the integral $\iint_B \gamma(z)\,|f(z)|^2\,d\sigma$ is finite, where $\gamma(z)$ is a non-negative function measurable in B.

In Section 3.1 we shall state the fundamental facts relating to the theory of orthonormal systems in $\mathcal{A}(B)$. Introducing in $\mathcal{L}^2(B)$ the scalar product by the formula

$$(f, g) = \iint_B f(z)\,\overline{g(z)}\,d\sigma$$

we shall transform $\mathcal{L}^2(B)$ into a Hilbert space (see Chapter V, Sections 120 and 121, of Smirnov [2]). It is remarkable that $\mathcal{A}(B)$ then proves to be a closed sub-space of $\mathcal{L}^2(B)$ and is therefore also a Hilbert space (Smirnov [2] p. 393). The closedness of $\mathcal{A}(B)$ in $\mathcal{L}^2(B)$ follows from the fact that the convergence of the sequence of functions $f_n(z) \in \mathcal{A}(B)$ in the norm of the space $\mathcal{L}^2(B)$ implies the uniform convergence of this series on the compact sub-sets of B (Theorem 2, Section 3.1.1). Note also that in $\mathcal{A}(B)$, as opposed to $\mathcal{L}^2(B)$, equivalent functions are necessarily identical so that the elements of $\mathcal{A}(B)$ are individual regular functions not equivalence classes of functions. Many theorems relating to orthonormal systems in $\mathcal{A}(B)$ (and therefore in $\mathcal{A}(\gamma, B)$ and also in Chapter 4 in class $\mathcal{E}_2$) are simple corollaries of the general theory of orthonormal systems in Hilbert space (see Section 121 of Smirnov [2]). However, bearing in mind the reader who is not familiar with this theory, we intend to state the properties of orthonormal systems in the classes of functions indicated without making use of the concept of Hilbert space.

In this chapter, unless otherwise stated, we shall take a domain to be an open, bounded set of points consisting of a finite number of finitely-connected domains.

In reading this chapter the reader who is not familiar with the Lebesque integral can use the following concept of an integral over a domain. If the function $f(z)$ is continuous in B, $|f(z)|$ is a bounded function and the domain B is bounded and measurable, then the integral $\iint_B f(z)\,d\sigma$ can be taken as the ordinary Riemann integral.

Let us now assume that for continuous $f(z)$ the above conditions for $|f(z)|$ or B are not fulfilled. Let B_n, $n=1, 2, \ldots$, be the sequence of domains exhausting B, e.g. B_n are measurable domains $\overline{B}_n \subset B$, $n=1, 2, \ldots$, and every closed set contained in B belongs to all B_n, starting from some n. Let us assume by definition

$$\iint_B |f(z)|\,d\sigma = \lim_{n\to\infty} \iint_{B_n} |f(z)|\,d\sigma$$

if the limit on the right is finite. It is known that this limit does not depend on the choice of sequence of domains B_n exhausting the domain B (see Goluzin [1] p. 35).

In the case of a measurable domain B and bounded $|f(z)|$ the Riemann integral clearly coincides with the integral just defined.

In this chapter we shall use the following result: let

$$f(z) = c_0 + c_1 z + c_2 z^2 + \ldots$$

be a function, regular and bounded in the disc $|z|<1$ e.g. there is a positive number M such that $|f(z)| \leqslant M$ for $|z|<1$. Let us assume

$$s_k(z) = c_0 + c_1 z + \ldots + c_k z^k$$

and

$$\sigma_n(z) = \frac{s_0(z) + s_1(z) + \ldots + s_k(z)}{n+1}$$

Then $|\sigma_n(z)| \leqslant M$ for $|z| \leqslant 1$ and any n (see Titchmarsh [1] p. 268) and $\sigma_n(z) \rightrightarrows f(z)$ on the compact sub-sets of $|z|<1$.

Let $f(z)$ now be an arbitrary function, regular in the disc $|z|<1$. Let us assume

$$M(r) = \sup_{|z|\leqslant r} |f(z)|, \quad 0<r<1$$

As before let us introduce the sums $\sigma_n(z)$ for the function $f(z)$. Applying the previous result to the function $f(rz)$, $0<r<1$, we have $|\sigma_n(rz)| \leqslant M(r)$ in the disc $|z| \leqslant 1$ and $\sigma_n(rz) \rightrightarrows f(rz)$ as $n\to\infty$ on $|z|=1$. Hence we conclude that

$$|\sigma_n(z)| \leqslant M(|z|), \quad |z|<1$$

and $\sigma_n(z) \rightrightarrows f(z)$ as $n\to\infty$ on the compact sub-sets of the disc $|z|<1$.

3.1 THEORY OF FUNCTIONS ORTHOGONAL WITH RESPECT TO A DOMAIN

3.1.1 Class $\mathcal{A}$ and convergence in the mean

We shall consider basically the class $\mathcal{A}(B)$ (or simply $\mathcal{A}$) of functions regular in the domain B such that

$$\iint_B |f(z)|^2 \, d\sigma < +\infty$$

Let us note certain properties of functions of class $\mathcal{A}$. Clearly if $f(z) \in \mathcal{A}$ and c is some constant, then $cf(z) \in \mathcal{A}$. If $f_1(z) \in \mathcal{A}$ and $f_2(z) \in \mathcal{A}$ then from the inequality

$$|f_1(z) + f_2(z)|^2 \leqslant (|f_1(z)| + |f_2(z)|)^2 \leqslant 2(|f_1(z)|^2 + |f_2(z)|^2)$$

it follows that also $f_1(z) + f_2(z) \in \mathcal{A}$.

From these properties it follows that if $f_k(z) \in \mathcal{A}$, $k = 1, 2, \ldots, m$, and c_k are arbitrary constants, then the sum

$$c_1 f_1(z) + c_2 f_2(z) + \ldots + c_m f_m(z)$$

belongs to $\mathcal{A}$. Further if $f_1(z) \in \mathcal{A}$ and $f_2(z) \in \mathcal{A}$, then from the inequality

$$|f_1(z) f_2(z)| \leqslant \frac{1}{2} (|f_1(z)|^2 + |f_2(z)|^2)$$

it follows that the product $f_1(z) f_2(z)$ is integrable over the domain B in the above sense. The same is obviously also true for the product $f_1(z) f_2(z)$. The following also holds as shown in the lemma.

Lemma 1 For functions $f(z) \in \mathcal{A}$ the inequality

$$\iint_B |f(z)|^2 \, d\sigma \geqslant \pi \rho_{z_0}^2 \, |f(z_0)|^2 \tag{1}$$

holds, where z_0 is any point from B and ρ_{z_0} is the distance from z_0 to the boundary of B.

PROOF Let us write the expansion of the function $f(z)$ by powers of $z - z_0$

$$f(z) = \sum_{k=0}^{\infty} a_k (z - z_0)^k, \quad a_0 = f(z_0)$$

This series converges absolutely and uniformly in the disc $|z-z_0| \leqslant \rho$ where $0 < \rho < \rho_{z_0}$. Putting here $z = z_0 + re^{i\varphi}$ we have

$$\iint_{|z-z_0|\leqslant\rho} |f(z)|^2 d\sigma = \int_0^\rho \int_0^{2\pi} \left|\sum_{k=0}^{\infty} a_k r^k e^{ik\varphi}\right|^2 d\varphi\, r dr = 2\pi \sum_{k=0}^{\infty} \frac{|a_k|^2}{2(k+1)} \rho^{2(k+1)}$$

Hence

$$\iint_B |f(z)|^2 d\sigma \geqslant \iint_{|z-z_0|\leqslant\rho} |f(z)|^2 d\sigma \geqslant \pi |a_0|^2 \rho^2, \quad a_0 = f(z_0)$$

If ρ then tends to ρ_{z_0}, in the limit we have inequality (1).

DEFINITION The sequence of functions $\psi_k(z) \in \mathcal{A}$ is said to converge in the mean to the function $\psi(z) \in \mathcal{A}$ if

$$\lim_{k\to\infty} \iint_B |\psi_k(z) - \psi(z)|^2 d\sigma = 0$$

It is not difficult to prove that the limiting function $\psi(z)$ is unique. Let us assume that besides $\psi(z)$ there is a limiting function $\varphi(z) \in \mathcal{A}$ and prove that $\psi(z) \equiv \varphi(z)$ in B. We have

$$|\psi(z) - \varphi(z)| \leqslant |\psi(z) - \psi_k(z)| + |\varphi(z) - \psi_k(z)|$$

and

$$\iint_B |\psi(z) - \varphi(z)|^2 d\sigma$$

$$\leqslant 2 \iint_B |\psi(z) - \psi_k(z)|^2 d\sigma + 2 \iint_B |\varphi(z) - \psi_k(z)|^2 d\sigma$$

The right-hand side of this inequality tends to zero as $k \to \infty$. Therefore

$$\iint_B |\psi(z) - \varphi(z)|^2 d\sigma = 0$$

Hence it follows that $\psi(z) \equiv \varphi(z)$ in B.

Theorem 1 If the sequence of functions $\psi_k(z) \in \mathcal{A}$ converges in the mean to the function $\psi(z) \in \mathcal{A}$ then this sequence converges uniformly on the compact sub-sets of B to the function $\psi(z)$

In fact for any point z_0 of the closed set $E \subset B$ we have

$$\pi\rho_E^2 |\psi_k(z_0) - \psi(z_0)|^2 \leqslant \iint_B |\psi_k(z) - \psi(z)|^2 d\sigma$$

where ρ_E is the distance of the set E to the contour of the domain B. The theorem follows directly from the inequality obtained.

DEFINITION The sequence of functions $\psi_k(z) \in \mathcal{A}$ is said to be a fundamental sequence in the mean if

$$\iint_B |\psi_k(z) - \psi_l(z)|^2 \, d\sigma \to 0 \quad \text{as} \quad k, l \to \infty \tag{2}$$

Then as before we have

$$|\psi_k(z) - \psi_l(z)| \rightrightarrows 0 \quad \text{as} \quad k, l \to \infty$$

on the compact sub-sets of B. At the same time the sequence $\psi_k(z)$ converges to some limiting function $\psi(z)$ uniformly on the compact sub-sets of B, and $\psi(z)$ is regular in B. By Fatou's theorem (see Smirnov [2] p. 165) it follows from (2) that for any $\varepsilon > 0$ there is an m such that

$$\iint_B |\psi(z) - \psi_l(z)|^2 \, d\sigma \leqslant \varepsilon \quad \text{as} \quad l > m, \tag{3}$$

and this means that the function $\psi(z) - \psi_l(z) \in \mathcal{A}$. But then $\psi(z) \in \mathcal{A}$ also. From (3) it therefore follows that $\psi_k(z)$ converges in the mean to $\psi(z)$, e.g. we have:

Theorem 2 If $\psi_k(z)$ is a fundamental sequence in the mean, then this sequence converges uniformly on the compact sub-sets of B and in the mean to some function $\psi(z) \in \mathcal{A}$.

Let us introduce another class $\mathcal{A}_1(B)$ ($\mathcal{A}_1$ for short) of functions $f(z)$ regular in the domain B and such that

$$\iint_B |f(z)| \, d\sigma < +\infty$$

If $f(z) \in \mathcal{A}_1$ and $\varphi(z)$ is any function regular and bounded in the domain B, then $f(z)\varphi(z) \in \mathcal{A}_1$. Any finite linear combination

$$\sum_{k=1}^{m} c_k f_k(z)$$

where c_k are constants and $f_k(z) \in \mathcal{A}_1$ thus belongs to $\mathcal{A}_1$. If $f(z) \in \mathcal{A}$ (domain B is bounded), then $f(z) \in \mathcal{A}_1$. This follows from Schwartz's inequality:

$$\iint_B |f(z)| \, d\sigma \leqslant \left(\iint_B |f(z)|^2 \, d\sigma \cdot \iint_B d\sigma \right)^{\frac{1}{2}} < +\infty$$

3.1.2 *Orthogonal functions - Bessel inequality*

The theory of quadratic approximation with respect to a domain is closely connected with the theory of functions which are orthogonal with respect to a domain. Let us introduce the fundamental concepts relating to this theory.

DEFINITION The sequence of functions

$$\varphi_0(z),\ \varphi_1(z),\ \dots \tag{1}$$

from $\mathcal{A}$ is said to be an orthogonal sequence with respect to the domain B if

$$\iint_B \varphi_n(z)\overline{\varphi_m(z)}\,d\sigma = 0 \quad \text{for} \quad n \neq m$$

In the following we shall assume that the sequence (1) is infinite. We assume, of course, that amongst the functions (1) there are no functions identically equal to zero. If we multiply each of the functions (1) by some complex number (differing from zero), then we again have a sequence which is orthogonal with respect to B. These multipliers can, for example, be chosen such that the integral of the square of the modulus of each of the functions (1) will be equal to one and, retaining the same notation for the functions, we have

$$\iint_B \varphi_n(z)\overline{\varphi_m(z)}\,d\sigma = \delta_{n,m} \tag{2}$$

where $\delta_{n,m} = 0$ for $n \neq m$ and $\delta_{n,n} = 1$. Such a system of functions is said to be an orthonormalised system. Note that conditions (2) are maintained if each of the functions $\varphi_k(z)$ are multiplied by the quantity $e^{i\omega_k}$ where ω_k is any real number. We shall assume that the system (1) is orthonormalised. If the function $f(z)$ is any function from the class $\mathcal{A}$, then the quantity

$$a_k = \iint_B f(z)\overline{\varphi_k(z)}\,d\sigma \tag{3}$$

is called the Fourier coefficient of the function $f(z)$ with respect to the orthonormalised system (1). The function $f(z)$ compares with its Fourier series

$$f(z) \sim \sum_{k=0}^{\infty} a_k\varphi_k(z) \tag{4}$$

For the moment we know nothing about the convergence of this series, and if it does converge, we cannot say that its sum is equal to $f(z)$.

Let us form a linear combination of the first $n+1$ of the functions (1)

$$\sum_{k=0}^{n} d_k \varphi_k(z)$$

where d_k are arbitrary coefficients, and let us investigate the square deviation of this sum from $f(z)$

$$I_n = \iint_B \left| f(z) - \sum_{k=0}^{n} d_k \varphi_k(z) \right|^2 d\sigma$$

$$= \iint_B |f(z)|^2 d\sigma - \sum_{k=0}^{n} d_k \iint_B \overline{f(z)} \varphi_k(z)\, d\sigma$$

$$- \sum_{k=0}^{n} \overline{d_k} \iint_B f(z) \overline{\varphi_k(z)}\, d\sigma + \sum_{k=0}^{n} |d_k|^2$$

$$= \iint_B |f(z)|^2 d\sigma - \sum_{k=0}^{n} d_k \overline{a_k} - \sum_{k=0}^{n} \overline{d_k} a_k$$

$$+ \sum_{k=0}^{n} |d_k|^2 + \sum_{k=0}^{n} |a_k|^2 - \sum_{k=0}^{n} |a_k|^2$$

$$= \iint_B |f(z)|^2 d\sigma - \sum_{k=0}^{n} |a_k|^2 + \sum_{k=0}^{n} |a_k - d_k|^2$$

Thus

$$I_n = \iint_B \left| f(z) - \sum_{k=0}^{n} d_k \varphi_k(z) \right|^2 d\sigma$$

$$= \iint_B |f(z)|^2 d\sigma - \sum_{k=0}^{n} |a_k|^2 + \sum_{k=0}^{n} |a_k - d_k|^2$$

Hence it follows that the integral in question reaches a minimum for $d_k = a_k$ $(k = 1, 2, \ldots, n)$. We then have

$$\iint_B \left| f(z) - \sum_{k=0}^{n} a_k \varphi_k(z) \right|^2 d\sigma = \iint_B |f(z)|^2 d\sigma - \sum_{k=0}^{n} |a_k|^2 \tag{5}$$

Since the left-hand side is non-negative, we therefore find that

$$\sum_{k=0}^{n} |a_k|^2 \leqslant \iint_B |f(z)|^2 \, d\sigma$$

This inequality is valid for any n, and consequently the series with terms $|a_k|^2$ converges and

$$\sum_{k=0}^{\infty} |a_k|^2 \leqslant \iint_B |f(z)|^2 \, d\sigma \tag{6}$$

This is the so-called Bessel inequality. We have proved the following theorem:

Theorem 1 Amongst the generalised polynomials

$$P_n(z) = \sum_{k=0}^{n} d_k \varphi_k(z)$$

the best square approximation to the function $f(z)$ in the domain B is given by the polynomial with coefficients d_k which are Fourier coefficients of the function $f(z)$ $(d_k = a_k)$ and the measure of approximation will then be

$$\iint_B \left| f(z) - \sum_{k=0}^{n} a_k \varphi_k(z) \right|^2 d\sigma = \iint_B |f(z)|^2 \, d\sigma - \sum_{k=0}^{n} |a_k|^2 \tag{7}$$

Let us assume that the equal sign holds in Formula (6), e.g.

$$\iint_B |f(z)|^2 \, d\sigma = \sum_{k=0}^{\infty} |a_k|^2 \tag{8}$$

This equation is called Parseval's equation and Parseval's equation is said to hold for the function $f(z)$ with respect to the system (1).

From (7) it follows directly that Parseval's equation (8) is equivalent to the fact that the sums of the Fourier series of function $f(z)$

$$S_n(z) = \sum_{k=0}^{n} a_k \varphi_k(z)$$

converge in the mean to $f(z)$ and by virtue of Theorem 1, Section 3.1.1, we have the following theorem.

Theorem 2 If Parseval's equation holds for the function $f(z) \in \mathcal{A}$ then the Fourier series of the function $f(z)$ converges uniformly in B to $f(z)$.

Let us now consider a series of the form

$$\sum_{k=0}^{\infty} d_k \varphi_k(z) \tag{9}$$

where d_k are given complex numbers, and let us introduce the partial sum of this series

$$\sigma_n(z) = \sum_{k=0}^{n} d_k \varphi_k(z)$$

From the formula

$$\iint_B |\sigma_{n+p}(z) - \sigma_n(z)|^2 \, d\sigma = \sum_{k=n+1}^{n+p} |d_k|^2$$

it follows directly that the sequence $\sigma_n(z)$ (or series (9)) is a fundamental sequence in the mean if and only if the series

$$\sum_{k=0}^{\infty} |d_k|^2 \tag{10}$$

converges. If this is the case then by virtue of Theorem 2, Section 3.1.1, as $n \to \infty$ the partial sum $\sigma_n(z)$ tends to some function $\omega(z) \in \mathcal{A}$ uniformly on the compact sub-sets of B and in the mean. Denoting the Fourier coefficients of $\omega(z)$ by a_k, we have

$$\iint_B |\omega(z) - \sigma_n(z)|^2 \, d\sigma = \iint_B |\omega(z)|^2 \, d\sigma - \sum_{k=0}^{n} |a_k|^2 + \sum_{k=0}^{n} |d_k - a_k|^2$$

and from the fact that the left-hand side tends to zero as $n \to \infty$ it follows that $d_k = a_k$ for all k and Parseval's equation holds for $\omega(z)$. In this way we obtain the following theorem:

Theorem 3 If the series (10) converges, e.g. the series (9) is fundamental, then the following applies:

(1) series (9) converges uniformly on the compact sub-sets of B to some functions $\omega(z) \in \mathcal{A}$;

(2) d_k are the Fourier coefficients for $\omega(z)$;

(3) Parseval's equation holds for the function $\omega(z)$.

NOTE If the function $f(z)$ does not belong to the class $\mathcal{A}$ and the functions $\varphi_k(z)$ of the orthonormal system are such that all the integrals (3) defining the Fourier coefficients of the function $f(z)$ exist, then we can compare function $f(z)$ with its Fourier series in this case also.

3.1.3 Completeness and closure

If a_k are the Fourier coefficients of some function $f(z) \in \mathcal{A}$ then the series with the general term $|a_k|^2$ converges, and to the Fourier series

$$\sum_{k=0}^{\infty} a_k \varphi_k(z) \tag{1}$$

one can apply the Theorem 3, Section 3.1.2, just proved. But its sum $\omega(z)$ may differ from $f(z)$. As proved $\omega(z) \in \mathcal{A}$ and the series (1) is the Fourier series of $\omega(z)$, e.g. $\omega(z)$ has the same Fourier coefficients as the function $f(z)$. Hence it follows directly that the difference $f(z) - \omega(z)$ has Fourier coefficients which are all equal to zero.

$$\iint_B [f(z) - \omega(z)]\,\overline{\varphi_k(z)}\,d\sigma = 0, \quad k = 0, 1, \ldots \tag{2}$$

for example, the difference $f(z) - \omega(z)$ belonging to class $\mathcal{A}$ is orthogonal to all $\varphi_k(z)$. If $\omega(z)$ coincides with $f(z)$ e.g. $f(z) - \omega(z) \equiv 0$, equations (2) are trivial. If $f(z) - \omega(z) \not\equiv 0$ the function $f(z) - \omega(z)$ from $\mathcal{A}$ not identically equal to zero is orthogonal to all $\varphi_k(z)$.

DEFINITION The orthonormal system $\varphi_k(z)$ is said to be complete in the class $\mathcal{A}$ if there is no function of $\mathcal{A}$ differing from identically zero which is orthogonal to all functions $\varphi_k(z)$.

From the above it follows that if the system $\varphi_k(z)$ is complete, for any chosen $f(z)$ from $\mathcal{A}$ the function $\omega(z)$ mentioned above coincides with $f(z)$.

Bearing in mind the properties of $\omega(z)$ proved above, we obtain the following theorem:

Theorem 1 For every complete system $\varphi_k(z)$ the Fourier series of any function $f(z) \in \mathcal{A}$ converges uniformly on the compact sub-sets of B to $f(z)$ and Parseval's equation holds for $f(z)$.

COROLLARY The system of functions $\varphi_k(z)$ cannot be complete in the class $\mathcal{A}$ if all functions of the system become zero at some point in the domain B.

NOTE 1 From Theorem 3, Section 3.1.2, and the Bessel inequality it follows that the series (1) will be the Fourier series of some function of $\mathcal{A}$ if, and only if, the series with the general term $|a_k|^2$ converges. If the system is complete, different functions from $\mathcal{A}$ correspond to different Fourier series, e.g. to Fourier series with different coefficients a_k. Thus if the system $\varphi_k(z)$ is complete, there is a one-to-one correspondence between the functions $f(z) \in \mathcal{A}$ and those sequences of complex numbers a_k, $k = 0, 1, \ldots$, for which the series with the general term $|a_k|^2$ converge. These quantities a_k are Fourier coefficients of the function $f(z)$ with respect to the system $\varphi_k(z)$.

The concept of the closedness of a system is often introduced side by side with the concept of completeness.

DEFINITION The orthonormalised system $\varphi_k(z)$ is said to be closed in the class $\mathcal{A}$ if Parseval's equation holds for any function from $\mathcal{A}$.

Theorem 2 The concepts of completeness and closedness are equivalent in the class $\mathcal{A}$.

PROOF Let us assume that the system $\varphi_k(z)$ is complete. From Theorem 1 it follows that Parseval's equation then holds for every function from $\mathcal{A}$. Let us now assume that the system is closed and prove that it is complete, e.g. if $\psi(z) \in \mathcal{A}$ and is orthogonal to all $\varphi_k(z)$

$$\iint_B \psi(z)\overline{\varphi_k(z)}\, d\sigma = 0, \quad k = 0, 1, \ldots,$$

then it is identically equal to zero. From this it follows that all Fourier coefficients for $\psi(z)$ are zero. But since it was assumed that the system is closed, Parseval's equation must hold for $\psi(z)$ e.g.

$$\iint_B |\psi(z)|^2\, d\sigma = 0$$

and by virtue of the continuity of $|\psi(z)|^2$ the function $\psi(z) \equiv 0$ in B. The theorem is proved.

NOTE 2 We have seen that Parseval's equation for $f(z)$ is equivalent to saying that the partial sums of the Fourier series of the function $f(z)$

$$S_n(z) = \sum_{k=0}^{n} a_k \varphi_k(z)$$

converge in the mean to $f(z)$. On the other hand, we know that if instead of $S_n(z)$ we choose any linear combination of the functions $\varphi_k(z)$, $k=0, 1, \ldots, n$

$$\sigma_n(z)=\sum_{k=0}^{n} d_k\varphi_k(z) \tag{3}$$

then the mean square deviation of $\sigma_n(z)$ from $f(z)$ will be a minimum for $d_k=a_k$, $k=0, 1, \ldots, n$. Hence it follows that the completeness and closedness of the system are equivalent to the following property: for any function $f(z)\in\mathcal{A}$ and for any given $\varepsilon>0$ there is a generalised polynomial (3) such that

$$\iint_B |f(z)-\sigma_n(z)|^2\, d\sigma \leqslant \varepsilon$$

NOTE 3 Let $f_1(z)$ and $f_2(z)$ be two functions of the class $\mathcal{A}$ and let Parseval's equation hold for each, e.g. with respect to any $\varepsilon>0$ there are generalised polynomials $\sigma_{n,1}(z)$ and $\sigma_{n,2}(z)$ of form (3) such that

$$\iint_B |f_1(z)-\sigma_{n,1}(z)|^2\, d\sigma<\varepsilon, \quad \iint_B |f_2(z)-\sigma_{n,2}(z)|^2\, d\sigma<\varepsilon$$

If C_1 and C_2 are constants, then

$$\iint_B |(C_1f_1(z)+C_2f_2(z))-(C_1\sigma_{n,1}(z)+C_2\sigma_{n,2}(z))|^2\, d\sigma$$

$$\leqslant 2|C_1|^2\iint_B |f_1(z)-\sigma_{n,1}(z)|^2\, d\sigma+2|C_2|^2\iint_B |f_2(z)-\sigma_{n,2}(z)|^2\, d\sigma$$

$$\leqslant 2(|C_1|^2+|C_2|^2)\varepsilon$$

and in view of the arbitrary nature of ε Parseval's equation will hold for the function $C_1f_1(z)+C_2f_2(z)$. Clearly this assertion extends to any finite number of functions from class.

NOTE 4 It is not difficult to see that if $a_k^{(1)}$ and $a_k^{(2)}$ are Fourier coefficients of the functions $f_1(z)$ and $f_2(z)$ then $C_1a_k^{(1)}+C_2a_k^{(2)}$ are Fourier coefficients of the function $C_1f_1(z)+C_2f_2(z)$ and for this function Parseval's equation is of the form

$$\iint_B |C_1f_1(z)+C_2f_2(z)|^2\, d\sigma=\sum_{k=0}^{\infty}|C_1a_k^{(1)}+C_2a_k^{(2)}|^2$$

NOTE 5 Let Parseval's equation

$$\iint_B |f_1(z)|^2\, d\sigma=\sum_{k=0}^{\infty}|a_k^{(1)}|^2, \quad \iint_B |f_2(z)|^2\, d\sigma=\sum_{k=0}^{\infty}|a_k^{(2)}|^2 \tag{4}$$

hold for functions $f_1(z)$ and $f_2(z)$ from A, where $a_k^{(1)}$ and $a_k^{(2)}$ are the Fourier coefficients of $f_1(z)$ and $f_2(z)$. Let us show the following equation holds

$$\iint_B f_1(z)\overline{f_2(z)}\,d\sigma = \sum_{k=0}^{\infty} a_k^{(1)}\overline{a_k^{(2)}} \tag{5}$$

this being called the generalised Parseval's equation.

If $f_2(z)\equiv f_1(z)$ the latter equation becomes Parseval's equation. Note that from the inequality $|xy|\leqslant\frac{1}{2}(|x|^2+|y|^2)$ it follows that the series (5) converges absolutely. As we indicated in Note 2, Parseval's equation holds for $f_1(z)+f_2(z)$ and $f_1(z)+if_2(z)$, e.g.

$$\iint_B |f_1(z)+f_2(z)|^2\,d\sigma = \sum_{k=1}^{\infty} |a_k^{(1)}+a_k^{(2)}|^2$$

$$\iint_B |f_1(z)+if_2(z)|^2\,d\sigma = \sum_{k=1}^{\infty} |a_k^{(1)}+ia_k^{(2)}|^2$$

Bearing in mind that the square of the modulus of a complex number is equal to the product of this number and its complex conjugate, and taking (4) into consideration, we have

$$\operatorname{Re}\iint_B f_1(z)\overline{f_2(z)}\,d\sigma = \operatorname{Re}\sum_{k=1}^{\infty} a_k^{(1)}\overline{a_k^{(2)}}$$

$$\operatorname{Im}\iint_B f_1(z)\overline{f_2(z)}\,d\sigma = \operatorname{Im}\sum_{k=1}^{\infty} a_k^{(1)}\overline{a_k^{(2)}}$$

whence (5) follows.

NOTE 6 We have seen (Theorem 3, Section 3.1.2) that the Fourier series of any function $f(z)\in\mathcal{A}$ converges uniformly on the compact sub-sets of B. In the following we shall see that there are series (1) which converge uniformly on the compact sub-sets of B but are not Fourier series of a function from $\mathcal{A}$.

Note finally, without going into the proof, that for every domain B there are orthonormal systems of functions which are complete in class $\mathcal{A}(B)$ (see Bochner [1]).

3.1.4 Example

Let the domain B be the disc $|z|<1$. The system $\psi_k(z)=z^k$, $k=0, 1, 2, \ldots$, is orthogonal in it:

$$\iint_{|z|<1} \varphi_k(z)\overline{\varphi_l(z)}\, d\sigma = \iint_{|z|<1} z^k \bar{z}^l\, d\sigma = \int_0^1\int_0^{2\pi} r^{k+l} e^{i\varphi(k-l)}\, r\, dr\, d\varphi = 0$$

But this system is not orthonormalised since

$$\iint_{|z|<1} |\varphi_k(z)|^2\, d\sigma = \iint_{|z|<1} |z|^{2k}\, ds = \int_0^1\int_0^{2\pi} r^{2k}\, d\varphi r\, dr = \frac{\pi}{k+1}$$

Hence it follows that the system of functions

$$\varphi_k(z) = \sqrt{\frac{k+1}{\pi}}\, z^k, \quad k=0, 1, 2, \ldots$$

is an orthonormalised system. Let $f(z)$ be an arbitrary function from $\mathcal{A}$ in the circle $|z|<1$. This function can be expanded in a Maclaurin series

$$f(z) = \sum_{k=0}^{\infty} c_k z^k$$

Let us find the Fourier coefficients a_k of this function by using the fact that the integrals for calculating a_k do not depend on the choice of sequence B_n of domains exhausting the domain B $(|z|<1)$

$$\begin{aligned} a_k &= \iint_{|z|<1} f(z)\overline{\varphi_k(z)}\, d\sigma = \iint_{|z|<1} \sum_{l=0}^{\infty} c_l z^l \sqrt{\frac{k+1}{\pi}}\, \bar{z}^k\, d\sigma \\ &= \lim_{\rho\to 1-0} \iint_{|z|<\rho} \sum_{l=0}^{\infty} c_l z^l \sqrt{\frac{k+1}{\pi}}\, \bar{z}^k\, d\sigma \qquad (1) \\ &= \lim_{\rho\to 1-0} \sum_{l=0}^{\infty} c_l \sqrt{\frac{k+1}{\pi}} \iint_{|z|<\rho} z^l \bar{z}^k\, d\sigma = \sqrt{\frac{\pi}{k+1}}\, c_k \end{aligned}$$

This means the Fourier series of the function $f(z)$ with respect to $\varphi_k(z)=\sqrt{\frac{k+1}{\pi}}\, z^k$ will be

$$\sum_{k=0}^{\infty} \sqrt{\frac{\pi}{k+1}}\, c_k \sqrt{\frac{k+1}{\pi}}\, z^k = \sum_{k=0}^{\infty} c_k z^k$$

This series coincides with the Maclaurin series, e.g. the Fourier series for $f(z)$ converges, and does so uniformly, on the compact sub-sets of $|z|<1$ to $f(z)$. Hence it follows that every function $f(z)\in \mathcal{A}$ in the disc $|z|<1$ can be expanded into a Fourier series with respect to the orthonormalised functions (polynomials) $\sqrt{\frac{k+1}{\pi}}z^k$, $k=0, 1, \ldots$ and this series coincides with the Maclaurin series. By the same token the system of polynomials $\sqrt{\frac{k+1}{\pi}}\ z^k$ orthonormalised in the disc is closed (complete) in class $\mathcal{A}$ and Parseval's equation

$$\iint\limits_{|z|<1} |f(z)|^2\,d\sigma=\sum_{k=0}^{\infty}|a_k|^2=\pi\sum_{k=0}^{\infty}\frac{|c_k|^2}{k+1}<+\infty \tag{2}$$

holds. Conversely, if the series $\sum\limits_{k=0}^{\infty}\frac{|c_k|^2}{k+1}$ converges, then the function $f(z)=\sum\limits_{k=0}^{\infty}c_k z^k\in\mathcal{A}$ since in this case the series

$$\sum_{k=0}^{\infty}\sqrt{\frac{\pi}{k+1}}\,c_k\sqrt{\frac{k+1}{\pi}}\,z^k$$

is the Fourier series of the function $f(z)$. In the case of the disc $|z|<1$ every function, regular in this disc, and not only functions from $\mathcal{A}$, can be expanded into a series with respect to orthonormal system of polynomials $\sqrt{\frac{k+1}{\pi}}z^k$, e.g. with respect to powers of z.

In the following we shall prove an analogous property for a certain class of domains B (see Theorem 2, Section 3.2.4).

3.1.5 *Fundamental bi-linear series*

Let $\varphi_k(z)$, $k=0, 1, \ldots$ be an orthonormal system of functions from $\mathcal{A}$ for the domain B. Let us consider the sum

$$K_n(z, \zeta)=\sum_{k=0}^{n}\overline{\varphi_k(\zeta)}\,\varphi_k(z), \quad \zeta\in B,\ z\in B \tag{1}$$

We can show that as a function of z, $K_n(z, \zeta)$ belongs to $\mathcal{A}$ and

$$\iint\limits_{B}|K_n(z, \zeta)|^2\,d\sigma_z=\sum_{k=0}^{n}|\varphi_k(\zeta)|^2=K_n(\zeta, \zeta)\geqslant 0 \tag{2}$$

By using inequality (1), Section 3.1.1, we have

$$K_n(\zeta, \zeta) = \sum_{k=0}^{n} |\varphi_k(\zeta)|^2 \leqslant \frac{1}{\pi\rho_\zeta^2}$$

where ρ_ζ is the distance from the point ζ to the boundary of the domain B, and consequently at any point $\zeta \in B$ the series

$$\sum_{k=0}^{\infty} |\varphi_k(\zeta)|^2 \leqslant \frac{1}{\pi\rho_\zeta^2}$$

converges.

Let E be any compact sub-set of B. Denoting the distance of E to the boundary of B by ρ_E, we have

$$\sum_{k=0}^{\infty} |\varphi_k(\zeta)|^2 \leqslant \frac{1}{\pi\rho_E^2}, \quad \zeta \in E \tag{3}$$

Applying Theorem 3, Section 3.1.2, for $d_k = \overline{\varphi_k(\zeta)}$ we obtain the following theorem.

Theorem 1 For any fixed ζ from B, the bi-linear series

$$\sum_{k=0}^{\infty} \overline{\varphi_k(\zeta)}\,\varphi_k(z)$$

converges uniformly on the compact sub-sets of B to some regular function $K(z, \zeta)$

$$K(z, \zeta) = \sum_{k=0}^{\infty} \overline{\varphi_k(\zeta)}\,\varphi_k(z)$$

$\overline{\varphi_k(\zeta)}$ are Fourier coefficients of the function $K(z, \zeta)$ and Parseval's equation

$$\iint_B |K(z, \zeta)|^2 \, d\sigma = \sum_{k=0}^{\infty} |\varphi_k(\zeta)|^2 = K(\zeta, \zeta) \tag{4}$$

holds.

Let us assume that B is a simply-connected domain and establish the connection between the bi-linear series and the function $w = \varphi(z, z_0)$, $\varphi(z_0, z_0) = 0$, $\psi'(0, z_0) > 0$ where z_0 is a fixed point from B, mapping one-to-one conformally the domain B on to the disc $|w| < 1$. Note that the function $\varphi_z'(z, z_0) \in \mathcal{A}$ since the integral

$$\iint_B |\varphi_z'(z, z_0)|^2 \, d\sigma_z$$

is the area of the image when the domain B is mapped by the function $w=\varphi(z, z_0)$ and is consequently equal to π.

Theorem 2 If $\varphi_k(z)$, $k=0, 1, \ldots$, is an arbitrary, orthonormal system of functions from $\mathcal{A}$ in the simply-connected domain B, then

$$K(z, z_0)=\frac{1}{\pi}\varphi_z'(z_0, z_0)\varphi_z'(z, z_0); \quad \varphi_z'(z, z_0)=\sqrt{\frac{\pi}{K(z_0, z_0)}}K(z, z_0) \tag{5}$$

PROOF Let us denote by $z=\psi(w, z_0)$ the inverse function to $w=\varphi(z, z_0)$. Let us determine the Fourier coefficients of the function $\varphi_z'(z, z_0)\in\mathcal{A}$

$$a_k=\iint_B \varphi_z'(z, z_0)\overline{\varphi_k(z)}\,d\sigma_z$$

or

$$\overline{a_k}=\iint_B \varphi_k(z)\overline{\varphi_z'(z, z_0)}\,d\sigma_z$$

If B_r, $0<r<1$ is the image of the disc $|w|<r$ when mapped by the function $z=\psi(w, z_0)$, then

$$\overline{a_k}=\lim_{r\to 1-0}\iint_{B_r}\frac{\varphi_k(z)}{\varphi_z'(z, z_0)}|\varphi_z'(z, z_0)|^2\,d\sigma_z$$

Introducing the variable $w=\varphi(z, z_0)$ we have

$$\overline{a_k}=\lim_{r\to 1-0}\iint_{|w|<r}\frac{\varphi_k(\psi(w, z_0))}{\varphi_z'(\psi(w, z_0), z_0)}\,d\sigma_w$$

Under the integral sign is a function which is regular in the disc $|w|<1$. If its power series expansion is of the form $\sum_{k=0}^{\infty} b_k w^k$ then putting $w=\rho e^{i\theta}$ we have

$$\overline{a_k}=\lim_{r\to 1-0}\int_0^r\int_0^{2\pi}\sum_{k=0}^{\infty} b_k\rho^k e^{i\theta k}\,d\theta\,\rho\,d\rho=\frac{1}{2}2\pi b_0=\pi b_0$$

and b_0 is equal to the value of the function under the integral sign for $w=0$

$$b_0=\frac{\varphi_k(z_0)}{\varphi_z'(z_0, z_0)}$$

e.g.

$$a_k = \pi \frac{\overline{\varphi_k(z_0)}}{\varphi_z'(z_0, z_0)}$$

and

$$\varphi_z'(z, z_0) = \frac{\pi}{\varphi_z'(z_0, z_0)} \sum_{k=0}^{\infty} \overline{\varphi_k(z_0)}\varphi_k(z) = \frac{\pi}{\varphi_z'(z_0, z_0)} K(z, z_0)$$

Hence and from the formula

$$K(z_0, z_0) = \frac{1}{\pi}[\varphi_z'(z_0, z_0)]^2$$

which is obtained from the above for $z = z_0$ we obtain (5). The theorem is proved.

COROLLARY The function $K(z, \zeta)$ does not depend on the choice of the closed, orthonormal system $\varphi_k(z)$.

We have proved this corollary for simply-connected domains but it also holds for multiply-connected domains (see the appendix of Shiffer to the book by Courant [1]).

Theorem 3 If $\varphi_k(z)$, $k = 0, 1, \ldots,$ is an arbitrary closed, orthonormal system of functions from $\mathcal{A}$ in the domain B and $f(z) \in \mathcal{A}$ then for any z_0 from B

$$f(z_0) = \frac{1}{\pi} \int_B \int \overline{K(z, z_0)} f(z)\, d\sigma, \quad z_0 \in B \tag{6}$$

PROOF The integral in question exists since $K(z, z_0)$ and $f(z) \in \mathcal{A}$. Let us write it in the form

$$\frac{1}{\pi} \int_B \int \overline{K(z, z_0)} f(z)\, d\sigma = \frac{1}{\pi} \int_B \int \overline{K_n(z, z_0)} f(z)\, d\sigma$$

$$+ \frac{1}{\pi} \int_B \int [\overline{K(z, z_0)} - \overline{K_n(z, z_0)}] f(z)\, d\sigma$$

The first integral on the right is equal to $\sum_{k=0}^{n} a_k \varphi_k(z_0)$ where a_k are the Fourier coefficients of the function $f(z)$. As $n \to \infty$ this integral tends to $f(z_0)$ due to the completeness of the system of functions $\varphi_k(z)$. If we show that the second integral on the right tends to zero as $n \to \infty$ the theorem will be proved. By using Schwarz's inequality and Theorem

3, Section 3.1.2, we have

$$\left|\iint_B [\overline{K(z, z_0)} - \overline{K_n(z, z_0)}]\, f(z)\, d\sigma_z\right|$$

$$\leqslant \iint_B \left|\sum_{k=n+1}^{\infty} \varphi_k(z_0)\,\overline{\varphi_k(z)}\right| \cdot |f(z)|\, d\sigma_z$$

$$\leqslant \sqrt{\iint_B \left|\sum_{k=n+1}^{\infty} \varphi_k(z_0)\,\overline{\varphi_k(z)}\right|^2 d\sigma_z \cdot \iint_B |f(z)|^2\, d\sigma}$$

$$= \sqrt{\sum_{k=n+1}^{\infty} |\varphi_k(z_0)|^2 \iint_B |f(z)|^2\, d\sigma}$$

The first factor under the root tends to zero as $n \to \infty$ since the series $\sum_{k=1}^{\infty} |\varphi_k(z_0)|^2$ converges, and the second factor is finite, e.g. the integral in question in fact tends to zero as $n \to \infty$ and the theorem is proved.

NOTE 1 From the proof of this theorem and Theorem 3, Section 3.1.2, it follows that if the system $\varphi_k(z)$ is not closed in class $\mathcal{A}$ it is closed in a sub-class of functions from $\mathcal{A}$ which are represented by the Formula (6).

NOTE 2 For a system of polynomials orthonormalised in the disc $|z| < 1$ the bi-linear series and Formula (6) have the form

$$K(z, z_0) = \frac{1}{\pi} \sum_{k=0}^{\infty} (k+1)\, \bar{z}_0^k z^k = \frac{1}{\pi (1 - \bar{z}_0 z)^2}$$

$$f(z) = \frac{1}{\pi} \iint_{|z|<1} \frac{f(z)}{(1 - \bar{z}_0 z)^2}\, d\sigma_z$$

Let us assume that the domain B is simply-connected and for some $z_0 \in B$ the function $K(z, z_0)$ is bounded in B. On the strength of (4) this means that the function $\varphi_z'(z, z_0)$ is bounded in B. But for different choices of the point z_0 from B, the functions $w = \varphi(z, z_0)$ are obtained one from another by means of bi-linear transformations mapping the unit discs on to itself.

$$\varphi(z, z_*) = \frac{\varphi(z, z_0) - \varphi(z_*, z_0)}{1 - \overline{\varphi(z_*, z_0)}\, \varphi(z, z_0)}, \quad z_0 \in B, \; z_* \in B$$

Hence it follows that if the domain B is simply-connected and the function $K(z, z_0)$ is bounded for some choice of z_0 from B, then it is also bounded for every other choice of z_0 from B.

Supplement to Theorem 3 If the domain B is simply-connected and the function $K(z, z_0)$ is bounded in B for some fixed $z_0 \in B$ then Formula (6) holds for every function $f(z) \in \mathcal{A}_1$ for every $z_0 \in B$.

PROOF Under the circumstances of the supplement to Theorem 3 the integral (6) exists as an improper integral. By using Theorem 2 we have

$$\iint_B \overline{K(z, z_0)} f(z)\, d\sigma_z = \frac{1}{\pi} \varphi_z'(z_0, z_0) \iint_B f(z) \overline{\varphi_z'(z, z_0)}\, d\sigma_z$$

If B_ρ, $0 < \rho < 1$ is the image of the disc $|w| < \rho$ when mapped by the function $z = \varphi(w, z_0)$ then (see proof of Theorem 2)

$$\iint_B \overline{K(z, z_0)} f(z)\, d\sigma_z$$

$$= \frac{1}{\pi} \varphi_z'(z_0, z_0) \lim_{r \to 1-0} \iint_{B_r} \frac{f(z)}{\varphi_z'(z, z_0)} |\varphi_z'(z, z_0)|^2\, d\sigma_z$$

$$= \frac{1}{\pi} \varphi_z'(z_0, z_0) \lim_{r \to 1-0} \iint_{|w| < r} \frac{f(\psi(w, z_0))}{\varphi_z'(\psi(w, z_0), z_0)}\, d\sigma_w$$

$$= \frac{1}{\pi} \varphi_z'(z_0, z_0) \frac{\pi f(z_0)}{\varphi_z'(z_0, z_0)} = f(z_0)$$

and the supplement is proved.

3.1.6 *The connection between conformal mapping and functions orthonormalised with respect to a domain*

Let B_z be a bounded domain of the z plane, $w = \varphi(z)$ be some function mapping one-to-one conformally the domain B_z on to some bounded domain B_w, and $z = \psi(w)$ be its inverse function. If $f(z) \in \mathcal{A}(B_z)$ then $f_1(w) = f[\psi(w)]\psi'(w) \in \mathcal{A}(B_w)$ and if $f_1(w) \in \mathcal{A}(B_w)$ then

$$f(z) = f_1(w)\psi'(w)^{-1} = f_1[\varphi(z)]\varphi'(z) \in \mathcal{A}(B_z)$$

This follows directly from

$$\iint\limits_{B_w} |f[\psi(w)]|^2 |\psi'(w)|^2 d\sigma_w = \iint\limits_{B_z} |f(z)|^2 d\sigma_z$$

$$\iint\limits_{B_z} |f_1[\varphi(z)]|^2 |\varphi'(z)|^2 d\sigma_z = \iint\limits_{B_w} |f_1(w)|^2 d\sigma_w$$

Let us assume that there is a system of functions $\varphi_k(z) \in \mathcal{A}_z$ orthonormalised in B_z. Then the functions

$$\psi_k(w) = \varphi_k[\psi(w)]\,\psi'(w) \tag{1}$$

belong to $\mathcal{A}(B_w)$ and form a system which is orthonormal in B_w. This follows directly from the following equations:

$$\iint\limits_{B_w} \psi_n(w)\,\overline{\psi_m(w)}\, d\sigma_w = \iint\limits_{B_w} \varphi_m[\psi(w)]\,\overline{\varphi_m[\psi(w)]}\,|\psi'(w)|^2 d\sigma_w$$

$$= \iint\limits_{B_z} \varphi_n(z)\,\overline{\varphi_m(z)}\, d\sigma_z = \delta_{n,m}$$

Analogously it can be proved that the Fourier coefficients of an arbitrary function $f(z) \in \mathcal{A}_z$ with respect to the system $\varphi_k(z)$ and the function $f_1(w) = f(\psi(w))\,\psi'(w) \in \mathcal{A}_w$ with respect to the system $\psi_k(w)$ (1) are identical.

Theorem 1 If the system $\varphi_k(z)$ is complete in B_z, then the system $\psi_k(w)$ (1) is complete in B_w.

PROOF Let us assume the reverse, that the system (1) is incomplete in B_w, e.g. there is a function not identically equal to zero $\psi(w) \in \mathcal{A}_w$ which is orthogonal to all $\psi_k(w)$.

It is easy to verify that then the function $\psi(\varphi(z))\,\varphi'(z) \in \mathcal{A}_z$ not identically equal to zero, is orthogonal to all functions $\varphi_k(z)$. But this contradicts the condition for the completeness of the system $\varphi_k(z)$. The theorem is proved.

NOTE 1 Note that in accordance with (1) we have

$$\varphi_k(z) = \psi_k(\varphi(z))\,\varphi'(z) \tag{2}$$

Let B be an arbitrary, simply-connected, bounded domain of the z plane and $w = \varphi(z, z_0)$, $\varphi(z_0, z_0) = 0$, $\varphi'_z(z_0, z_0) > 0$, $z \in B$ be the function mapping one-to-one conformally the domain

B on to the disc $K: |w| < 1$. Since the system of functions $\psi_k(w) = \sqrt{\frac{k+1}{\pi}} w^k, k = 0, 1, \ldots,$ is a closed orthonormal system in the class $\mathcal{A}(K)$ the system of functions

$$\varphi_k(z) = \sqrt{\frac{k+1}{\pi}} (\varphi(z, z_0))^k \varphi'_z(z, z_0), \quad k = 0, 1, \ldots,$$

is also a closed orthonormal system in the class $\mathcal{A}$ for the domain B.

Thus for every simply-connected domain B there are closed, orthonormal systems of functions.

EXAMPLE (CASSINI'S OVAL) Cassini's oval is the name given to the geometrical locus of points such that the product of their distances from two fixed points is constant.

Let us consider the curve

$$|z - 1| \cdot |z + 1| = \alpha, \quad \alpha > 0$$

If $\alpha \leqslant 1$ this curve defines two bounded domains which are symmetric with respect to the imaginary axis. Let us denote either of them by B. The function $w = \frac{1}{\alpha}(z^2 - 1)$ maps B one-to-one on to the disc $|w| < 1$. Starting from the complete orthonormal system of polynomials $\sqrt{\frac{n+1}{\pi}} w^n$, $n = 0, 1, \ldots,$ in the disc $|w| < 1$ we obtain on the z plane a complete orthonormal system of polynomials

$$P_n(z) = \frac{2\sqrt{n+1}}{\sqrt{\pi}\,\alpha^{n+1}} z(z^2 - 1)^n, \quad n = 0, 1, \ldots,$$

with respect to the domain B.

3.1.7 *Fundamental extremal problem*

Let $\varphi_k(z)$ be a system of functions from $\mathcal{A}$ orthonormalised with respect to the domain B; $\mathcal{A}^{(n)}$, $n = 0, 1, \ldots,$ be the class of functions $\omega_n(z) = \sum_{k=0}^{n} a_k \varphi_k(z)$, where a_k are arbitrary numbers; $\mathcal{A}^{\infty}$ be the class of functions $\omega(z) = \sum_{k=0}^{\infty} a_k \varphi_k(z)$, where a_k are arbitrary numbers satisfying the condition $\sum_{k=0}^{\infty} |a_k|^2 < +\infty$.

Clearly $\mathcal{A}^{(n)} \subset \mathcal{A}^{\infty} \subset \mathcal{A}$, and if the system $\varphi_k(z)$ is closed in class $\mathcal{A}$, then $\mathcal{A}^{\infty} = \mathcal{A}$ (see Theorems 1 and 2, Section 3.1.3).

Theorem 1 Let there be amongst the functions $\varphi_k(z)$, $k = 0, 1, \ldots, n$ functions which are not equal to zero at some point z_0 from B. Let us denote

$$I_n = \iint_B |\omega_n(z)|^2 \, d\sigma$$

Then for functions $\omega_n(z) \in \mathcal{A}^{(n)}$ satisfying the condition $\omega_n(z_0) = \alpha$ where α is a fixed number, we have

$$\mu_n(z_0) = \inf_{\omega_n} I_n = \frac{|\alpha|^2}{\sum_{k=0}^{n} |\varphi_k(z_0)|^2} = \frac{|\alpha|^2}{K_n(z_0, z_0)}$$

and $I_n = \mu_n(z_0)$ only for the function

$$\omega_n(z) = \alpha \frac{\sum_{k=0}^{n} \overline{\varphi_k(z_0)} \varphi_k(z)}{\sum_{k=0}^{n} |\varphi_k(z_0)|^2} = \alpha \frac{K_n(z, z_0)}{K_n(z_0, z_0)}$$

PROOF For the function $\omega_n(z) = \sum_{k=0}^{n} a_k \varphi_k(z)$

$$I_n = \sum_{k=0}^{n} |a_k|^2$$

According to Cauchy's inequality we have

$$|\alpha|^2 = \left| \sum_{k=0}^{n} a_k \varphi_k(z_0) \right|^2 \leqslant \sum_{k=0}^{n} |a_k|^2 \cdot \sum_{k=0}^{n} |\varphi_k(z_0)|^2 = I_n K_n(z_0, z_0)$$

e.g.

$$I_n \geqslant \frac{|\alpha|^2}{K_n(z_0, z_0)}$$

The equal sign will hold in this inequality when, and only when $a_k = \beta \overline{\varphi_k(z_0)}$ where β is a constant which we find from the condition $\omega_n(z_0) = \alpha$

$$\alpha = \omega_n(z_0) = \sum_{k=0}^{n} \beta \overline{\varphi_k(z_0)} \varphi_k(z_0) = \beta K_n(z_0, z_0)$$

whence

$$\beta = \frac{\alpha}{K_n(z_0, z_0)}, \quad a_k = \alpha \frac{\overline{\varphi_k(z_0)}}{K_n(z_0, z_0)}$$

Then

$$\omega_n(z)=\alpha\frac{\sum_{k=0}^{n}\overline{\varphi_k(z_0)}\,\varphi_k(z)}{K_n(z_0,z_0)}=\alpha\frac{K_n(z,z_0)}{K_n(z_0,z_0)}$$

In fact for this function

$$I_n=\frac{|\alpha|^2}{K_n(z_0,z_0)}$$

and the theorem is proved.

In an analogous manner using the convergence of the series $\sum_{k=1}^{\infty}|\varphi_k(z_0)|^2$ we can prove the following theorem.

Theorem 2 Amongst the functions $\varphi_k(z)$, $k=0, 1, \ldots,$ let there be functions not equal to zero at some point z_0 from B. Let us denote

$$I_\infty=\iint_B|\omega(z)|^2\,d\sigma$$

Then for the functions $\omega(z)\in\mathcal{A}^\infty$ satisfying the condition $\omega(z_0)=\alpha$ where α is a fixed number, we have

$$\mu_\infty(z_0)=\inf_\omega I_\infty=\frac{|\alpha|^2}{\sum_{k=0}^{\infty}|\varphi_k(z_0)|^2}=\frac{|\alpha|^2}{K(z_0,z_0)}$$

and $I_\infty=\mu_\infty(z_0)$ only for the function

$$\omega(z)=\alpha\frac{\sum_{k=0}^{\infty}\overline{\varphi_k(z_0)}\,\varphi_k(z)}{\sum_{k=0}^{\infty}|\varphi_k(z_0)|^2}=\alpha\frac{K(z,z_0)}{K(z_0,z_0)}\tag{1}$$

Let

$$I=\iint_B|f(z)|^2\,d\sigma\tag{2}$$

For functions $f(z)\in\mathcal{A}$ satisfying the condition $f(z_0)=\alpha$ where α is a fixed number, let us put

$$\mu(z_0)=\inf_f I$$

From the above it follows that if the system of functions $\varphi_k(z)$ is closed in the class $\mathcal{A}$, then $\mu(z_0)=\mu_\infty(z_0)$ only for function (1). From the above and Theorem 3, Section 3.1.5, there also follows the theorem overleaf.

Theorem 3 For functions $f(z)$ regular in the simply-connected domain B and satisfying the condition $f(z_0)=\alpha$, $z_0 \in B$, the minimum of the integral (2) is equal to

$$\mu(z_0) = \frac{|\alpha|^2}{[\varphi_z'(z_0, z_0)]^2}$$

and $I=\mu(z_0)$ only for the function

$$f(z) = \alpha \cdot \frac{\varphi'(z, z_0)}{\varphi'(z_0, z_0)}$$

Here $w=\varphi(z, z_0)$ is the function mapping the domain B one-to-one conformally on to the disc $|w|<1$ and such that

$$\varphi(z_0, z_0)=0, \quad \varphi'(z_0, z_0)>0$$

If the system $\varphi_k(z)$ is not closed, then $\mu(z_0) \leqslant \mu_\infty(z_0)$ Let us specify this more exactly. Let the system $\varphi_k(z)$ not be closed in class $\mathcal{A}$ and the function $\psi(z) \in \mathcal{A}$, $\psi(z) \not\equiv 0$ in B, be orthogonal to all functions $\varphi_k(z)$, $k=1, 2, \ldots$, and normalised by the condition

$$\iint\limits_B |\psi(z)|^2 \, d\sigma = 1$$

Then for the system $\varphi_k(z)$, $\psi(z)$ we have

$$\mu(z_0) \leqslant \frac{|\alpha|^2}{\sum\limits_{k=0}^{\infty} |\varphi_k(z_0)|^2 + |\psi(z_0)|^2} \leqslant \frac{|\alpha|^2}{\sum\limits_{k=0}^{\infty} |\varphi_k(z_0)|^2} = \mu_\infty(z_0)$$

The equals sign then holds in the last inequality when, and only when, $\psi(z_0)=0$. Since the roots of the function are isolated

$$\mu(z_0) < \mu_\infty(z_0)$$

in any closed domain $\bar{B}' \subset B$ apart perhaps from a finite number of points where $\mu(z_0)=\mu_\infty(z_0)$.

3.1.8 The Schmidt orthogonalisation process - orthogonal polynomials

Let us take a given system of functions from

$$\psi_0(z), \ \psi_1(z), \ \ldots, \ \psi_n(z), \ \ldots$$

and let these functions be linearly independent, for example, the identity

$$C_0^{(n)}\psi_0(z)+C_1^{(n)}\psi_1(z)+\ \dots\ +C_n^{(n)}\psi_n(z)\equiv 0, \quad n=0,\ 1,\ 2,\ \dots,$$

- where $C_k^{(n)}$, $k=0, 1, \dots n$, are constant - is possible only when all $C_k^{(n)}=0$.

In the following for the sake of brevity let us put

$$(F,\ f)=\iint\limits_B F(z)\overline{f(z)}\,d\sigma$$

The problem is to construct an orthonormalised system of functions $\varphi_k(z)$, $k=0, 1, \dots,$ such that every function $\varphi_n(z)$ of this system is represented linearly by functions $\psi_k(z)$, $k=0, 1, \dots, n$, and conversely $\psi_n(z)$ are represented linearly by $\varphi_k(z)$, $k=0, 1, \dots, n$.

The process of constructing the system $\varphi_k(z)$, $k=0, 1, \dots,$ with respect to the system $\psi_k(z)$, $k=0, 1, \dots,$ is called the process of orthogonalisation of the system $\psi_k(z)$, $k=0, 1, \dots$

The process of orthogonalisation proposed by Schmidt consists of the following:

1. Let us assume

$$\varphi_0(z)=\frac{e^{i\omega_0}}{\sqrt{(\psi_0,\ \psi_0)}}\psi_0(z)$$

where ω_0 is an arbitrary real number.

2. Let us denote by $\chi_1(z)=\psi_1(z)-(\psi_1,\varphi_0)\varphi_0(z)$ ($\chi_1(z)\not\equiv 0$ and does not depend on ω_0) and put

$$\varphi_1(z)=\frac{e^{i\omega_1}}{\sqrt{(\chi_1,\ \chi_1)}}\chi_1(z)$$

where ω_1 is a real number.

3. Let us denote $\chi_2(z)=\psi_2(z)-\sum\limits_{k=0}^{1}(\psi_2,\varphi_k)\varphi_k(z)$ ($\chi_2(z)\not\equiv 0$ and does not depend on ω_0 and ω_1) and put

$$\varphi_2(z)=\frac{e^{i\omega_2}}{\sqrt{(\chi_2,\ \chi_2)}}\chi_2(z)$$

where ω_2 is a real number.

....

$n+1$. Let us denote $\chi_n(z)=\psi_n(z)-\sum\limits_{k=0}^{n-1}(\psi_n,\varphi_k)\varphi_k(z)$ ($\chi_n(z)\not\equiv 0$

and does not depend on $\omega_0, \omega_1, \ldots, \omega_{n-1}$) and put

$$\varphi_n(z) = \frac{e^{i\omega_n}}{\sqrt{(\chi_n, \chi_n)}} \chi_n(z)$$

where ω_n is a real number.

....

By construction the linear expression of $\varphi_n(z)$ by $\psi_k(z)$, $k=0, 1, \ldots, n$, contains $\psi_n(z)$, and $\psi_n(z)$ can be represented by $\varphi_k(z)$, $k=0, 1, \ldots, n$. Clearly the functions $\varphi_n(z)$, $n=0, 1, \ldots$, are normalised: $(\varphi_n, \varphi_n)=1$. Let us prove that the functions $\varphi_k(z)$, $k=0, 1, \ldots$, form an orthogonal system. Assuming that the functions $\varphi_k(z)$, $k=0, 1, \ldots, n-1$ are mutually orthogonal, i.e. $(\varphi_k, \varphi_l)=0$, $k \neq l$, $k, l=0, 1, \ldots, n-1$, let us consider (φ_n, φ_m), $m < n$. We have

$$(\varphi_n, \varphi_m) = \frac{e^{i\omega_n}}{\sqrt{(\chi_n, \chi_n)}} \left(\psi_n - \sum_{k=0}^{n-1} (\psi_n, \varphi_k)\varphi_k, \varphi_m\right)$$

$$= \frac{e^{i\omega_n}}{\sqrt{(\chi_n, \chi_n)}} \left[(\psi_n, \varphi_m) - \sum_{k=0}^{n-1} (\psi_n, \varphi_k)(\varphi_k, \varphi_m)\right]$$

$$= \frac{e^{i\omega_n}}{\sqrt{(\chi_n, \chi_n)}} [(\psi_n, \varphi_m) - (\psi_n, \varphi_m)] = 0$$

e.g. $(\varphi_n, \varphi_m)=0$ which is what we required to prove.

In the known way this process of orthogonalisation determines a unique orthonormal system. In particular, the function $\varphi_n(z)$ is defined in this process with an accuracy of up to $e^{i\omega_n}$. The factors $e^{i\omega_n}$ are called phase factors.

Note that $\varphi_n(z)$ is orthogonal to every function $\psi_k(z)$, $k=0, 1, \ldots, n-1$, since $\psi_k(z)$ is represented linearly by $\varphi_l(z)$, $l=0, 1, \ldots, k-1$.

Let us prove that the system of orthonormalised functions $\varphi_n(z)$ with the properties

$$\varphi_n(z) = \sum_{k=0}^{n} a_k^{(n)} \psi_k(z) \text{ and } a_n^{(n)} > 0$$

is unique.

Let us assume that there are two such systems $\Phi_n(z)$ and $\varphi_n(z)$. It must be shown that $\Phi_n(z) \equiv \varphi_n(z)$, $n=0, 1, \ldots$ From the condition of normalisation it follows that

$$\Phi_0(z) = A_0 \varphi_0(z)$$

where $A_0 > 0$ and $A_0^2 = 1$, e.g. $A_0 = 1$ and $\Phi_0(z) \equiv \varphi_0(z)$. For $n > 0$ we can expand $\Phi_n(z)$ with respect to $\varphi_k(z)$, $k = 0, 1, \ldots, n$

$$\Phi_n(z) = \sum_{k=0}^{n} A_k \varphi_k(z)$$

whereby on the strength of the fact that the system $\varphi_k(z)$ is orthonormalised, we have

$$A_k = \iint_B \Phi_n(z) \overline{\varphi_k(z)}\, d\sigma, \quad k = 0, 1, \ldots, n$$

and by virtue of the fact that the function $\Phi_n(z)$ is orthogonal to any function $\psi_k(z)$, $k = 0, 1, \ldots, n-1$ we have $A_k = 0$, $k = 0, 1, \ldots, n-1$, so that $\Phi_n(z) = A_n \varphi_n(z)$, $A_n > 0$ and, as for $n = 1$, we have $A_n = 1$.

A clear picture of orthonormalised functions $\varphi_n(z)$ can be given by means of determinants.

Let us denote

$$c_{j,k} = \iint_B \psi_j(z) \overline{\psi_k(z)}\, d\sigma$$

It is clear that $c_{j,j} > 0$ and $c_{k,j} = \overline{c_{j,k}}$. Let us introduce the determinant

$$D_0 = c_{0,0}, \quad D_n = \begin{vmatrix} c_{0,0}, & c_{1,0}, & \ldots, & c_{n,0} \\ c_{0,1}, & c_{1,1}, & \ldots, & c_{n,1} \\ \cdot & \cdot & \cdot & \cdot \\ c_{0,n}, & c_{1,n}, & \ldots, & c_{n,n} \end{vmatrix} \tag{1}$$

All of the determinants D_n are real since when the rows are interchanged with the columns the value of the determinant is not changed and secondly all the elements of the determinant obtained are changed to conjugate elements, e.g. we obtain the determinant $\overline{D_k}$. Let us now construct the function

$$Q_n(z) = \begin{vmatrix} c_{0,0}, & c_{1,0}, & \ldots, & c_{n-1,0}, & c_{n,0} \\ c_{0,1}, & c_{1,1}, & \ldots, & c_{n-1,1}, & c_{n,1} \\ \cdot & \cdot & \cdot & \cdot & \cdot \\ c_{0,n-1}, & c_{1,n-1}, & \ldots, & c_{n-1,n-1}, & c_{n,n-1} \\ \psi_0(z), & \psi_1(z), & \ldots, & \psi_{n-1}(z), & \psi_n(z) \end{vmatrix} = \sum_{k=0}^{n} \alpha_k^{(n)} \psi_k(z) \tag{2}$$

Here, the coefficient $\alpha_n^{(n)}$ of $\psi_n(z)$ is equal to D_{n-1} and for

$Q_n(z)$ we have

$$\iint_B Q_n(z)\overline{\psi_k(z)}\,d\sigma = \begin{cases} 0, & k=0,\ 1,\ \dots,\ n-1 \\ D_n, & k=n, \end{cases} \tag{3}$$

since the integral under consideration is equal to the determinant which is obtained from the determinant D_n by changing the last row to the row $c_{0,k}, c_{1,k}, \dots, c_{n,k}$. From this formula follows

$$\iint_B |Q_n(z)|^2\,d\sigma = \iint_B Q_n(z)\sum_{k=0}^{n}\overline{\alpha_k^{(n)}}\,\overline{\psi_k(z)}\,d\sigma = \overline{\alpha_n^{(n)}}\,D_n = D_{n-1}D_n \tag{4}$$

Let us show that $D_n > 0,\ n=0,\ 1,\ \dots$ Assuming that $D_n=0$ we have $\iint_B |Q_n(z)|^2\,d\sigma = 0$. This means $Q_n(z)\equiv 0$. But then the coefficient of $\psi_n(z)$ in (2) would be equal to zero, e.g. $D_{n-1}=0$. Similarly we find $D_{n-2}=0,\ D_{n-3}=0,\ \dots,\ D_0=0$. But $D_0=c_{0,0}>0$. This contradiction shows that $D_n\neq 0$. From (4) and $D_0>0$ it follows that all D_n are positive.

On the strength of (3) and (4) we have

$$\iint_B Q_n(z)\overline{Q_m(z)}\,d\sigma = \begin{cases} 0, & k=0,\ 1,\ \dots,\ n-1,\ n+1,\ \dots, \\ D_nD_{n-1}, & k=n, \end{cases}$$

e.g. the functions $Q_n(z)$ are orthogonal. Normalising them, we find the desired orthonormalised functions

$$\varphi_n(z) = \frac{1}{\sqrt{D_nD_{n-1}}}\,Q_n(z) = \sqrt{\frac{D_{n-1}}{D_n}}\,\psi_n(z) + \dots \tag{5}$$

We take the root as positive.

Applying the above to powers $z^k,\ k=0,\ 1,\ \dots,$ we obtain a system of polynomials $P_k(z),\ k=0,\ 1,\ \dots,$ orthogonal with respect to the domain B, $P_k(z)$ being a polynomial of degree k, the highest coefficient of which is positive and $P_0(z)$ being a positive constant.

These orthogonal polynomials are connected with the following extremal problem: to determine a polynomial $Q_n(z) = z^n + c_1z^{n-1} + \dots + c_n$ of degree n with the highest coefficient equal to unity, for which the integral

$$I_n = \iint_B |Q_n(z)|^2\,d\sigma$$

has its minimum value. The polynomial under the integral sign can be written in the form of a linear combination of orthogonal polynomials $P_k(z)$, $k=0, 1, \ldots, n$

$$z^n + c_1 z^{n-1} + \ldots + c_{n-1} z + c_n = a_n^{-1} P_n(z) + \sum_{k=0}^{n-1} d_k P_k(z)$$

where a_n is the leading coefficient of the polynomial $P_n(z)$. For the integral I_n we find

$$I_n = a_n^{-2} + \sum_{k=0}^{n-1} |d_k|^2$$

and the minimum of I_n is reached only in the case where $d_k = 0$, $k=0, 1, \ldots, n-1$, for example, for the polynomial $Q_n(z) = a_n^{-1} P_n(z)$ and then $I_n = a_n^{-2}$.

3.1.9 Completeness and closedness of non-orthonormalised systems of functions

DEFINITION 1 A system of linearly independent functions $\psi_k(z)$, $k=0, 1, \ldots,$ of class $\mathcal{A}$ is said to be complete if there is no function from class $\mathcal{A}$ differing from identically zero which is orthogonal to all functions $\psi_k(z)$.

DEFINITION 2 A system of linearly independent functions $\psi_k(z)$, $k=0, 1, \ldots,$ of class $\mathcal{A}$ is said to be closed in class $\mathcal{A}$ if for any function $f(z) \in \mathcal{A}$ there is a sequence of generalised polynomials $\sigma_n(z) = \sum_{k=0}^{n} d_k^{(n)} \psi_k(z)$, $n=0, 1, \ldots,$ such that

$$\lim_{n\to\infty} \iint_B |f(z) - \sigma_n(z)|^2 d\sigma = 0 \tag{1}$$

This equation will be called the closure equation for the function $f(z)$ with respect to the system $\psi_k(z)$.

For a system of linearly independent functions $\psi_k(z)$, $k=0, 1, \ldots,$ of class $\mathcal{A}$ an orthonormalised system of functions $\omega_k(z)$, $k=0, 1, \ldots$ can be constructed by the Schmidt orthogonalisation process. Since the functions $\omega_k(z)$, $k=0, 1, \ldots,$ are linearly represented by $\psi_l(z)$, $l=0, 1, \ldots, k$, and $\psi_k(z)$ is linearly represented by $\omega_l(z)$, $l=0, 1, \ldots, k$, as the system $\psi_k(z)$ is complete, it follows that the orthonormalised system $\omega_k(z)$ is complete and vice-versa. Similar arguments apply to $\psi_k(z)$ and $\omega_k(z)$. Completeness and closedness are equivalent for non-orthonormalised systems.

Lemma 1 If the closure equation holds for the functions $\psi_k(z)\in\mathcal{A}$, $k=1, 2, \ldots, n$, with respect to the system $\varphi_k(z)\in\mathcal{A}$, then the closure equation with respect to the system $\varphi_k(z)$ holds for every function $f_n(z)=\sum_{k=0}^{n} c_k\psi_k(z)$ where c_k are constants.

The proof is first carried out for $n=2$. For a given $\varepsilon>0$ there are generalised polynomials $P_m^{(1)}(z)=\sum_{k=0}^{m} c_k^{(1)}\varphi_k(z)$ and $P_m^{(2)}(z)=\sum_{k=0}^{m} c_k^{(2)}\varphi_k(z)$ such that

$$\iint_B |\psi_\nu(z)-P_m^{(\nu)}(z)|^2\,d\sigma<\varepsilon, \quad \nu=1, 2$$

For the function $f_2(z)=c_1\psi_1(z)+c_2\psi_2(z)$ and the polynomial $P_m(z)=c_1P_m^{(1)}(z)+c_2P_m^{(2)}(z)$ we have

$$\iint_B |f_2(z)-P_m(z)|^2\,d\sigma\leqslant 2|c_1|^2\iint_B |\psi_1(z)-P_m^{(1)}(z)|^2\,d\sigma$$

$$+2|c_2|^2\iint_B |\psi_2(z)-P_m^{(2)}(z)|^2\,d\sigma<2(|c_1|^2+|c_2|^2)\varepsilon$$

By virtue of the arbitrary smallness of ε, the closure equation with respect to the system $\varphi_k(z)$ holds for $f_2(z)$. The general case can be proved by the method of induction.

Lemma 2 If the closure equation with respect to the system $\varphi_k(z)\in\mathcal{A}$ holds for the functions $f_n(z)\in\mathcal{A}$, $n=0, 1, \ldots,$ and the function $f(z)\in\mathcal{A}$ is such that

$$\lim_{n\to\infty}\iint_B |f(z)-f_n(z)|^2\,d\sigma=0$$

then the closure equation with respect to the system $\varphi_k(z)$ also holds for the function $f(z)$.

PROOF For given $\varepsilon>0$ there is n such that

$$\iint_B |f(z)-f_n(z)|^2\,d\sigma<\frac{\varepsilon}{4}$$

and a generalised polynomial $P_{n,m}(z)=\sum_{k=0}^{m} c_k^{(n)}\varphi_k(z)$ such that

$$\iint_B |f_n(z)-P_{n,m}(z)|^2\,d\sigma<\frac{\varepsilon}{4}$$

Hence

$$\iint_B |f(z)-P_{n,m}(z)|^2 d\sigma \leqslant 2\iint_B |f(z)-f_n(z)|^2 d\sigma + 2\iint_B |f_n(z)-P_{n,m}(z)|^2 d\sigma < \varepsilon$$

In view of the arbitrary smallness of ε, the closure equation with respect to the system $\varphi_k(z)$ holds for $f(z)$ and the lemma is proved.

Theorem 1 Let the system of functions

$$\psi_k(z) \in \mathcal{A},\ k=0,\ 1,\ \ldots,$$

be closed in the class $\mathcal{A}$ and $\varphi_k(z)$, $k=0, 1, \ldots,$ be some other system of functions of class $\mathcal{A}$. For the system $\varphi_k(z)$ to be closed in class $\mathcal{A}$, it is necessary and sufficient that the closure equation with respect to the system $\varphi_k(z)$ should hold for every function of the system $\psi_k(z)$.

PROOF The necessity is obvious.

SUFFICIENCY Let $f(z) \in \mathcal{A}$. According to the premises there is a sequence of generalised polynomials

$$f_n(z)=\sum_{k=0}^{n} c_k^{(n)} \psi_k(z),\ n=0,\ 1,\ \ldots,$$

such that

$$\lim_{n\to\infty} \iint_B |f(z)-f_n(z)|^2 d\sigma = 0$$

On the strength of Lemma 1 the closure equation with respect to the system $\varphi_k(z)$ holds for functions $f_n(z)$, $n=0, 1, \ldots,$ and consequently, by virtue of Lemma 2, the closure equation with respect to the system $\varphi_k(z)$ holds for the function $f(z)$ and the theorem is proved.

EXAMPLE Let B be the disc K: $|z|<1$. We know (see Section 3.1.3) that the system of functions z^k, $k=0, 1, \ldots,$ is closed in the class $\mathcal{A}(K)$. For the system of functions $\varphi_k(z) \in \mathcal{A}(K)$, $k=0, 1, \ldots,$ to be closed in the class $\mathcal{A}(K)$ it is necessary and sufficient that the closure equation with respect to the system $\varphi_k(z)$ should hold for every function z^k, $k=0, 1, \ldots$

3.1.10 Unitary transformation

Let $\psi_n(z)$, $n=0, 1, \ldots$, be a closed, orthonormal system of functions of the class $\mathcal{A}$. Let us construct the series

$$\varphi_k(z)=\sum_{n=0}^{\infty} a_{k,n}\psi_n(z), \quad k=0, 1, \ldots, \tag{1}$$

where the infinite array of coefficients $a_{k,n}$ satisfies the condition

$$\sum_{n=0}^{\infty} |a_{k,n}|^2 < +\infty, \quad k=0, 1, \ldots \tag{2}$$

As we know the series (1) converge uniformly on the compact sub-sets of B, their sums $\varphi_k(z)$ belong to the class $\mathcal{A}$ and these series are the Fourier series of the functions $\varphi_k(z)$ with respect to the closed, orthonormal system $\psi_n(z)$. Let us investigate the condition which the array of coefficients $a_{k,n}$ must satisfy if the function $\varphi_k(z)$ is to form a closed, orthonormal system. Note first that the choice of coefficients $a_{k,n}$ under condition (2) is equivalent to the choice of a system of functions $\varphi_k(z)$ from class $\mathcal{A}$. Then $a_{k,n}$ are the Fourier coefficients of $\varphi_k(z)$ with respect to the system of functions $\psi_n(z)$ and the series (1) are the Fourier series of $\varphi_k(z)$.

From the generalised Parseval's equation

$$\iint_B \varphi_k(z)\overline{\varphi_l(z)}\,d\sigma=\sum_{n=0}^{\infty} a_{k,n}\bar{a}_{l,n}, \quad k, l=0, 1, \ldots,$$

it follows that for the system $\varphi_k(z)$ to be orthonormal it is necessary and sufficient that the conditions

$$\sum_{n=0}^{\infty} a_{k,n}\bar{a}_{l,n}=\delta_{k,l}, \quad k, l=0, 1, \ldots, \tag{3}$$

be fulfilled, where $\delta_{k,l}=0$ for $k\neq l$ and $\delta_{k,l}=1$ for $k=l$.

For the orthonormal system $\varphi_k(z)$ to be closed it is necessary and sufficient, as we have seen, that the closure equation with respect to the system $\varphi_n(z)$ should hold for all functions $\psi_k(z)$, $k=0, 1, \ldots$. The Fourier coefficients of $\psi_k(z)$ with respect to the system $\varphi_n(z)$ are the integrals of the product $\psi_k(z)\overline{\varphi_n(z)}$. The integral of this product is the Fourier coefficient $a_{n,k}$ of the function $\varphi_n(z)$ e.g.

$$\iint_B \psi_k(z)\overline{\varphi_n(z)}\,d\sigma=\bar{a}_{n,k} \tag{4}$$

and the closure equation of normalised functions $\psi_k(z)$ with respect to the system $\varphi_n(z)$ is expressed by the formulae

$$\sum_{n=0}^{\infty} |\overline{a}_{n,k}|^2 = 1; \qquad \sum_{n=0}^{\infty} |a_{n,k}|^2 = 1, \quad k = 0, 1, \ldots, n \tag{5}$$

We arrive at the following result.

Theorem 1 Let $\psi_n(z)$ be a closed, orthonormal system of functions of class $\mathcal{A}$, $\varphi_k(z)$ be some system of functions of class $\mathcal{A}$ and (1) be Fourier series of functions $\varphi_k(z)$ with respect to functions $\psi_n(z)$. Then conditions (3) are necessary and sufficient for system (1) to be orthonormalised, and condition (5) for this system to be closed. If conditions (3) and (5) are fulfilled, then the Fourier series of $\psi_k(z)$ with respect to the system $\varphi_n(z)$ are of the form

$$\psi_k(z) = \sum_{n=0}^{\infty} \overline{a}_{n,k}\varphi_n(z) \tag{6}$$

Let conditions (3) and (5) be fulfilled, e.g. let the system $\varphi_n(z)$ be a closed, orthonormal system. Then we can choose $\varphi_n(z)$ instead of $\psi_k(z)$ as the basic closed, orthonormal system. Taking into account expansion (6) for functions of the closed, orthonormal system $\psi_k(z)$ with respect to $\varphi_n(z)$ and applying condition (3) to the Fourier coefficients $\overline{a}_{n,k}$ we have

$$\sum_{n=0}^{\infty} a_{n,k}\overline{a}_{n,l} = \delta_{k,l} \tag{7}$$

Let us now introduce the concept of unitary transformation. Let U be a matrix of infinite order, the elements of which are the quantities $a_{k,n}$, $k, n = 0, 1, \ldots,$ satisfying the conditions (3) and (5). Taking the first subscript of $a_{k,n}$ as the number of the row and the second as the number of the column, we call conditions (3) the conditions for orthonormalisation by rows. If these conditions and (5) are satisfied, we call the matrix U with elements $a_{k,n}$ the unitary matrix. As we have shown, from the conditions for orthonormalisation by rows and (5) follow the conditions (7) for orthonormalisation by columns and we can define the unitary matrix by the orthonormalisation of its elements by rows and columns. It is not difficult to see that if the matrix $\{a_{k,n}\}$ is unitary, then the matrices $\{a_{n,k}\}$, $\{\overline{a}_{k,n}\}$ and $\{\overline{a}_{n,k}\}$ are also unitary.

Let us introduce the linear transformation corresponding

to the unitary matrix U:

$$y_k = \sum_{n=0}^{\infty} a_{k,n} x_n, \quad k = 0, 1, \ldots, \tag{8}$$

where the sequence of numbers x_n, $n = 0, 1, \ldots,$ satisfies the condition

$$\sum_{n=0}^{\infty} |x_n|^2 < +\infty \tag{9}$$

From this condition and from conditions (2) by virtue of the inequality $|a_{k,n} x_n| \leqslant \frac{1}{2}(|a_{k,n}|^2 + |x_n|^2)$ it follows that series (8) converge absolutely.

Let us introduce the function $f(z) \in \mathcal{A}$ which has the numbers x_n as Fourier coefficients with respect to the orthonormal system $\psi_k(z)$:

$$f(z) = \sum_{n=0}^{\infty} x_n \psi_n(z)$$

and determine its Fourier coefficients with respect to the system $\varphi_k(z)$ which is a closed, orthonormal system since the matrix $\{a_{k,n}\}$ is unitary:

$$x'_k = \iint_B f(z) \overline{\varphi_k(z)} \, d\sigma$$

Applying to this integral the generalised Parseval's equation with respect to the system $\psi_n(z)$ we have

$$x'_k = \sum_{n=0}^{\infty} \overline{a_{k,n}} x_n \tag{10}$$

If we now write Parseval's equation for $f(z)$ with respect to the systems $\varphi_k(z)$ and $\psi_n(z)$ we find

$$\iint_B |f(z)|^2 \, d\sigma = \sum_{k=0}^{\infty} |x'_k|^2, \quad \iint_B |f(z)|^2 \, d\sigma = \sum_{k=0}^{\infty} |x_k|^2$$

that is

$$\sum_{k=0}^{\infty} |x'_k|^2 = \sum_{k=0}^{\infty} |x_k|^2$$

Instead of (8) we can write

$$\overline{y_k} = \sum_{n=0}^{\infty} \overline{a}_{k,n} \overline{x_n}$$

and from the above directly follows

$$\sum_{k=0}^{\infty} |y_k|^2 = \sum_{n=0}^{\infty} |x_n|^2 \tag{11}$$

e.g. the relation (11) holds for the linear unitary transformation (8).

In consequence it follows directly that the infinite homogeneous system

$$\sum_{n=0}^{\infty} a_{k,n} x_n = 0, \quad k = 0, 1, \ldots, \tag{12}$$

where $\{a_{k,n}\}$ is a unitary matrix, has only zero solutions $x_1 = x_2 = \ldots = 0$ if solutions are limited to those satisfying condition (9). By using the conditions that the matrix $\{a_{k,n}\}$ be unitary, it is not difficult to prove that if y_k are any given complex numbers satisfying the condition

$$\sum_{k=0}^{\infty} |y_k|^2 < +\infty$$

then the infinite system of Equations (8) with unknown x_n has a unique solution satisfying the condition (9), and this solution is of the form

$$x_n = \sum_{k=0}^{\infty} \bar{a}_{k,n} y_k, \quad n = 0, 1, 2, \ldots \tag{13}$$

3.1.11 Orthogonal functions with a weight function

In this section we shall briefly consider quadratic approximation with respect to a domain of the complex variable with a weight function.

In the domain B let there be given a measurable real function $\gamma(z) \geqslant 0$ such that there exists the finite or infinite integral

$$\iint_B \gamma(z)\, d\sigma > 0 \tag{1}$$

We shall mainly consider the class $\mathcal{A}(\gamma, B)$ of functions $f(z)$ regular in the domain B with the condition

$$\iint_B \gamma(z) |f(z)|^2\, d\sigma < +\infty$$

The measure of quadratic approximation to the function $f(z)$

of class $\mathcal{A}(\gamma, B)$ by means of function $P(z)$ of class $\mathcal{A}(\gamma, B)$ will be defined by the formula

$$\delta = \iint_B \gamma(z)\,|f(z) - P(z)|^2\, d\sigma$$

In addition let the function $\gamma(z)$ satisfy the following condition: for any compact sub-set E of B there is a number $\rho_E > 0$ such that for any point z_0 from E and any function $f(z)$ from $\mathcal{A}(\gamma, B)$ the inequality

$$\iint_B \gamma(z)\,|f(z)|^2\, d\sigma \geqslant \rho_E\,|f(z_0)|^2 \tag{2}$$

holds.

Such a function $\gamma(z)$ will be called the weight function or simply the weight.

Let us show that the function $\gamma(z)$ satisfying condition (1) will be a weight function if for every compact set $E \subset B$ there is a closed partially smooth Jordan curve $l \subset B$ containing E in its interior and a quantity $\delta > 0$ such that (a) the lesser of the distances from l to E and from l to the boundary of the domain B is not less than δ; (b) the function $\gamma(z)$ is bounded below by a positive number on the set $\overline{G}$ containing all the points not further than $\delta/2$ from l.

In point of fact for every point $\zeta \in l$ and every function $f(z) \in \mathcal{A}(\gamma, B)$ we have

$$\iint_B \gamma(z)\,|f(z)|^2\, d\sigma \geqslant \iint_{|z-\zeta| \leqslant \frac{\delta}{2}} \gamma(z)\,|f(z)|^2\, d\sigma \geqslant m_l \iint_{|z-\zeta| \leqslant \frac{\delta}{2}} |f(z)|^2 d\sigma$$

where $m_l = \inf\limits_{z \in G} \gamma(z) > 0$. By using inequality (1), Section 3.1.1, we find

$$\iint_B \gamma(z)\,|f(z)|^2\, d\sigma \geqslant \pi \left(\frac{\delta}{2}\right)^2 m_l\,|f(\zeta)|^2$$

If we put $M = \max\limits_{\zeta \in l} |f(\zeta)|$ then from this inequality we have

$$\iint_B \gamma(z)\,|f(z)|^2\, d\sigma \geqslant \pi \left(\frac{\delta}{2}\right)^2 m_l M^2 \tag{3}$$

Now let z_0 be an arbitrary point of E. Then

$$f(z_0) = \frac{1}{2\pi i} \int_l \frac{f(\zeta)}{\zeta - z_0}\, d\zeta$$

and consequently

$$|f(z_0)| \leqslant \frac{M\lambda}{2\pi\frac{\delta}{2}} = \frac{M\lambda}{\pi\delta}$$

where λ is the length of the curve l. Hence and from (3) we find

$$\iint_B \gamma(z)\,|f(z)|^2\,d\sigma \geqslant \frac{1}{4}\frac{\pi^3\delta^4}{\lambda^2}\,m_l\,|f(z_0)|^2, \quad z_0 \in E$$

As the compact set $E \subset B$ and the function $f(z) \in \mathcal{A}(\gamma, B)$ are arbitrary, we conclude that $\gamma(z)$ is a weight function. Hence, in particular, it easily follows that $\gamma(z) = |D(z)|$ where $D(z)$ is a regular function in B of class $\mathcal{A}(B)$ not becoming identically zero, is a weight function.

Retaining all the definitions given earlier but changing $d\sigma$ to $\gamma(z)\,d\sigma$ everywhere in the integrals, we can easily prove that all the theorems proved above with the exception of certain ones which will be indicated below, remain in force for an arbitrary weight function $\gamma(z)$ when the formulation has been altered in the obvious way. The condition for orthonormalisation, the formulae for the Fourier coefficients and Bessel's inequality are in the form

$$\iint_B \gamma(z)\,\varphi_k(z)\,\overline{\varphi_l(z)}\,d\sigma = \delta_{k,l}, \quad a_k = \iint_B \gamma(z)\,f(z)\,\overline{\varphi_k(z)}\,d\sigma$$

$$\sum_{k=0}^{\infty} |a_k|^2 \leqslant \iint_B \gamma(z)\,|f(z)|^2\,d\sigma.$$

In an analogous manner to that used before, we introduce the notation $A_1(\gamma(z), B)$.

Let us briefly list the exceptions.

Inequality (2) will correspond to inequality (1), Section 3.1.1. There is no analogue to Section 3.1.4. There is no analogous theorem to Theorem 2, Section 3.1.5 or to the corollary to Theorem 3, Section 3.1.5.

Let us consider Section 3.1.6. By retaining the same notation we shall have: if $f(z) \in \mathcal{A}(\gamma(z), B_z)$, then

$$f_1(w) = f(\psi(w))\,\psi'(w) \in \mathcal{A}(\gamma(\psi(w)), B_w)$$

and

$$f_2(w) = f(\psi(w)) \in \mathcal{A}(\gamma(\psi(w))\,|\psi'(w)|^2, B_w)$$

Besides this, functions $\psi_k(w) = \varphi_k(\psi(w))\psi'(w)$ belong to the class $\mathcal{A}(\gamma(\psi(w)), B_w)$ and form a closed, orthonormal system in this class, and the functions $\psi_k^*(z) = \varphi_k(\psi(w))$ belong to the class $\mathcal{A}(\gamma(\psi(w))|\psi'(w)|^2, B_w)$ and also form a closed, orthonormal system in this class.

There is no analogue to the example in Section 3.1.6 (Cassini's oval), and Theorem 3, Section 3.1.7 is not applicable. Let us denote by $\Phi(z, z_0)$ the function giving the minimum of the integral

$$\iint_B \gamma(z)|f(z)|^2 d\sigma$$

in the class of functions $f(z)$ regular in the domain B and normalised by the condition $f(z_0) = \alpha$, $z_0 \in B$. By using the analogy to Theorem 2, Section 3.1.7, we find that

$$\alpha \frac{K(z, z_0)}{K(z_0, z_0)} = \Phi(z, z_0)$$

and

$$\iint_B \gamma(z)|\Phi(z, z_0)|^2 d\sigma = \frac{|\alpha|^2}{K(z_0, z_0)}$$

There is no analogue to the example in Section 3.1.9.

3.2 POLYNOMIALS ORTHONORMALISED WITH RESPECT TO A DOMAIN IN THE PRESENCE OF AN ANALYTIC WEIGHT FUNCTION

3.2.1 *Construction of generalised polynomials*

Let B be a bounded domain of the z plane and $D(z)$ be some function given in B such that the integral

$$\iint_B |D(z)|^2 d\sigma < +\infty$$

exists. In this section we shall assume that $D(z)$ is regular and not identically equal to zero in B. The weight $\gamma(z) = |D(z)|^2$ will be called an analytic weight.

Let us consider the system of functions $\psi_k(z) = D(z)z^k$, $k = 0, 1, \ldots$ By applying to this Schmidt's method of orthogonalisation (Section 3.1.8), we find the system of functions orthonormalised in the domain B

$$\varphi_k(z) = D(z)\, q_k(z), \quad k = 0, 1, \ldots, \tag{1}$$

where $q_k(z)$ is a polynomial of degree k. We choose the phase factors $e^{i\omega_k}$ such that the coefficient of z^k in the polynomial $q_k(z)$ is positive. The functions $\varphi_k(z)$ will be called orthonormalised generalised polynomials with respect to the domain B for $D(z)$. If $D(z) = 1$ then we have polynomials $P_k(z) = q_k(z)$, $k = 0, 1, \ldots$ orthonormalised with respect to the domain B.

We obtain an explicit expression for orthonormalised generalised polynomials $\varphi_k(z)$ by means of determinants as explained in Section 3.1.8, putting $\psi_k(z) = D(z)\, z^k$.

For the system of functions $\varphi_k(z)$ we have

$$\iint_B D(z)\, q_k(z)\, \overline{D(z)}\, \overline{q_m(z)} d\sigma = \delta_{n,m}$$

Or, assuming $|D(z)|^2 = \gamma(z)$, we have

$$\iint_B \gamma(z)\, q_k(z)\, \overline{q_m(z)}\, d\sigma = \delta_{n,m}, \quad n, m = 0, 1, \ldots \tag{2}$$

Thus we can consider the polynomials $q_k(z)$ as polynomials orthonormalised with respect to the domain B in the presence of a weight $\gamma(z) = |D(z)|^2$.

The system of functions (1) cannot be closed (complete) in the class $\mathcal{A}(B)$ if the function $D(z)$ becomes zero even at only one point z_0 in B since at every such point all functions of the system (1) become zero (see Corollary to Theorem 1, Section 3.1.3).

If the function $D(z) \in \mathcal{A}(B)$ is such that the system of functions (1) is closed in class $\mathcal{A}(B)$ (and consequently $D(z) \neq 0$ in B), then the class of series of the form

$$f(z) = \sum_{k=0}^{\infty} a_k D(z)\, q_k(z) \tag{3}$$

where

$$\sum_{k=0}^{\infty} |a_k|^2 < +\infty \tag{4}$$

is a class of Fourier series of functions $f(z)$ of class $\mathcal{A}(B)'$, the series converge uniformly on the compact sub-sets of B and Parseval's equation holds for $f(z)$.

Let us write (3) in the form

$$\frac{f(z)}{D(z)} = F(z) = \sum_{k=0}^{\infty} a_k q_k(z)$$

Bearing in mind the above, we can prove the following: if the function $D(z) \in \mathcal{A}(B)$ is such that system (1) is closed in class $\mathcal{A}(B)$ then the class of series of the form

$$F(z) = \sum_{k=0}^{\infty} a_k q_k(z) \tag{5}$$

where a_k satisfy condition (4), is a class of Fourier series with respect to orthonormalised polynomials $q_k(z)$ with weight $\gamma(z) = |D(z)|^2$ for functions $F(z)$ such that $F(z)D(z) \in \mathcal{A}(B)$, e.g. such that $F(z) \in \mathcal{A}(\gamma, B)$ and for a_k we have the formulae

$$a_k = \iint_B f(z)\overline{D(z)}\,\overline{q_k(z)}\, d\sigma = \iint_B \gamma(z) F(z) \overline{q_k(z)}\, d\sigma \tag{6}$$

The series (5) converge uniformly on the compact sub-sets of the domain B and Parseval's equation

$$\iint_B \gamma(z) |F(z)|^2 d\sigma = \sum_{k=0}^{\infty} |a_k|^2$$

holds for $F(z)$.

The system of functions $\varphi_k(z) = D(z) q_k(z)$ cannot be closed in the class $\mathcal{A}(B)$ if B is a multiply-connected domain. In fact, if the system of functions $\varphi_k(z)$ is closed in the class $\mathcal{A}(B)$ then the system of polynomials $q_k(z)$ is closed in the class $\mathcal{A}(\gamma, B)$, $\gamma(z) = |D(z)|^2$ and consequently the Fourier series for any function $f(z) \in \mathcal{A}(\gamma, B)$ with respect to the system of polynomials $q_k(z)$ will converge uniformly on the compact sub-sets of B to $f(z)$. On the basis of the maximum modulus principle, this series will converge uniformly on the compact sub-sets of the smallest simply-connected domain B' containing B, and the function $f(z)$ must be regular in the domain B' whilst at the same time there are clearly functions $f(z) \in \mathcal{A}(\gamma, B)$ not regular in B'.

In the remainder of this section we shall assume that the domain is simply-connected.

There are simply-connected domains B for which, for some choice of function $D(z) \in \mathcal{A}(B)$ ($D(z) \neq 0$ in B), the orthonormalised system of generalised polynomials $\varphi_k(z) = D(z) q_k(z)$ will not be closed in the class $\mathcal{A}(B)$. Let us consider, for example, the domain B which is the disc $|z|<1$ excluding the radius $0\leqslant z<1$ and let us put $D(z)\equiv$. The orthonormalised polynomials for this domain will be the same as for the disc $|z|<1$ since the exclusion of the radius has no effect on the double integral of the product $z^j\bar{z}^k$. But if the function $f(z)\in\mathcal{A}(B)$ has singular points on the excluded radius (for example $f(z)=\sqrt{z}$) this function cannot be expanded with respect to the orthonormal polynomials $\sqrt{\frac{k+1}{\pi}}z^k$, i.e. into a Maclaurin series, in the domain B.

Let us consider another example. Let us denote by Δ a class of simply-connected, bounded domains, the boundary of which consists of two Jordan curves with one common point. There is a domain $B_1\in\Delta$ for which the system of polynomials with weight $\gamma(z)\equiv 1$ is closed in the class $\mathcal{A}(B_1)$ and there is a domain $B_2\in\Delta$ for which the system of polynomials is not closed in the class $\mathcal{A}(B_2)$. The works of Keldysh [1], Mergelian [3] and Shaginian [1 and 2] deal with questions of this type.

Keldysh [2] proved that if the domain B is the disc $|z|<1$, and $D(z)=e^{\frac{1}{2}\frac{z+1}{z-1}}$ the orthonormalised system of generalised polynomials $\varphi_k(z)=D(z)q_k(z)$ is not closed in the class $\mathcal{A}(B)$ in particular, Parseval's equation does not hold for the function $f(z)\equiv 1$.

These examples demonstrate that the closedness of the orthonormalised system of generalised polynomials $\varphi_k(z)=D(z)q_k(z)$ in the class $\mathcal{A}(B)$ depends both on the form of the domain B and on the function $D(z)$.

Let the function $F(z)\in\mathcal{A}(B)$ be such that the orthonormalised system of generalised polynomials $\varphi_k(z)=D(z)q_k(z)$ is closed in class $\mathcal{A}(B)$ and let $w=\varphi(z)$ be a function mapping the domain B one-to-one conformally on to the disc $|w|<1$. Then the function

$$w=\varphi(z,\ z_0)=\frac{|\varphi'(z_0)|}{\varphi'(z_0)}\,\frac{\varphi(z)-\varphi(z_0)}{1-\overline{\varphi(z_0)}\,\varphi(z)}$$

maps the domain B one-to-one conformally on to the disc $|w|<1$ and $\varphi(z_0,\ z_0)=0$ and

$$\varphi'(z_0,\ z_0)=\frac{|\varphi'(z_0)|}{1-|\varphi(z_0)|^2}>0$$

Applying the Formula (5), Section 3.1.5, for the sums of a

bi-linear series to the system $\varphi_k(z)$, we find

$$\sum_{k=0}^{\infty} \overline{q_k(z_0)}\, q_k(z) = \frac{1}{\pi} \frac{|\varphi'(z_0)|^2}{[1-\overline{\varphi(z_0)}\,\varphi(z)]^2} \frac{1}{\overline{D(z_0)}D(z)} \tag{7}$$

Let us return to the orthonormalised system (1) and let

$$q_n(z) = a_n z^n + a_1^{(n)} z^{n-1} + \dots + a_n^{(n)} \qquad (a_n > 0)$$

Theorem 1 Amongst the polynomials $Q_n(z) = z^n + \alpha_1 z^{n-1} + \dots + \alpha_n$ of degree n the integral

$$I_n = \iint_B |D(z)|^2 |Q_n(z)|^2 \, d\sigma$$

has a minimum only for the polynomial $Q_n^{(0)}(z) = a_n^{-1} q_n(z)$ and $\inf I_n = a_n^{-2}$.

PROOF Let us expand the polynomial $Q_n(z)$ with respect to polynomials $q_k(z)$, $k = 0, 1, \dots, n$

$$Q_n(z) = c_0 q_0(z) + c_1 q_1(z) + \dots + c_n q_n(z), \qquad c_n = a_n^{-1}$$

Then

$$I_n = |c_0|^2 + |c_1|^2 + \dots + |c_n|^2$$

Clearly $\min I_n = |c_n|^2$ for $Q_n^{(0)}(z) = c_n q_n(z)$ and the theorem is proved.

COROLLARY The roots of the polynomial

$$q_n(z), \; n = 1, 2, \dots$$

belong to a convex hull of the domain B (e.g. the smallest closed convex domain $\overline{G}$ containing B).

In fact, if there were a root of $q_n(z)$ outside $\overline{G}$, then we could displace it in such a way that it was closer to all points of the domain B, which would entail a decrease in the integral

$$I_n = \iint_B |D(z)|^2 |a_n^{-1} q_n(z)|^2 \, d\sigma$$

and this contradicts the assertion of the theorem.

3.2.2 Case of a disc

In this section we shall consider an orthonormalised system of generalised polynomials for the case when the domain B is the circle $K : |z| < 1$.

Theorem 1 For the orthonormalised system of generalised polynomials $\varphi_k(z) = D(z) q_k(z)$ for the disc K to be closed in the class $\mathcal{A}(K)(D(z) \in \mathcal{A}(K))$ it is necessary and sufficient that one of the following equivalent conditions be fulfilled:

1. There exists a sequence of polynomials $u_n(z)$, $n = 0, 1, \ldots$, such that

$$\lim_{n\to\infty} \iint_K |1 - D(z) u_n(z)|^2 d\sigma = 0 \tag{1}$$

2. $$\sum_{k=0}^{\infty} |D(0)|^2 |q_k(0)|^2 = \frac{1}{\pi} \tag{2}$$

3. $$\mu = \pi |D(0)|^2 \tag{3}$$

where

$$\mu = \lim_{n\to\infty} \mu_n, \quad \mu_n = \min_{d_k} \iint_K |D(z)|^2 |1 + d_1 z + \ldots + d_n z^n|^2 d\sigma$$
$$k = 1, 2, \ldots, n.$$

PROOF The function $D(z) u_n(z)$ can be represented as a linear combination of the functions $\varphi_k(z)$ and by virtue of Note 1 to Theorem 2, Section 3.1.3, condition (1) is equivalent to the assertion that Parseval's equation holds for the function $f(z) \equiv 1$ and the necessity of condition (1) is obvious.

SUFFICIENCY From (1) it follows that

$$\lim_{n\to\infty} \iint_K |z^m - D(z) z^m u_n(z)|^2 d\sigma = 0, \quad m = 0, 1, \ldots$$

Hence as above we conclude that Parseval's equation holds for the functions z^m, $m = 0, 1, \ldots$ On the strength of the theorem in Section 3.1.9 (see also example given) the closure equation holds for every function of class (K), and the sufficiency of condition (1) is proved.

Let us now calculate the Fourier coefficients of the function $f(z) \equiv 1$ with respect to the system of functions $\varphi_k(z)$

$$a_k = \iint_K \overline{D(z) q_k(z)}\, d\sigma = \pi \overline{D(0) q_k(0)}$$

and the Bessel inequality gives

$$\sum_{k=0}^{\infty} \pi^2 |D(0)|^2 |q_k(0)|^2 \leqslant \pi$$

or

$$\sum_{k=0}^{\infty} \pi |D(0)|^2 |q_k(0)|^2 \leqslant 1 \tag{4}$$

Since (1) expresses the closure equation for $f(z) \equiv 1$ it follows that condition (2) is equivalent to (1).

Let us put

$$I_n = \iint_K |\omega_n(z)|^2\, d\sigma$$

where $\omega_n(z)$ is a linear combination of n functions $\varphi_k(z)$

$$\omega_n(z) = \sum_{k=0}^{n} b_k D(z) q_k(z) = D(z) \sum_{k=0}^{n} d_k z^k$$

satisfying the condition $\omega_n(0) = D(0)$, e.g. $d_0 = 1$. According to Theorem 1, Section 3.1.7,

$$\mu_n = \frac{|D(0)|^2}{\sum_{k=0}^{n} |D(0) q_k(0)|^2} = \frac{1}{\sum_{k=0}^{n} |q_k(0)|^2}$$

and

$$\mu = \lim_{n\to\infty} \mu_n = \frac{1}{\sum_{k=0}^{\infty} |q_k(0)|^2} \tag{5}$$

Hence it follows that condition (2) is equivalent to condition (3) and the theorem is proved.

DEFINITION Function $D(z)$ of class $\mathcal{A}(K)$ is said to be maximal in the class $\mathcal{A}(K)$ if condition (1) is fulfilled for it.

NOTE 1 In proving the theorem we could have assumed that $D(z) \neq 0$ in the disc $|z| < 1$ (see Corollary to Theorem 1, Section 3.1.3). The proof carried out above shows that it follows from condition (1) that $D(z) \neq 0$ for $|z| < 1$.

NOTE 2 If $D_1(z)$ and $D_2(z)$ are two maximal functions of class $\mathcal{A}(K)$ and one of them, for example $D_1(z)$, is bounded in the disc K, then the product $D_1(z) D_2(z)$ is a maximal function of the class $\mathcal{A}(K)$. This follows from the definition of a maximal function and the inequality

$$\iint_K |1 - D_1(z) D_2(z) u_{n,1}(z) u_{m,2}(z)|^2 d\sigma$$

$$\leqslant 2\iint_K |1 - D_1(z) u_{n,1}(z)|^2 d\sigma + 2\iint_K |D_1(z) u_{n,1}(z)|^2 |1 - D_2(z) u_{m,2}(z)|^2 d\sigma$$

$$\leqslant 2\iint_K |1 - D_1(z) u_{n,1}(z)|^2 d\sigma + 2M_n \iint_K |1 - D_2(z) u_{m,2}(z)|^2 d\sigma \quad (6)$$

where

$$M_n = \max_{z \in K} |D_1(z) u_{n,1}(z)|^2$$

$u_{n,j}(z)$, $n = 1, 2, \ldots$, $j = 1, 2$, is a sequence of polynomials such that as $n \to \infty$

$$I_{n,j} = \iint_K |1 - D_j(z) u_{n,j}(z)|^2 d\sigma \to 0$$

For given $\varepsilon > 0$ let us first fix n such that $I_{n,1} < \frac{\varepsilon}{4}$ and then fix m such that $I_{m,2} < \frac{\varepsilon}{4M_n}$. Hence and from (6) follows

$$\iint_K |1 - D_1(z) D_2(z) u_{n,1}(z) u_{m,2}(z)|^2 d\sigma \leqslant \varepsilon$$

The theorem is proved.

Theorem 2 For the function $D(z)$ of class $\mathcal{A}(K)$ to be maximal in class $\mathcal{A}(K)$ it is necessary and sufficient that there be a sequence of functions $g_l(z)$, $l = 1, 2, \ldots$, regular in the disc $|z| < 1$ and continuous in $|z| \leqslant 1$ such that

$$\lim_{l \to \infty} \iint_{|z|<1} |1 - D(z) g_l(z)|^2 d\sigma = 0 \quad (7)$$

PROOF Necessity follows from Theorem 1 (see (1)).

SUFFICIENCY Let (7) be fulfilled and $u(z)$ be an arbitrary polynomial for the time being. We have

$$\iint_{|z|<1} |1 - D(z) u(z)|^2 d\sigma$$

$$= \iint_{|z|<1} |[1 - D(z) g_l(z)] + D(z)[g_l(z) - u(z)]|^2 d\sigma$$

$$\leqslant 2 \iint_{|z|<1} |1 - D(z) g_l(z)|^2 d\sigma + 2 \iint_{|z|<1} |D(z)|^2 |g_l(z) - u(z)|^2 d\sigma$$

Here the first term on the right-hand side can be made as small as is desired by the choice of l. Having fixed l, we can make the second term on the right-hand side as small as is desired by the choice of $u(z)$ (see Equation (1), Section 3.1.6). The sufficiency of (7) hence follows and the theorem is proved.

Theorem 2 can be formulated as follows:

Theorem 2′ For the function $D(z)$ of class $\mathcal{A}(K)$ to be a maximal function of class $\mathcal{A}(K)$ it is necessary and sufficient that there be a sequence of functions $g_l(z)$, $l = 1, 2, \ldots$, regular in the disc $|z| < 1$ and continuous in $|z| \leqslant 1$ uniformly converging on the compact sub-sets of the disc $|z| < 1$ to $\frac{1}{D(z)}$ and such that

$$\lim_{l\to\infty} \iint\limits_{|z|<1} |D(z)\, g_l(z)|^2\, d\sigma = \pi \tag{8}$$

PROOF: NECESSITY If $D(z)$ is a maximal function of class $\mathcal{A}(K)$ then (7) holds and consequently

$$D(z)\, g_l(z) \rightrightarrows 1 \quad \text{of} \quad g_l(z) \rightrightarrows \frac{1}{D(z)}$$

in the disc $|z| < 1$ as $l \to \infty$. Hence by virtue of Fatou's theorem (Smirnov [2] p. 165) it follows that

$$\varliminf_{l\to\infty} \iint\limits_{|z|<1} |D(z)\, g_l(z)|^2\, d\sigma \geqslant \pi$$

and from Minkovsky's inequality (see Smirnov [2] p. 191)

$$\left(\iint\limits_{|z|<1} |D(z)\, g_l(z)|^2\, d\sigma\right)^{\frac{1}{2}} \leqslant \left(\iint\limits_{|z|<1} |1 - D(z)\, g_l(z)|^2\, d\sigma\right)^{\frac{1}{2}} + \left(\iint\limits_{|z|<1} 1\, d\sigma\right)^{\frac{1}{2}}$$

we have

$$\varlimsup_{l\to\infty} \iint\limits_{|z|<1} |D(z)\, g_l(z)|^2\, d\sigma \leqslant \pi$$

whence follows (8).

SUFFICIENCY From the fact that $g_l(z) \rightrightarrows \frac{1}{D(z)}$ in $|z| < 1$ it follows that $D(z)$ does not become zero inside the disc

$|z|<1$ and for every r_0, $0<r_0<1$

$$\lim_{l\to\infty}\iint_{|z|<r_0}|D(z)\,g_l(z)|^2\,d\sigma=\pi r_0^2$$

and by virtue of (8)

$$\lim_{l\to\infty}\iint_{r_0<|z|<1}|D(z)\,g_l(z)|^2\,d\sigma=\pi(1-r_0^2)$$

From the above it follows that in the inequality

$$\iint_{|z|<1}|1-D(z)\,g_l(z)|^2\,d\sigma$$

$$\leqslant\iint_{|z|<r_0<1}|1-D(z)\,g_l(z)|^2\,d\sigma+2\iint_{r_0\leqslant|z|<1}d\sigma+2\iint_{r_0\leqslant|z|<1}|D(z)g_l(z)|^2 d\sigma$$

the right-hand side can be made as small as desired by the choice of r_0 and l, and therefore (7) holds.

COROLLARY If the function $D(z)$ is regular in the disc $|z|<1, D(z)\neq 0$ for $|z|<1$ and

$$\lim_{r\to 1-0}\iint_{|z|<1}\left|\frac{D(z)}{D(rz)}\right|^2 d\sigma=\pi$$

then the function $D(z)$ is a maximal function of class $\mathcal{A}(K)$.

Let us denote by $\mathcal{A}^{(0)}(K)$ a sub-class of functions $D(z)\in\mathcal{A}(K)$ for which $D(z)\neq 0$ for $|z|<1$ and

$$\lim_{r\to 1-0}\iint_{|z|<r}\left|\frac{D(z)}{m(|z|)}\right|^2 d\sigma<+\infty,\quad m(r)=\min_{|z|=r}|D(z)| \tag{9}$$

Note that if $D(z)\in\mathcal{A}(K)$ and $|D(z)|\geqslant m>0$ in $|z|<1$ then $D(z)\in\mathcal{A}^{(0)}(K)$.

Theorem 3 The function $D(z)\in\mathcal{A}^{(0)}(K)$ is a maximal function of class $\mathcal{A}(K)$.

PROOF By the conditions of the theorem the function $D(z)$ does not become zero in the disc $|z|<1$. Let us put

$$u_n(z)=\frac{1}{n+1}\sum_{k=0}^{n}S_k(z)$$

where $S_k(z)$ is the k-th partial sum of the Maclaurin series of the function $\frac{1}{D(z)}$. We know (see introduction) that the sequence of functions $u_n(z)$ converges uniformly on the compact sub-sets of the disc $|z|<1$ to $\frac{1}{D(z)}$ and

$$|u_n(z)| \leqslant \sup_{|\zeta|=|z|} \frac{1}{|D(\zeta)|} = \frac{1}{m(|z|)}$$

For the function $u_n(z)$ let us transform and estimate from above the integral of Formula (1)

$$\iint\limits_{|z|<1} |D(z)|^2 \left|\frac{1}{D(z)} - u_n(z)\right|^2 d\sigma$$

$$\leqslant \iint\limits_{|z|<r<1} |D(z)|^2 \left|\frac{1}{D(z)} - u_n(z)\right|^2 d\sigma + 2 \iint\limits_{r\leqslant|z|<1} |D(z)|^2 \left|\frac{1}{D(z)}\right|^2 d\sigma$$

$$+ 2 \iint\limits_{r\leqslant|z|<1} |D(z)|^2 |u_n(z)|^2 d\sigma$$

$$\leqslant \iint\limits_{|z|<r} |D(z)|^2 \left|\frac{1}{D(z)} - u_n(z)\right|^2 d\sigma + 2\pi(1-r^2)$$

$$+ 2 \iint\limits_{r\leqslant|z|<1} \left|\frac{D(z)}{m(|z|)}\right|^2 d\sigma.$$

By the choice of r, $0<r<1$ the two last terms can be made as small as desired and the first term on the right-hand side of the inequality tends to zero as $n\to\infty$ for any choice of $r, 0<r<1$. Hence it follows that (1) holds and the theorem is proved.

Theorem 4 If for a function $D(z)$ of class $\mathcal{A}(K)$ there is a sequence of functions $D_l(z)$, $l=1, 2, \ldots$, of class $\mathcal{A}^{(0)}(K)$ such that

$$\lim_{l\to\infty} \iint\limits_{|z|<1} \frac{|D_l(z)-D(z)|^2}{[m_l(|z|)]^2} d\sigma = 0$$

$$m_l(r) = \min_{|z|=r} |D_l(z)|$$

then $D(z)$ is a maximal function of class $\mathcal{A}(K)$.

PROOF Assuming

$$u_{n,l}(z) = \frac{1}{n+1} \sum_{k=0}^{n} S_{k,l}(z)$$

where $S_{k,l}(z)$ is the k-th partial sum of the Maclaurin series of the function $\frac{1}{D_l(z)}$ we have

$$\iint\limits_{|z|<1} |1-D(z)\, u_{n,l}(z)|^2\, d\sigma$$
$$= \iint\limits_{|z|<1} |(1-D_l(z)\, u_{n,l}(z)) + (D_l(z)-D(z))\, u_{n,l}(z)|^2\, d\sigma$$
$$\leqslant 2 \iint\limits_{|z|<1} |1-D_l(z)\, u_{n,l}(z)|^2\, d\sigma + 2 \iint\limits_{|z|<1} \frac{|D_l(z)-D(z)|^2}{[m_l(|z|)]^2}\, d\sigma$$

By choosing l the second term on the right can be made as small as desired and the first term on the right tends to zero as $n \to \infty$ (see proof of Theorem 3) for every fixed l. Hence the theorem is proved.

Theorem 5 If for a function $D(z)$ of class $\mathcal{A}(K)$ there is a sequence of maximal functions $D_l(z)$, $l=1, 2, \ldots,$ of class $\mathcal{A}(K)$ such that

1. $|D_l(z)| \geqslant |D(z)|, \quad |z|<1, \quad l=1, 2, \ldots,$

2. $\lim\limits_{l\to\infty} |D_l(0)| = |D(0)| \neq 0$

then $D(z)$ is also a maximal function of class $\mathcal{A}(K)$.

PROOF For the function $D_l(z)$ the quantities μ_n and μ (see Theorem 1) are denoted respectively by $\mu_n^{(l)}$ and $\mu^{(l)}$. From Theorem 1 it follows that $\mu^{(l)} = \pi |D_l(0)|^2$, $l=1, 2, \ldots$ Besides this

$$\mu_n^{(l)} = \min_{d_k} \iint\limits_{|z|<1} |D_l(z)|^2\, |1+d_1 z + \ldots + d_n z^n|^2\, d\sigma \geqslant \mu_n$$

and consequently $\mu \leqslant \mu^{(l)} = \pi |D_l(0)|^2$. In the limit we find $\mu \leqslant \pi |D(0)|^2$. Hence and from (4) and (5) follows (3) which means that $D(z)$ is also a maximal function of class $\mathcal{A}(K)$.

Let $z_k, |z_k| \geqslant 1, k=1, 2, \ldots,$ be distinct points; $\alpha_k, \alpha_k > -1$ be real numbers amongst which there is only a finite number of negative ones and such that $\sum\limits_{k=1}^{\infty} \alpha_k < +\infty$. Let us introduce the function

$$\Phi(z) = \sum_{k=1}^{\infty} \alpha_k \ln\left(1 - \frac{z}{z_k}\right) \tag{10}$$

where by $\ln\left(1-\frac{z}{z_k}\right)$ is understood that branch regular in the disc $|z|<1$ which becomes zero for $z=0$. The series (10) converges uniformly on the compact sub-sets of the disc $|z|<1$.

Theorem 6 The function

$$D(z)=Ce^{\Phi(z)}=C\prod_{k=1}^{\infty}\left(1-\frac{z}{z_k}\right)^{\alpha_k} \tag{11}$$

where C is a constant and $\Phi(z)$ is the function defined above, is a maximal function of the class $\mathcal{A}(K)$.

PROOF Let us first show that $D(z)\in\mathcal{A}(K)$. Let α_k, $k=1, 2, \ldots, m$ be negative numbers and α_k, $k=m+1$, $m+2, \ldots$, be positive $\alpha=\sum_{k=1}^{m}\alpha_k$, $\beta=\sum_{k=m+1}^{\infty}\alpha_k$. Let us cover every point z_k, $k=1, 2, \ldots, m$ with a disc R_k with its centre at the point z_k and radius equal to ε which is small enough for none of these discs to have common points. Let us denote by G the part of the disc $|z|<1$ not covered with discs. Clearly

$$|D(z)|\leqslant|C|\,2^{\beta}\varepsilon^{\alpha},\quad z\in G,$$

$$|D(z)|\leqslant|C|\,2^{\beta}\varepsilon^{\alpha-\alpha_k}\left(\frac{|z-z_k|}{|z_k|}\right)^{\alpha_k},\quad z\in R_k,\ k=1, 2, \ldots, m$$

Consequently

$$\iint_{|z|<1}|D(z)|^2\,d\sigma\leqslant\iint_{G}|D(z)|^2\,d\sigma+\sum_{k=1}^{m}\iint_{R_k}|D(z)|^2\,d\sigma$$

$$\leqslant|C|^2\,2^{2\beta}\varepsilon^{2\alpha}\pi+\sum_{k=1}^{m}|C|^2\,2^{2\beta}\varepsilon^{2\alpha-2\alpha_k}\frac{\varepsilon^{2\alpha_k+2}\pi}{(2\alpha_k+2)\,|z_k|^{2\alpha_k}}$$

$$=|C|^2\,2^{2\beta}\varepsilon^{2\alpha}\pi\left(1+\varepsilon^2\sum_{k=1}^{m}\frac{1}{2\,(1+\alpha_k)\,|z_k|^{2\alpha_k}}\right)<+\infty$$

Thus $D(z)\in\mathcal{A}(K)$.

In proving that condition (1) is fulfilled we shall assume that all polynomials $u_n(z)$ contain the factor

$$\prod_{k=1}^{m}\left(1-\frac{z}{z_k}\right)$$

Let us transfer this factor from $u_n(z)$ to $D(z)$. We can then also assume that all α_k in (10) are positive.

Let $\rho_l > 1$ and $\rho_l \to 1$ as $l \to \infty$. Let us put

$$D_l(z) = C\rho_l^{\sum_{k=1}^{\infty} \alpha_k} e^{\Phi\left(\frac{z}{\rho_l}\right)}$$

In the disc $|z| < 1$ the modulus of the function $D_l(z)$ is bounded above by the quantity $|C|(2\rho_l)^{\sum_{k=1}^{\infty} \alpha_k}$ and below by the quantity $|C|(\rho_l - 1)^{\sum_{k=1}^{\infty} \alpha_k}$. By virtue of Theorem 3 $D_l(z)$ is a maximal function of class $\mathcal{A}(K)$. Clearly in the disc $|z| < 1$ the conditions of Theorem 5 are fulfilled for the sequence $D_l(z)$ and the theorem is proved.

3.2.3 *Case of a general domain*

Let B be a bounded, simply-connected domain and the function $D(z) \in \mathcal{A}(B)$. Let us investigate the conditions under which the orthonormalised system of generalised polynomials $\varphi_k(z) = D(z) q_k(z)$, $k = 0, 1, \ldots$, which was introduced in Section 3.2.1 ($q_k(z)$ is a polynomial of degree k), is closed in the class $\mathcal{A}(B)$.

Let $w = \varphi(z)$ be a function mapping the domain B conformally on to the disc $K: |w| < 1$ and let $z = \psi(w)$ be its inverse function.

Theorem 1 For the orthonormalised system of generalised polynomials $\varphi_k(z) = D(z) q_k(z)$, $k = 0, 1, \ldots$, for the domain B to be closed in the class $\mathcal{A}(B)$ it is necessary and sufficient that Parseval's equation should hold for the functions $\varphi(z)^m \varphi'(z)$, $m = 0, 1, \ldots$

PROOF The question of whether the system $\varphi_k(z)$ is closed in the class $\mathcal{A}(B)$ is equivalent to the question of whether the system of functions (see Section 3.1.6)

$$R_k(w) = D(\psi(w))\, q_k(\psi(w))\, \psi'(w),$$

$$k = 0, 1, \ldots, \tag{1}$$

belonging to class $\mathcal{A}(K)$ and orthonormalised in it, is closed.

For the system (1) to be closed in class $\mathcal{A}(K)$ (see example for Theorem 1, Section 3.1.9) it is necessary and sufficient that Parseval's equation hold for functions w^m, $m=0, 1, \ldots$:

$$\iint\limits_{|w|<1} |w|^{2m}\, d\sigma = \sum_{k=0}^{\infty} |a_k^{(m)}|^2, \quad m=0, 1, \ldots,$$

where

$$a_k^{(m)} = \iint\limits_{|w|<1} w^m\, \overline{D(\psi(w))}\, \overline{q_k(\psi(w))}\, \overline{\psi'(w)}\, d\sigma$$

Changing the variable by using $w=\varphi(z)$ we have

$$\iint\limits_{B} |\varphi(z)^m \varphi'(z)|^2\, d\sigma = \sum_{k=0}^{\infty} |a_k^{(m)}|^2 \tag{2}$$

and

$$a_k^{(m)} = \iint\limits_{B} \varphi(z)^m \varphi'(z)\, \overline{D(z)}\, \overline{q_k(z)}\, d\sigma$$

e.g. Parseval's equation must hold for the function $\varphi(z)^m$ $\varphi'(z)$, $m=0, 1, \ldots$ and the theorem is proved.

COROLLARY Let the domain B be such that for every $m=0, 1, \ldots$ there is a sequence of polynomials $u_{m,n}(z)$ $n=0, 1, \ldots$, such that

$$\lim_{n\to\infty} \iint\limits_{B} |\varphi(z)^m \varphi'(z) - u_{m,n}(z)|^2\, d\sigma = 0$$

Then for the system of generalised polynomials $\varphi_k(z) = D(z) q_k(z)$ to be closed in the class $\mathcal{A}(B)$ it is necessary and sufficient that there be a sequence of polynomials $u_n(z)$, $n=0, 1, \ldots$, such that

$$\lim_{n\to\infty} \iint\limits_{B} |1 - D(z) u_n(z)|^2\, d\sigma = 0 \tag{3}$$

In fact, by virtue of Lemma 2, Section 3.1.9, to prove that the system $\varphi_k(z)$ is closed it is sufficient to prove that the closure equation holds for an arbitrary polynomial and by virtue of Lemma 1, Section 3.1.9, it is sufficient to prove that the closure equation holds for the functions z^m, $m=0, 1, \ldots$ Proceeding as in the proof of Theorem 1, Section 3.2.2, we prove the sufficiency. The necessity is obvious.

In this section we shall also assume that B is a Caratheodory domain, that is, a bounded, simply-connected domain the boundary of which coincides with the boundary of the domain amongst those complementary to B which contains the point ∞. This complementary domain is denoted by B^1. A simply-connected, bounded Jordan domain (domain bounded by a closed Jordan curve) is a Caratheodory domain, but we know that there are Caratheodory domains with boundaries which are not closed Jordan curves.

Let the function $w=\Phi(z)$, $\Phi(\infty)=0$, $\Phi'(\infty)>0$ be a function mapping the domain B^1 conformally on to the domain $|w|>1$ and let L_ρ be the inverse image of the circle $|w|=\rho>1$ for this mapping. The bounded domain with the contour L_ρ is denoted by B_ρ. Let ρ_n, $n=0,1,\ldots$, be a decreasing sequence of real numbers such that $\lim_{n\to\infty}\rho_n=1$. The sequence of domains B_{ρ_n}, $n=0,1,\ldots$, has as its kernel* with respect to the arbitrary point $z_0\in B$ the domain B and converges to B (see Goluzin [1] p. 62).

Let z_0 be a point of the Caratheodory domain B. Let us denote by $w=\varphi_n(z, z_0)$ the function mapping the domain B_{ρ_n} on to the disc $|w|<1$ and normalised by the conditions $\varphi_n(z_0, z_0)=0$, $\varphi_n'(z_0, z_0)>0$ and by $w=\varphi(z, z_0)$, $\varphi(z_0, z_0)=0$, $\varphi'(z_0, z_0)>0$ the function mapping the domain B conformally on to the disc $|w|<1$ and let $z=\psi(w, z_0)$ denote its inverse function. From Caratheodory's theorem (Goluzin [1] p. 62) it follows that as $n\to\infty$ we have, in B,

$$\varphi_n(z, z_0)\rightrightarrows\varphi(z, z_0) \text{ and } \varphi_n'(z, z_0)\rightrightarrows\varphi'(z, z_0)$$

Lemma 1 For the Caratheodory domain B

$$\lim_{n\to\infty}\int\int_B|\varphi(z, z_0)^m\varphi'(z, z_0)-\varphi_n(z, z_0)^m\varphi_n'(z, z_0)|^2\,d\sigma=0 \qquad (4)$$
$$m=0,\ 1,\ \ldots$$

PROOF Note that the sequence of functions

$$f_n(z)=|\varphi(z, z_0)^m\varphi'(z, z_0)-\varphi_n(z, z_0)^m\varphi_n'(z, z_0)|^2$$

*The kernel of a sequence of simply-connected domains B_n, $n=1,2,\ldots$, containing some disc $|z-z_0|<\rho, \rho>0$, with respect to the point z_0, is the name given to the largest domain B, containing the point z_0 and having the property that for every closed set E belonging to B there is a number N such that E belongs to domains B_n for $n>N$. The largest domain is understood to mean that it contains any other domain having the same property. We shall say that the sequence of domains B_n, $n=1,2,\ldots$, converges to the kernel B if any sequence from B_n has B as its kernel.

converges uniformly to zero on the compact sub-sets of B and

$$f_n(z) \leqslant 2|\varphi(z, z_0)^m|^2 |\varphi'(z, z_0)|^2 + 2|\varphi_n(z, z_0)^m|^2 |\varphi_n'(z, z_0)|^2$$
$$\leqslant 2(|\varphi'(z, z_0)|^2 + |\varphi_n'(z, z_0)|^2) = f_n^*(z)$$

and the function $f_n^*(z)$ is integrable in B since the integral

$$\iint_B |\varphi'(z, z_0)|^2 \, d\sigma \quad \left(u \iint_B |\varphi_n'(z, z_0)|^2 \, d\sigma\right)$$

is equal to the area of the image of the domain B when mapped by the function $w = \varphi(z, z_0)$ (respectively $w = \varphi_n(z, z_0)$).

Let us now break up the integral (4) into two integrals with respect to the domain B_r, $0 < r < 1$ defined by the inequality $|\varphi(z, z_0)| \leqslant r$ and with respect to the domain $\bar{B} - B_r$. The first of these integrals obviously tends to zero as $n \to \infty$. The second does not exceed the sum of the integrals

$$2 \iint_{\bar{B}-B_r} |\varphi'(z, z_0)|^2 \, d\sigma + 2 \iint_{\bar{B}-B_r} |\varphi_n'(z, z_0)|^2 \, d\sigma$$

The first of these integrals is equal to twice the area of the image of the set $\bar{B} - B_r$ when mapped by the function $w = \varphi(z, z_0)$, e.g. $2\pi(1 - r^2)$. Let $0 < r_1 < r$. Since the image of the contour line $|\varphi(z, z_0)| = r$ when mapped by the function $w = \varphi_n(z, z_0)$ for sufficiently large n lies in the ring $r_1 < |w| < 1$ and consequently the image of $\bar{B} - B_r$ lies in the same ring, for such n the second of these integrals does not exceed $2\pi(1 - r_1^2)$. Hence in view of the arbitrary choice of r and r_1, $0 < r_1 < r < 1$ the lemma is easily proved.

Let B_* be the complement of $\bar{B}^1$ (note that $\bar{B}_*$ may not coincide with $\bar{B}$ (see Goluzin [1] p. 42)). The functions $\varphi_n(z, z_0)^m \varphi_n'(z, z_0)$ are regular in $\bar{B}_*$ and consequently any of them can be uniformly approximated as well as is desired by means of polynomials (see Section 1.7.4) on $\bar{B}_*$. Hence we conclude that for functions $\varphi_n(z, z_0)^m \varphi_n'(z, z_0)$ Parseval's equation holds in $\bar{B}^*$ with respect to the system of functions z^k, $k = 0, 1, \ldots$ By virtue of Lemma 1 and Lemma 2, Section 3.1.9, the conditions of the Corollary to Theorem 1 are fulfilled and therefore the following theorem is valid.

Theorem 2 (Markushevich [1] *)** Let B be a Caratheodory domain and $\varphi_k(z)=D(z)q_k(z)$, $k=0, 1, \ldots$, be an orthonormalised system of generalised polynomials with respect to the domain B. For the system $\varphi_k(z)$ to be closed in the class $\mathcal{A}(B)$ it is necessary and sufficient that there be a sequence of polynomials $u_n(z)$, $n=0, 1, \ldots$, such that

$$\lim_{n\to\infty}\iint_B |1-D(z)\,u_n(z)|^2\,d\sigma=0 \tag{5}$$

In particular, if $D(z)\equiv 1$ in B, we have the following corollary.

COROLLARY 1 The orthonormalised system of polynomials with respect to a Caratheodory domain B is closed in the class $\mathcal{A}(B)$.

COROLLARY 2 Let B be a Caratheodory domain. If $f(z)\in\mathcal{A}(B)$ and

$$\iint_B f(z)\,\bar{z}^m\,d\sigma=0, \quad m=0, 1, \ldots, \tag{6}$$

then $f(z)\equiv 0$ in B.

In fact, in view of the fact that the orthonormalised system of polynomials $P_k(z)$, $k=0, 1, \ldots$, is closed in the class $\mathcal{A}(B)$ the system $P_k(z)$ is also complete in $\mathcal{A}(B)$ and for this it is necessary and sufficient that the system z^m, $m=0, 1, \ldots$ be complete.

The question arises: if $f(z)\in\mathcal{A}_1(B)$ and (6) holds, will $f(z)\equiv 0$? Let us prove the following theorem.

Theorem 3 Let B be a Caratheodory domain and let the function $w=\varphi(z)$ mapping the domain B conformally on to the disc $|w|<1$ have a derivative bounded in B. Then if $f(z)\in\mathcal{A}_1(B)$ and condition (6) is fulfilled, $f(z)\equiv 0$ in B.

PROOF Let us retain the notation introduced earlier when considering Caratheodory domains. In our case the function

$$w=\varphi(z, z_0)=e^{-i\alpha}\frac{\varphi(z)-\varphi(z_0)}{1-\overline{\varphi(z_0)}\,\varphi(z)}, \quad \alpha=\arg\varphi'(z_0), \quad z_0\in B$$

We have $|\varphi'(z, z_0)|\leqslant M_{z_0}$ where M_{z_0} is some positive number and $z\in B$. By virtue of the Corollary to Theorem 3,

* See also Mergelian [3].

Section 3.1.5

$$f(z_0) = \frac{\varphi'(z_0, z_0)}{\pi} \iint_B f(z) \overline{\varphi'(z, z_0)}\, d\sigma_z$$

The theorem will be proved if we can show that the integral in this formula is equal to zero for every z_0 from B. Let us put

$$v_m(w) = \frac{1}{m+1} \sum_{k=0}^{m} S_k(w)$$

where $S_k(w)$ is the k-th partial sum of the Maclaurin series of $\varphi'(\psi(w, z_0), z_0)$. The sequence of functions $v_m(w)$ converges uniformly on the compact sub-sets of the disc $|w| < 1$ to $\varphi'(\psi(w, z_0), z_0)$ and $|v_m(w)| \leqslant M_{z_0}$ for $|w| < 1$ and consequently the sequence $v_m(\varphi(z, z_0))$ converges uniformly on the compact sub-sets of B to $\varphi'(z, z_0)$ and $|v_m(\varphi(z, z_0))| \leqslant M_{z_0}$ in B. Thus

$$|f(z) v_m(\varphi(z, z_0))| \leqslant M_{z_0} |f(z)|, \quad z \in B$$

and the function $M_{z_0}|f(z)|$ is integrable in $B (f(z) \in \mathcal{A}_1)$. Since $f(z)\overline{v_m(\varphi(z, z_0))} \rightrightarrows f(z)\overline{\varphi'(z, z_0)}$ in B, by virtue of the theorem on taking the limit under the integral sign (Smirnov [2] p. 162) for $\varepsilon > 0$ there is m such that

$$\left| \iint_B f(z) \overline{\varphi'(z, z_0)}\, d\sigma_z - \iint_B f(z) \overline{v_m(\varphi(z, z_0))}\, d\sigma_z \right| < \frac{\varepsilon}{3}$$

Note that as $n \to \infty$ the sequence $v_m(\varphi_n(z, z_0))$ uniformly converges to $v_m(\varphi(z, z_0))$ on the compact sub-sets of B and $|v_m(\varphi_n(z, z_0))| \leqslant M_{z_0}$ in B. Hence, as above, it follows that there is n such that

$$\left| \iint_B f(z) \overline{v_m(\varphi(z, z_0))}\, d\sigma_z - \iint_B f(z) \overline{v_m(\varphi_n(z, z_0))}\, d\sigma_z \right| < \frac{\varepsilon}{3}$$

The function $v_m(\varphi_n(z, z_0))$ is regular on $\overline{B}_*$ (in particular regular in B_{ρ_n}) and consequently there is a polynomial $u(z)$

such that

$$|v_m(\varphi_n(z, z_0)) - u(z)| < \frac{\varepsilon}{3M}, \quad z \in \bar{B}_*, \quad M = \int_B\int |f(z)|\, d\sigma$$

and this means

$$\left| \int_B\int f(z)\overline{v_m(\varphi_n(z, z_0))}\, d\sigma - \int_B\int f(z)\overline{u(z)}\, d\sigma \right| < \frac{\varepsilon}{3}$$

From the estimates obtained we have

$$\left| \int_B\int f(z)\overline{\varphi'(z, z_0)}\, d\sigma - \int_B\int f(z)\overline{u(z)}\, d\sigma \right| < \varepsilon$$

and on the strength of (6)

$$\left| \int_B\int f(z)\overline{\varphi'(z, z_0)}\, d\sigma \right| < \varepsilon$$

In view of the arbitrary nature of ε this integral is equal to zero and the theorem is proved.

As before in the case of a disc let us introduce the concept of a maximal function. The function $D(z)$ of class $\mathcal{A}(B)$ is said to be a maximal function in class $\mathcal{A}(B)$ if condition (5) is fulfilled for it.

NOTE 1 If $D_1(z)$ and $D_2(z)$ are two maximal functions of class $\mathcal{A}(B)$ and one of them is bounded in B, then the product $D_1(z)D_2(z)$ is a maximal function in class $\mathcal{A}(B)$. This is proved in the same way as in the corresponding note in Section 3.2.2.

Theorem 4 For the function $D(z)$ of class $\mathcal{A}(B)$, where B is a Caratheodory domain, to be a maximal function of class $\mathcal{A}(B)$ it is necessary and sufficient that there should be a sequence of functions $g_l(z)$, $l = 1, 2, \ldots$, regular in internal points of $\bar{B}_*$ and continuous on $\bar{B}_*$ such that

$$\lim_{l\to\infty} \int_B\int |1 - D(z)g_l(z)|^2\, d\sigma = 0 \tag{7}$$

Theorem 4 can also be formulated

Theorem 4′ For the function $D(z) \in \mathcal{A}(B)$ to be a maximal function in class $\mathcal{A}(B)$ it is necessary and sufficient that there should be a sequence of functions $g_l(z)$, $l = 1, 2, \ldots,$

regular at internal points of $\bar{B}_*$ and continuous on $\bar{B}_*$, uniformly converging on the compact sub-sets of B to $\frac{1}{D(z)}$ and such that

$$\lim_{l\to\infty}\iint_B |D(z)g_l(z)|^2\,d\sigma = s$$

where s is the internal area of the domain B.

The proof of Theorems 4 and 4′ is analogous to the proof of Theorems 2 and 2′, Section 3.2.2.

Further we shall indicate some sufficient conditions such that $D(z)\in\mathcal{A}(B)$ is a maximal function in class $\mathcal{A}(B)$ if they are fulfilled.

Let the sequence ρ_n, $n=1, 2, \ldots$, $\rho_n > 1$ tend to one as $n\to\infty$ and let $w=\varphi_n(z, z_0)$, $z_0\in B$ be a function mapping the domain B_{ρ_n} conformally on to the disc $|w|<1$ and normalised by the conditions $\varphi_n(z_0, z_0)=0$, $\varphi_n'(z_0, z_0)>0$ and also let $w=\varphi(z, z_0)$ and $z=\psi(w, z_0)$ be the functions introduced earlier. From the theorem stated we have the following corollary.

COROLLARY If the function $D(z)$ is regular in the Caratheodory domain B and

$$\lim_{n\to\infty}\iint_B \left|\frac{D(z)}{D(\psi(\varphi_n(z, z_0), z_0))}\right|^2 d\sigma = s$$

where s is the internal area of the domain B, then $D(z)$ is a maximal function of class $\mathcal{A}(B)$.

Let us introduce the contour lines $L^r_{z_0}$, $z_0\in B$, $0<r<1$ of the domain B: $L^r_{z_0}$ is the image of the circle $|w|=r$ when mapped by the function $z=\psi(w, z_0)$. Let us denote by $B^r_{z_0}$ the bounded domain with boundary $L^r_{z_0}$. Let us denote by $\mathcal{A}^{(z_0)}(B)$ the sub-class of functions $D(z)\in\mathcal{A}(B)$ $(D(z)\not\equiv 0)$ for which

$$\lim_{r\to 1-0}\iint_{B^r_{z_0}} \left|\frac{D(z)}{m(|\varphi(z, z_0)|)}\right|^2 d\sigma < +\infty, \quad m(r)=\min_{z\in L^r_{z_0}}|D(z)| \qquad (8)$$
$$0<r<1.$$

Note that if $D(z)\in\mathcal{A}(B)$ and $|D(z)|\geqslant m>0$ in B, then $D(z)\in\mathcal{A}^{(z_0)}(B)$.

The following two theorems analogous to Theorems 3 and 4, Section 3.2.2, can easily be proved.

Theorem 5 The function $D(z)$ belonging to $\mathcal{A}^{(z_0)}(B)$, $z_0 \in B$ where B is a Caratheodory domain, is a maximal function of class $\mathcal{A}(B)$.

Theorem 6 If for a function $D(z)$ of the class $\mathcal{A}(B)$ where B is a Caratheodory domain, there is a sequence of functions $D_l(z)$, $l=1, 2, \ldots$, of class $\mathcal{A}^{(z_0)}(B)$, $z_0 \in B$ such that

$$\lim_{l\to\infty} \int_B\int \left| \frac{D_l(z)-D(z)}{m_l(|\varphi(z, z_0)|)} \right|^2 d\sigma = 0, \quad m_l(r) = \inf_{z \in L^r_{z_0}} |D_l(z)|, \; 0<r<1$$

then the orthonormalised system of generalised polynomials $\varphi_k(z) = D(z)q_k(z)$, $k=0, 1, \ldots$, for the domain B is closed in $\mathcal{A}(B)$.

Further, we can extend Theorem 6, Section 3.2.2, to a domain of the Caratheodory type. To do this we require the following lemma.

Lemma 2 Let B be a Caratheodory domain and L be the boundary of B; $\alpha > 0$ and $\delta > 0$ be given numbers. Then for any point $\zeta \in L$ accessible from B^1 and integer α, for any point $\zeta \in L$ there is a function $g_\zeta(z)$ regular at internal points of $\bar{B}_*$ and continuous on $\bar{B}_*$, such that

$$|g_\zeta(z)| \leqslant \frac{c_1}{\delta^\alpha}, \quad z \in \bar{B}_* \tag{9}$$

$$|1-(z-\zeta)^\alpha g_\zeta(z)| \leqslant c_2 \frac{\delta^2}{|z-\zeta|^2}, \quad z \in \bar{B}_* \tag{10}$$

where $\arg(z-\zeta)^\alpha$ will be indicated below and c_1 and c_2 depend only on α.

PROOF In the δ neighbourhood of the point ζ let us take an arbitrary Jordan arc $\mathscr{l}$ with diameter $d \geqslant 1/2\,\delta$ and belonging to that domain B^1 of those complementary to B which contains the point ∞, whereby if α is not an integer, then one of the end points of l must coincide with the point ζ (the existance of such an arc l follows easily from the definition of a Caratheodory domain). Let us denote by G the domain which is obtained if we exclude the arc l from the z plane. Let

$$z = z(w) = aw + b + \frac{a_1}{w} + \frac{a_2}{w^2} + \ldots, \quad a > 0$$

be the function mapping the domain $|w| > 1$ conformally onto the domain G.

Since the function $\frac{z(w)-\zeta}{w}$ is regular in the domain $|w| > 1$ continuous in $|w| \geqslant 1$ and $\left|\frac{z(w)-\zeta}{w}\right| \leqslant \delta$ for $|w| = 1$

$$|w| \geqslant \frac{1}{\delta}|z(w)-\zeta|, \quad |w| \geqslant 1 \tag{11}$$

The function

$$f(\xi) = \frac{a}{z\left(\frac{1}{\xi}\right)-\beta} = \xi - \frac{b-\beta}{a}\xi^2 + \ldots$$

where $\beta \in L$ is regular and schlicht in the disc $|\xi| < 1$ and belongs to the class S (see Goluzin [1] p. 138). Let the point $\gamma \in l$ be the furthest from the point β, e.g. the point $\frac{a}{\gamma-\beta}$ be the nearest to the origin of coordinates of the boundary points of the image of $|\xi| < 1$ when mapped by the function $u = f(\xi)$. According to Koebe's theorem (see Goluzin [1] p. 56)

$$\frac{a}{|\gamma-\beta|} \geqslant \frac{1}{4}$$

or assuming that β is chosen such that $|\gamma-\beta| = d$ we find

$$a \geqslant \frac{1}{4}d \geqslant \frac{1}{8}\delta \tag{12}$$

Besides, for the coefficient of ξ^2 in the expansion of the function $f(\xi)$ of class S we have the bound (see Goluzin [1] p. 56)

$$\frac{|b-\beta|}{a} \leqslant 2$$

hence

$$\frac{|b-\zeta|}{a} - \frac{|\zeta-\beta|}{a} \leqslant 2$$

and on the strength of (12)

$$\frac{|b-\zeta|}{a} \leqslant 2 + \frac{\delta}{a} \leqslant 2 + 8 = 10 \tag{13}$$

Let us denote the inverse function to $z = z(w)$ by $w = w(z)$ and put

$$g_\zeta(z) = \frac{1}{a^\alpha w(z)^\alpha}\left[1 - \alpha\frac{b-\zeta}{a}\frac{1}{w(z)}\right]$$

Any branch of the function $g_\zeta(z)$ is regular in the domain B_* and continuous in $\bar{B}_*$. By using (12) and (13), on the arc l we have

$$|g_\zeta(z)| \leqslant \frac{1}{\left(\frac{1}{8}\delta\right)^\alpha}(1+10\alpha) = \frac{c_1}{\delta^\alpha}$$

This estimate holds in G but to obtain (9) it is sufficient to apply it to $\bar{B}_*$.

To prove the inequality (10) let us consider the function $(z-\zeta)^\alpha g_\zeta(z)$. Any branch of it is regular for $z \in G$ or as a function of w is regular in the domain $|w|>1$. Let us choose that branch which is equal to one for $w=\infty$. It is easy to see that the function $h(z)=[1-(z-\zeta)^\alpha g_\zeta(z)]\, w(z)^2$ is regular in G, for such a choice, and on l we have

$$|h(z)| \leqslant 1+(8\delta)^\alpha \frac{1+10\alpha}{\delta^\alpha} = 1+(1+10\alpha)\,8^\alpha = c_2$$

Consequently

$$|1-(z-\zeta)^\alpha g_\zeta(z)| \leqslant \frac{c_2}{w(z)^2}, \quad z \in G$$

Hence, by using (11), we obtain (10) and the lemma is proved.

Let B be a Caratheodory domain; α_k, $\alpha_k > -1$, $k=1, 2, \ldots,$ be real numbers amongst which there is only a finite number of negative ones and such that $\sum_{k=1}^{\infty} \alpha_k < +\infty$; z_k be points not belonging to B whereby if $z_k \in L$ either α_k is an integer or the point z_k is accessible from B^1; a be some point of B. Let us form the function

$$\Phi(z) = \sum_{k=1}^{\infty} \alpha_k \ln\left(1-\frac{z-a}{z_k-a}\right)$$

where by $\ln\left(1-\frac{z-a}{z_k-a}\right)$ is understood that branch which is regular in B and becomes zero for $z=a$ ($\Phi(z)$ is regular in B).

Theorem 7 The function

$$D(z) = Ce^{\Phi(z)} = C\prod_{k=1}^{\infty}\left(\frac{z-z_k}{a-z_k}\right)^{\alpha_k}$$

where C is a constant and $\Phi(z)$ is the function defined above, is a maximal function of class $\mathcal{A}(B)$.

PROOF It is easy to prove that the function $D(z) \in \mathcal{A}(B)$ (see proof of Theorem 6, Section 3.2.2). Let us first prove the theorem for the case when

$$\Phi(z) = \alpha \ln\left(1 - \frac{z-a}{\zeta - a}\right) \tag{14}$$

where ζ is a point lying on the boundary L of the domain B and $\alpha > 0$. We can then assume that $D(z) = (z - \zeta)^{\alpha}$. By virtue of Lemma 2 there is a function $g_{\zeta}(z)$ regular in the domain B_* and continuous on $\bar{B}_*$, for which inequalities (9) and (10) are fulfilled. For such a function $g_{\zeta}(z)$ we have

$$I = \iint_B |1 - (z-\zeta)^{\alpha} g_{\zeta}(z)|^2 \, d\sigma \leqslant \iint_{\substack{|z-\zeta| > \delta \\ z \in B}} |1 - (z-\zeta)^{\alpha} g_{\zeta}(z)|^2 \, d\sigma$$

$$+ 2 \iint_{\substack{|z-\zeta| \leqslant \delta \\ z \in B}} 1 \, d\sigma + 2 \iint_{\substack{|z-\zeta| \leqslant \delta \\ z \in B}} |(z-\zeta)^{\alpha} g_{\zeta}(z)|^2 \, d\sigma$$

$$\leqslant \iint_{|z-\zeta| > \delta} c_2^2 \frac{\delta^4}{|z-\zeta|^4} \, d\sigma + 2\pi\delta^2 + 2 \iint_{|z-\zeta| \leqslant \delta} c_1^2 \frac{|z-\zeta|^{2\alpha}}{\delta^{2\alpha}} \, d\sigma$$

Assuming $z = \zeta + re^{i\vartheta}$ we obtain

$$I \leqslant \int_{\delta}^{\infty} \int_0^{2\pi} c_2^2 \frac{\delta^4}{r^4} \, d\theta r \, dr + 2\pi\delta^2 + 2 \int_0^{\delta} \int_0^{2\pi} c_1^2 \frac{r^{2\alpha}}{\delta^{2\alpha}} \, d\theta r \, dr$$

$$= \pi \left(c_2^2 + 2 + \frac{c_1^2}{1+\alpha}\right) \delta^2$$

Because of the choice of δ (that is because of the choice of $g_{\zeta}(z)$) we can assume that $I < \frac{\varepsilon}{4}$, $\varepsilon > 0$. Since the function $g_{\zeta}(z)$ is regular in the domain B_* and continuous in $\bar{B}_*$, there is a polynomial (see Theorem 3, Section 1.7.4) $u_n(z)$ such that

$$|g_{\zeta}(z) - u_n(z)| \leqslant \sqrt{\frac{\varepsilon}{4M}}$$

$$z \in \bar{B}_*, \quad M = \iint_B |z - \zeta|^{2\alpha} \, d\sigma$$

and consequently

$$\iint_B |1-(z-\zeta)^\alpha u_n(z)|^2\, d\sigma$$

$$=\iint_B |[1-(z-\zeta)^\alpha g_\zeta(z)]+(z-\zeta)^\alpha [g_\zeta(z)-u_n(z)]|^2\, d\sigma$$

$$\leqslant 2\iint_B |1-(z-\zeta)^\alpha g_\zeta(z)|^2\, d\sigma$$

$$+2\iint_B |z-\zeta|^{2\alpha}\, |g_\zeta(z)-u_n(z)|^2\, d\sigma$$

$$\leqslant 2\cdot\frac{\varepsilon}{4}+2\frac{\varepsilon}{4M}\iint_B |z-\zeta|^{2\alpha}\, d\sigma=\varepsilon$$

The theorem is proved for the case (14).

Let us now consider the case when there is only a finite number of points z_k. If amongst the points z_k there is a point lying outside B, then we represent $D(z)$ in the form of the product $D_1(z)\, D_2(z)$,

$$D_1(z)=e^{\Phi_1(z)}, \quad \Phi_1(z)=\sum{}'\alpha_k \ln\left(1-\frac{z-a}{z_k-a}\right)$$

where the prime by the sign $\sum$ denotes that this sum only includes all those terms in which z_k belongs to the contour L of the domain B. Clearly there are two numbers m and M such that

$$0<m\leqslant|D_2(z)|\leqslant M<+\infty, \quad z\in B$$

and by virtue of Theorem 5, $D_2(z)$ is a maximal function of class $\mathcal{A}(B)$ and by virtue of the earlier note on the definition of a maximal function it is sufficient to prove that the function $D_1(z)$ is a maximal function of class $\mathcal{A}(B)$. On the strength of the same note this follows from the fact that the function $(z-\zeta)^\alpha$, $\zeta\in L$, $\alpha>0$ is a maximal function of class $\mathcal{A}(B)$. If $\alpha<0$ then, proceeding as in the proof of Theorem 6, Section 3.2.2, we reduce this case to the previous one.

Let us now consider the general case. Let us write the function $D(z)$ in the form

$$D(z)=D_l(z)\, d_l(z), \;\; d_l(z)=Ce^{\Phi_l(z)}, \;\; \Phi_l(z)=\sum_{k=l+1}^{\infty}\alpha_k \ln\left(1-\frac{z-a}{z_k-a}\right)$$

We shall consider only l for which α_k are positive for $k > l$.

Note that $|d_l(z)| \leqslant M$ in $\overline{B}$ where M is some positive number. In fact, if Δ is the diameter of the domain B and δ is the distance of the point a to the contour L, then for $z \in \overline{B}$ we have

$$|d_l(z)| = |C| \prod_{k=l+1}^{\infty} \left|1 - \frac{z-a}{z_k - a}\right|^{\alpha_k} \leqslant |C| \prod_{k=l+1}^{\infty} \left(1 + \frac{|z-a|}{|z_k - a|}\right)^{\alpha_k}$$

$$\leqslant |C| \prod_{k=l+1}^{\infty} \left(1 + \frac{\Delta}{\delta}\right)^{\alpha_k} \leqslant |C| \left(1 + \frac{\Delta}{\delta}\right)^{\sum_{k=1}^{\infty} |\alpha_k|} = M$$

In the domain B itself $d_l(z) \rightrightarrows 1$ as $l \to \infty$. From the above, $D_l(z)$ is a maximal function of class $\mathcal{A}(B)$ and consequently there is a sequence of polynomials $u_{l,n}(z)$, $n = 0, 1, \ldots$, $l = 1, 2, \ldots$, such that

$$\lim_{n \to \infty} \iint_B |1 - D_l(z) u_{l,n}(z)|^2 \, d\sigma = 0$$

Let us denote the area of the domain B by S and the area of B_r, $0 < r < 1$ by S_r. Given $\varepsilon > 0$, $\varepsilon < \frac{1}{2} S$ and let N_l be such that for $n > N_l$ the inequality

$$\iint_B |1 - D_l(z) u_{l,n}(z)|^2 \, d\sigma < \varepsilon$$

is fulfilled. Further let r, $0 < r < 1$ be such that $S - S_r < \varepsilon$, and finally let N be such that for $l > N$, $z \in B_r$

$$|1 - d_l(z)|^2 < \varepsilon$$

For $n > N_l$ we have

$$\iint_B |D_l(z) u_{l,n}(z)|^2 \, d\sigma \leqslant 2 \iint_B |1 - D_l(z) u_{l,n}(z)|^2 \, d\sigma$$

$$+ 2 \iint_B 1 \, d\sigma \leqslant 2\varepsilon + 2S < 3S$$

$$\iint_{B - B_r} |D_l(z) u_{l,n}(z)|^2 \, d\sigma \leqslant 2\varepsilon + 2(S - S_r) < 4\varepsilon$$

and hence for $l > N$ and $n > N_l$

$$\iint_B |1 - D(z)\, u_{l,n}(z)|^2 \, d\sigma = \iint_B |[1 - D_l(z)\, u_{l,n}(z)]$$

$$+ D_l(z)\, u_{l,n}(z)[1 - d_l(z)]|^2 d\sigma \leqslant 2 \iint_B |1 - D_l(z)\, u_{l,n}(z)|^2 d\sigma$$

$$+ 2 \iint_{B_r} |D_l(z)\, u_{l,n}(z)|^2 \, |1 - d_l(z)|^2 \, d\sigma$$

$$+ \iint_{B-B_r} |D_l(z)\, u_{l,n}(z)|^2 \, |1 - d_l(z)|^2 \, d\sigma \leqslant 2\varepsilon + 2 \cdot 3\varepsilon \cdot S + 2 \cdot 4 \cdot \varepsilon (1 + M)^2$$

In view of the arbitrary smallness of ε we hence conclude that $D(z)$ is a maximal function of class $\mathcal{A}(B)$. The theorem is proved.

The special case of this theorem when α_k are positive integers was proved in a different way by Shaginian [3].

COROLLARY If B is a Caratheodory domain and the function $D(z)$ is regular in B_ρ, $\rho > 1$ and does not become zero in B, then $D(z)$ is a maximal function in class $\mathcal{A}(B)$.

We obtain the proof if we write $D(z)$ in the form

$$D(z) = D_1(z)\, D_2(z)$$

where $D_1(z)$ is a regular function not becoming zero in B_{ρ_1}, $1 \leqslant \rho_1 < \rho$ and $D_2(z)$ is some polynomial, and apply Note 1.

We have considered the closedness of orthonormalised systems of generalised polynomials in Caratheodory domains. To conclude this section let us formulate without proof one of the criteria for the closedness of a system of polynomials for arbitrary, bounded, simply-connected domains.

Theorem 8 (Mergelian [4]*)* Let B be a bounded simply-connected domain and S be a circle enclosing B. The system of polynomials is closed in the domain B if there is an enumerable set of points $z_1, z_2, \ldots, z_k, \ldots$, the closure of which contains the boundary of the domain B and having the following property: to each point z of this set there corresponds a quantity $\delta_0(z)$ such that for any $\delta < \delta_0(z)$ the point z can be joined to S by a straight line l such that

$$\text{mes}\, B_\delta(l) < \exp\left\{-\exp 5\, \frac{\text{lth.}\, l}{\delta}\right\}$$

Here $B_\delta(l)$ is the part of the domain B contained in the δ neighbourhood of the line l; mes $B_\delta(l)$ is the area of $B_\delta(l)$; lth. l is the length of the line l.

A series of other criteria can be found in a review article by Mergelian [3].

3.2.4 *Investigation of series of orthonormalised generalised polynomials*

Let $\varphi_k(z)=D(z)\,q_k(z)$, $k=0, 1, \ldots,$ be a system of generalised polynomials orthonormalised with respect to a Caratheodory domain B which is closed in the class $\mathcal{A}(B)$ and F be a closed sub-set of B. Assuming that in Formula (7), Section 3.2.1, $z=z_0$, $z_0\in F$ we have

$$\sum_{k=0}^{\infty}|q_k(z_0)|^2=\frac{1}{\pi}\frac{|\varphi'(z_0)|}{(1-|\varphi(z_0)|^2)^2}\cdot\frac{1}{|D(z_0)|^2}$$

whence

$$|q_k(z)|\leqslant\sqrt{M_F},\quad z\in F,\quad M_F=\sup_{z\in F}\frac{1}{\pi}\frac{|\varphi'(z)|}{(1-|\varphi(z)|^2)^2}\frac{1}{|D(z)|}\tag{1}$$

Lemma 1 For every simply-connected, bounded Caratheodory domain B and $\sigma>1$ there is a C such that

$$|q_k(z)|\leqslant C\sigma^k,\quad k=0, 1, \ldots, z\in\overline{B}\tag{2}$$

PROOF Let $0<r<1$. For the domain B_r let us construct an external contour line L_r^σ, $\sigma>1$ and let us denote the bounded domain with boundary L_r^σ by B_r^σ. As $r\to 1$ the variable domain B_r converges to the kernel B, for fixed σ the domains B_r^σ converge to the domain B_σ bounded by the external contour line L_σ for B (in accordance with Caratheodory's theorem, see Goluzin [1] p. 62). But $\overline{B}\subset B_\sigma$ and consequently, for r sufficiently close to unity, $\overline{B}\subset B_r^\sigma$. Formula (1) gives

$$|q_k(z)|\leqslant C_r,\quad C_r=\sqrt{M_{\overline{B}_r}},\quad z\in\overline{B}_r$$

and according to Lemma 4, Section 1.2

$$|q_k(z)|\leqslant C_r\sigma^k,\quad z\in B_r^\sigma,\quad k=0, 1, \ldots$$

As $\overline{B}\subset B_r^\sigma$ the lemma is proved.

Remember: if $\varphi_k(z)=D(z)q_k(z)$, $k=0, 1, \ldots,$ is a system of generalised polynomials orthonormalised with respect to the domain B and closed in the class $\mathcal{A}(B)$ then $q_k(z)$ is a system of polynomials orthonormalised with respect to the domain B with the weight $\gamma(z)=|D(z)|^2$ and closed in class $\mathcal{A}(\gamma(z), B)$ and vice-versa.

Theorem 1 Let B be a Caratheodory domain and $q_k(z)$, $k=0,1, \ldots,$ be a system of polynomials orthonormalised with respect to the domain B for the weight $\gamma(z)=|D(z)|^2$ where $D(z)$ is regular in B, the system of being closed in class $\mathcal{A}(\gamma(z), B)$. Let the function $f(z)$ be regular in the domain B_ρ, $\rho>1$ and have a singular point on L_ρ.

1. It can be expanded in a Fourier series by the polynomials $q_k(z)$

$$f(z)=\sum_{k=0}^{\infty} a_k q_k(z) \tag{3}$$

uniformly converging on the compact sub-sets of B_ρ. This series diverges in B_ρ^1.

2. $$\overline{\lim_{k\to\infty}} \sqrt[k]{|a_k|}=\frac{1}{\rho} \tag{4}$$

3. Conversely, if (4) holds, then the series (3) uniformly converges on the compact sub-sets of B_ρ, diverges outside $\overline{B}_\rho$ and is a Fourier series of its sum $f(z)$ and this function is regular in B_ρ and has a singular point on L_ρ.

PROOF By virtue of the closedness of the system of polynomials $q_k(z)$ we have expansion (3), uniformly converging on the compact sub-sets of B.

Let us estimate a_k. According to the Corollary to Theorem 1, Section 1.3.1, there is a sequence of polynomials $Q_n(z)$, $n=0, 1, \ldots,$ such that

$$|f(z)-Q_n(z)|^2 \leqslant \frac{M_{\rho_1}}{\rho_1^{n+1}} \quad z\in\overline{B} \tag{5}$$

where $1<\rho_1<\rho$, and M_{ρ_1} is some constant. We have

$$a_k=\iint_B \gamma(z) f(z)\overline{q_k(z)}\,d\sigma=\iint_B \gamma(z)(f(z)-Q_{k-1}(z))\overline{q_k(z)}\,d\sigma + \iint_B \gamma(z)Q_{k-1}(z)\overline{q_k(z)}\,d\sigma$$

The second term of the right-hand side is equal to zero

since the polynomial $q_k(z)$ is orthogonal to any polynomial of degree less than k, and by virtue of (5) and (2)

$$|a_k| \leqslant \iint_B \gamma(z)|f(z) - Q_{k-1}(z)||q_k(z)|\,d\sigma$$

$$\leqslant \frac{M_{\rho_1}}{\rho_1^k} C\sigma^k \iint_B \gamma(z)\,d\sigma = T\left(\frac{\sigma}{\rho_1}\right)^k$$

whence

$$\sqrt[k]{|a_k|} \leqslant \sqrt[k]{T}\,\frac{\sigma}{\rho_1}, \qquad \overline{\lim_{k\to\infty}} \sqrt[k]{|a_k|} \leqslant \frac{\sigma}{\rho_1}$$

In view of the arbitrary nature of ρ_1 and σ, where $1 < \rho_1 < \rho$, $\sigma > 1$, it follows

$$\overline{\lim_{k\to\infty}} \sqrt[k]{|a_k|} \leqslant \frac{1}{\rho} \tag{6}$$

Let us assume

$$\overline{\lim_{k\to\infty}} \sqrt[k]{|a_k|} = q < \frac{1}{\rho} \tag{7}$$

and arbitrarily fix ρ_1, $\rho < \rho_1 < \frac{1}{q}$. In the domain B_{ρ_1} on the strength of Lemma 1 and Lemma 4, Section 1.2, we have

$$\overline{\lim_{k\to\infty}} \sqrt[k]{|a_k q_k(z)|} < \sigma\rho_1 q, \quad z \in B_{\rho_1}, \quad B_\rho \subset B_{\rho_1}$$

Since $\rho_1 q < 1$, σ can be chosen such that $\sigma\rho_1 q < 1$. Hence it follows that the series (3) converges uniformly in B_{ρ_1} to a function regular in B_{ρ_1} which coincides with $f(z)$ in B_ρ, that is $f(z)$ is regular in B_{ρ_1}. But this is impossible since, according to the premises, $f(z)$ has a singular point on $L_\rho \subset B_{\rho_1}$. Consequently our assumption (7) is invalid and on the strength of (6) Formula (4) is proved.

Let us now prove that the third assertion of the theorem holds. If (4) holds, then as before it can be proved that the series (3) converges uniformly on the compact sub-sets of B to a function regular in B_ρ. By the same token it converges uniformly in the domain B and is therefore the Fourier series of its sum $f(z)$. If we assume that the function $f(z)$ has no singular points on L_ρ, then on the basis of the first (proved) part of the theorem we shall have

$$\overline{\lim_{k\to\infty}} \sqrt[k]{|a_k|} < \frac{1}{\rho}$$

which contradicts (4) and the theorem is proved.

NOTE 1 From the proof of the theorem and the expression for M_{ρ_1} (see Note 5, Section 1.2) it follows that: if the function $f(z)$ is regular in B_ρ, $\rho > 1$ and is bounded there with respect to modulus by the number M, then for every ρ_1, $1 < \rho_1 < \rho$ there is a constant C not dependent on $f(z)$ such that

$$|a_k| \leqslant \frac{CM}{\rho_1^k}, \quad k = 0, 1, \ldots$$

We proved above (see Theorem 1, Section 3.1.3 and Theorem 1, Section 3.1.11) that if $f(z) \in \mathcal{A}(\gamma(z), B)$ and the system of polynomials $q_k(z)$ is closed, then $f(z)$ can be expanded in B in a uniformly converging Fourier series of these polynomials and Parseval's equation holds. We shall now consider the cases when $f(z)$ does not belong to the above-mentioned class. Let us assume that $f(z) \in \mathcal{A}_1(\gamma(z), B)$ whereby $D(z)$ is a function regular in B and the system of polynomials is complete in $\mathcal{A}(\gamma(z), B)$. We can then determine the coefficients of the function $f(z)$

$$a_k = \iint_B \gamma(z) f(z) \overline{q_k(z)}\, d\sigma, \quad k = 0, 1, \ldots \tag{8}$$

The question arises: does the Fourier series of $f(z)$ converge to $f(z)$ or not in this case? If the function $f(z)$ is regular in the domain, but does not belong to the class $\mathcal{A}_1(\gamma(z), B)$, then the integrals (8) may not exist. In this case the question arises: is it impossible to find numbers a_k such that the series $\sum_{k=0}^{\infty} a_k q_k(z)$ converges in B to $f(z)$.

These questions have not been solved for the arbitrary simply-connected domain B. We shall investigate the solution to these problems in the case where the domain B is bounded by a closed analytic Jordan curve.

We shall require two lemmas.

Lemma 2 Let B be a Caratheodory domain and $\varphi_k(z) = D(z) q_k(z)$, $k = 0, 1, \ldots$, be an orthonormalised system of generalised polynomials, the function $D(z)$ being regular in some domain B_ρ, $\rho > 1$. Then for every $\sigma > 1$ and ρ_1, $1 < \rho_1 < \rho$ there is a constant C_1 such that

$$|D(z) q_k(z)| \leqslant C_1 \sigma^k \rho_1^k, \quad k = 0, 1, \ldots, \quad z \in \overline{B}_{\rho_1} \tag{9}$$

This is proved by noting that $D(z)$ is bounded on $\overline{B}_{\rho_1}$ and applying Lemma 1 and then Lemma 4 of Section 1.2.

Let B be a domain bounded by a closed analytic curve, $z_0 \in B$ and $z = \psi(w, z_0)$ be the function mapping the disc $|w| < 1$ conformally on to B, this function being normalised by the conditions $\psi(0, z_0) = z_0$, $\psi'(0, z_0) > 0$, and let $w = \varphi(z, z_0)$ be the inverse function. The function $w = \varphi(z, z_0)$ is regular and schlicht in some domain containing $\bar{B}$ (see Goluzin [1] p. 51), and its inverse function $z = \psi(w, z_0)$ is regular and schlicht in some disc $|w| < \rho$, $\rho > 1$. Let us denote by B_r, $0 < r < \rho$ the image of the circle $|w| < r$ when mapped by the function $z = \psi(w, z_0)$.

Lemma 3 Let B be a domain bounded by a closed analytic curve, and $\varphi_k(z) = D(z) q_k(z)$, $k = 0, 1, \ldots$ be an orthonormalised system of generalised polynomials with respect to the domain B, $D(z)$ being regular on $\bar{B}$. Then for every $r, 0 < r < 1$ there is $\rho_r > 1$ and constant C_r such that

$$|D(z) q_k(z)| \leqslant \frac{C_r}{\rho_r^k}, \quad k = 0, 1, \ldots, \quad z \in \bar{B}_r \tag{10}$$

PROOF For every $\zeta \in \bar{B}_r$ the function

$$w = \varphi(z, \zeta) = e^{i\alpha} \frac{\varphi(z, z_0) - \varphi(\zeta, z_0)}{1 - \overline{\varphi(\zeta, z_0)}\, \varphi(z, z_0)} \tag{11}$$

where α is a real number such that $\varphi'(\zeta, \zeta) > 0$ is regular and schlicht in the domain B_{r_1}, $1 < r_1 < \frac{1}{r}$, $r_1 < \rho$. From (11) follows

$$|\varphi(z, \zeta)| \leqslant \frac{r + r_1}{1 - r r_1} = M, \quad z \in B_{r_1}$$

$$\varphi'(\zeta, \zeta) \leqslant \frac{|\varphi'(\zeta, z_0)|}{1 - r^2} \leqslant \frac{\sup\limits_{\zeta \in \bar{B}_r} |\varphi'(\zeta, z_0)|}{1 - r^2} = M' \tag{12}$$

From Formulae (1) and (5) of Section 3.1.5 we obtain the following expression for the Fourier coefficients of the function $\varphi(z, \zeta)$:

$$d_k = \frac{\pi}{\varphi'(\zeta, \zeta)} D(\zeta) q_k(\zeta), \quad k = 0, 1, \ldots$$

Since $B_{r_1} \supset \bar{B}$ there is $\rho_2 > 1$ such that $\bar{B}_{\rho_2} \subset B_{r_1}$ and the function $D(z)$ is regular and bounded on $\bar{B}_{\rho_2}$, and by virtue of Note 1

$$\frac{\pi}{\varphi'(\zeta, \zeta)} |D(\zeta) q_k(\zeta)| \leqslant \frac{CM}{\rho_2^n}, \quad \zeta \in B_r$$

where C is a constant. Hence and from (12) follows (10) and the lemma is proved.

Theorem 2 Let the domain B be bounded by a closed analytic curve L and $\varphi_k(z)=D(z)q_k(z)$, $k=0, 1, \ldots$, be an orthonormalised system of generalised polynomials with respect to the domain B, $D(z)$ being regular on $\overline{B}$ and differing from zero in B. Then every function $f(z)$, regular in the domain B, can be expanded on to a series

$$f(z)=\sum_{k=0}^{\infty} a_k D(z) q_k(z)$$

uniformly converging on the compact sub-sets of B.

PROOF We make use of the notation of Lemmas 2 and 3. By virtue of the Corollary to Theorem 7, Section 3.2.3, the system $\varphi_k(z)$ is closed in the class $\mathcal{A}(B)$ and in the disc $|w|<1$ we have the closed orthonormalised system of functions

$$Q_k(w)=D(\psi(w, z_0))\,q_k(\psi(w, z_0))\,\psi'(w, z_0), \quad k=0, 1, \ldots$$

From Lemma 2 it follows that in some disc $|w|<\rho_1$, $1<\rho_1<\rho$ the following values hold

$$|Q_n(w)|\leqslant C_3\sigma^n\rho_1^n \qquad n=0, 1, \ldots, \quad |w|<\rho_1 \tag{13}$$

where C_3 is some constant. Let us expand $Q_n(w)$ and $f(\psi(w, z_0))\psi'(w, z_0)$ as a Maclaurin series:

$$\begin{aligned} Q_n(w)&=\sum_{k=0}^{\infty} c_k^{(n)} w^k, \quad |w|<\rho_1 \\ f(\psi(w, z_0))\,\psi'(w, z_0)&=\sum_{k=0}^{\infty} b_k w^k, \quad |w|<1 \end{aligned} \tag{14}$$

For the coefficients $c_k^{(n)}$ and b_k we have the estimates

$$|c_k^{(n)}|\leqslant \frac{C_3\sigma^n\rho_1^n}{\rho_1^k}, \quad |b_k|\leqslant\frac{M_{r_0}}{r_0^k}, \quad \frac{1}{\rho_1}<r_0<1 \tag{15}$$

where

$$M_{r_0}=\sup_{|w|=r_0}|f(\psi(w, z_0))\,\psi'(w, z_0)|$$

Substituting the Fourier expansion of w^k

$$w^k = \pi \sum_{n=0}^{\infty} \frac{\overline{c_k^{(n)}}}{k+1} Q_n(w)$$

in (14) and changing the variable to z, we obtain

$$f(z) = \pi \sum_{k=0}^{\infty} \sum_{n=0}^{\infty} b_k \frac{\overline{c_k^{(n)}}}{k+1} D(z) q_n(z) \tag{16}$$

By using values (15) and Lemma 3, we arrive at the following estimate for the general term of this series for $z \in B_r$, $0 < r < 1$:

$$\left| b_k \frac{\overline{c_k^{(n)}}}{k+1} D(z) q_n(z) \right| \leqslant M_{r_0} C_3 C_2 \left(\frac{\sigma\rho_1}{\rho_2}\right)^n \left(\frac{1}{r_0\rho_1}\right)^k \tag{17}$$

Let us choose ρ_1 and σ so close to one that $\frac{\sigma\rho_1}{\rho_2} < 1$. Then by virtue of the inequality $\frac{1}{r_0\rho_1} < 1$ it follows from the last estimate that the series (16) converges absolutely and uniformly on $\overline{B}_r$, $r < 1$. By altering the order of summation, we obtain a series uniformly converging on $\overline{B}_r$

$$f(z) = \sum_{n=0}^{\infty} a_n D(z) q_n(z), \quad a_n = \pi \sum_{k=0}^{\infty} \frac{b_k \overline{c_k^{(n)}}}{k+1} \tag{18}$$

In view of the arbitrary choice of r, $0 < r < 1$ the series (18) converges uniformly on the compact sub-sets of B and the theorem is proved.

Let us introduce the concept of the 'principal value of an integral'. Let B be a simply-connected bounded domain; z_0 a fixed point from B; B_r, $0 < r < 1$ be the image of the disc $|w| < r$ when mapped by the function $z = \psi(w, z_0)$, $\psi(0, z_0) = z_0$ mapping the disc $|w| < 1$ conformally on to B. Let $f(z)$ be a function measurable in the domain B. If the finite limit

$$\lim_{r \to 1-0} \iint_{B_r} f(z)\, d\sigma$$

exists, then we denote this limit by

$$\iint_{B}^{z_0} f(z)\, d\sigma \tag{19}$$

and say that the integral $\int\limits_B\int f(z)\,d$ exists as a principal value - with respect to the point z_0 - or that the integral (19) exists.

Now let $\varphi_k(z)$, $k=0, 1, \ldots$, be an orthonormal system of functions in the class $\mathcal{A}(B)$ and $f(z)$ be a function regular in B. Let us put

$$a_n = \int\limits_B\int^{z_0} f(z)\overline{\varphi_k(z)}\,d\sigma, \quad k=0, 1, \ldots$$

If all these integrals exist, then the function $f(z)$ can be compared with the series

$$\sum_{k=0}^{\infty} a_k\varphi_k(z)$$

This series is called the special Fourier series of the function $f(z)$ with respect to the point z_0 and the quantities a_k are called the Fourier coefficients of the special Fourier series. Analogously we can introduce a special Fourier series with respect to an orthonormalised system of polynomials with a weight function.

The question arises: under what conditions for the domain B, the function $f(z)$ and the functions of the orthonormalised system does the special Fourier series of the function $f(z)$ converge to $f(z)$ in B?

Let the conditions of Theorem 2 be fulfilled. Let us prove that the series (18) obtained in proving this theorem is the special Fourier series of $f(z)$. To do this let us evaluate the Fourier coefficients of the special Fourier series of the function $f(z)$

$$a_n = \int\limits_B\int^{z_0} f(z)\,D(z)\,q_n(z)\,d\sigma$$

$$= \lim_{r\to 1-0}\int\limits_{|w|<r}\int f(\psi(w, z_0))\psi'(w, z_0)\overline{Q_n(w)}\,d\sigma_w = \pi\lim_{r\to 1-0}\sum_{k=0}^{\infty}\frac{b_k\overline{c_k^{(n)}}}{k+1}r^{2k} \qquad (20)$$

Then from (13) and (15) it follows that

$$\left|\frac{b_k c_k^{(n)}}{k+1}\right| \leqslant C_1 M_{r_0}\sigma^n\rho_1^n\left(\frac{1}{r_0\rho_1}\right)^k, \quad \frac{1}{r_0\rho_1}<1$$

whence it follows that the limit (20) exists and

$$a_n = \pi\sum_{k=0}^{\infty}\frac{b_k\overline{c_k^{(n)}}}{k+1}$$

Thus from Theorem 2 we have the following corollaries.

COROLLARY 1 The series (18) is the special Fourier series of the function $f(z)$.

COROLLARY 2 Under the conditions of the theorem, if $f(z) \in \mathcal{A}_1(B)$ then the series (18) is the series of function $f(z)$. This follows from that fact that in this case the Fourier coefficients of the function $f(z)$ exist and they can be determined by Formulae (20) for any choice of z_0, e.g. the Fourier series of the function $f(z)$ coincides with series (18).

3.2.5 Orthonormalised polynomials with a general weight function

Up to now we have considered an orthonormalised system of polynomials $q_k(z)$, $k=0, 1, \ldots$, for the weight $\gamma(z)= |D(z)|^2$, where $D(z)$ is a function regular in the domain B. Let us now turn to the general case and let $\gamma(z)$ be an arbitrary weight function in the simply-connected domain B (see Section 3.1.11):

$$\iint_B \gamma(z)\, q_n(z)\, \overline{q_m(z)}\, d\sigma = \delta_{n,m}, \quad n, m = 0, 1, \ldots \tag{1}$$

If this system of polynomials $q_k(z)$ is closed, the class of series of the form

$$f(z) = \sum_{k=0}^{\infty} a_k q_k(z) \tag{2}$$

where

$$\sum_{k=0}^{\infty} |a_k|^2 < \infty$$

is a class of Fourier series, for functions $f(z) \in \mathcal{A}(\gamma, B)$ the series (2) converge uniformly on the compact sub-sets of B and Parseval's equation

$$\iint_B \gamma(z) |f(z)|^2 d\sigma = \sum_{k=0}^{\infty} |a_k|^2$$

holds.

NOTE 1 If two weight functions $\gamma_1(z)$ and $\gamma_2(z)$ coincide outside some closed domain $\bar{B}' \subset B$, then the classes $\mathcal{A}(\gamma_1, B)$ and $\mathcal{A}(\gamma_2, B)$ coincide and the systems of orthonormalised polynomials with the given weights are simultaneously either closed or not closed in the given classes of functions.

The domain B' can clearly be considered measurable. The first assertion follows directly from the inequalities

$$\iint_B \gamma_j(z)|f(z)|^2 d\sigma = \iint_{B'} \gamma_j(z)|f(z)|^2 d\sigma + \iint_{B-B'} \gamma_j(z)|f(z)|^2 d\sigma, \quad j=1, 2$$

the second terms on the right-hand side of which are identical for $j=1$ and $j=2$, and the first terms are finite for every function $f(z)$ regular in B. The second assertion follows directly from the analogous inequalities

$$\iint_B \gamma_j(z)|f(z)-u_n(z)|^2 d\sigma = \iint_{B'} \gamma_j(z)|f(z)-u_n(z)|^2 d\sigma + \iint_{B-B'} \gamma_j(z)|f(z)-u_n(z)|^2 d\sigma, \quad j=1, 2$$

Let us indicate certain sufficient conditions for $\gamma(z)$ in the disc K such that the systems of polynomials $q_k(z)$ are closed.

Theorem 1 The system of polynomials $q_k(z)$, $k=0, 1, \ldots,$ is closed in the class $\mathcal{A}(\gamma, K)$ if there is r_0, $0<r_0<1$ such that the weight function $\gamma(z)$ has constant values on all circles $|z|=r$, $r_0<r<1$.

PROOF On the basis of the note given above we can assume that the function $\gamma(z)$ takes on constant values on all circles $|z|=r<1$. Let

$$f(z) = \sum_{k=0}^{\infty} a_k z^k \in \mathcal{A}(\gamma, K)$$

e.g.

$$\iint_{|z|<1} \gamma(z)|f(z)|^2 d\sigma = \int_0^1 \int_0^{2\pi} \gamma(z) \left| \sum_{k=0}^{\infty} a_k r^k e^{i\theta k} \right|^2 d\theta\, r\, dr = 2\pi \sum_{k=0}^{\infty} |a_k|^2 \int_0^1 \gamma(r) r^{2k+1} dr < \infty$$

Hence it follows that the quantity

$$I_n = \iint_{|z|<1} \gamma(z) \left| f(z) - \sum_{k=0}^{n} a_k z^k \right|^2 d\sigma = 2\pi \sum_{k=n+1}^{\infty} |a_k|^2 \int_0^1 \gamma(r) r^{2k+1} dr$$

tends to zero as $n \to \infty$ and the theorem is proved.

Theorem 2 (Dzhrbashian [1]) The system of polynomials $q_k(z)$, $k=0, 1, \ldots,$ is closed in the class $\mathcal{A}(\gamma, K)$ if there are two numbers $A>0$ and r_0, $0<r_0<1$ such that

$$\gamma(z) \leqslant A\gamma(rz), \quad r_0 \leqslant r < 1, \quad r_0 \leqslant |z| < 1$$

PROOF Let $f(z) \in \mathcal{A}(\gamma, K)$ and $P(z)$ be a polynomial which is arbitrary for the moment. Let $r_0^2 < r_1 < r^2 < 1$. We have

$$I = \iint_{|z|<1} \gamma(z)|f(z)-P(z)|^2 d\sigma \leqslant 2 \iint_{|z|<\sqrt{r_1}} \gamma(z)|f(z)-f(rz)|^2 d\sigma$$

$$+2 \iint_{\sqrt{r_1}<|z|<1} \gamma(z)|f(z)-f(rz)|^2 d\sigma + 2 \iint_{|z|<1} \gamma(z)|f(rz)-P(z)|^2 d\sigma$$

We have to prove that the quantity I can be made as small as is desired by the choice of the polynomial $P(z)$. In view of the regularity of the function $f(z)$ for any fixed r_1 the first term on the right-hand side can be made as small as is desired by the choice of r, and since $f(rz)$ is a function which is regular in the closed disc $|z| \leqslant 1$ the last term can also be made as small as is desired by the choice of the polynomial $P(z)$. The theorem will be proved if we prove that the middle term can be made as small as is desired by the choice of r_1. We have

$$\iint_{\sqrt{r_1}<|z|<1} \gamma(z)|f(z)-f(rz)|\, d\sigma \leqslant 2 \iint_{\sqrt{r_1}<|z|<1} \gamma(z)|f(z)|^2 d\sigma$$

$$+2 \iint_{\sqrt{r_1}<|z|<1} \gamma(z)|f(rz)|^2 d\sigma \leqslant 2 \iint_{\sqrt{r_1}<|z|<1} \gamma(z)|f(z)|^2 d\sigma$$

$$+2A \iint_{\sqrt{r_1}<|z|<1} \gamma(rz)|f(rz)|^2 d\sigma = 2 \iint_{\sqrt{r_1}<|z|<1} \gamma(z)|f(z)|^2 d\sigma$$

$$+\frac{2A}{r} \iint_{r\sqrt{r_1}<|z|<r} \gamma(z)|f(z)|^2 d\sigma \leqslant 2\left(1+\frac{A}{r_0}\right) \iint_{r_1<|z|<1} \gamma(z)|f(z)|^2 d\sigma.$$

Since $f(z) \in \mathcal{A}(\gamma, K)$ the right-hand side of this inequality can be made as small as is desired by the choice of r_1 and the theorem is proved.

Let us return to the case of the domain B and, as above, let $z=\psi(w)$ be a function mapping the disc $K(|w|<1)$ on to the domain B and let $w=\varphi(z)$ be the inverse function to this.

We shall say that the weight $\gamma(z)$ satisfies condition I or condition II if the system of functions w^m, $m=0, 1, \ldots$, is closed in the class $\mathcal{A}(\gamma(\psi(w)), K)$ or respectively in the class $\mathcal{A}(\gamma(\psi(w))|\psi'(w)|^2, K)$.

If $q_k(z)$ is an orthonormalised system of polynomials in the class $\mathcal{A}(\gamma(z), B)$ then the systems of functions

$$R_k(w)=q_k(\psi(w))\psi'(w) \text{ and } R_k^*(w)=q_k(\psi(w)), \quad k<0, 1, \ldots,$$

are orthonormalised systems respectively in the classes

$$\mathcal{A}(\gamma(\psi(w)), K) \text{ and } \mathcal{A}(\gamma(\psi(w))|\psi'(w)|^2, K)$$

By using Parseval's equation and returning to the variable z, we arrive at the following result.

Theorem 3 If $\gamma(z)$ satisfies condition I or condition II, then for the system of polynomials $q_k(z)$ with weight $\gamma(z)$ to be closed it is necessary and sufficient that Parseval's equation hold for the functions $\varphi(z)^m \varphi'(z)$, $m=0, 1, \ldots$, or respectively for the function $\varphi(z)^m$, $m=0, 1, \ldots$

If B is a Caratheodory domain, then, as in Section 3.2.2, we have the following theorem.

Theorem 4 If the weight $\gamma(z)$ is bounded in the Caratheodory domain B and satisfies condition I or II, then for the polynomials $q_k(z)$ to be closed in the class $\mathcal{A}(\gamma, B)$ it is necessary and sufficient that there be a sequence of polynomials $u_n(z)$ such that

$$\lim_{n\to\infty} \iint_B \gamma(z)|1-u_n(z)|^2 d\sigma=0$$

In the case where the weight function $\gamma(z)$ satisfies the conditions of Theorem 4, we have valid analogues of the following results: Theorem 6, the note and corollary relating to it, Lemma 2, Theorem 7 and Theorem 8 from Section 3.2.3. All we have to do is to take account of the presence of the weight function (and change $d\sigma$ to $\gamma(z)d\sigma$ everywhere in Section 3.2.3).

For an arbitrary, bounded, simply-connected domain B we have the following theorem.

Theorem 5 If the weight function $\gamma(z)>0$ is bounded in the domain B, satisfies condition I and the inequality

$$\lim_{\delta \to 0} \inf_{\delta(z)=\delta} [\delta(z)]^2 \ln \ln \frac{1}{\gamma(z)} > 0$$

where $\delta(z)$ is the distance of the point z from the boundary of the domain B, then the system of polynomials is closed in the domain B for the weight $\gamma(z)$.

This result was obtained by Tamadian [1] improving on a result of Keldysh [2]. A whole series of results in this field have been obtained by Shaginian (e.g., see [5] and [7]).

Note that up to now we have not found a necessary and sufficient condition which must be satisfied by the weight function $\gamma(z)$ if the system of orthonormalised polynomials $q_k(z)$ with weight $\gamma(z)$ is to be closed in the class $\mathcal{A}(\gamma, B)$ even for the case where the domain B is the disc $|z| < 1$ and the weight $\gamma(z)$ is the square of the modulus of a function regular in the disc $|z| < 1$.

Note also that instead of a system of polynomials orthonormalised with weight $\gamma(z)$ we could consider a system of generalised polynomials $D(z)q_k(z)$ orthonormalised with weight $\gamma(z)$ where $D(z)$ is a function which is regular in the domain B.

3.3 ASYMPTOTIC REPRESENTATION OF POLYNOMIALS ORTHONORMALISED WITH RESPECT TO A DOMAIN

Our task is the asymptotic representation of the polynomials orthonormalised with respect to the domain

$$P_k(z) = \alpha_k z^k + a_1^{(k)} z^{k-1} + \dots + a_k^{(k)}, \quad k = 0, 1, \dots,$$

for large values of k. It is assumed that the boundary L of the domain B is an analytic curve. This representation was established by Carleman [1].

Let $z = \Psi(w) = w + c_0 + \frac{c_1}{w} + \dots$ be a function mapping the domain $|w| > r_0$ conformally on to the domain $B^1 = B^1_{r_0}$ complementary to B with respect to the extended plane (we shall use the same normalisation of the mapping function here as we used in Chapter 2), and let $w = \Phi(z)$ be the inverse function to this. Let us denote by ρ the smallest number such that the function $z = \Psi(w)$ is regular and schlicht in the ring $\rho < |w| < \infty$. As the contour L is an analytic curve,

$\rho < r_0$ (see Goluzin [1] p. 51). We shall denote by B_r^1 (as in Chapter 2) the image of the domain $|w| > r$, $r > \rho$ (when mapped by the function $z = \Psi(w)$); by B_r the complement to $\bar{B}_r^1$ and by L_r the boundary of B_r^1 (and B_r).

3.3.1 *Asymptotic value of the leading coefficient* a_k

First let us consider the function $\Phi(z)^k \Phi'(z)$, $k = 0, 1, \ldots$ The sum of the terms with positive powers of z in the Laurent expansion of this function in the neighbourhood of the point $z = \infty$ is a polynomial $\Pi_k(z)$ of degree k with the leading coefficient equal to one, and we have

$$\Pi_k(z) = \Phi(z)^k \Phi'(z) + \omega_k(z) \tag{1}$$

where the function $\omega_k(z)$ is regular in B_ρ^1 and $\omega_k(\infty) = 0$. We can interpret $\Pi_k(z)$ as a generalised Faber polynomial (see Section 3.3) for the function $R(t) = \Phi'(\Psi(t))$ and for $z \in B_r^1$, $r > \rho$ the following estimate holds (see Section 2.2.1)

$$|\omega_k(z)| < G(r) r^k, \quad z \in B_r^1, \tag{2}$$

where $G(r)$ is constant and depends only on the choice of r and the form of the domain B. Let us now derive this estimate without using the theory of generalised Faber polynomials. Let us put $r' = 1/2\,(r + \rho)$ and represent $\omega_k(z)$ in B_r^1 by means of Cauchy's formula:

$$\omega_k(z) = \frac{1}{2\pi i} \int_{L_{r'}} \frac{\omega_k(\xi)}{\xi - z}\, d\xi, \quad z \in \bar{B}_r^1$$

By using Formula (1) we have

$$\omega_k(z) = -\frac{1}{2\pi i} \int_{L_{r'}} \frac{\Phi(\xi)^k \Phi'(\xi)}{\xi - z}\, d\xi, \quad z \in \bar{B}_r^1$$

whence follows

$$|\omega_k(z)| \leqslant \frac{r'^k G_0(r')\, l_{r'}}{2\pi\delta_r} < \frac{G_0(r')\, l_{r'}}{2\pi\delta_r} r^k$$

where $G_0(r') = \sup_{z \in L_{r'}} |\Phi'(z)|$, $l_{r'}$ is the length of $L_{r'}$ and δ_r is the shortest distance between L_r and $L_{r'}$. The estimate obtained is in the form of (2).

Thus on L_r, $r > \rho$ we have the estimate

$$|\Pi_k(z)| \leqslant |\Phi(z)^k \Phi'(z)| + |\omega_k(z)| < G_0(r) r^k + G(r) r^k = G_1(r) r^k \quad (3)$$

where $G_1(r)$ is a constant of the same nature as $G(r)$. Clearly (3) is valid in $\bar{B}_r$ also.

In the following we shall denote by $G(r)$ (with indices) different constants depending only on the choice of r and the form of the domain B.

NOTE Let $P_n(z) = a_0 z^n + \ldots + a_n$ be a polynomial of degree n. Let us expand the function $P_n(\Psi(w))\Psi'(w)$ in a power series

$$P_n(\Psi(w))\Psi'(w) = a_0 w^n + A_1 w^{n-1} + \ldots + A_n + \sum_{\nu=2}^{\infty} \frac{\alpha_{-\nu}}{w^{\nu}}$$

$$\rho < |w| < \infty.$$

The summation with respect to ν starts from 2 and not 1 since the primitive function for $P_n(\Psi(w))\Psi'(w)$ is regular in the domain $\rho < |w| < \infty$. In particular by virtue of (1) we have the expansion

$$\Pi_k(\Psi(w))\Psi'(w) = w^k + \sum_{\nu=2}^{\infty} \frac{\alpha_{-\nu}^{(k)}}{w^{\nu}}, \quad \rho < |w| < +\infty$$

where

$$\omega_k(\Psi(w))\Psi'(w) = \sum_{\nu=2}^{\infty} \frac{\alpha_{-\nu}^{(k)}}{w^{\nu}}, \quad \rho < |w| < +\infty$$

Let us now find the asymptotic value of α_k for large k. The following lemma holds.

Lemma 1 The following estimates are valid

$$\frac{\pi}{k+1}\left(r_0^{2k+2} - r_1^{2k+2}\right) \leqslant \alpha_k^{-2} \leqslant \int_B\int |\Pi_k(z)|^2 d\sigma \quad (4)$$

$$\leqslant \frac{\pi}{k+1}\left(r_0^{2k+2} + G_2(r_1) r_1^{2k+2}\right), \quad \rho < r_1 < r_0$$

PROOF The polynomial $\Pi_k(z)$ is a polynomial of degree k with leading coefficient equal to one, and on the basis of Theorem 1, Section 3.2.2

$$\alpha_k^{-2} \leqslant \iint_B |\Pi_k(z)|^2 d\sigma$$
$$= \iint_{B_{r_1}} |\Pi_k(z)|^2 d\sigma + \iint_{B-B_{r_1}} |\Pi_k(z)|^2 d\sigma, \quad \rho < r_1 < r_0. \tag{5}$$

By using Formula (3) we have

$$\iint_{B_{r_1}} |\Pi_k(z)|^2 d\sigma \leqslant G_1(r_1)^2 S(r_1) r_1^{2k} \tag{6}$$

where $S(r_1)$ is the area of the domain B_{r_1}. Operating on the basis of the note made above, we obtain

$$\iint_{B-B_{r_1}} |\Pi_k(z)|^2 d\sigma = \iint_{r_1 \leqslant |w| \leqslant r_0} |\Pi_k(\Psi(w))\Psi'(w)|^2 d\sigma$$
$$= \int_{r_1}^{r_0} \left[\int_0^{2\pi} \left| r^k e^{ik\theta} + \sum_{\nu=2}^{\infty} \frac{\alpha_{-\nu}^{(k)}}{r^\nu e^{i\nu\theta}} \right|^2 d\theta \right] r\, dr$$
$$= 2\pi \int_{r_1}^{r_0} \left(r^{2k} + \sum_{\nu=2}^{\infty} \frac{|\alpha_{-\nu}^{(k)}|^2}{r^{2\nu}} \right) r\, dr$$
$$= \frac{\pi}{k+1}(r_0^{2k+2} - r_1^{2k+2}) + 2\pi \sum_{\nu=2}^{\infty} \frac{|\alpha_{-\nu}^{(k)}|^2}{2(\nu-1)} \left(\frac{1}{r_1^{2(\nu-1)}} - \frac{1}{r_0^{2(\nu-1)}} \right)$$

The last term does not exceed

$$\pi \sum_{\nu=2}^{\infty} |\alpha_{-\nu}^{(k)}|^2 \frac{1}{r_1^{2(\nu-1)}}$$
$$= \frac{r_1^2}{2} \int_0^{2\pi} |\omega_k(\Psi(r_1 e^{i\theta}))\Psi'(r_1 e^{i\theta})|^2 d\theta \leqslant \pi r_1^2 G_1(r_1) G_*(r_1) r_1^{2k}$$

where $G_*(r_1) = \sup_{|w|=r_1} |\Psi'(w)|$ and therefore

$$\iint_{B-B_r} |\Pi_k(z)|^2 d\sigma \leqslant \frac{\pi}{k+1}(r_0^{2k+2} - r_1^{2k+2}) + \pi G(r_1) G_*(r_1) r_1^{2k+2}$$

Hence and also from (5) and (6) we obtain the upper bound for α_k^{-2}

To find the lower bound α_k^{-2} let us introduce the function

$$R_k(w) = P_k(\Psi(w))\Psi(w) = \alpha_k w^k + \sum_{\nu=0}^{k-1} A_\nu^{(k)} w + \sum_{\nu=2}^{\infty} \frac{\beta_{-\nu}^{(k)}}{w^\nu}$$

regular in the domain $\rho < |w| < \infty$. Assuming $w = re^{i\theta}$ we find

$$1 = \iint_B |P_k(z)|^2 d\sigma > \iint_{B-B_{r_1}} |P_k(z)|^2 d\sigma = \int_{r_1}^{r_0} \left[\int_0^{2\pi} |R_k(w)|^2 d\theta \right] r\, dr$$

$$= 2\pi \int_{r_1}^{r_0} \left[\alpha_k^2 r^{2k+1} + \sum_{\nu=0}^{k-1} |A_\nu|^2 r^{2\nu+1} + \sum_{\nu=2}^{\infty} \frac{|\beta_{-\nu}^{(k)}|^2}{r^{2\nu-1}} \right] dr$$

$$\geqslant \alpha_k^2 \frac{\pi}{k+1} [r_0^{2k+2} - r_1^{2k+2}]$$

Hence we find the lower bound for α_k^{-2}. The lemma is proved.

COROLLARY The following asymptotic formula holds:

$$\alpha_k = \sqrt{\frac{k+1}{\pi}}\, r_0^{-(k+1)} (1 + \delta_k), |\delta_k| \leqslant G_3(r_1) \left(\frac{r_1}{r_0}\right)^{2k+2}, \ \rho < r_1 < r_0 \quad (7)$$

3.3.2 *Connection between orthonormalised polynomials $P_k(z)$ and generalised Faber polynomials $\pi_k(z)$*

Theorem 1 The following formulae hold

$$P_k(z) = \sqrt{\frac{k+1}{\pi}}\, r_0^{-(k+1)} \Pi_k(z) + \omega_k^*(z), \quad k = 0, 1, \ldots, \quad (1)$$

where for the polynomial $\omega_k^*(z)$ we have the estimates

$$|\omega_k^*(z)| \leqslant \begin{cases} G_4(r)(k+1)\left(\frac{r}{r_0}\right)^{k+1}, & \rho < r < r_0, \ z \in \bar{B}, \quad (2) \\ G_4(r)(k+1)\left(\frac{r}{r_0}\right)^{k+1} \left(\frac{|\Phi(z)|}{r_0}\right)^k, & \rho < r < r_0, \ z \in \bar{B}^1 \quad (3) \end{cases}$$

PROOF Let us express the polynomial $\omega_k^*(z)$ of degree not greater than k by the polynomials $\Pi_s(z)$, $s = 0, 1, \ldots, k$

$$\omega_k^*(z) = \sum_{s=0}^{k} \lambda_s^{(k)} \Pi_s(z) \quad (4)$$

By virtue of the note in Section 3.3.1 we have

$$\Pi_s(\Psi(w))\Psi'(w) = w^s + \sum_{\nu=2}^{\infty} \frac{\alpha_{-\nu}^{(s)}}{w^\nu}$$

Multiplying this equation by $\lambda_s^{(k)}$ and summing with respect to s, we arrive at the equation

$$\omega_k^*(\Psi(w))\,\Psi'(w)=\sum_{s=0}^{k}\lambda_s^{(k)}w^s+\sum_{\nu=2}^{\infty}\frac{\mu_{-\nu}^{(k)}}{w^\nu} \tag{5}$$

Let us integrate the square of the modulus of both sides of this equation with respect to the domain $r_1<|w|<r_0$, $\rho<r_1<r_0$ and change the variable on the left-hand side to z:

$$\iint_{B-B_{r_1}}|\omega_k^*(z)|^2\,d\sigma=\pi\sum_{s=0}^{k}\frac{|\lambda_s^{(k)}|^2}{s+1}\left(r_0^{2s+2}-r_1^{2s+2}\right)$$
$$+\pi\sum_{\nu=2}^{\infty}\frac{|\mu_{-\nu}^{(k)}|^2}{\nu-1}\left(\frac{1}{r_1^{2\nu-2}}-\frac{1}{r_0^{2\nu-2}}\right)\leqslant\iint_B|\omega_k^*(z)|^2\,d\sigma$$

hence, taking (1) into account, we have

$$I_k=\sum_{s=0}^{k}\frac{|\lambda_s^{(k)}|^2}{s+1}\left(r_0^{2s+2}-r_1^{2s+2}\right)\leqslant\frac{1}{\pi}\iint_B|P_k(z)|^2\,d\sigma$$
$$-2\,\mathrm{Re}\,\frac{1}{\pi}\sqrt{\frac{k+1}{\pi}}\,r_0^{-(k+1)}\iint_B P_k(z)\,\overline{\Pi_k(z)}\,d\sigma$$
$$+\frac{k+1}{\pi^2}\,r_0^{-(2k+2)}\iint_B|\Pi_k(z)|^2\,d\sigma$$

Bearing in mind the fact that the polynomial $P_k(z)$ is orthogonal to any polynomial of degree less than k, we can change $\Pi_k(z)$ to $\frac{1}{\alpha_k}P_k(z)$ in the integral under the sign Re:

$$I_k\leqslant\frac{1}{\pi}\left(1-2\sqrt{\frac{k+1}{\pi}}\,r_0^{-(k+1)}\frac{1}{\alpha_k}\right)\iint_B|P_k(z)|^2\,d\sigma$$
$$+\frac{k+1}{\pi^2}\,r_0^{-(2k+2)}\iint_B|\Pi_k(z)|^2\,d\sigma$$

Further let us make use of the estimates (4), Section 3.3.1, for the lower bound of $1/\alpha_k$ and for the upper bound of the last integral

$$I_k\leqslant\frac{1}{\pi}\left(1-2\sqrt{1-\left(\frac{r_1}{r_0}\right)^{2k+2}}\right)$$
$$+\frac{k+1}{\pi^2}\,r_0^{-(2k+2)}\cdot\frac{\pi}{k+1}\left(r_0^{2k+2}+G_2(r_1)\,r_1^{2k+2}\right)$$

or

$$I_k = \sum_{s=0}^{k} \frac{|\lambda_s^{(k)}|^2}{s+1} (r_0^{2s+2} - r_1^{2s+2}) \leqslant \frac{2}{\pi} \left(1 - \sqrt{1 - \left(\frac{r_1}{r_0}\right)^{2k+2}}\right. + \frac{1}{\pi} G_2(r_1) \left(\frac{r_1}{r_0}\right)^{2k+2} \leqslant G_5(r_1) \left(\frac{r_1}{r_0}\right)^{2k+2} \tag{6}$$

Cauchy's inequality gives

$$|\omega_k^*(z)|^2 = \left| \sum_{s=0}^{k} \lambda_s^{(k)} \Pi_s(z) \right|^2 \leqslant \sum_{s=0}^{k} \frac{|\lambda_s^{(k)}|^2}{s+1} (r_0^{2s+2} - r_1^{2s+2}) \sum_{s=0}^{k} \frac{s+1}{r_0^{2s+2} - r_1^{2s+2}} |\Pi_s(z)|^2$$

or, by virtue of (3), Section 3.3.1, and (6), for $z \in L_{r_0}$

$$|\omega_k^*(z)|^2 \leqslant G_5(r_1) \left(\frac{r_1}{r_0}\right)^{2k+2} G_1(r_0)^2 \sum_{s=0}^{k} \frac{s+1}{r_0^2 \left(1 - \left(\frac{r_1}{r_0}\right)^{2s+2}\right)} \leqslant \frac{(k+1)(k+2)}{2(r_0^2 - r_1^2)} G_5(r_1) G_1(r_0)^2 \left(\frac{r_1}{r_0}\right)^{2k+2}$$

Thus the estimate (2) holds on L and consequently also in $\bar{B}$. Since $\omega_k^*(z)$ is a polynomial of degree not greater than k, for $z \in \bar{B}^1$ we have estimate (3) (see Lemma 4, Section 1.2).

3.3.3 *Asymptotic representation of orthonormalised polynomials $P_k(z)$ for large k*

Theorem 1 For $z \in \bar{B} - \bar{B}_r$, $\rho < r < r_0$ the formulae

$$P_k(z) = \sqrt{\frac{k+1}{\pi}}\, r_0^{-(k+1)} \Phi(z)^k \Phi'(z) + \delta_k(z), \quad k = 0, 1, \ldots, \tag{1}$$

hold where

$$|\delta_k(z)| \leqslant G_6(r)(k+1)\left(\frac{r}{r_0}\right)^{k+1}$$

$$\rho < r < r_0, \quad k = 0, 1, \ldots, \quad z \in \bar{B} - \bar{B}_r$$

Theorem 2 For $z \in B^1$ the formulae

$$P_k(z) = \sqrt{\frac{k+1}{\pi}}\, r_0^{-(k+1)} \Phi(z)^k \Phi'(z)(1 + \delta_k^*(z)), \quad k = 0, 1, \ldots, \tag{2}$$

hold where

$$|\delta_k^*(z)| \leqslant G_7(r)\sqrt{k+1}\left(\frac{r}{r_0}\right)^{k+1}, \quad \rho < r < r_0, \quad k=0, 1, \ldots, z \in \bar{B}^1$$

PROOF From Formulae (1), Section 3.3.1, and (1), Section 3.3.2, follows (1) where

$$\delta_k(z) = \sqrt{\frac{k+1}{\pi}}\, r_0^{-(k+1)}\omega_k(z) + \omega_k^*(z)$$

and for $z \in \bar{B} - \bar{B}_r$ by using the estimates (2), Section 3.3.1, and (2), Section 3.3.2, we find

$$|\delta_k(z)| \leqslant \sqrt{\frac{k+1}{\pi}}\, r_0^{-(k+1)} G(r) r^k + G_4(r)(k+1)\left(\frac{r}{r_0}\right)^{k+1}$$
$$\leqslant G_6(r)(k+1)\left(\frac{r}{r_0}\right)^{k+1}$$

and Theorem 1 is proved.

From (1) follows Formula (2) where

$$\delta_k^*(z) = \frac{\omega_k(z)}{\Phi(z)^k \Phi'(z)} + \sqrt{\frac{\pi}{k+1}}\, r_0^{k+1} \frac{\omega_k^*(z)}{\Phi(z)^k \Phi'(z)}$$

Putting $m = \min|\Phi'(z)|$ and using the estimates (2), Section 3.3.1, and (3), Section 3.3.2, for $z \in \bar{B}^1$ we find

$$|\delta_k^*(z)| \leqslant \frac{G(r) r^k}{r_0^k m} + \sqrt{\frac{\pi}{k+1}}\, r_0^{k+1} \frac{G_4(r)(k+1)\left(\frac{r}{r_0}\right)^{k+1}}{r_0^k m}$$
$$\leqslant G_7(r)\sqrt{k+1}\left(\frac{r}{r_0}\right)^{k+1}$$

Theorem 2 is proved.

From Theorems 1 and 2 we obtain the following corollary.

COROLLARY As $k \to \infty$

$$\frac{P_{k+1}(z)}{P_k(z)} \to \frac{\Phi(z)}{r_0}, \qquad \sqrt[k]{|P_k(z)|} \to \frac{|\Phi(z)|}{r_0}$$

uniformly inside B_ρ^1. If $z \in L_r$ then $|\Phi(z)| = r$ and consequently

$$\lim_{k\to\infty} \sqrt[k]{|P_k(z)|} = \frac{r}{r_0}, \quad z \in L_r, \quad r > \rho \tag{3}$$

3.3.4 *Asymptotic behaviour of orthonormalised polynomials with a weight function*

Let us assume as before that the boundary of the domain B is an analytic curve and let the weight function $\gamma(z)$ satisfy the following two conditions:

(1) $\gamma(z)$ is non-negative and integrable in the domain B;

(2) there is ρ_0, $\rho < \rho_0 < r_0$ such that in the domain $B - B_{\rho_0}$ the function $\gamma(z) = |d(z)|^2 \neq 0$ where $d(z)$ is a function which is regular in the domain $B^1_{\rho_0}$ with the possible exception of the point $z = \infty$ the expansion of which in the neighbourhood of the point ∞ is of the form

$$d(z) = \sum_{k=m}^{\infty} \frac{a_k}{z^k}, \quad a_m = a > 0$$

where m is some integer. The case $m = 0$ was investigated by Korovkin [3].

Proceeding as in Sections 3.3.1, 2 and 3, we obtain the asymptotic representation for the polynomials orthonormalised with respect to the domain

$$B\ P_k(z) = \alpha_k z^k + a_1^{(k)} z^{k-1} + \ldots + a_k^{(k)}, \quad k = 0, 1, \ldots,$$

for the weight $\gamma(z)$. As opposed to the above we have to put

$$R(t) = \frac{a\Phi'(\Psi(t))}{d(\Psi(t))\, t^m}$$

Instead of Formula (1), Section 3.3.1, we find

$$\Pi_k(z) = \frac{a\Phi(z)^{k-m}}{d(z)} \Phi'(z) + \omega_k(z), \qquad k = 0, 1, \ldots,$$

where for $\omega_k(z)$ we have an estimate of the form (2), Section 3.3.1, for $r > \rho_0$ whereby ρ is replaced by ρ_0 throughout. Instead of Formula (1), Section 3.3.2, we have the formula

$$P_k(z) = \sqrt{\frac{k - m + 1}{\pi}}\, r_0^{-(k-m+1)} \Pi_k(z) + \omega_k^*(z), \qquad k \geqslant m - 1$$

where

$$|\omega_k^*(z)| \leqslant \begin{cases} G_8(r)(k+1)\left(\frac{r}{r_0}\right)^k, & \rho_0 < r < r_0, \quad z \in \overline{B}, \\ G_8(r)(k+1)\left(\frac{r}{r_0}\right)^k \left(\frac{|\Phi(z)|}{r_0}\right)^k, & \rho_0 < r < r_0, \quad z \in \overline{B}^1 \end{cases}$$

Instead of Formula (1), Section 3.3.3, we have the formula

$$P_k(z) = \sqrt{\frac{k-m+1}{\pi}}\, r_0^{-(k-m+1)} \frac{\Phi(z)^{k-m}}{d(z)} \Phi'(z) + \delta_k(z), \quad z \in \bar{B} - \bar{B}_r$$

where

$$|\delta_k(z)| \leqslant G_9(r)(k+1)\left(\frac{r}{r_0}\right)^k, \quad \rho_0 < r < r_0, \quad z \in \bar{B} - \bar{B}_r$$

and instead of (2), Section 3.3.3.,

$$P_k(z) = \sqrt{\frac{k-m+1}{\pi}}\, r_0^{-(k-m+1)} \frac{\Phi(z)^{k-m}}{d(z)} \Phi'(z)\left[1 + \delta_k^*(z)\right]$$

where

$$|\delta_k^*(z)| \leqslant G_{10}(r)\sqrt{k+1}\left(\frac{r}{r_0}\right)^{k+1}, \quad \rho_0 < r < r_0, \quad z \in \bar{B}^1$$

If we use the estimates for the generalised Faber polynomials given in Section 1.2.1, the expression for $G(r)$ can be found.

3.3.5 *Expansion of a function, regular in $\bar{B}_\rho$, by means of orthonormal polynomials*

We shall retain the notation introduced at the beginning of this section.

Theorem 1 If the function $f(z)$ is regular in the domain B_r, $\rho < r < \infty$ and there is a singular point on L_r, then the following applies:

1. The function $f(z)$ can be expanded in a series with respect to polynomials $P_k(z)$

$$f(z) = \sum_{k=0}^{\infty} c_k P_k(z) \tag{1}$$

uniformly converging on the compact sub-sets of B_r. This series diverges in B_r^1;

2. $$\overline{\lim_{k \to \infty}} \sqrt[k]{|c_k|} = \frac{r_0}{r} \tag{2}$$

3. The expansion of the function $f(z)$ in a series of form (1), uniformly converging in some domain B_{r_1}, $\rho < r_1 < \infty$, is unique;

4. If (2) holds, then the series (1) converges uniformly on the compact sub-sets of B_r and diverges in B_r^1, and the function $f(z)$ defined by the series (1), is regular in the domain B_r and has a singular point on the contour line L_r.

PROOF The case for $r > r_0$ follows directly from Theorem 1, Section 3.2.4, and the corollary to Theorem 2, Section 3.2.2. Let us consider only the case $\rho < r \leqslant r_0$. The function can be expanded in a series of generalised Faber polynomials $\Pi_k(z)$ (see Section 2.2.4):

$$f(z) = \sum_{k=0}^{\infty} a_k \Pi_k(z), \qquad z \in B_r \tag{3}$$

and

$$\overline{\lim_{k\to\infty}} \sqrt[k]{|a_k|} = \frac{1}{r} \tag{4}$$

By using Formula (1), Section 3.3.2, let us write the series (3) in the form

$$f(z) = \sum_{k=0}^{\infty} \sqrt{\frac{\pi}{k+1}}\, r_0^{k+1} a_k P_k(z) + \sum_{k=0}^{\infty} \sqrt{\frac{\pi}{k+1}}\, r_0^{k+1} a_k \omega_k^*(z) \tag{5}$$

By using (4) and (3), Section 3.3.2, we find

$$\overline{\lim_{k\to\infty}} \sqrt[k]{\sqrt{\frac{\pi}{k+1}}\, r_0^{k+1} |a_k| |P_k(z)|} \leqslant r_0 \frac{1}{r} \cdot \frac{r_1}{r_0} = \frac{r_1}{r} < 1$$
$$z \in \bar{B}_{r_1}, \quad \rho < r_1 < r.$$

Hence we conclude that the first series in (5) converges uniformly on the compact sub-sets of B_r to a function regular in B_r.

Let us now consider the second series in (5). By using the estimate (3), Section 3.3.2, we conclude that

$$|\omega_k^*(z)| \leqslant G_5(r_1)(k+1)\left(\frac{r_1}{r_0}\right)^{k+1}\left(\frac{r''}{r_0}\right)^k, \quad z \in B_{r''}, \quad \rho < r_1 < r_0, \quad r'' > r_0 \tag{6}$$

By using (4) we find

$$\overline{\lim_{k\to\infty}} \sqrt[k]{\sqrt{\frac{\pi}{k+1}}\, r_0^{k+1} |a_k| |\omega_k^*(z)|} \leqslant r_0 \frac{1}{r} \frac{r_1}{r_0} \frac{r''}{r_0} = \frac{r_1 r''}{r r_0}$$
$$z \in B_{r''}, \quad r'' > r_0.$$

Since we can choose r_1 as close to ρ as we like, the limit in question does not exceed $\frac{\rho r''}{rr_0}$. This quantity is less than one for $r'' < \frac{rr_0}{\rho}$. Hence we conclude that the second series in (5) converges uniformly on the compact sub-sets of the domain $B_{\frac{rr_0}{\rho}}$ to a function regular in this domain. Since $\frac{r}{\rho} > 1$ the domain $B_{\frac{rr_0}{\rho}}$ contains the domain $\bar{B} = \bar{B}_{r_0}$ and consequently this function can be expanded in a Fourier series

$$\sum_{k=0}^{\infty} b_k P_k(z)$$

uniformly converging on the compact sub-sets of $B_{\frac{rr_0}{\rho}}$. Then

$$\overline{\lim_{k\to\infty}} \sqrt[k]{|b_k|} \leqslant \frac{r_0}{\frac{rr_0}{\rho}} = \frac{\rho}{r}$$

Thus for the function $f(z)$ we have the expansion

$$f(z) = \sum_{k=0}^{\infty} c_k P_k(z), \quad c_k = \sqrt{\frac{\pi}{k+1}}\, r_0^{k+1} a_k + b_k$$

uniformly converging on the compact sub-sets of B_r, and bearing (4) in mind, we obtain

$$\overline{\lim_{k\to\infty}} \sqrt[k]{|c_k|} = \frac{r_0}{r}$$

By using Formula (3), Section 3.3.3, and proceeding as usual we can prove that: if (2) holds, then the series (1) converges uniformly on the compact sub-sets of B_r and diverges in B_r^1.

Let us now prove assertion (3) of Theorem 1. Let us assume that there are two expansions of the function $f(z)$

$$f(z) = \sum_{k=0}^{\infty} c_k' P_k(z), \quad f(z) = \sum_{k=0}^{\infty} c_k'' P_k(z) \tag{7}$$

and the series converge uniformly in B_{r_1}. By virtue of (3), Section 3.3.3, for $c_k = c_k'$ and $c_k = c_k''$ we have

$$\overline{\lim_{k\to\infty}} \sqrt[k]{|c_k P_k(z)|} = \overline{\lim_{k\to\infty}} \sqrt[k]{|c_k|} \cdot \frac{r_1}{r_0} \leqslant 1, \quad z \in L_{r_1}$$

whence

$$\overline{\lim_{k\to\infty}} \sqrt[k]{|c_k'|} \leqslant \frac{r_0}{r_1}, \quad \overline{\lim_{k\to\infty}} \sqrt[k]{|c_k''|} \leqslant \frac{r_0}{r_1} \tag{8}$$

From (7) and (8) we have

$$\sum_{k=0}^{\infty} a_k P_k(z) \equiv 0, \quad a_k = c_k' - c_k'', \quad z \in B_{r_1} \tag{9}$$

and

$$\overline{\lim_{k\to\infty}} \sqrt[k]{|a_k|} \leqslant \frac{r_0}{r_1}$$

From (9) we find by using (1), Section 3.3.2,

$$\sum_{k=0}^{\infty} \sqrt{\frac{k+1}{\pi}} \frac{a_k}{r_0^{k+1}} \Pi_k(z) = -\sum_{k=0}^{\infty} a_k \omega_k^*(z), \quad z \in B_{r_1} \tag{10}$$

Proceeding as above, we show that the series on the left in this equation converges uniformly on the compact sub-sets of B_{r_1} and the series on the right uniformly converges in the domain $B_{\frac{r_1 r_0}{\rho}} \supset B_{r_0}$ to a function regular in this domain. From this we conclude that the sum of the series on the left in (10) is a regular function in the domain $B_{\frac{r_1 r_0}{\rho}}$ and by virtue of the uniqueness of the expansion of this function in a series of polynomials $\Pi_k(z)$ (see Theorem 1, Section 2.2.4) we have

$$\overline{\lim_{k\to\infty}} \sqrt[k]{\sqrt{\frac{k+1}{\pi}} \frac{|a_k|}{r_0^{k+1}}} \leqslant \frac{1}{\frac{r_1 r_0}{\rho}} = \frac{\rho}{r_1 r_0}$$

Consequently

$$\overline{\lim_{k\to\infty}} \sqrt[k]{|a_k|} \leqslant \frac{\rho}{r_1}$$

Hence it follows that the series (9) converges uniformly on the compact sub-sets of $B_{\frac{r_1 r_0}{\rho}} \supset \overline{B_{r_0}}$ to a function regular in this domain identically equal to zero in B_{r_1}. By virtue of the uniqueness of analytic continuation, this function is identically equal to zero in $B_{\frac{r_1 r_0}{\rho}}$ in particular in $\overline{B}$. Multiplying equation (9) by $\overline{P_m(z)}$, $m = 0, 1, \ldots,$ and integrating with respect to the domain B, we find

$$a_m = c_k' - c_k'' = 0, \quad k = 0, 1, \ldots$$

Thus $c_k' = c_k''$, $k = 0, 1, \ldots,$ and the series (7) coincide in spite of our assumption. Assertion (3) of the theorem is proved.

To complete the proof of the theorem it must be shown that the function $f(z)$ defined by the series (1) (for condition (2)) is regular in B_r and has a singular point on the contour line L_r. From (2) and (3), Section 3.3.3, it follows that the series (1) converges uniformly on the compact sub-sets of B_r and consequently $f(z)$ is regular in B_r. If the function $f(z)$ did not have a singular point on L_r, then the function $f(z)$ would be regular in some domain $B_{r'}$, $r' > r$ and by virtue of the uniqueness of the expansion, would have the same expansion (1), whereby the limit (2) would not exceed $\frac{r_0}{r'} < \frac{r_0}{r}$. The theorem is proved.

SUPPLEMENTARY LITERATURE

In addition to the works mentioned in the text we should like to recommend the book by Goncharov [1], Chapter 5, and articles by Mergelian [3, 6], where a long list of references is indicated. We should also like to mention: Aronshain [1], Berman [1], Dzhrbashian [2], Dzhrbashian and Khachatrian [1], Kalashnikov [1], Rosenbloom and Warshawski [1], Suetin [5], and Shaginian [1 to 7].

CHAPTER FOUR

Functions orthogonal with respect to a contour

In Chapter 3 we investigated the theory of analytical functions which were orthogonal with respect to a domain. In the present chapter we shall deal with functions which are orthogonal with respect to the contour of a domain. Throughout this chapter, apart from the supplementary remarks at the end, we shall consider a simply-connected, bounded domain B, the boundary L of which is a rectifiable Jordan curve. An important part will be played by the integrals along L of the limiting values on L of functions regular in B. In this connection we shall require the results relating to limiting values of various classes of functions regular in B. In stating these results and the theory of orthogonal functions we shall make use of the Lebesque integral and fundamental facts of the theory of functions of a real variable. Questions relating to the boundary values of analytic functions are discussed in the well-known monograph by Privalov [1]. For the sake of clarity we shall state in this introduction the fundamental facts relating to boundary values of analytic functions and to certain other questions from the theory of analytic functions.

Let us begin with the case of the unit disc $|z| < 1$. Let us introduce the class $\mathscr{H}_\delta (\delta > 0)$ of functions $f(z)$ which are regular for $|z| < 1$ and such that

$$\int_0^{2\pi} |f(re^{i\theta})|^\delta d\theta \leqslant C, \quad 0 < r < 1$$

where C is a constant dependent only on the choice of $f(z)$. Note that these integrals do not diminish as r increases. If $f(z) \in \mathscr{H}_\delta$ (Privalov [1]), then almost everywhere on the circle $|z|=1$ there exist angular limiting values* of the function $f(e^{i\theta})$, belonging to $\mathscr{L}_\delta$, i.e.

$$\int_0^{2\pi} |f(e^{i\theta})|^\delta d\theta < +\infty$$

and the formulae

$$\lim_{r\to 1-0} \int_0^{2\pi} |f(re^{i\theta})|^\delta d\theta = \int_0^{2\pi} |f(e^{i\theta})|^\delta d\theta$$

$$\lim_{r\to 1-0} \int_0^{2\pi} |f(e^{i\theta}) - f(re^{i\theta})|^\delta d\theta = 0$$

hold. The class $\mathscr{H}_1$ coincides with the class of functions $f(z)$, represented in the disc $|z|<1$ by means of its limiting values on the circle $|z|=1$ by Cauchy's formula:

$$f(z) = \frac{1}{2\pi i} \int_{|z'|=1} \frac{f(z')}{z'-z} dz' = \frac{1}{2\pi} \int_0^{2\pi} \frac{f(e^{i\vartheta})\, e^{i\vartheta}}{e^{i\vartheta} - z} d\vartheta \quad (|z|<1,\ z'=e^{i\vartheta})$$

where the integral is a Lebesque integral.

The class $\mathscr{H}_2$ coincides with the class of functions

$$f(z) = c_0 + c_1 z + c_2 z^2 + \dots \quad (|z|<1)$$

such that

$$\sum_{k=0}^{\infty} |c_k|^2 < +\infty$$

*Let L be a closed, rectifiable Jordan curve, $f(z)$ be a function regular inside L, $\zeta \in L$ and at the point ζ let there be a tangent to L (as is known, this is so for almost all points of L). The quantity A is called the angular limit of the function $f(z)$ at the point ζ if, for any sequence of points z_n, belonging to the inside of L and tending to the ζ remaining in the angle with its vertex at ζ and its bisector along the normal to L with span less than π, $\lim_{n\to\infty} f(z_n) = A$ always.

U

If there are two functions from $\mathscr{H}_2$:

$$f_1(z)=\sum_{k=0}^{\infty} a_k z^k, \quad f_2(z)=\sum_{k=0}^{\infty} b_k z^k$$

then

$$\frac{1}{2\pi}\int_0^{2\pi} f_1(e^{i\vartheta})\,\overline{f_2(e^{i\vartheta})}\,d\vartheta=\sum_{k=1}^{\infty} a_k \overline{b}_k \tag{1}$$

Let us now assume that B is a simply-connected, bounded domain, the boundary L of which is a rectifiable Jordan curve, and let us introduce the class $\mathscr{E}_\delta$, analogous to classes $\mathscr{H}_\delta$ for a circle. Let L_r be the images of the circles $|w|=r\,(0<r<1)$ when a function $z=\psi(w)$ maps the disc $|w|<1$ conformally on to the domain B. Let us introduce the class $\mathscr{E}_\delta$ of functions regular in B and such that

$$\int_{L_r} |f(z)|^\delta\, ds \leqslant C \qquad (0<r<1)$$

where C is a constant depending only on the choice of $f(z)$, and ds is the element of length of the arc L_r. The indicated property of the integrals with respect to L_r does not depend on the choice of conformal transformation $z=\psi(w)$, and $f(z)$ belongs to the class $\mathscr{E}_\delta$ if and only if

$$f[\psi(w)]\,[\psi'(w)]^{\frac{1}{\delta}} \in \mathscr{H}_\delta$$

for some conformal transformation $z=\psi(w)$, and if this holds for one of them, then it will also hold for any other choice of $z=\psi(w)$. The class $\mathscr{E}_\delta$ can be defined, not depending on the conformal transformation, in this way: $f(z)\in\mathscr{E}_\delta$, if there is a sequence of closed, rectifiable Jordan curves L_n, in B and such that B is the kernel of the bounded domains B_n with boundaries L_n, and the integrals

$$\int_{L_n} |f(z)|^\delta\, ds$$

remain bounded as $n\to\infty$.

It is easy to show that if $f(z)\in\mathscr{E}_\delta$ for $\delta=\delta_0$, then $f(z)\in\mathscr{E}_\delta$ for every δ satisfying the condition $0<\delta<\delta_0$. If $f(z)\in\mathscr{E}_\delta$, then almost everywhere on L there exist angular limits $f(z')$, where z' is a variable point on L, whereby

$$\lim_{r\to 1-0}\int_{L_r} |f(z)|^{\delta}\, ds = \int_L |f(z')|^{\delta}\, ds < +\infty$$

The statement $f(z)\in \mathcal{E}_1$ is equivalent to any of the following three statements: (1) $f(z)$ has almost everywhere on L, angular limits $f(z')$ integrable on L, and is expressed by them by Cauchy's formula; (2) $f(z)$ has almost everywhere on L, angular limits $f(z')$, integrable on L, and the formulae

$$\int_L f(z')\, z'^n\, dz' = 0 \qquad (n = 0,\ 1,\ 2, \ldots) \tag{2}$$

hold; (3) the primitive function of $f(z)$ is continuous right up to L and is absolutely continuous on L as a function of s.

The following important statement is also true: if a complex, integrable function $f(z')$, satisfying the relation (2), is defined on L then $f(z')$ is equivalent to the limiting values of some function $f(z)$ from $\mathcal{E}_1$ in B. Clearly all the above is true for a circle changing $\mathcal{E}_\delta$ to $\mathcal{H}_\delta$.

Let B_z and B_w be two simply-connected domains of the type indicated above on the z and w planes with boundaries L_z and L_w, $w = \varphi(z)$ be a conformal transformation of B_z on to B_w, and $z = \psi(w)$ be the inverse function. Then $\varphi(z)$ is continuous in the closed domain B_z, absolutely continuous on L_z and gives a one-one transformation of L_z to L_w. The derivative $\varphi'(z)\in\mathcal{E}_1$ in B_z. Analogous statements are, of course, also true for $\psi(w)$. In changing from integration with respect to L_z to integration with respect to L_w, the element of length ds_z has to be altered to the expression $|\psi'(w)|\, ds_w$.

If $f(z)\in\mathcal{E}_\delta$ in B_z, then $f[\psi(w)]\,[\psi'(w)]^{\frac{1}{\delta}}\in\mathcal{E}_\delta$ in B_w, and the formula

$$F(w) = f[\psi(w)]\,[\psi'(w)]^{\frac{1}{\delta}}$$

gives all functions of $\mathcal{E}_\delta$ in B_w, if we take all functions $f(z)$ from $\mathcal{E}_\delta$ in B_z.

If the function $z = \psi(w)$ maps the disc $|w| < 1$ conformally on to B, then the length l_r of the image L_r of the circle $|w| = r\ (r < 1)$ is expressed by the formula

$$l_r = \int_0^{2\pi} |\varphi'(re^{i\vartheta})|\, d\vartheta$$

and $l_r \to l$, as $r \to 1$, where l is the length of L.

In the following an important part is played by the class $\mathcal{E}_2$. The statement $f(z) \in \mathcal{E}_2$ is equivalent to one of the two following statements: (1) $f(z)$ almost everywhere on L has an angular limit $f(z')$, $|f(z')|^2$ is integrable, and $f(z')$ is expressed through $f(z')$ by Cauchy's formula; (2) $f(z)$ almost everywhere on L has an angular limit $f(z)$, $|f(z)|^2$ is integrable and $f(z')$ satisfies conditions (2).

Let us introduce the so-called parametric representation of functions of class $\mathcal{H}_\delta$. Let $p(\vartheta) = |F(e^{i\vartheta})|$, where $F(w)$ is some function of the class $\mathcal{H}_\delta$ and $F(e^{i\vartheta})$ is its limiting value. Then, as is known, the function $[p(\vartheta)]^\delta$ is integrable on the interval $(0, 2\pi)$. It can be shown that the function $\lg p(\vartheta)$ is also an integrable function. Let us now assume that in the interval $(0, 2\pi)$ we have a given function $p(\vartheta)$, which is positive almost everywhere, whereby $[p(\vartheta)]^\delta$ and $\lg p(\vartheta)$ are integrable functions. It can be proved that there is an infinite class of functions from class $\mathcal{H}_\delta$, for which the modulus of the limiting values is equal to $p(\vartheta)$, and all these functions are obtained from the formula

$$F(w) = e^{i\lambda} w^m b(w) \exp \frac{1}{2\pi} \int_0^{2\pi} \lg p(\vartheta) \frac{e^{i\vartheta} + w}{e^{i\vartheta} - w} d\vartheta \times \exp \frac{1}{2\pi} \int_0^{2\pi} \frac{e^{i\vartheta} + w}{e^{i\vartheta} - w} dq(\vartheta) \tag{3}$$

where $\exp \alpha = e^\alpha$, λ is a real number, m is a non-negative integer, $q(\vartheta)$ is an arbitrary, real non-increasing bounded function, the derivative of which is almost everywhere equal to zero, $b(w)$ is a so-called Blaschke function

$$b(w) = \prod_n \frac{w - \alpha_n}{\bar{\alpha}_n w - 1} \cdot \frac{|\alpha_n|}{\alpha_n}$$

where $\alpha_n \neq 0$ are the roots of $F(w)$ in the disc $|w| < 1$. If they are infinite in number, then the series with the general term $(1 - |\alpha_n|)$ must be convergent. If $F(w)$ does not have roots apart, perhaps, from the root $w = 0$, then in Formula (3) we must put $b(w) \equiv 1$. If $w = 0$ is not a root, then we must put $m = 0$. If $b(w) \not\equiv 1$, then $|b(w)| < 1$ for $|w| < 1$, and the limiting values of $|b(e^{i\vartheta})|$ are equivalent to one.

Formula (3) gives all the functions of class $\mathcal{H}_\delta$ if in an arbitrary manner we choose a real number λ, a non-negative integer m, a real, non-increasing bounded function

$q(\vartheta)$ with a derivative almost everywhere equal to zero, a real function $p(\vartheta)$ positive almost everywhere and such that $[p(\vartheta)]^{\delta}$ and $\lg p(\vartheta)$ are integrable, and numbers α_n for the Blaschke function. The function $p(\vartheta)$ from Formula (3) is equivalent to the modulus of the limiting values of $F(w)$. The function

$$D(w) = \exp \cdot \frac{1}{2\pi} \int_0^{2\pi} \lg p(\vartheta) \cdot \frac{e^{i\vartheta} + w}{e^{i\vartheta} - w} \, d\vartheta \tag{4}$$

is called the maximal function of the class $\mathscr{H}_\delta$ with the given modulus $p(\vartheta)$ of the limiting values. If $F(w)$ is any function with a modulus $p(\vartheta)$ for the limiting values and such that the ratio $F(w) : D(w)$ is not constant then

$$|F(w)| < |D(w)| \quad \text{for} \quad |w| < 1$$

Let B be a simply-connected domain, bounded by the rectifiable Jordan curve L, $z = \psi(w)$ be the conformal transformation of the disc $|w| < 1$ on to B and $w = \varphi(z)$ be the inverse function. Taking into account the fact mentioned above that the following two assertions are equivalent: $f(z) \in \mathfrak{E}_\delta$ and $f[\psi(w)][\psi'(w)]^{\frac{1}{\delta}} \in \mathscr{H}_\delta$, we see that from $f(z) \in \mathfrak{E}_\delta$ it follows that $|f(z')|^{\delta}$ and $|\varphi'(z')| \lg |f(z')|$ must be functions which are integrable on L. Conversely, if we are given a function $p(z')$ which is positive almost everywhere on L such that $[p(z')]^{\delta}$ and $|\varphi'(z')| \lg p(z')$ are integrable on L for some choice of $z = \psi(w)$, then this will also be true for any choice of $z = \varphi(w)$, and $p(z')$ is equivalent to the limiting values of the modulus of certain functions from $\mathfrak{E}_\delta$. All these functions can be obtained in the following manner: Let us construct on the interval $(0, 2\pi)$ a function $p_1(\vartheta) = p[\psi(e^{i\vartheta})] |\psi'(e^{i\vartheta})|^{\frac{1}{\delta}}$, having the properties that $[p_1(\vartheta)]^{\delta}$ and $\lg p_1(\vartheta)$ are integrable on the interval $(0, 2\pi)$. The choice of $z = \psi(w)$ makes no difference. Let us also construct $F(w)$ in accordance with Formula (3) by changing $p(\vartheta)$ to $p_1(\vartheta)$ and let us assume that $f(z) = F[\varphi(z)][\varphi'(z)]^{\frac{1}{\delta}}$. If we choose the maximal function $F(w)$, defined by Formula (4) for $p(\vartheta) = p_1(\vartheta)$, then we obtain the maximal function $f(z)$ with the given limiting value of the modulus: $p(z')$.

Let us introduce another class of functions which is regular in the disc $|w| < 1$, namely the class of functions $\mathscr{D}$

defined by the parametric representation (3) in which m, $b(w)$ and $q(\vartheta)$ have the same properties as for class $\mathcal{H}_\delta$. Besides this it is assumed that $p(\vartheta) \geqslant 0$ almost everywhere and $\lg p(\vartheta)$ is an integrable function. No assumptions are made about the integrability of $p(\vartheta)$ or any power of this function. Let us denote by the symbol $\lg^+\alpha$, where $\alpha > 0$, the function equal to $\lg \alpha$ for $\alpha \geqslant 1$ and equal to zero for $\alpha < 1$. The following statement is true: $F(w) \in \mathcal{D}$ when, and only when, $F(w)$ is regular in the disc $|w| < 1$, has almost everywhere on the circle $|w| = 1$ an angular limit $F(e^{i\theta})$ such that $\ln^+|F(e^{i\theta})|$ is integrable on the interval $(0, 2\pi)$ and

$$\lim_{r \to 1-0} \int_0^{2\pi} \ln^+|F(re^{i\theta})|\, d\theta = \int_0^{2\pi} \ln^+|F(e^{i\theta})|\, d\theta$$

From the parametric representations of $\mathcal{H}_\delta$ and $\mathcal{D}$ the following statements follow directly: if $F(w) \in \mathcal{D}$ and $|F(e^{i\theta})|^\delta$, $\delta > 0$, is integrable on the interval $(0, 2\pi)$, then $F(w) \in \mathcal{H}_\delta$.

In this chapter we shall require the following result. Let $f(z) = c_0 + c_1 z + c_2 z^2 + \ldots$ be a regular function in the disc $|z| < 1$, and $|f(z)| \leqslant M$, $M > 0$, for $|z| < 1$. Let us put $s_k(z) = c_0 + c_1 z + \ldots c_k z^k$ and $\sigma_n(z) = \frac{1}{n+1} \sum_{k=0}^{n} s_k(z)$. Then $|\sigma_n(z)| \leqslant M$ for $|z| \leqslant 1$ and any n (see Titchmarsh [1] p. 268), and on the circle $|z| = 1$, $\sigma_n(e^{i\theta}) \to f(e^{i\theta})$ almost everywhere as $n \to \infty$ (see Zygmund [1]).

In the first section of this chapter we shall follow the exposition in the first section of Chapter 3, omitting proofs which do not differ essentially from those in Chapter 3.

Unless otherwise stated, we shall assume throughout this chapter that B is a simply-connected, bounded domain, the boundary L of which is a rectifiable Jordan curve.

4.1 THEORY OF FUNCTIONS ORTHOGONAL WITH RESPECT TO THE BOUNDARY OF A SIMPLY-CONNECTED DOMAIN

4.1.1 *Class $\mathfrak{E}_2$ and convergence in the mean*

Functions of class $\mathfrak{E}_2$ in the domain B have the property that any finite linear combination of functions $f_k(z)$ from $\mathfrak{E}_2$

$$c_1 f_1(z) + c_2 f_2(z) + \ldots + c_m f_m(z)$$

also belongs to $\mathfrak{E}_2$. Also if $f_1(z)\in\mathfrak{E}_2$ and $f_2(z)\in\mathfrak{E}_2$, then the products $f_1(z')f_2(z')$ and $f_1(z')\overline{f_2(z')}$ are integrable with respect to L. Besides this, the following important inequality holds for the functions $f(z)\in\mathfrak{E}_2$

$$\int_L |f(z')|^2\,ds \geqslant \frac{4\pi^2}{l}\,\rho_{z_0}^2\,|f(z_0)|^2 \tag{1}$$

where z_0 is any point from B, ρ_{z_0} is the distance from z_0 to L and l is the length of L. Here and in the following z' denotes a variable point on L.

To prove this let us write Cauchy's formula

$$f(z_0)=\frac{1}{2\pi i}\int_L \frac{f(z')}{z'-z_0}\,dz'$$

and apply Schwarz's inequality

$$|f(z_0)|^2\leqslant\frac{1}{4\pi^2}\int_L |f(z')|^2\,ds\int_L\frac{ds}{|z'-z_0|^2}$$

whence we obtain (1).

DEFINITION The sequence of functions $\psi_k(z)\in\mathfrak{E}_2$ is said to converge in the mean to the function $\psi(z)\in\mathfrak{E}_2$, if

$$\lim_{k\to\infty}\int_L |\psi_k(z')-\psi(z')|^2\,ds=0$$

It is not difficult to prove that the limiting function $\psi(z')$ is unique. The uniqueness of the limiting function $\psi(z')$ from L_2 on L within the accuracy of the equivalent function is known (Smirnov [2] p. 168), and the function $\psi(z)$ from $\mathfrak{E}_2$ is defined by its limiting values in a unique fashion (Cauchy's formula).

Theorem 1 If the sequence of functions $\psi_k(z)\in\mathfrak{E}_2$ converges in the mean to the function $\psi(z)\in\mathfrak{E}_2$, then this sequence converges uniformly on the compact sub-sets of B to this function $\psi(z)$.

The theorem is proved immediately if inequality (1) is applied to the difference $\psi_k(z)\in\mathfrak{E}_2$.

DEFINITION The sequence of functions $\psi_k(z)$. is said to converge in the mean in itself if

$$\int_L |\psi_k(z')-\psi_l(z')|^2\,ds\to 0 \text{ as } k \text{ and } l\to\infty \tag{2}$$

As is known (Smirnov [2] p. 169) there is a function $\psi(z)$ from L_2 on L, such that

$$\lim_{k\to\infty}\int_L |\psi(z')-\psi_k(z')|^2\,ds=0 \tag{3}$$

From $\psi_k(z)\in\mathscr{E}_2$ it follows that

$$\int_L \psi_k(z')\,z'^n\,dz'=0 \qquad (n=0,\ 1,\ \dots)$$

By virtue of (2) in these equations we can pass to the limit under the integral sign (Smirnov [2] p. 162)

$$\int_L \psi(z')\,z'^n\,dz'=0$$

whence it follows (see introduction) that $\psi(z')$ is equivalent to the limiting values of some function $\psi(z)\in\mathscr{E}_2$. Bearing in mind (3) and Theorem 1, we obtain the following theorem.

Theorem 2 If the sequence of functions $\psi_k(z)\in\mathscr{E}_2$ converges in the mean to itself, then this sequence converges in the mean and uniformly on the compact sub-sets of B to some function $\psi(z)\in\mathscr{E}_2$.

4.1.2 Orthogonal functions - Bessel inequality

Let us begin with the fundamental definition for functions which are orthogonal with respect to a contour.

DEFINITION The sequence of functions

$$\varphi_1(z),\ \varphi_2(z),\ \dots \tag{1}$$

from $\mathscr{E}_2$ is said to be a sequence orthogonal with respect to the contour L if

$$\int_L \varphi_n(z')\,\overline{\varphi_m(z')}\,ds=0 \qquad (n\neq m)$$

We assume, of course, that there are no functions amongst functions (1) which are identically equal to zero. Multiplying these functions by numbers chosen in an appropriate manner, it can be assumed that, retaining the earlier notation, the following equation holds

$$\frac{1}{l}\int_L \varphi_n(z')\overline{\varphi_m(z')}\,ds = \delta_{n,m} \tag{2}$$

where l, is the length of L, as before, and $\delta_{n,m}=0$ for $n \neq m$ and $\delta_{n,n}=1$. Such a system of functions $\varphi_k(z)$ from $\mathfrak{E}_2$ will be said to be orthonormalised. Note that conditions (2) remain if we multiply every function $\varphi_k(z)$ by the quantity $e^{i\omega_k}$, where ω_k is some real number. The normalisation (2) differs in character from that which we used for the integral with respect to a domain in Chapter 3. We have to alter $d\sigma$ to $\frac{1}{l}\,ds$ and the double integral with respect to B to the integral with respect to L. The later formulae will look slightly different as a result.

We shall assume that the system (1) is orthonormalised. If $f(z)$ is some function from $\mathfrak{E}_2$, then

$$a_k = \frac{1}{l}\int_L f(z')\overline{\varphi_k(z')}\,ds \tag{3}$$

are called the Fourier coefficients of $f(z)$ with respect to the orthonormal system (1). The function $f(z)$ is comparable with its Fourier series

$$f(z) \sim \sum_{k=1}^{\infty} a_k\varphi_k(z) \tag{4}$$

Let us form a linear combination of the first n of these functions (1)

$$\sum_{k=1}^{n} d_k\varphi_k(z)$$

where d_k are arbitrary coefficients, and let us consider the quadratic deviation of this sum from $f(z)$

$$I_n = \frac{1}{l}\int_L \left|f(z') - \sum_{k=1}^{n} d_k\varphi_k(z')\right|^2 ds$$

As in Chapter 3 it can be shown that this integral is a minimum for $d_k = a_k$ $(k=1,2,\ldots,n)$ and

$$\frac{1}{l}\int_L \left|f(z') - \sum_{k=1}^{n} a_k\varphi_k(z')\right|^2 ds = \frac{1}{l}\int_L |f(z')|^2\,ds - \sum_{k=1}^{n}|a_k|^2 \tag{5}$$

Whence, as in Chapter 3, follows the Bessel inequality

$$\sum_{k=1}^{\infty} |a_k|^2 \leqslant \frac{1}{l} \int_L |f(z')|^2 ds \tag{6}$$

Let us formulate the result obtained.

Theorem 1 Amongst the finite linear combinations of the function $\varphi_k(z)$

$$P_n(z) = \sum_{k=1}^{n} d_k \varphi_k(z)$$

the best quadratic approximation to the function $f(z) \in \mathscr{E}_2$ on L is given by that for which the coefficients d_k are the Fourier coefficients of the function $f(z) (d_k = a_k)$, and the measure of approximation will then be

$$\frac{1}{l} \int_L |f(z') - \sum_{k=1}^{n} a_k \varphi_k(z')|^2 ds = \frac{1}{l} \int_L |f(z')|^2 ds - \sum_{k=1}^{n} |a_k|^2 \tag{7}$$

Let us assume that the equals sign holds in Formula (6), for example

$$\frac{1}{l} \int_L |f(z')|^2 ds = \sum_{k=1}^{\infty} |a_k|^2 \tag{8}$$

This equation is called Parseval's equation and Parseval's equation is said to hold for the function $f(z)$. It follows directly from (7) that Parseval's equation (8) is equivalent to saying that the partial sums of the Fourier series of the functions $f(z)$

$$S_n(z) = \sum_{k=1}^{n} a_k \varphi_k(z)$$

converge in the mean to $f(z)$, and on the strength of Theorem 1, Section 4.1.1, we have the following theorem.

Theorem 2 If Parseval's equation holds for $f(z) \in \mathscr{E}_2$ then the Fourier series of $f(z)$ converges to $f(z)$ uniformly on the compact sub-sets of B.

Let us now consider a series of the form

$$\sum_{k=1}^{\infty} b_k \varphi_k(z) \tag{9}$$

where b_k are given complex numbers, and let us introduce the partial sum of this series

$$\sigma_n(z) = \sum_{k=1}^{n} b_k \varphi_k(z)$$

The convergence of this sequence in the mean (or of sequence (9) in itself) is equivalent to the convergence of the series

$$\sum_{k=1}^{\infty} |b_k|^2 \tag{10}$$

The following theorem can be proved exactly as in Chapter 3.

Theorem 3 If the series (10) converges, e.g. the series (9) in the mean converges in itself, then the following applies:

1. The series (9) converges uniformly on the compact sub-sets of B to some function $\omega(z) \in \mathcal{E}_2$;
2. b_k are the Fourier coefficients of $\omega(z)$;
3. Parseval's equation holds for the function $\omega(z)$.

4.1.3 Completeness and closure

The discussion is exactly the same as in Section 3.1.3, so we shall merely formulate the results.

DEFINITION The orthonormalised system $\varphi_k(z)$ is said to be complete in the class $\mathcal{E}_2$, if there is no function from class $\mathcal{E}_2$, which differs from identically zero and is orthogonal to all functions $\varphi_k(z)$.

Theorem 1 For every complete system $\varphi_k(z)$ the Fourier series of any function $f(z) \in \mathcal{E}_2$ converges uniformly to $f(z)$ on the compact sub-sets of B and Parseval's equation holds for $f(z)$.

DEFINITION The orthonormal system $\varphi_k(z)$ is said to be closed if Parseval's equation holds for any function from $\mathcal{E}_2$.

Theorem 2 The concepts of completeness and closedness in class $\mathcal{E}_2$ are equivalent.

All the remarks in Section 3.1.3 hold. In particular, the generalised Parseval's equation is of the form

$$\frac{1}{l} \int_L f_1(z') \overline{f_2(z')}\, ds = \sum_{k=1}^{\infty} a_k^{(1)} \overline{a}_k^{(2)} \tag{1}$$

where $a_k^{(1)}$ and $a_k^{(2)}$ are Fourier coefficients of $f_1(z)$ and $f_2(z)$ with respect to a closed orthonormal system.

4.1.4 *Example*

Let B be a disc. The system of functions

$$\psi_k(z) = z^k \ (k = 0, 1\ 2, \ldots)$$

is orthonormal since

$$\frac{1}{2\pi}\int_{|z'|=1} z'^k \overline{z'^l}\, d\vartheta = \frac{1}{2\pi}\int_0^{2\pi} e^{i(k-l)\vartheta}\, d\vartheta = \delta_{k,l}$$

For any function

$$f(z) = \sum_{k=0}^{\infty} c_k z^k$$

from $\mathscr{H}_2$ the coefficients c_k are the Fourier coefficients

$$c_k = \frac{1}{2\pi}\int_0^{2\pi} f(e^{i\vartheta})\, \overline{z^k}\, d\vartheta, \quad z = e^{i\vartheta}$$

and Parseval's equation (see introduction to chapter)

$$\frac{1}{2\pi}\int_0^{2\pi} |f(e^{i\vartheta})|^2\, d\vartheta = \sum_{k=1}^{\infty} |c_k|^2$$

holds, e.g. the orthonormal system of functions z^k is closed.

From the above it follows that functions from $\mathscr{H}_2$ can be expanded in the disc $|z| < 1$ by means of the process z^k, i.e. into a Maclaurin series. But, as is known, this is a property of every function regular in a disc $|z| < 1$. In the following we shall prove an analogous property, as in the previous chapter, for domains B, with analytic boundaries.

4.1.5 *Fundamental bi-linear series*

Let $\varphi_k(z)$ $(k = 1, 2, \ldots)$ be an orthonormal system of functions from $\mathfrak{E}_2$ in the domain B and let F be a compact sub-set of B. Exactly as in Section 3.1.5 we can prove that

$$\sum_{k=1}^{\infty} |\varphi_k(\zeta)|^2 \leqslant \frac{l}{4\pi^2\rho_F^2}, \quad \zeta \in F$$

where ρ_F is the distance from F to L, and the following theorem holds.

Theorem 1 For any fixed ζ from B, the bi-linear series

$$\sum_{k=1}^{\infty} \overline{\varphi_k(\zeta)}\, \varphi_k(z)$$

converges uniformly on the compact sub-sets of B to some regular function $K(z, \zeta) \in \mathfrak{E}_2$; $\overline{\varphi_k(\zeta)}$ are the Fourier coefficients of the function $K(z, \zeta)$ and Parseval's equation

$$\frac{1}{l} \int_L |K(z', \zeta)|^2\, ds = \sum_{k=1}^{\infty} |\varphi_k(\zeta)|^2 = K(\zeta, \zeta) \tag{1}$$

Let us now establish the connection between the bi-linear series and the function mapping the domain B conformally on to the disc $|w| < 1$. Let us assume that $w = \varphi(z_0, z_0)$ is the function amongst those just indicated which is normalised by the conditions $\varphi(z_0, z_0) = 0$ and $\varphi_z'(z_0, z_0) > 0$, where z_0 is a fixed point of B. Note that $\varphi_z'(z, z_0) \in \mathfrak{E}_1$ and consequently $\sqrt{\varphi'(z, z_0)} \in \mathfrak{E}_2$, assuming $\sqrt{\varphi_z'(z_0, z_0)} > 0$.

Theorem 2 If $\varphi_k(z)$ $(k = 1, 2, \ldots)$ is an arbitrary closed system of functions from $\mathfrak{E}_2$ in the domain B, then

$$\begin{aligned} K(z, z_0) &= \frac{l}{2\pi} \sqrt{\varphi_z'(z_0, z_0)} \cdot \sqrt{\varphi_z'(z, z_0)} \\ \varphi_z'(z, z_0) &= \frac{2\pi}{lK(z_0, z_0)} K(z, z_0)^2 \end{aligned} \tag{2}$$

PROOF Let $z = \psi(w)$ be the inverse function to $w = \varphi(z, z_0)$, such that $\psi(0) = z_0$ and $\psi'(0) > 0$. Let us define the Fourier coefficients of the function $\sqrt{\varphi'(z, z_0)} \in \mathfrak{E}_2$ by:

$$a_k = \frac{1}{l} \int_L \sqrt{\varphi'(z', z_0)}\, \overline{\varphi_k(z')}\, ds \quad \text{of} \quad \bar{a}_k = \frac{1}{l} \int_L \varphi_k(z') \overline{\sqrt{\varphi'(z', z_0)}}\, ds$$

Let us change the integration variable by means of $z' = \psi(w')$ where here and in the following we assume that $w' = e^{i\vartheta}$.

$$\bar{a}_k = \frac{1}{l} \int_0^{2\pi} \varphi_k [\psi(w')] \overline{\sqrt{\varphi' [\psi(w'), z_0]}} \, |\psi'(w')| \, d\vartheta$$

$$= \frac{1}{l} \int_0^{2\pi} \varphi_k [\psi(w')] \frac{|\psi'(w')|}{\overline{\sqrt{\psi'(w')}}} \, d\vartheta$$

Putting $w' = e^{i\vartheta}$, we find

$$\bar{a}_k = \frac{1}{li} \int_{|w'|=1} \frac{\varphi_k [\psi(w')] \sqrt{\psi'(w')}}{w'} \, dw'$$

whence, since $\varphi_k [\psi(w)] \sqrt{\psi'(w)} \in \mathscr{H}_2$ and is therefore expressed by Cauchy's formula, we have

$$\bar{a}_k = \frac{2\pi}{l} \varphi_k (z_0) \sqrt{\psi'(0)} = \frac{2\pi}{l} \frac{\varphi_k (z_0)}{\sqrt{\varphi'(z_0, z_0)}}$$

Bearing in mind that the system $\varphi_k(z)$ is closed, we obtain the expansion

$$\sqrt{\varphi'(z, z_0)} = \frac{2\pi}{l \sqrt{\varphi'(z_0, z_0)}} \sum_{k=1}^{\infty} \overline{\varphi_k (z_0)} \varphi_k (z)$$

uniformly converging on the compact sub-sets of B, whence follows the first of Formulae (2). For $z = z_0$ this gives

$$\varphi'(z_0, z_0) = \frac{2\pi}{l} K(z_0, z_0) \tag{3}$$

whence

$$K(z, z_0) = \sqrt{\frac{l}{2\pi}} \sqrt{K(z_0, z_0)} \cdot \sqrt{\varphi'_z(z, z_0)}$$

which leads to the second of Formulae (2).

Theorem 3 If $\varphi_k(z)$ $(k = 1, 2, \ldots)$ is an arbitrary closed orthonormal system of functions from $\mathscr{E}_2$ and $f(z) \in \mathscr{E}_2$, then for any z_0 from B

$$f(z_0) = \frac{1}{l} \int_L \overline{K(z', z_0)} f(z') \, ds \tag{4}$$

PROOF Note that both factors under the integral sign belong to $\mathscr{L}_2$ on L and the integral exists. By virtue of (2) we have

$$\frac{1}{l}\int_L \overline{K(z', z_0)}\, f(z')\, ds = \frac{\sqrt{\varphi'(z_0, z_0)}}{2\pi}\int_L \overline{\sqrt{\varphi'(z', z_0)}}\, f(z')\, ds$$

or, changing the variable to $w' = e^{i\vartheta}$,

$$\frac{1}{l}\int_L \overline{K(z', z_0)}\, f(z')\, ds = \frac{1}{2\pi\sqrt{\psi'(0)}}\int_{|z'|=1} \frac{f[\psi(w')]}{\overline{\sqrt{\psi'(w')}}}\,|\psi'(w')|\,\frac{dw'}{iw'}$$

$$= \frac{1}{2\pi i\sqrt{\psi'(0)}}\int_{|w'|=1} \frac{f[\psi(w')]\sqrt{\psi'(w')}}{w'}\, dw'$$

The function $f[\psi(w)]\sqrt{\psi'(w)} \in \mathscr{H}_2$, and on the basis of Cauchy's formula we can show that the last expression is equal to $f(z_0)$, which proves Formula (4).

In the case of the disc $|z| < 1$ we have for the closed, orthonormalised system z^k $(k = 0, \mathrm{i}, \ldots)$

$$K(z, z_0) = \sum_{k=0}^{\infty} \bar{z}_0^k z^k = \frac{1}{1 - \bar{z}_0 z}$$

and Formula (4) is in the form

$$f(z_0) = \frac{1}{2\pi i}\int_{|z'|=1} \frac{1}{z' - z_0} f(z')\, dz'$$

e.g. it is transformed into the Cauchy formula and holds for the class $\mathscr{H}_1$. The following is easily proved.

Supplement to Theorem 3 If $K(z, z_0)$ for some fixed $z_0 \in B$ is bounded in B, then Formula (4) holds for $f(z) \in \mathfrak{E}_1$ for every $z_0 \in B$.

Note first of all that saying $K(z, z_0)$ is bounded is equivalent to saying that $\varphi'(z, z_0)$ is bounded in B and that, under the given assumptions, the function under the integral sign in Formula (4) is the product of a bounded and an integrable function, i.e. is an integrable function. Further it follows from $f(z) \in \mathfrak{E}_1$ that (see introduction) $f[\psi(w)]\psi'(w) \in \mathscr{H}_1$, and from the formula

$$f[\psi(w)]\sqrt{\psi'(w)} = f[\psi(w)]\psi'(w)\sqrt{\varphi'(z, z_0)}$$

and the fact that $\sqrt{\varphi'(z, z_0)}$ is bounded, it follows that $f[\psi(w)]\sqrt{\psi'(w)} \in \mathscr{H}_1$, also, i.e. Cauchy's formula can be applied to this function, and Formula (4) can be proved exactly as before. For different choices of z_0, the functions $w=\varphi(z, z_0)$ can be obtained from each other by means of bi-linear transformations preserving the unit circle. Hence it follows that if $K(z, z_0)$ is bounded for some choice of z_0 from B, then it is also bounded for every other choice of z_0 from B.

4.1.6 *Connection between conformal mapping and functions orthonormalised with respect to the boundary of a domain*

Let B_z be a bounded, simply-connected domain of the z plane, the boundary L_z of which is a rectifiable Jordan curve, and let B_w be a domain of the same type on the w plane with boundary L_w. Let $w=\varphi(z)$ be the function conformally mapping B_z on to B_w, and $z=\psi(w)$ be the inverse function. We indicated the fundamental properties of $\varphi(z)$ and $\psi(w)$ in the introduction. We shall now formulate results which are proved in an exactly analogous manner to those obtained in the corresponding context in Chapter 3.

If there is an orthonormal system $\varphi_k(z)\in \mathcal{E}_2^{(z)}$, then the functions

$$\psi_k(w)=\sqrt{\frac{l_w}{l_z}}\,\varphi_k[\psi(w)]\sqrt{\psi'(w)} \tag{1}$$

where l_z and l_w are the lengths of L_z and L_w form an orthonormal system in B_w. If the system $\varphi_k(z)$ is closed in B_z, the system $\psi_k(w)$ is closed in B_w. Note that from (1) follows

$$\varphi_k(z)=\sqrt{\frac{l_z}{l_w}}\,\psi_k[\varphi(z)]\sqrt{\varphi'(z)} \tag{2}$$

If B is a simply-connected domain and $w=\varphi(z, z_0)$ is the function mapping B conformally on to the disc $|w|<1$, as before, then the system of functions

$$\varphi_k(z)=[\varphi(z, z_0)]^k\sqrt{\varphi'(z, z_0)} \qquad (k=0, 1, \dots)$$

is a closed orthonormal system.

4.1.7 *Extremal problems*

Let $\varphi_k(z)\in \mathcal{E}_2$ be an orthonormal system of functions, $\mathcal{E}^n$ be a class of functions of the form $\omega_n(z)=\sum_{k=1}^{n} a_k\varphi_k(z)$, where a_k

are arbitrary numbers, and $\mathcal{E}^{\infty}$ be a class of functions $\sum_{k=1}^{\infty} a_k \varphi_k(z)$, where a_k are arbitrary numbers satisfying the condition that $\sum_{k=1}^{\infty} |a_k|^2 < +\infty$. Clearly $\mathcal{E}^n \subset \mathcal{E}^{\infty} \subset \mathcal{E}_2$. If the system is closed, then $\mathcal{E}^{\infty} = \mathcal{E}_2$. Just as in Section 3.1.7, the following theorems hold.

Theorem 1 Amongst the functions $\varphi_k(z)$ $(k = 1, 2, \ldots, n)$ let there be functions not equal to zero at some point $z_0 \in B$. Let us denote

$$I_n = \frac{1}{l} \int_L |\omega_n(z)|^2 \, ds \tag{1}$$

Then for the functions $\omega_n(z) \in \mathcal{E}^n$ satisfying the condition $w_n(z_0) = \alpha$, where α is a number different from zero, we have

$$\mu_n(z_0) = \min I_n = \frac{|\alpha|^2}{\sum_{k=1}^{n} |\varphi_k(z_0)|^2} \tag{2}$$

and $I_n = \mu_n(z_0)$ only for the function

$$\omega_n(z) = \alpha \frac{\sum_{k=1}^{n} \overline{\varphi_k(z_0)}\, \varphi_k(z)}{\sum_{k=1}^{\infty} |\varphi_k(z_0)|^2} = \alpha \frac{K_n(z, z_0)}{K_n(z_0, z_0)} \tag{3}$$

Theorem 2 Amongst the functions $\varphi_k(z)$ $(k = 1, 2, \ldots)$ let there be functions which are not equal to zero at some point $z_0 \in B$. Let us denote

$$I_{\infty} = \frac{1}{l} \int_L |\omega(z)|^2 \, ds \tag{4}$$

Then for the functions $\omega(z) \in \mathcal{E}^{\infty}$ and satisfying the condition $\omega(z_0) = \alpha$, where α is a number differing from zero, we have

$$\mu_{\infty}(z_0) = \min I_{\infty} = \frac{|\alpha|^2}{\sum_{k=1}^{\infty} |\varphi_k(z_0)|^2} \tag{5}$$

and $I_{\infty} = \mu_{\infty}(z_0)$ only for the function

$$\omega(z) = \alpha \frac{\sum_{k=1}^{\infty} \overline{\varphi_k(z_0)}\, \varphi_k(z)}{\sum_{k=1}^{\infty} |\varphi_k(z_0)|^2} = \alpha \frac{K(z, z_0)}{K(z_0, z_0)} \tag{6}$$

If the system $\varphi_k(z)$ is closed, then $\mathcal{E}^{\infty}$ coincides with $\mathcal{E}_2$, and we can formulate the following result: the expression

$$I = \frac{1}{l} \int_L |f(z)|^2\, ds \tag{7}$$

in the class of functions $f(z) \in \mathcal{E}_2$ and satisfying the condition $f(z_0) = \alpha$ has a minimum equal to the expression (5) and only reaches it for the function (6). Note that if the system $\varphi_k(z)$ is closed amongst these functions there are always functions which differ from zero at any point $z_0 \in B$. This follows, for example, from the fact that any constant should be expandable into a Fourier series with respect to the functions $\varphi_k(z)$ (see Section 3.1.7).

The Schmidt orthogonalisation process and the formulae from Section 3.1.8 are still applicable if the integrals over B are altered to integrals along L, divided by l.

Analogously all the results of Sections 3.1.9 and 10 are preserved. In the following section we shall consider orthogonal polynomials in the presence of weight.

Note one other extremal problem which we investigated for the integral with respect to a domain in the previous chapter. We shall only give the result.

Let $P_k(z)$ $(k = 0, 1, \ldots)$ be a system of polynomials which is orthogonal on the boundary L of the domain B, and $\alpha_n > 0$ be the leading coefficient of the polynomial $P_n(z)$. The polynomials $P_k(z)$ are obtained by orthogonalisation of the powers z^k $(k = 0, 1, \ldots)$ with respect to L. The extremal problem consists of the following: to find the polynomial $Q_n(z) = z^n + c_1 z^{n-1} + \ldots + c_n$ of degree n with leading coefficient one, for which the integral

$$I_n = \frac{1}{l} \int_L |Q_n(z)|^2\, ds$$

has its smallest value. This problem has a unique solution

$$Q_n(z) = \alpha_n^{-1} P_n(z)$$

and for this $I_n = \alpha_n^{-2}$.

4.2 POLYNOMIALS ORTHOGONAL WITH RESPECT TO THE BOUNDARY OF A DOMAIN

4.2.1 *Polynomials orthonormalised on the unit circle with respect to a weight function*

There are considerable difficulties in investigating the closedness of a system of polynomials $P_k(z)$ orthonormalised with respect to the boundary L of a domain B. As we shall see the type of system $P_k(z)$ in question is not closed for every domain B. We shall start by considering systems of polynomials orthonormalised on the circle $|z|=1$ in the presence of a weight function $p(\vartheta)$ with definite properties. This will then be used to consider the closedness of systems $P_k(z)$ for any domain B, the boundary of which is a rectifiable Jordan curve.

On the interval $0 \leqslant \vartheta \leqslant 2\pi$ let there be a function $p(\vartheta)$, having the following properties: (1) $p(\vartheta) > 0$ almost everywhere; (2) $p(\vartheta)$ and $\lg p(\vartheta)$ are integrable in the interval $[0, 2\pi]$. Further let $D(z)$ be a maximal function of $\mathscr{H}_2$ in the disc $|z|<1$ which, on the circle $|z|=1$, has limiting values of the modulus equal to $\sqrt{p(\vartheta)}$ e.g.

$$D(z) = \exp \frac{1}{2\pi} \int_0^{2\pi} \lg \sqrt{p(\vartheta)} \, \frac{e^{i\vartheta}+z}{e^{i\vartheta}-z} \, d\vartheta \tag{1}$$

Note that

$$D(0) = \exp \frac{1}{2\pi} \int_0^{2\pi} \lg \sqrt{p(\vartheta)} \, d\vartheta$$

is a real, positive number. In the right-hand part of (1) we could add a numerical factor with a modulus of one. This would have no effect on the following discussion but instead of $[D(0)]^2$ we would have to write $|D(0)|^2$.

Applying the orthogonalisation process to the functions $D(z)z^k$ $(k=0, 1, \ldots)$ we obtain a system of functions orthonormalised in the disc $|z|<1$

$$\varphi_k(z) = D(z)\, q_k(z) \qquad (k=0, 1, \ldots) \tag{2}$$

where $q_k(z)$ is a polynomial of degree k with a positive coefficient for z^k. The functions $\varphi_k(z) \in \mathscr{H}_2$, and we have

$$\frac{1}{2\pi}\int_0^{2\pi} D(z')\,q_n(z')\,\overline{D(z')\,q_m(z')}\,d\vartheta = \delta_{n,m}. \quad (z' = e^{i\vartheta})$$

or

$$\frac{1}{2\pi}\int_0^{2\pi} p(\vartheta)\,q_n(e^{i\vartheta})\,\overline{q_m(e^{i\vartheta})}\,d\vartheta = \delta_{n,m}, \tag{3}$$

since $|D(z')|^2 = p(\vartheta)$. We can regard $q_k(z)$ as polynomials orthonormalised on the circle $|z| = 1$ in the presence of a weight $p(\vartheta)$. In the following we shall base our discussions on the orthonormalised system (2) and prove that it is closed.

To do this it is sufficient to prove that Parseval's equation holds for the functions z^m, $m = 0, 1, \ldots$ (see Section 3.1.9). A linear combination of the first $n+1$ functions (2) is a function of the form $D(z)S_n(z)$, where $S_n(z)$ is a polynomial of degree not greater than n, and whereby the polynomial $S_n(z)$ can be any polynomial if a suitable combination of functions (2) is chosen. Bearing this in mind, we see that to prove that the system of functions (2) is closed it is sufficient for us to prove the following: there is a sequence of polynomials $u_{n,m}(z)$, such that

$$\lim_{n\to\infty}\int_0^{2\pi} |z'^m - D(z')\,u_{n,m}(z')|^2\,d\vartheta = 0 \qquad (m = 0, 1, \ldots) \tag{4}$$

where $z' = e^{i\vartheta}$. But it is clearly sufficient to prove this for $m = 0$. If $u_n(z)$ is a corresponding sequence of polynomials for which the condition (4) is fulfilled for $m = 0$, then for z^m the sequence of polynomials satisfying condition (4) will be $u_{n,m}(z) = z^m u_n(z)$, since $|z'| = 1$. Thus proving the system of functions (2) is closed reduces to proving the following.

Lemma 1 There is a sequence of polynomials $u_n(z)$, such that

$$\lim_{n\to\infty}\int_0^{2\pi} |1 - D(z')\,u_n(z')|^2\,d\vartheta = 0 \tag{5}$$

Let us determine the Fourier coefficients a_k of the function $f(z) \equiv 1$ as follows on the next page.

$$\overline{a}_k = \frac{1}{2\pi}\int_0^{2\pi} D(z')\, q_k(z')\, d\vartheta = \frac{1}{2\pi i}\int_{|z'|=1} \frac{D(z')\, q_k(z')}{z'}\, dz'$$

The function $D(z)\, q_k(z) \in \mathscr{H}_2$ and Cauchy's formula can be applied to it, and from the above formula it follows that $\overline{a}_k = D(0)\, q_k(0)$, e.g.

$$a_k = D(0)\, \overline{q_k(0)} \tag{6}$$

Let us put $D(z)\, u_n(z)$ equal to the partial sum of the Fourier series of the function $f(z) \equiv 1$, e.g.

$$u_n(z) = \sum_{k=0}^{n} D(0)\, \overline{q_k(0)}\, q_k(z) \tag{7}$$

Then as we know

$$\frac{1}{2\pi}\int_0^{2\pi} |1 - D(z')\, u_n(z')|^2\, d\vartheta = 1 - \sum_{k=0}^{n} |a_k|^2$$

and it remains to show that

$$\sum_{k=0}^{\infty} |a_k|^2 = 1 \tag{8}$$

e.g. on the strength of (6) that

$$\sum_{k=0}^{\infty} |q_k(0)|^2 = \frac{1}{[D(0)]^2} \tag{9}$$

Note that the constant $q_0(z)$ is different from zero. First let us prove Formula (5) for the case when $p(\vartheta)$ is not less than some positive number m almost everywhere, e.g. $p(\vartheta) \geqslant m > 0$. Then

$$|D(z)| = \exp \frac{1}{2\pi}\int_0^{2\pi} \lg \sqrt{p(\vartheta)}\, \frac{1 - r^2}{1 - 2r\cos(\varphi - \vartheta) + r^2}\, d\vartheta$$

$$\geqslant e^{\lg \sqrt{m}} = \sqrt{m} \quad (z = re^{i\varphi})$$

and $\dfrac{1}{|D(z)|} \leqslant \dfrac{1}{\sqrt{m}}$. Let us re-write the integral on the left-hand side of Formula (5) in the form

$$\int_0^{2\pi} |D(z')|^2 \left| \frac{1}{D(z')} - u_n(z') \right|^2 d\vartheta \tag{10}$$

and let us assume that $u_n(z')$ are the means $\sigma_n(z')$ of the partial sums of the Maclaurin series for the function $\frac{1}{D(z)}$, regular and bounded in the disc. The function under the integral sign in (10) then does not exceed the integrable function $\frac{2|D(z')|^2}{m}$ and it is possible to pass to the limit under the integral sign as $n \to \infty$. But we know that $u_n(z') \to \frac{1}{D(z')}$ almost everywhere on the circle $|z'| = 1$, and consequently the integral (10) tends to zero as $n \to \infty$, which leads to Parseval's equation (8) since altering $\sigma_n(z')$ to $u_n(z)$, as defined by Formula (7), can only decrease the integral (5).

Before going on to the general case $p(\vartheta) > 0$, let us recall the extremal problem investigated above. Let us put

$$I_n = \frac{1}{2\pi} \int_0^{2\pi} |\tau_n(z')|^2 d\vartheta, \quad z' = e^{i\vartheta}$$

where $\tau_n(z)$ is a linear combination of the first n of functions (2)

$$\tau_n(z) = \sum_{k=0}^{n} d_k \varphi_k(z) = D(z) \sum_{k=0}^{n} d'_k z^k$$

and let μ_n be the minimum of I_n for the condition $\sigma_n(0) = D(0)$, e.g. $d'_0 = 1$. We know that

$$\mu_n = \frac{[D(0)]^2}{\sum_{k=0}^{w} [D(0)]^2 |q_k(0)|^2} = \frac{1}{\sum_{k=0}^{n} |q_k(0)|^2}$$

Let us denote

$$\mu = \lim_{n \to \infty} \mu_n = \frac{1}{\sum_{k=0}^{\infty} |q_k(0)|^2} \tag{11}$$

Formula (9) can be written in the form

$$\mu = [D(0)]^2 \qquad (12)$$

Let us alter the function $p(\vartheta)$ to the function $p^{(\varepsilon)}(\vartheta) = p(\vartheta) + \varepsilon$, where ε is some positive number. It is not difficult to see that $p(\vartheta) + \varepsilon$ and $\lg[p(\vartheta) + \varepsilon]$ are integrable in the interval $[0, 2\pi]$. Let us denote by $D^{(\varepsilon)}(z)$, $\mu_n^{(\varepsilon)}$ and $\mu^{(\varepsilon)}$ the magnitudes of $D(z)$, μ_n and μ obtained as a result of this alteration. We clearly have:

$$|D^{(\varepsilon)}(z)| = \exp \frac{1}{2\pi} \int_0^{2\pi} \lg[p(\vartheta) + \varepsilon] \frac{1 - r^2}{1 - 2r\cos(\varphi - \vartheta) + r^2} d\vartheta > |D(z)|$$

$$\mu_n^{(\varepsilon)} = \min \frac{1}{2\pi} \int_0^{2\pi} |D^{(\varepsilon)}(z')|^2 |1 + d_1 z' + d_2 z'^2 + \ldots + d_n z'^n|^2 d\vartheta \geqslant \mu_n$$

and consequently $|D^{(\varepsilon)}(z')| \geqslant |D(z')|$ and $\mu^{(\varepsilon)} \geqslant \mu$. But $p^{(\varepsilon)}(\vartheta) \geqslant \varepsilon$ and consequently, in accordance with the above, Parseval's Formula (9) holds, i.e. $\mu^{(\varepsilon)} = |D^{(\varepsilon)}(0)|$ whence it follows that

$$\mu \leqslant [D^{(\varepsilon)}(0)]^2 = \exp \frac{1}{2\pi} \int_0^{2\pi} \lg[p(\vartheta) + \varepsilon] d\vartheta$$

If positive ε tends monotonically to zero, the function $\lg[p(\vartheta) + \varepsilon]$ diminishes, and in the last integral it is possible to pass to the limit under the integral sign (Smirnov [2] p. 162)

$$\mu \leqslant \exp \frac{1}{2\pi} \int_0^{2\pi} \lg p(\vartheta) d\vartheta = [D(0)]^2$$

But Bessel's inequality

$$\sum_{k=0}^{\infty} [D(0)]^2 |q_k(0)|^2 \leqslant 1$$

on the strength of (11) leads to the opposite inequality $\mu \geqslant [D(0)]^2$, whence it follows that: $\mu = [D(0)]^2$, i.e. Parseval's equation (9). The lemma is proved.

The proof of this lemma has a lot in common with the contents of Section 3.2.2. It can be found, in essence, in the works of Szegö [2] (see also Smirnov [5]).

From this lemma we have the following theorem.

Theorem 1 The system of functions (2) is closed, and the class of series of the form

$$\sum_{k=0}^{\infty} a_k D(z) q_k(z) \tag{13}$$

where

$$\sum_{k=0}^{\infty} |a_k|^2 < +\infty \tag{14}$$

is a class of Fourier series of functions of class $\mathscr{H}_2$:

$$f(z) = \sum_{k=0}^{\infty} a_k D(z) q_k(z) \qquad (f(z) \in \mathscr{H}_2) \tag{15}$$

the series converge uniformly on the compact sub-sets of the disc $|z| < 1$ and Parseval's equation holds for $f(z)$.

Let us write (15) in the form

$$\frac{f(z)}{D(z)} = F(z) = \sum_{k=0}^{\infty} a_k q_k(z)$$

Bearing the above statements in mind, we can prove the following.

Theorem 2 The class of series of the form

$$\sum_{k=0}^{\infty} a_k q_k(z) \tag{16}$$

where a_k satisfy condition (14), is a class of Fourier series for orthonormalised polynomials $q_k(z)$ with weight $p(\vartheta) = |D(e^{i\vartheta})|^2$ of class $\mathscr{H}_D$ for functions $F(z)$, regular in the circle $|z| < 1$ and such that $F(z)D(z) \in \mathscr{H}_2$. Different Fourier series (16) correspond to different functions $F(z)$ from $\mathscr{H}_D$. These series converge uniformly in the circle $|z| < 1$ to the function from $\mathscr{H}_D$ of which they are the Fourier series, and Parseval's equation

$$\frac{1}{2\pi} \int_0^{2\pi} p(\vartheta) |F(e^{i\vartheta})|^2 \, d\vartheta = \sum_{k=0}^{\infty} |a_k|^2$$

holds for $F(z)$. The following formulae hold for a_k.

$$a_k = \frac{1}{2\pi} \int_0^{2\pi} p(\vartheta) F(e^{i\vartheta}) \overline{q_k(e^{i\vartheta})} \, d\vartheta \tag{17}$$

Let us apply the formula for the sum of bi-linear series to the system (2). In the given case $w = \varphi(z, z_0)$ is the conformal transformation of the disc $|z| < 1$ on to the disc $|w| < 1$ for the conditions $\varphi(z_0, z_0) = 0$ and $\varphi'_z(z_0, z_0) > 0$. It is not difficult to see that

$$\varphi(z, z_0) = \frac{z - z_0}{1 - \bar{z}_0 z}$$

$$\varphi'_z(z, z_0) = \frac{1 - |z_0|^2}{(1 - \bar{z}_0 z)^2}$$

and we obtain

$$\sum_{k=0}^{\infty} \overline{q_k(z_0)}\, q_k(z) = \frac{1}{1 - \bar{z}_0 z} \cdot \frac{1}{\overline{D(z_0)}\, D(z)} \tag{18}$$

Let us also prove the following theorem.

Theorem 3 If $D_1(z)$ is not a maximal function of $\mathscr{H}_2$, then for the system of orthonormalised functions

$$D_1(z)\, q_k(z) \tag{19}$$

Parseval's equation is not fulfilled for $f(z) \equiv 1$.

Let $\sqrt{p(\vartheta)}$ be the limiting values of the modulus of $D_1(z)$, and $D(z)$ be a maximal function of $\mathscr{H}_2$ with the same limiting values of the modulus. Bearing in mind that $|D_1(e^{i\vartheta})| = |D(e^{i\vartheta})|$ we can show that the polynomials $q_k(z)$ in the orthonormalised systems (2) and (19) are identical. If $D_1(z)$ becomes zero at some $z = \alpha$ $(|\alpha| < 1)$, then the system (19) is not closed since all functions of this system become zero for $z = \alpha$. Let us assume that $D_1(z)$ does not have roots in the disc $|z| < 1$. From (9) and $|D_1(0)| < |D(0)|$ it follows that

$$\sum_{k=0}^{\infty} |q_k(0)|^2 < \frac{1}{|D_1(0)|^2}$$

i.e. Parseval's equation is not fulfilled for $f(z) \equiv 1$, and the theorem is proved.

In the case when $D(\alpha) = 0$, Parseval's equation is also not fulfilled for $f(z) \equiv 1$, since it would follow from Parseval's equation for $f(z) \equiv 1$ that the system (19) was closed.

4.2.2 *Case of simply-connected domains with rectifiable boundary*

Let us return to the system of polynomials $P_k(z)$, orthonormalised on the boundary L of a domain B, and consider the question of the closedness of this system. Let $z = \psi(w)$ be a function mapping the disc $|w| < 1$ conformally onto the domain B. The function $\psi(w)$ is continuous right up to the circle $|w| = 1$ and $\psi'(w) \in \mathscr{H}_1$. As we have seen (Section 4.1.6), the question of whether the system $P_k(z)$ is closed in the class $\mathfrak{E}_2$ is equivalent to the question as to whether the system of functions

$$R_k(w) = \sqrt{\frac{2\pi}{l}}\, P_k[\psi(w)] \sqrt{\psi'(w)} \qquad (k = 0, 1, \ldots) \tag{1}$$

belonging to $\mathscr{H}_2$ in the disc $|w| < 1$ and orthonormalised on the circle $|w| = 1$ is closed.

Lemma 1 In the case of system (1), Parseval's equation holds for any function of the form $f(w) = \sigma(w)\sqrt{\psi'(w)}$, where $\sigma(w)$ is a polynomial.

Note that the function $f(w)$ of the type indicated belongs to $\mathscr{H}_2$. We have to prove the equation

$$\frac{1}{2\pi}\int_0^{2\pi} |\sigma(e^{i\vartheta})|^2 |\psi'(e^{i\vartheta})|\, d\vartheta = \sum_{k=0}^{\infty} |a_k|^2 \tag{2}$$

where

$$a_k = \frac{1}{2\pi}\int_0^{2\pi} \sigma(e^{i\vartheta})\sqrt{\psi'(e^{i\vartheta})}\sqrt{\frac{2\pi}{l}}\,\overline{P_k[\psi(w')]}\,\overline{\sqrt{\psi'(w')}}\, d\vartheta$$

or

$$a_k = \frac{1}{\sqrt{2\pi l}}\int_0^{2\pi} \sigma(e^{i\vartheta})\overline{P_k[\psi(w')]}\, |\psi'(w')|\, d\vartheta \tag{3}$$

Converting to the z plane and introducing, as usual, the function $z = \psi(w)$ inverse to $w = \varphi(z)$, we obtain instead of (2) and (3)

$$\frac{1}{2\pi}\int_L |\sigma[\varphi(z')]|^2\, ds = \sum_{k=0}^{\infty} |a_k|^2 \tag{4}$$

$$a_k = \frac{1}{\sqrt{2\pi l}} \int_L \sigma[\varphi(z')] \overline{P_n(z')}\, ds \tag{5}$$

where z' is an arbitrary point on L. Let us introduce the Fourier coefficients of the function $\sigma[\varphi(z)]$ with respect to the orthonormalised system $P_n(z)$:

$$b_k = \frac{1}{l} \int_L \sigma[\varphi(z')] \overline{P_n(z')}\, ds = \sqrt{\frac{2\pi}{l}}\, a_k$$

and the Equation (4) can be written in the form

$$\frac{1}{l} \int_L |\sigma[\varphi(z')]|^2\, ds = \sum_{k=0}^{\infty} |b_k|^2 \tag{6}$$

Hence it can be seen that the Equation (4) is equivalent to Parseval's equation for the function $\sigma[\varphi(z)]$ with respect to the orthonormalised system $P_k(z)$. But the function $\sigma[\varphi(z)]$ is regular in B and continuous up to L, and consequently for any given $\varepsilon > 0$ there is a polynomial $Q(z)$, such that

$$|\sigma[\varphi(z)] - Q(z)| < \varepsilon$$

in the whole of the closed domain $\bar{B}$, whence

$$\frac{1}{l} \int_l |\sigma[\varphi(z')] - Q(z')|^2\, ds \leqslant \varepsilon^2$$

But the polynomial $Q(z)$ can be represented in the form of a finite, linear combination of polynomials $P_k(z)$ and consequently Parseval's equation (4) holds for $\sigma[\varphi(z)]$ and the lemma is proved.

Let us consider the closedness of the orthonormalised system (1). To establish that it is closed it is sufficient to prove that Parseval's equation holds for functions w^m ($m = 0, 1, \ldots$), and for this, on the basis of Lemma 1, it is sufficient to prove (Section 3.1.9) that there is a sequence of polynomials $\sigma_{n,m}(w)$ such that

$$\lim_{n \to \infty} \int_0^{2\pi} |w'^m - \sigma_{n,m}(w') \sqrt{\psi'(w')}|^2\, d\vartheta = 0, \quad m = 0, 1, 2, \ldots$$

As in Section 4.2.1, it is sufficient to prove that there is a sequence of polynomials $\sigma_n(w)$, such that

$$\lim_{n\to\infty} \int_0^{2\pi} |1-\sigma_n(w')\sqrt{\psi'(w')}|^2\, d\vartheta = 0 \tag{7}$$

The function $\psi'(w) \in \mathscr{H}_1$ and does not become zero in the disc $|w| < 1$. Let $\psi'(w)$ be a maximal function of the class $\mathscr{H}_1$, e.g.

$$\psi'(w) = e^{i\lambda} \exp \int_0^{2\pi} \lg p(\vartheta) \frac{e^{i\vartheta}+w}{e^{i\vartheta}-w}\, d\vartheta \tag{8}$$

where λ is a real number. The factor $e^{i\lambda}$ is absent if $\psi(w)$ is normalised such that $\psi'(0) > 0$. This is possible by means of the transformation $w = e^{-i\lambda}w'$, i.e. by rotating the circle $|w| < 1$. Then, in accordance with Lemma 1, Section 4.2.1, the sequence of polynomials $\sigma_n(w)$, having the properties (7), exists, and by the same respect the system (1) is closed. Since the closedness of the system (1) is equivalent to the closedness of the system $P_n(z)$, we have the following theorem.

Theorem 1 If for a function $z = \varphi(w)$, mapping the disc $|w| < 1$ conformally on to the domain B, $\psi'(w)$ is a maximal function of class $\mathscr{H}_1$, then the system of polynomials $P_k(z)$ is closed in the class $\mathfrak{E}_2$.

NOTE The Formula (8) is equivalent to the following:

$$\lg|\psi'(w)| = \frac{1}{2\pi} \int_0^{2\pi} \lg|\psi'(e^{i\vartheta})| \frac{1-r^2}{1-2r\cos(\theta-\vartheta)+r^2}\, d\vartheta \qquad (w = re^{i\theta}) \tag{9}$$

i.e. is equivalent to the fact that the harmonic function $\lg|\psi'(w)|$ can be represented by a Poisson integral.

In the monograph by Privalov [1] already mentioned, it is shown that if the condition (9) is fulfilled for some conformal transformation $z = \varphi(w)$ then it is also fulfilled for every other transformation. He introduces the example of a domain B, the boundary of which is a rectifiable Jordan curve and for which the condition (8) is not fulfilled. This example is taken from the work of Keldysh and Lavrentev [1]. Let us prove that in this case the system (1) is not closed, and that the system of polynomials $P_k(z)$ is not closed.

Theorem 2 If $\psi'(w)$ is not a maximal function of class $\mathscr{H}_1$, then Parseval's equation with respect to the system (1) does not hold for the function $f(w) \equiv 1$.

Let us prove this from the converse. Let the equation indicated hold, e.g. for any given $\varepsilon > 0$ there is m, such that

$$\int_0^{2\pi} \left| 1 - \sqrt{\frac{2\pi}{l}}\, \psi'(w') \sum_{k=0}^{m} a_k P_k [\psi(w)] \right|^2 d\vartheta \leqslant \frac{\varepsilon}{4} \tag{10}$$

where a_k are the Fourier coefficients of one with respect to the system (1). The function

$$S(w) = \sqrt{\frac{2\pi}{l}} \sum_{k=0}^{m} a_k P_k [\psi(w)]$$

is regular in the disc $|w| < 1$ and continuous in the closed disc $w| \leqslant 1$, and consequently there is a polynomial $Q(w)$, such that $|S(w) - Q(w)|^2 \leqslant \frac{\varepsilon}{4l}$ in the closed disc $|w| \leqslant 1$, such that

$$\int_0^{2\pi} |S(w') - Q(w')|\, |\psi'(w')|\, d\vartheta < \frac{\varepsilon}{4l} \int_0^{2\pi} |\psi'(w')|\, d\vartheta = \frac{\varepsilon}{4} \tag{11}$$

Bearing in mind (10), (11) and the obvious inequality

$$\int_0^{2\pi} |1 - Q(w') \sqrt{\psi'(w')}\, |^2 d\vartheta$$

$$\leqslant 2 \int_0^{2\pi} |1 - S(w') \sqrt{\psi'(w')}\, |^2 d\vartheta + 2 \int_0^{2\pi} |^2 S(w') - Q(w')|^2 |\psi'(w')|\, d\vartheta \tag{12}$$

we find

$$\int_0^{2\pi} |1 - Q(w') \sqrt{\psi'(w')}\, |^2 d\vartheta \leqslant \varepsilon$$

But since the function $\psi'(w)$ is not maximal, this contradicts what was said in Section 4.2.1 in view of the arbitrary nature of ε, and the theorem is proved.

Note that it follows from Theorems 1 and 2 that if a function $z = \psi(w)$ satisfies condition (9), then every other function, mapping the disc $|w| < 1$ conformally on to the domain B, also satisfies condition (9).

The following also follows directly from Theorems 1 and 2.

Theorem 3 The condition (9) is sufficient and necessary if the system of polynomials $P_k(z)$ is to be closed.

If condition (9) is fulfilled and by the same token the system of polynomials $P_k(z)$ is closed, then all the statements proved above for closed systems hold for this system also.

The class of domains for which the condition (9) holds is called class C. Sufficient conditions for a domain B to belong to class C are given in the monograph by Privalov [1] pp. 250-257. Also see Smirnov [4 and 5]. In particular, domains bounded by piecewise smooth curves and, more generally, by rectifiable curves, belong to class C.

4.2.3 *Special case*

Let us assume that the domain B is such that $\sqrt{\psi'(w)}$ is not a maximal function of class $\mathscr{H}_1$. We shall have

$$\sqrt{\frac{2\pi}{l}}\sqrt{\psi'(w)} = \exp\frac{1}{2\pi}\int_0^{2\pi} \lg\sqrt{p(\vartheta)}\,\frac{e^{i\vartheta}+w}{e^{i\vartheta}-w}\,d\vartheta\cdot\pi(w) \tag{1}$$
$$= D(w)\cdot\pi(w)$$

where

$$\pi(w) = \exp\frac{1}{2\pi}\int_0^{2\pi}\frac{e^{i\vartheta}+w}{e^{i\vartheta}-w}\,dq(\vartheta)$$

and $q(\vartheta)$ is a non-increasing, non-constant, bounded function, the derivative of which is equal to zero almost everywhere. We have $0<|\pi(w)|<1$ for $|w|<1$ and $|\pi(w')|=1$ almost everywhere on the circle $|w'|=1$. Let $q_k(w)$ be polynomials defined by the conditions (3), Section 4.2.1, where $p(\vartheta)=|D(e^{i\vartheta})|^2=\frac{2\pi}{l}|\psi'(e^{i\vartheta})|$. On L they satisfy the conditions

$$\frac{1}{l}\int_L q_n[\varphi(z)]\; q_m[\varphi(z)]\,ds = \delta_{n,m} \tag{2}$$

where $w=\varphi(z)$, as before, is the inverse function to $z=\psi(w)$. We have seen (Section 4.2.1) that the series

$$F(w) = \sum_{k=0}^{\infty} a_k q_k(w) \tag{3}$$

with the condition

$$\sum_{k=0}^{\infty} |a_k|^2 < +\infty \tag{4}$$

gives a class of functions $F(w)$ such that

$$D(w)\, F(w) = \sqrt{\psi'(w)}\, \frac{F(w)}{\pi(w)} \in \mathscr{H}_2$$

i.e.

$$\int\limits_{|w|=r} \left| \sqrt{\psi'(w)} \cdot \frac{F(w)}{\pi(w)} \right|^2 d\vartheta < C$$

or, transferring to the plane $z = \psi(w)$:

$$r \int\limits_{|\varphi(z)|<r} \left| \frac{F[\varphi(z)]}{\pi[\varphi(z)]} \right|^2 ds \leqslant C \tag{5}$$

Thus we can state that the series

$$f(z) = \sum_{k=0}^{\infty} a_k q_k [\varphi(z)] \tag{6}$$

under condition (4) gives a class of functions such that

$$\frac{f(z)}{\pi[\varphi(z)]} \in \mathscr{E}_2 \tag{7}$$

Let us prove that the series with respect to polynomials $P_k(z)$

$$f(z) = \sum_{k=0}^{\infty} a_k P_k(z) \tag{8}$$

also give the same class.

Note that the existence of the Fourier expansions (6) and (8) for functions from $\mathscr{E}_2$ is equivalent to the corresponding Parseval's equation for $f(z)$. Thus we need to prove that if Parseval's equation holds for functions $f(z)$ with respect to the system $P_k(z)$ then it also holds with respect to the system $q_k[\varphi(z)]$, and vice-versa.

Bearing in mind that the functions $q_k[\varphi(z)]$ are regular in B, and continuous in the closed domain $\overline{B}$, we can state that for any $\varepsilon > 0$ there is a polynomial $Q(z)$, such that

$$|q_k[\varphi(z)] - Q(z)| \leqslant \varepsilon \quad \text{for} \quad z \in \overline{B}.$$

But $Q(z)$ is a finite, linear combination of $P_k(z)$, and hence it follows that closure with respect to $q_k[\varphi(z)]$ involves closure with respect to $P_k(z)$. It is sufficient to use an inequality analogous to inequality (12), Section 4.2.2.

Let us now choose $P_k(z) = P_k[\psi(w)]$. These functions are regular in the disc $|w| < 1$ and continuous in the closed disc $|w| \leqslant 1$ and therefore for any $\varepsilon > 0$ there is a polynomial $R(w)$, such that

$$|P_k[\psi(w)] - R(w)| \leqslant \varepsilon \quad \text{for} \quad |w| \leqslant 1$$

But $R(w)$ is a finite, linear combination of $q_k(w)$, e.g.

$$\left| P_k(z) - \sum_{k=0}^{m} d_k q_k[\varphi(z)] \right| \leqslant \varepsilon \text{ for } z \in \bar{B}$$

whence as above it follows that closure with respect to $P_k(z)$ involves closure with respect to $q_k[\varphi(z)]$. We thus arrive at the following theorem.

Theorem 1 If $\sqrt{\psi'(w)}$ is expressed by Formula (1), then, under condition (4), the series (8) give a class of functions $f(z)$, satisfying condition (7).

NOTE 1 Bearing in mind that $|\pi[\varphi(z)]| < 1$, if $z \in B$, we can prove that it follows from (7) that $f(z) \in \mathscr{E}_2$. Theorem 3, Section 4.2.2, proves that (7) defines a class of functions which forms at least part of $\mathscr{E}_2$, if $\pi(z) \not\equiv 1$.

NOTE 2 If Formula (9), Section 4.2.2, holds, i.e. the system of polynomials $P_k(z)$ is closed, then

$$K(z, z_0) = \sum_{k=0}^{\infty} \overline{P_k(z_0)} P_k(z) = \frac{l}{2\pi} \sqrt{\overline{\varphi_z'(z_0, z_0)}} \sqrt{\varphi_z'(z, z_0)} \tag{9}$$

If $\sqrt{\psi'(w)}$ is expressed by Formula (1), then by using Theorem 1, it is not difficult to show that in this case Formula (9) becomes

$$\begin{aligned} K(z, z_0) &= \sum_{k=0}^{\infty} \overline{P_k(z_0)} P_k(z) \\ &= \frac{l}{2\pi} \sqrt{\overline{\varphi_z'(z_0, z_0)}} \sqrt{\varphi_z'(z, z_0)}\, \pi(0)\, \pi[\varphi(z, z_0)] \end{aligned} \tag{10}$$

In the following, when orthogonal polynomials are considered in the presence of a weight function, we shall prove a formula of which (10) is a particular case.

Let us now consider the class of functions regular and bounded in B and show that condition (5) is fulfilled for functions of this class. Let $|f(z)| \leqslant m$, where m is a positive number. We have

$$I_r = \int\limits_{|\varphi(z)|=r} \frac{|f(z)|^2}{|\pi[\varphi(z)]|^2} ds \leqslant m^2 \int\limits_{|\varphi(z)|=r} \frac{ds}{|\pi[\varphi(z)]|^2}$$

$$= m^2 r \int\limits_{|w|=r} \frac{|\psi'(w)|}{|\pi(w)|^2} d\vartheta$$

By using (1) we obtain

$$I_r \leqslant C \int\limits_{|w|=r} |D(w)|^2 d\vartheta$$

and the above integral remains bounded as $r \to 1$, since $D(w) \in \mathscr{H}_2$. We have proved the following theorem.

Theorem 2 Parseval's equation with respect to $P_k(z)$ holds in the class of functions which are regular and bounded in B.

COROLLARY Every function $f(z)$, regular and bounded in B, can be expanded into a Fourier series with respect to the polynomials $P_k(z)$, this series converging uniformly on the compact sub-sets of B, without the condition (9), Section 4.2.2, being fulfilled.

4.2.4 *Investigation of series of orthonormal polynomials*

We have the following formula

$$\sum_{k=1}^{\infty} |P_k(z)|^2 \leqslant \frac{l}{4\pi^2 d^2} \qquad (z \in F)$$

where F is any closed sub-set of B and d is the distance of F from L. From this follows the estimate for $P_k(z)$ in F:

$$|P_k(z)| \leqslant \frac{\sqrt{l}}{2\pi d} \qquad (z \in F) \tag{1}$$

It is important to have an estimate in the closed domain $\bar{B}$ for the future. The following holds.

Lemma 1 For any $\sigma > 1$ there is $c > 0$ such that

$$|P_k(z)| \leqslant c\sigma^k \qquad (k = 0, 1, \ldots), \quad (z \in \bar{B}) \tag{2}$$

Proof of this lemma and the following theorem is literally the same as that in Section 3.2.4. We do not use condition (9), Section 4.2.2, either in this lemma or in Theorem 1. Let B^1 be the exterior of L, and

$$z = \psi(w) = \tau w + c_0 + \frac{c_1}{w} + \frac{c_2}{w^2} + \ldots \qquad (|w| > 1, \quad \tau > 0)$$

be the function mapping the domain $|w| > 1$ conformally onto the domain B^1. Let us introduce the inverse function which in the neighbourhood of $z = \infty$ has the expansion

$$w = \Phi(z) = \frac{1}{\tau} z + b_0 + \frac{b_1}{z} + \frac{b_2}{z^2} + \ldots$$

Let us denote the image of the circle $|w| = \rho$, $\rho > 1$, on the z plane by L_ρ (external contour lines for B) and the interior of L_ρ by B_ρ and exterior of L_ρ by B_ρ^1.

Let us assume that $f(z)$ is regular in the closed domain $\bar{B}$. Hence it is bounded in $\bar{B}$, and can therefore be expanded into a Fourier series by the polynomials $P_k(z)$

$$f(z) = \sum_{k=0}^{\infty} a_k P_k(z) \tag{3}$$

converging uniformly on the compact sub-sets of B. As in Section 3.2.4 the following holds.

Theorem 1 Let the function $f(z)$ be regular in B_ρ, $\rho > 1$, and have a singular point on L_ρ. Then

1. It can be expanded into the Fourier series (3) uniformly converging on the compact sub-sets of B_ρ and diverging in B_ρ^1;

2. $$\overline{\lim_{k \to \infty}} \sqrt[k]{|a_k|} = \frac{1}{\rho} \tag{4}$$

3. Conversely, if (4) holds, then the series (3) converges uniformly on the compact sub-sets of B_ρ, diverges in B_ρ^1, and is the Fourier series of its sum $f(z)$, and this function is regular in B_ρ and has a singular point on L_ρ.

Let us now assume that L is an analytic curve and let $w = \varphi(z, z_0)$ and $z = \psi(w, z_0)$ be the usual notation for a function mapping B on to a disc and its inverse function. The function $\psi(w, z_0)$ is regular in the closed disc $|w| \leqslant 1$ and in this $\psi'(w, z_0)$ is different from zero so that the condition (9), Section 4.2.2, is obviously fulfilled. Besides this, with the given assumptions, the function $\psi(w, z_0)$ is regular and schlicht in some disc $|w| < \rho^*$, where $\rho^* > 1$. From Theorem 1 and what was said in Section 4.1.5, we have the expansion

$$\frac{l}{2\pi} \sqrt{\overline{\varphi_z'(z_0, z_0)}} \sqrt{\varphi_z'(z, z_0)} = \sum_{k=0}^{\infty} \overline{P_k(z_0)}\, P_k(z) \tag{5}$$

uniformly converging for $|\varphi(z, z_0)| \leqslant 1$, where z_0 is any fixed point from B. Let us assume that the function $f(z) \in \mathscr{E}_1$ in B. Transferring to the conjugate quantities in Formula (5) and taking into account that the series appearing in this formula converges uniformly on L, we obtain

$$\frac{\sqrt{\varphi_z'(z_0, z_0)}}{2\pi} \int_L f(z') \overline{\sqrt{\varphi_z'(z', z_0)}}\, ds = \sum_{k=0}^{\infty} a_k P_k(z_0) \tag{6}$$

where a_k are the Fourier coefficients of the function $f(z)$:

$$a_k = \frac{1}{l} \int_L f(z') \overline{P_k(z')}\, ds$$

In the integral on the left in (6) let us change the variable to $w' = \varphi(z', z_0)$:

$$\frac{\sqrt{\varphi_z'(z_0, z_0)}}{2\pi} \int_{|w'|=1} f(\psi(w', z_0)) \overline{\sqrt{\psi_w'(w', z_0)}}\, |dw'|$$

$$= \frac{\sqrt{\varphi_z'(z_0, z_0)}}{2\pi i} \int_{|w'|=1} \frac{f(\psi(w', z_0)) \sqrt{\psi_w'(w', z_0)}}{w'}\, dw'$$

Clearly $f(\psi(w, z_0)) \sqrt{\psi'(w, z_0)} \in \mathscr{H}_1$ in the circle $|w| < 1$ whence it follows that

$$\frac{1}{2\pi i} \int_{|w'|=1} \frac{f(\psi(w', z_0)) \sqrt{\psi'(w, z_0)}}{w'}\, dw' = f(\psi(0, z_0)) \sqrt{\psi_w'(0, z_0)}$$

$$= f(z_0) \left[\sqrt{\varphi_z'(z_0, z_0)}\right]^{-1}$$

and Formula (6) leads to the Fourier expansion of $f(z)$ in B.

$$f(z_0) = \sum_{k=0}^{\infty} a_k P_k(z_0) \tag{7}$$

We have proved the following theorem.

Theorem 2 If the boundary L of a domain B is an analytic curve, then the Fourier series of any function $f(z) \in \mathfrak{E}_1$ converges in B to the function $f(z)$.

In the following we shall show that the series (7) converges uniformly on the compact sub-sets of B, and, as in Section 3.2.4, we can prove the theorem that it is possible to expand into a series of polynomials $P_k(z)$ any function, regular in B, and even any function, regular in a simply-connected domain $B' \subset B$, such that all points of $B - B'$ lie sufficiently close to the boundary L of the domain B.

4.3 ASYMPTOTIC REPRESENTATION OF POLYNOMIALS ORTHONORMALISED WITH RESPECT TO THE BOUNDARY OF A DOMAIN

In this section we shall obtain the asymptotic representation of polynomials orthonormalised with respect to the boundary of a domain

$$P_k(z) = \alpha_k z^k + a_1^{(k)} z^{k-1} + \ldots + a_k^{(k)}, \quad k = 0, 1, \ldots,$$

for large values of k. It will be assumed that the boundary L of the domain B is an analytic curve. This representation was established by Szego [3] in 1921.

Let $z = \Psi(w) = w + c_0 + \frac{c_1}{w} + \ldots$ be a function conformally mapping some domain $|w| > r_0$ on to the domain $B^1 = B^1_{r_0}$ complementary to B with respect to the extended plane (see Chapter 2) and $w = \Phi(z)$ be the inverse function. Let us denote by ρ the smallest number such that the function $z = \Psi(w)$ is regular and schlicht in the ring $\rho < |w| < +\infty$. Since L is an analytic curve, $\rho < r_0$. We shall denote by $\overline{B^1_r}$ the image of the domain $|w| > r$, $r > \rho$, when mapped by the function $z = \Psi(w)$, and by B_r the complement for B^1_r and by L_r the boundary of B^1_r (and B_r).

4.3.1 *Asymptotic value of the leading coefficient a_k*

Let us start by considering the function $\Phi(z)^k\sqrt{\Phi'(z)}$, $k = 0, 1, \ldots$, whereby we assume $\sqrt{\Phi'(\infty)} = 1$. The sum of the terms with non-negative powers of z in the expansion of the above function in the neighbourhood of the point $z=\infty$ is a polynomial $\Pi_k(z)$ of degree k with leading coefficient one, and we have

$$\Pi_k(z) = \Phi(z)^k \sqrt{\Phi'(z)} + \omega_k(z) \tag{1}$$

where the function $\omega_k(z)$ is regular in B^1_ρ and $\omega_k(\infty) = 0$. We can regard $\Pi_k(z)$ as a generalised Faber polynomial (see Section 2.3) for the function $R(t) = \sqrt{\Phi'(\Psi(t))}$, and for $z \in B^1_r$, $r > \rho$, we have the value (see Section 2.2.1)

$$|\omega_k(z)| < G(r)r^k, \qquad z \in B^1_r \tag{2}$$

where $G(r)$ is a constant depending only on the chaice of r and the form of the domain B. This estimate can be obtained without using generalised Faber polynomials as it was in Section 3.3.

Thus on L_r, $r > \rho$, we have the estimate

$$\begin{aligned}|\Pi_k(z)| &\leqslant |\Phi(z)^k \sqrt{\Phi'(z)}| + |\omega_k(z)| \\ &< G_0(r)r^k + G(r)r^k = G_1(r)r^k\end{aligned} \tag{3}$$

where $G_1(r)$ is a constant of the same type as $G(r)$. The estimate (3) is clearly also true in $\overline{B}_r$.

Let us now obtain the asymptotic value of α_k for large k. The following lemma holds.

Lemma 1 The following estimate holds

$$\frac{2\pi}{l} r_0^{2k+1} \leqslant \alpha_k^{-2} \leqslant \frac{1}{l}\int_L |\Pi_k(z)|^2\, ds \leqslant \frac{2\pi}{l} r_0^{2k+1} + G_2(r) r^{2k}$$

$$\rho < r < r_0 \tag{4}$$

PROOF Taking into account the extremal problem in Section 4.1.7, we find

$$\alpha_k^{-2} \leqslant \frac{1}{l} \int_L |\Pi_k(z)|^2 ds = \frac{1}{l} \int_L |\Phi(z)^k \sqrt{\Phi'(z)} + \omega_k(z)|^2 ds$$

$$= \frac{1}{l} \int_L |\Phi(z)|^{2k} |\Phi'(z)| ds + 2 \operatorname{Re} \frac{1}{l} \int_L \overline{\Phi(z)^k \sqrt{\Phi'(z)}}\, \omega_k(z) ds + \frac{1}{l} \int_L |\omega_k(z)|^2 ds$$

Changing to the variable w and proceeding as in Section 3.3.1, we have

$$\alpha_k^{-2} \leqslant \frac{1}{l} \int_L |\Pi_k(z)|^2 ds \leqslant \frac{2\pi}{l} r_0^{2k+1} + G_2(r) r^{2k}$$

To obtain the lower bound for α_k^{-2} let us introduce the functions

$$\Phi_k(w) = \sqrt{\frac{2\pi r_0}{l}} P_k(\Psi(w)) \sqrt{\Psi'(w)} = \sqrt{\frac{2\pi r_0}{l}} \alpha_k w^k + \sum_{\nu=-\infty}^{k-1} \alpha_\nu^{(k)} w^\nu$$

analogous to the functions in Section 2.3.1. The function $\Phi_k(w)$ is regular in the domain $|w| > \rho$, apart from the point $w = \infty$, where it has a pole of order $k \geqslant 1$. These functions form an orthonormalised system on the circle $|w| = r_0$ and consequently

$$1 = \frac{1}{2\pi r_0} \int_{|w|=r_0} |\Phi_k(w)|^2 ds = \frac{2\pi}{l} r_0^{2k+1} \alpha_k^2 + \sum_{\nu=-\infty}^{k-1} |\alpha_\nu^{(k)}| r_0^{2\nu}$$

whence

$$\alpha_k^{-2} \geqslant \frac{2\pi}{l} r_0^{2k+1}$$

The lemma is proved.

COROLLARY The following asymptotic formula holds

$$\alpha_k = \sqrt{\frac{l}{2\pi}} r_0^{-\left(k+\frac{1}{2}\right)} (1 + \delta_k), \quad 0 \leqslant \delta_k \leqslant G_3(r) \left(\frac{r}{r_0}\right)^{2k+1}, \quad \rho < r < r_0 \tag{5}$$

NOTE Formulae (4) and (5) also hold for $r = r_0$, but we shall not make any further use of this fact.

4.3.2 Connection between the orthonormal polynomials $P_k(z)$ and the generalised Faber polynomials $\Pi_k(z)$

Theorem 1 The following formulae hold

$$P_k(z)=\sqrt{\frac{l}{2\pi}}\, r_0^{-\left(k+\frac{1}{2}\right)}\Pi_k(z)+\omega_k^*(z), \quad k=0,1,\ldots, \tag{1}$$

whilst the estimate

$$|\omega_k^*(z)|\leqslant\begin{cases} G_4(r)\sqrt{k+1}\left(\frac{r}{r_0}\right)^k, & \rho<r<r_0, \quad z\in\overline{B} \quad (2)\\ G_4(r)\sqrt{k+1}\left(\frac{r}{r_0}\right)^k\left(\frac{|\Phi(z)|}{r_0}\right)^k, & \rho<r<r_0, \quad z\in\overline{B}^1 \quad (3)\end{cases}$$

holds for the polynomial $\omega_k^*(z)$.

PROOF As in Section 3.3.2, we have in succession

$$\omega_k^*(z)=\sum_{s=0}^{k}\lambda_s^{(k)}\Pi_s(z) \tag{4}$$

$$\Pi_s(\Psi(w))-w^s[\Psi'(w)]^{-\frac{1}{2}}=\sum_{\nu=1}^{\infty}\frac{d_\nu^{(s)}}{w^\nu}, \quad |w|>\rho$$

$$\Pi_s(\Psi(w))[\Psi'(w)]^{\frac{1}{2}}=w^s+\sum_{\nu=1}^{\infty}\frac{e_\nu^{(s)}}{w}$$

$$\omega_k^*(\Psi(w))[\Psi'(w)]^{\frac{1}{2}}=\sum_{s=0}^{k}\lambda_s^{(k)}w^s+\sum_{\nu=1}^{\infty}\frac{\mu_\nu^{(k)}}{w^\nu}$$

Let us integrate the square of the modulus of both sides of this equation along the circle $|w|=r_0$, changing to the variable z in the left-hand side and bearing in mind Formula (1):

$$\sum_{s=0}^{k}|\lambda_s^{(k)}|^2 r_0^{2s}\leqslant\frac{1}{2\pi r_0}\int_L|P_k(z)|^2\,ds-2\,\mathrm{Re}\sqrt{\frac{l}{2\pi}}\, r_0^{-\left(k+\frac{1}{2}\right)}$$
$$\times\frac{1}{2\pi r_0}\int_L P_k(z)\overline{\Pi_k(z)}\,ds+\frac{l}{2\pi}r_0^{-(2k+1)}\frac{1}{2\pi r_0}\int_L|\Pi_k(z)|^2\,ds$$

whence as in Section 3.3.2, we obtain

$$|\omega_k^*(z)|^2\leqslant G_5(r)\left(\frac{r}{r_0}\right)^{2k}(k+1)$$

Thus on L, and consequently also in $\bar{B}$, estimate (2) holds. From this, as also in Section 3.3.2, estimate (3) is obtained. The theorem is proved.

4.3.3 Asymptotic representation of the orthonormal polynomials $P_k(z)$ for large k

Theorem 1 For $z \in \bar{B} - \bar{B}_r$, $\rho < r < r_0$, the formulae

$$P_k(z) = \sqrt{\frac{l}{2\pi}}\, r_0^{-\left(k+\frac{1}{2}\right)} \Phi(z)^k \sqrt{\Phi'(z)} + \delta_k(z), \quad k = 0, 1, \ldots, \tag{1}$$

hold, whereby

$$|\delta_k(z)| \leqslant G_6(r) \sqrt{k+1} \left(\frac{r}{r_0}\right)^k, \quad \rho < r < r_0, \quad k = 0, 1, \ldots, \quad z \in \bar{B} - \bar{B}_r \tag{2}$$

Theorem 2 For $z \in \bar{B}_1$ the formulae

$$P_k(z) = \sqrt{\frac{l}{2\pi}}\, r_0^{-\left(k+\frac{1}{2}\right)} \Phi(z)^k \sqrt{\Phi'(z)} \left(1 + \delta_k^*(z)\right), \quad k = 0, 1, \ldots, \tag{3}$$

hold, whereby

$$|\delta_k^*(z)| \leqslant G_7(r) \sqrt{k+1} \left(\frac{r}{r_0}\right)^k, \quad \rho < r < r_0, \quad k = 0, 1, \ldots, \quad z \in \bar{B}^1 \tag{4}$$

PROOF Formula (1) follows from Formula (1), Section 4.3.1, and Formula (1), Section 4.3.2,

$$\delta_k(z) = \sqrt{\frac{l}{2\pi}}\, r_0^{-\left(k+\frac{1}{2}\right)} \omega_k(z) + \omega_k^*(z)$$

and for $z \in \bar{B} - B_r$ using the estimates (2), Section 4.3.1, and (2), Section 4.3.2, we obtain

$$|\delta_k(z)| \leqslant \sqrt{\frac{l}{2\pi}}\, r_0^{-\left(k+\frac{1}{2}\right)} G(r) r^k + G_4(r) \sqrt{k+1} \left(\frac{r}{r_0}\right)^k \leqslant G_6(r) \sqrt{k+1} \left(\frac{r}{r_0}\right)^k$$

and Theorem 1 is proved.

In Formula (3)

$$\delta_k^*(z) = \frac{\omega_k(z)}{\Phi(z)^k \sqrt{\Phi'(z)}} + \sqrt{\frac{2\pi}{l}}\, r_0^{k+\frac{1}{2}} \frac{\omega_k^*(z)}{\Phi(z)^k \sqrt{\Phi'(z)}}$$

Assuming $m = \inf_{z \in \overline{B}^1} |\Phi'(z)|$ and using the estimates (2), Section 4.3.1, and (3), Section 4.3.2, for $z \in \overline{B}^1$, we obtain

$$|\delta_k^*(z)| \leqslant \frac{G(r) r^k}{r_0^k \sqrt{m}} + \sqrt{\frac{2\pi}{l}} \frac{G_4(r)\sqrt{k+1}\left(\frac{r}{r_0}\right)^k}{\sqrt{m}} \leqslant G_7(r)\sqrt{k+1}\left(\frac{r}{r_0}\right)^k$$

Theorem 2 is proved.

From Theorems 1 and 2 we obtain the following corollary.

COROLLARY As $k \to \infty$ inside B_ρ^1

$$\frac{P_{k+1}(z)}{P_k(z)} \rightrightarrows \frac{\Phi(z)}{r_0}$$

$$\sqrt[k]{|P_k(z)|} \rightrightarrows \frac{|\Phi(z)|}{r_0}$$

If $z \in L_r$, then $|\Phi(z)| = r$, and consequently

$$\lim_{k \to \infty} \sqrt[k]{|P_k(z)|} = \frac{r}{r_0}, \quad z \in L_r, \quad r > \rho$$

In the article by Korovkin [1] an asymptotic formula for $P_k(z)$ was obtained assuming that the angle $\theta(s)$ formed between the tangent to the line L and a fixed direction, as a function of the length of the arc s satisfies a Lipschitz condition

$$|\theta(s_2) - \theta(s_1)| \leqslant A|s_2 - s_1|^\alpha, \quad 0 < \alpha \leqslant 1$$

We shall discuss the asymptotic behaviour of orthonormalised polynomials in the presence of a weight function in Section 4.4.

Note that theorems entirely analogous to those in Section 3.3.5 hold for functions regular in $\overline{B}_\rho$.

Theorem 3 If the function $f(z)$ is regular in the domain B_r $\rho < r < \infty$, and there is a singular point on the contour line L_r, then the following applies:

1. The function $f(z)$ can be expanded into a series of polynomials $P_k(z)$:

$$f(z) = \sum_{k=0}^{\infty} c_k P_k(z) \tag{5}$$

uniformly converging on the compact sub-sets of B_r. This

series diverges in B_r^1;

2. $\overline{\lim\limits_{k\to\infty}} \sqrt[k]{|c_k|} = \frac{r_0}{r}$ (6)

3. The expansion of $f(z)$ as a series of form (5) uniformly converging in $B_{r'}$, $\rho < r' < \infty$, is unique;
4. If (6) holds, then the series (5) converges uniformly on the compact sub-sets of B_r and diverges in B_r^1 and the function $f(z)$ defined by the series (5), is regular in B_r and has a singular point on the contour line L_r.

4.4 ORTHOGONAL POLYNOMIALS IN THE PRESENCE OF A WEIGHT FUNCTION

4.4.1 Basic concepts and expansion as a Fourier series

Let B, as before, be a simply-connected domain, the boundary L of which is a rectifiable Jordan curve, and let $p(z')$ be an integrable function positive nearly everywhere on L. Applying the process of orthogonalisation to the functions $\sqrt{p(z')}\, z'^k$, $k=0, 1, \ldots$, we obtain polynomials $P_k(z)$, $k=0, 1, \ldots$, orthogonal on L with weight $p(z')$:

$$\frac{1}{l}\int_L p(z')P_n(z')\overline{P_m(z')}\,ds = \delta_{n,m} \tag{1}$$

The coefficient α_n of z^n in the polynomial $P_n(z)$ is taken as positive. As in Section 4.2.1 the following holds: the minimum of the integral

$$\frac{1}{l}\int_L p(z')\,|z'^n + c_1 z'^{n-1} + \ldots + c_n|^2\,ds$$

is equal to α_n^{-2} and is attained only for the polynomial

$$z^n + c_1 z^{n-1} + \ldots + c_n = \alpha_n^{-1}P_n(z)$$

Let $f(z')$ be a complex function given on L such that $\sqrt{p(z')}\, f(z') \subset \mathscr{L}_2$ on L. We can then construct the Fourier coefficients

$$a_k = \frac{1}{l}\int_L p(z')f(z')\overline{P_k(z')}\,ds \tag{2}$$

and we shall have the following formula:

$$I_n = \frac{1}{l} \int_L p(z') \left| f(z') - \sum_{k=1}^{n} a_k P_k(z) \right|^2 ds$$

$$\leqslant \frac{1}{l} \int_L p(z') |f(z')|^2 ds - \sum_{k=1}^{n} |a_k|^2$$

from which, as before, follows Bessel's inequality

$$\sum_{k=0}^{\infty} |a_k|^2 \leqslant \frac{1}{l} \int_L p(z') |f(z')|^2 ds \tag{3}$$

The equal sign, i.e. Parseval's equation

$$\frac{1}{l} \int_L p(z') |f(z')|^2 ds = \sum_{k=1}^{\infty} |a_k|^2 \tag{4}$$

is equivalent to $I_n \to 0$ as $n \to \infty$. The function $f(z')$ is comparable with its Fourier series

$$f(z') \sim \sum_{k=0}^{\infty} a_k P_k(z') \tag{5}$$

whereby

$$\sum_{k=0}^{\infty} |a_k|^2 < +\infty \tag{6}$$

Let $w = \varphi(z)$ be the function conformally mapping B on to the disc $|w| < 1$ whereby $\varphi(z_0) = 0$, and $\varphi'(z_0) > 0$ and $z = \psi(w)$ be the inverse function. Let us assume that the function

$$|\varphi'(z')| \ln p(z') \tag{7}$$

is integrable on L. These conditions are equivalent to saying that the function $p(z')$ is equivalent to the limiting values on L of the square of the modulus of some family of functions belonging to $\mathfrak{E}_2$ in B. Let $\Phi_0(z)$ be a maximal function of this family. Let us introduce the function

$$p_1(w') = \frac{2\pi}{l} p(\psi(w')) |\psi'(w')|, \quad w' = e^{i\vartheta} \tag{8}$$

and let $D(w)$ be a maximal function of $\mathscr{H}_2$, the limiting values of the square of the modulus of which on the circle $|w'| = 1$ are equal to $p_1(w')$:

$$D(w) = \exp \frac{1}{2\pi} \int_0^{2\pi} \ln \sqrt{p_1(w')} \frac{w'+w}{w'-w} d\vartheta, \quad w' = e^{i\vartheta} \tag{9}$$

We have (see introduction to chapter)

$$\Phi_0(z) = \sqrt{\frac{l}{2\pi}} D(\varphi(z)) \sqrt{\varphi'(z)} \tag{10}$$

and we can prove that the functions $D(w)q_k(w)$, $k=0, 1, \ldots$, where $q_k(w)$ are polynomials defined in Section 4.2.2, form a system closed in the disc $|w|<1$. Hence, as in Section 4.2.3, it follows that the functions $D(w)P_k[\psi(w)]$ form a closed system and that every function $F(w) \in \mathscr{H}_2$ can be expanded into a Fourier series

$$F(w) = \sum_{k=0}^{\infty} a_k D(w) P_k[\psi(w)] \tag{11}$$

uniformly converging on the compact sub-sets of the disc $|w|<1$ whereby

$$a_k = \frac{1}{2\pi} \int_0^{2\pi} F(w') \overline{D(w')} \overline{P_k[\psi(w')]} d\vartheta \tag{12}$$

and Parseval's equation

$$\frac{1}{2\pi} \int_0^{2\pi} |F(w')|^2 d\vartheta = \sum_{k=0}^{\infty} |a_k|^2 \tag{13}$$

holds.

We can also prove that the class of series (11), where a_k satisfy the condition (6), is a class of Fourier series for functions $F(w) \in \mathscr{H}_2$, where the condition $F(w) \in \mathscr{H}_2$ is equivalent to boundedness of the integral

$$\int_{|\varphi(z)|=r} \left| F[\varphi(z)] \sqrt{\varphi'(z)} \right|^2 ds \quad (0<r<1)$$

In other words, the condition $F(w) \in \mathscr{H}_2$ is equivalent to $F[\varphi(z)] \sqrt{\varphi'(z)} \in \mathscr{E}_2$ in B.

Let us transfer to the z plane in Formulae (11), (12) and (13):

$$F[\varphi(z)] = \sum_{k=0}^{\infty} a_k D[\varphi(z)] P_k(z) \tag{14}$$

$$a_k = \frac{1}{l} \int_L p(z') \frac{F[\varphi(z')]}{D[\varphi(z')]} \overline{P_k(z')}\, ds \tag{15}$$

$$\frac{1}{l} \int_L p(z') \left| \frac{F[\varphi(z')]}{D[\varphi(z')]} \right|^2 ds = \sum_{k=0}^{\infty} |a_k|^2 \tag{16}$$

To prove the last two formulae we need to make use of the formula

$$|D(w')|^2 = p_1(w') = \frac{2\pi}{l} p(z') |\psi'(w')|$$

which follows directly from Formula (9).

Let us introduce the class $\mathscr{E}^{(p)}$ of functions $f(z)$, regular in B and such that $f(z)\Phi_0(z) \in \mathscr{E}_2$. From the above follows directly:

Theorem 1 If $p(z')$ is positive almost everywhere and integrable on L and the function (7) is also integrable, then the collection of series of the form

$$\sum_{k=0}^{\infty} a_k P_k(z) \tag{17}$$

where

$$\sum_{k=0}^{\infty} |a_k|^2 < +\infty \tag{18}$$

is a collection of Fourier series of functions of class $\mathscr{E}^{(p)}$, where different Fourier series correspond to different functions from $\mathscr{E}^{(p)}$. If condition (18) is fulfilled, the series (17) converges uniformly on the compact sub-sets of B, its sum is equal to the function $f(z)$ from $\mathscr{E}^{(p)}$ whose Fourier series it is, and Parseval's equation

$$\frac{1}{l} \int_L p(z') |f(z')|^2 ds = \sum_{k=0}^{\infty} |a_k|^2 \tag{19}$$

holds for this function.

If $\lg|\psi'(w)|$ is expressed by Poisson's formula, then in the presence of weight, i.e. for $p(z') \not\equiv 1$, we have

$$\Phi_0(z) \equiv \Phi_0[\psi(w)] \equiv \exp \frac{1}{2\pi} \int_0^{2\pi} \lg \sqrt{p[\psi(w')]}\, \frac{w'+w}{w'-w}\, d\vartheta$$

If $p(z') \equiv 1$ and $\lg|\psi'(w)|$ is expressed by Formula (1), Section 4.2.3, then $\Phi_0(z) = \Phi_0(\psi(w)) = \frac{1}{\pi(w)}$. Let us now consider the function

$$f(z) = \frac{\sqrt{\varphi'(z)}}{\Phi_0(z)} = \sqrt{\frac{2\pi}{l}} \frac{1}{D[\varphi(z)]}$$

obviously belonging to $\mathcal{E}^{(p)}$ and let us define its Fourier coefficients a_k with respect to $P_k(z)$:

$$\overline{a_k} = \sqrt{\frac{2\pi}{l}} \cdot \frac{1}{l} \int\limits_L \frac{p(z')}{D[\varphi(z')]} P_k(z')\, ds$$

Changing to the variable w', we find

$$\overline{a_k} = \sqrt{\frac{2\pi}{l}} \cdot \frac{1}{2\pi i} \int\limits_{|w'|=1} \frac{D(w')\, P_k[\psi(w')]}{w'}\, dw' = \sqrt{\frac{2\pi}{l}}\, D(0)\, P_k(z_0)$$

where $z_0 = \psi(0)$. Thus we have a Fourier expansion uniformly converging on the compact sub-sets of B:

$$\frac{\sqrt{\varphi'(z)}}{\Phi_0(z)} = \sqrt{\frac{2\pi}{l}}\, D(0) \sum_{k=0}^{\infty} \overline{P_k(z_0)}\, P_k(z)$$

Substituting $D(0) = D[\varphi(z_0)]$, we obtain

$$\frac{l}{2\pi} \frac{\sqrt{\varphi'(z)}\sqrt{\varphi'(z_0)}}{\Phi_0(z)\,\Phi_0(z_0)} = \sum_{k=0}^{\infty} \overline{P_k(z_0)}\, P_k(z) \tag{20}$$

If $p(z') \equiv 1$ and $\lg|\psi'(w)|$ is expressed by Poisson's formula, then (20) is in the form (2), Section 4.1.5, and if $p(z') \equiv 1$ and $\sqrt{\psi'(w)}$ is expressed by Formula (1), Section 4.2.3, then (20) is in the form (10), Section 4.2.3.

4.4.2 *Asymptotic expressions*

Let us assume that the weight $p(z')$ is strictly positive and continuous on L and that L is an analytic curve. The function $q(\vartheta) = p\{\Psi(r_0 e^{-i\vartheta})\}$ is strictly positive and continuous on the unit circle. To this corresponds a maximal function $D(w)$ of class $\mathcal{H}_2$, satisfying the condition $|D(e^{i\vartheta})|^2 = q(\vartheta)$, where $D(w)$ is continuous on the closed disc $|w| \leqslant 1$. Assuming $w = r_0[\Phi(z)]^{-1}$, we obtain the function

$$\Delta(z) = D\{r_0[\Phi(z)]^{-1}\}$$

regular outside L, including the point $z = \infty$, continuous right up to L, differing from zero and such that when z tends from without to the point z', $|\Delta(z)|$ tends to $p(z')$. Proceeding basically as in Section 4.3, we can obtain the asymptotic formula

$$P_k(z) = \sqrt{\frac{l}{2\pi}}[\Delta(z)]^{-1} r_0^{-\left(k+\frac{1}{2}\right)} \{\Phi'(z)\}^{\frac{1}{2}} [\Phi(z)]^k [1 + \varepsilon_k(z)] \qquad (1)$$

where $\varepsilon_k(z) \to 0$ uniformly outside L_r for any $r > r_0$.

If L is an analytic contour and $\Delta(z)$ is regular right up to L, then it is possible to prove an asymptotic formula analogous to Formula (3), Section 4.3.3,

$$P_k(z) = \sqrt{\frac{l}{2\pi}} [\Delta(z)]^{-1} [\Phi'(z)]^{\frac{1}{2}} [\Phi(z)]^k + O(h^k) \qquad (2)$$

where h, $0 < h < 1$, depends only on the curve L and the function $p(z')$. Formula (2) holds on L and in B sufficiently close to L. To prove the Formula (2) we make use of polynomials which are obtained by expanding the function

$$[\Delta(z)]^{-1} [\Phi'(z)]^{\frac{1}{2}} [\Phi(z)]^n$$

in the neighbourhood of the point $z = \infty$ into a power series. The asymptotic formula mentioned above was obtained by Szegö [1] Chapter XVI.

An asymptotic formula outside L with few assumptions relating to L and the weight $p(z')$ is derived in the work of Korovkin [2].

By using the asymptotic formulae it is possible to prove a theorem analogous to Theorem 3, Section 4.3.3. As a corollary to this theorem we find that the Fourier series of any function $f(z) \in \mathfrak{E}_1$ in B converges uniformly on the compact sub-sets of B to $f(z)$ (see Theorem 2, Section 4.2.4).

4.4.3 *Supplementary data*

Let us make some supplementary remarks and introduce new results relating to various of the above problems.

Let us begin with the case of polynomials orthogonalised

on the unit circle in the presence of a weight $p(\theta)$ assuming that $p(\theta)$ and $\lg p(\theta)$ are integrable functions. We have seen that Parseval's equation and the expansion theorem hold in the class $\mathcal{F}$ of functions $f(z)$, regular in the disc $|z|<1$ and such that $f(z)D(z)\in\mathcal{H}_2$, where $D(z)$ is a maximal function of the class $\mathcal{H}_2$, which satisfies the condition $|D(e^{i\vartheta})|^2 = p(\vartheta)$ almost everywhere in the interval $(0, 2\pi)$. The class $\mathcal{F}$ can be defined in the following manner: $f(z)\in\mathcal{D}$ (see introduction to chapter) and

$$\int_0^{2\pi} |D(e^{i\vartheta})|^2\, |f(e^{i\vartheta})|^2\, d\vartheta < +\infty \tag{1}$$

In fact it follows from $f(z)D(z)\in\mathcal{H}_2$ that $f(z)\in\mathcal{D}$. This follows directly from the parametric representation for classes $\mathcal{H}_2$ and $\mathcal{D}$. Let us now assume that $f(z)\in\mathcal{D}$ and condition (1) is fulfilled, and prove that $f(z)D(z)\in\mathcal{H}_2$. From the parametric representations mentioned above and $f(z)\in\mathcal{D}$ it follows that $f(z)D(z)\in\mathcal{D}$. But condition (1) then proves that $f(z)D(z)\in\mathcal{H}_2$ (see introduction). Analogously we can choose the case of a domain B, the boundary of which is a rectifiable Jordan curve C, and prove that the class $\mathcal{F}$ of functions $f(z)$, for which Parseval's equation and the expansion theorem hold with respect to polynomials orthonormalised in the presence of a weight $p(z')$, can be defined in the following manner $f[\psi(w)]\in\mathcal{D}$ and

$$\int_L p(z')\, |f(z')|^2\, ds < +\infty$$

It is assumed that $p(z')$ and $|\varphi'(z')|\lg p(z')$ are integrable functions on L.

Let us introduce the basic literature relating to the problems considered above and containing results which we have not mentioned.

Orthogonal polynomials in the presence of a weight function are investigated in the work of Korovkin [2] and in the monograph by Szegö [1].

In a monograph to be published in the near future, Geronimus [2] has studied in detail the polynomials $p_n(z)$, orthonormalised on the circle $|z|=1$ in the following sense:

$$\frac{1}{2\pi}\int_0^{2\pi} p_n(e^{i\vartheta})\,\overline{p_m(e^{i\vartheta})}\, d\sigma(\vartheta) = \begin{cases} 0 & \text{for} \quad n\neq m \\ 1 & \text{for} \quad n=m \end{cases}$$

where $\sigma(\vartheta)$ is a bounded function not vanishing in the interval $[0, 2\pi]$. If $\sigma(\vartheta)$ is an absolutely continuous function, then the polynomials considered above with weight $p(\vartheta)=\sigma'(\vartheta)$ are obtained. In this monograph and a number of previous works, Geronimus and other authors have also considered the case where $\lg\sigma'(\vartheta)$ is an integrable function and also the case when

$$\int_0^{2\pi} \lg\sigma'(\vartheta)\,d\vartheta = -\infty \tag{2}$$

In monographs by Szegö [2] and Geronimus [2] a connection is established between polynomials $P_n(z)$, orthonormalised on the unit circle $|z|=1$, and polynomials $p_n(x)$, orthogonal on the finite interval $a \leqslant x \leqslant b$ in the presence of a weight function. The famous results of Bernstein [2] pp. 7-106, for the asymptotic behaviour of the polynomials $p_n(x)$, are obtained here by a different method.

The theory of orthogonal polynomials is connected with mean quadratic approximation on the circle $|z|=1$ or on a rectifiable curve L. The question of mean approximation with a degree of approximation which is different from two has been investigated. This more general question has been considered in the work of Tumarkin [2] where a wide literature is listed. Let us introduce various results from this work. In the following we shall assume that $\sigma(s)$ is a non-diminishing function bounded on L, and shall denote by $\mathcal{L}_p(p>0)$ a class of complex functions $f(z')$, defined on L measurable on L with respect to $\sigma(s)$ and such that

$$\int_L |f(z')|^p\,d\sigma(s) < \infty \tag{3}$$

Theorem 1 For there to be for the measurable function $f(z')$, defined on L, a sequence of polynomials $P_k(z)$, such that

$$\lim_{n\to\infty}\int_L |f(z')-P_n(z')|^p\,d\sigma(s)=0 \qquad (p>0) \tag{4}$$

where $\sigma(s)$ is such that the integral

$$\int_L |\varphi'(z')|\lg\sigma'(s)\,ds \tag{5}$$

has a finite value, it is necessary and sufficient that:

$f(z') \in \mathscr{L}_p$ and $f(z')$ be equivalent on L to the limiting values of some function $f(z)$, regular inside and such that $f[\psi(w)] \in \mathscr{D}$.

Theorem 2 If the integral (5) has a finite value and $f(z') \in \mathscr{L}_p$, then the sequence of polynomials $P_n(z)$, satisfying condition (4), converges uniformly on the compact sub-sets inside L to the function $f(z)$, indicated at the end of the previous theorem.

Theorem 3 The necessary and sufficient condition that for any function $f(z') \in \mathscr{L}_p$ for any given $p > 0$ there shall be a sequence of polynomials $P_n(z)$, satisfying condition (4), is that

$$\int_L |\varphi'(z')| \lg \sigma'(s)\, ds = -\infty \tag{6}$$

NOTE If condition (6) is not fulfilled, the integral (5) has a finite value.

Theorem 4 Let $f(z) \in \mathscr{E}_\delta$ $(\delta > 0)$ in B and integral (5) be finite. Then if there is to be a sequence of polynomials $P_n(z)$, satisfying condition (4), it is necessary and sufficient that: $f(z') \in \mathscr{L}_p$ and $f(z)[g(z)]^{1/\delta} \in \mathscr{E}_\delta$, where

$$g(z) = \varphi'(z) \exp \frac{1}{2\pi} \int_0^{2\pi} \frac{e^{i\vartheta} + \varphi(z)}{e^{i\vartheta} - \varphi(z)} \lg |\psi'(e^{i\vartheta})|\, d\vartheta$$

NOTE If the function $\ln |\psi'(w)|$ can be represented by a Poisson integral, then $g(z) \equiv 1$ and the second of the above conditions drops out.

SUPPLEMENTARY LITERATURE

In addition to the works mentioned in the text, we should like to mention the following monographs and articles dealing with the material in the present chapter: Geronimus [1, 2], Goncharov [1] Chapter 5, Mergelian [6], Szegö [1] with review article by Geronimus. We should also like to mention the special articles: Dzhrbashian [5], Kuzmina [1], Rosenbloom and Warschawski [1], Smirnov [4, 5], Suetin [2, 3, 4, 7, 8], and Tumarkin [4].

CHAPTER FIVE

Questions related to the best uniform approximation

This chapter is devoted to the problem of the best uniform approximation of functions and to several related questions. Let us give a precise formulation of the problem of the best uniform approximation of a function by means of polynomials.

Suppose that a bounded function $f(z)$ is defined on a closed bounded set B of the complex z plane and let $\mathscr{P}_n$ denote the class of all polynomials $p_n(z)=c_0z^n+c_1z^{n-1}+\ldots+c_n$. We shall call the number

$$\mu(p_n(z))=\sup_{z\in B}|f(z)-p_n(z)|$$

the measure of uniform approximation of the function $f(z)$ on the set B by means of the polynomial $p_n(z)\in\mathscr{P}_n$.

Consider the greatest lower bound of the numbers $\mu(p_n(z))$ for $p_n(z)\in\mathscr{P}_n$, that is, the number

$$\mu_n=\inf_{p_n(z)\in\mathscr{P}_n}\mu(p_n(z))=\inf_{p_n(z)\in\mathscr{P}_n}\sup_{z\in B}|f(z)-p_n(z)|$$

We shall call this number the measure of the best uniform approximation of the function $f(z)$ on the set B by means of the polynomials $p_n(z)\in\mathscr{P}_n$.

We shall show later that there always exists a polynomial $\pi(z)\in\mathscr{P}_n$ such that $\mu(\pi(z))=\mu_n$, and, therefore, we shall say that the value μ_n is attained for the polynomial $\pi(z)$. We shall use the same term in analogous cases. We shall call such a polynomial $\pi(z)$ the polynomial of best approximation of the function $f(z)$ on B by means of polynomials $p_n(z) \in \mathscr{P}_n$. This problem is a special case of the problem of best approximation of the function $f(z)$ on the set B by means of the generalised polynomials

$$p_n(z) = c_1\varphi_1(z) + c_2\varphi_2(z) + \ldots + c_n\varphi_n(z)$$

where $\varphi_1(z), \ldots, \varphi_n(z)$ are functions defined on B and $c_1, \ldots, c_n$ are arbitrary constants. The latter problem is posed in an analogous manner to the above-stated problem of finding the best approximation by means of the polynomials $p_n(z) \in \mathscr{P}_n$.

The following problem reduces to the problem of best approximation of a function by means of polynomials. Among the polynomials $P_n(z) = z^n + c_1 z^{n-1} + \ldots + c_n$ (where the degree n of the polynomial is one or greater and the leading coefficient is one), find the polynomial whose maximum absolute value on the set B is minimum; that is, if we set

$$\mu_n = \inf_{P_n(z)} \sup_{z \in B} |P_n(z)|$$

then we have to find a polynomial $\Pi_n(z) = z^n + c_1^{(0)} z^{n-1} + \ldots + c_n^{(0)}$ for which the value μ_n is attained, i.e. such that

$$\sup_{z \in B} |\Pi_n(z)| = \mu_n$$

Such a polynomial $\Pi_n(z)$ is called a polynomial of least deviation from zero on the set B or the Chebyshev polynomial for the set B. If we represent the polynomial $P_n(z) = z^n + c_1 z^{n-1} + \ldots + c_n$ in the form $P_n(z) = z^n - p_{n-1}(z)$, where $p_{n-1}(z) = -c_1 z^{n-1} - \ldots - c_n \in \mathscr{P}_{n-1}$, we see that the Chebyshev polynomial for the set B is obtained by subtracting from z^n the polynomial of best approximation of the function $f(z) = z^n$ on B by means of the polynomials $p_{n-1}(z) \in \mathscr{P}_{n-1}$.

The problem of finding the Chebyshev polynomial for the set B is generalised as follows:

PROBLEM 1 Among the polynomials $p_n(z) = c_0 z^n + \ldots + c_n$ of degree $n \geqslant 1$ with coefficients satisfying the condition $\alpha_0 c_0 + \alpha_1 c_1 + \ldots + \alpha_n c_n = 1$, where $\alpha_0, \alpha_1, \ldots, \alpha_n$ are given numbers, find the polynomials of least deviation from 0 on the set B (that is, all those polynomials whose maximum absolute value on the set B is minimum) and find this minimum deviation μ_n.

NOTE In Problem 1, we may consider polynomials whose coefficients satisfy certain conditions of the type referred to in the problem.

We obtain Chebyshev's polynomial for the set B as a special case of this problem if the condition $p_0c_0 + \dots + p_nc_n = 1$ reduces to the condition $c_0 = 1$.

Problem 1 can be extended in an obvious manner to generalised polynomials. All these problems are easily reduced to the problem of best approximation of some function on B by means of generalised polynomials of a special kind.

Closely connected with Problem 1 is the following problem:

PROBLEM 2. Among the polynomials $p_n(z) = c_0z^n + \dots + c_n$ of degree $n \geqslant 1$ that satisfy the condition $\sup_{z \in B} |p_n(z)| \leqslant 1$, find the polynomials for which the quantity

$$\lambda_n = |\alpha_0 c_0 + \dots + \alpha_n c_n|, \qquad \sum_{k=0}^{n} |\alpha_k| \neq 0$$

where $\alpha_0, \alpha_1, \dots, \alpha_n$ are given numbers, assumes its greatest value.

Specifically, if $\Pi_n(z)$ is a polynomial solution to Problem 1, every polynomial of the form $R_n(z) = \frac{1}{\mu_n} e^{i\gamma}\Pi_n(z)$, for $0 \leqslant \gamma < 2\pi$, is a solution to Problem 2 (here $\lambda_n = \frac{1}{\mu_n}$) and, conversely, if $R_n(z)$ is a polynomial solution to Problem 2, the polynomial $\Pi_n(z) = \frac{1}{\lambda_n} e^{i\alpha} R_n(z)$ where α is a real number, correspondingly defined, is a solution to Problem 1 (and here, $\mu_n = \frac{1}{\lambda_n}$). We shall not stop to prove this almost obvious fact.

Each of Problems 1 and 2 we shall call the dual of the other. We shall also employ this term in connection with other analogous cases.

In the first two sections of this chapter, we shall consider certain particular cases of Problems 1 and 2 when B is the disc $|z| \leqslant R$, where $R > 0$. Other problems of this nature will also be solved. In the third section, we shall consider problems of approximating functions by means of polynomials and generalised polynomials. In Section 5.4, we shall treat Problems 1 and 2 in the general case.

5.1 POLYNOMIALS DEVIATING THE LEAST FROM ZERO

5.1.1 *Certain very simple problems*

PROBLEM 1 Among the polynomials $P_n(z) = z^n + c_1 z^{n-1} + \dots + c_n$ find the polynomial deviating the least from zero in the closed region $|z| \leqslant R$, where $R > 0$.

The solution to this problem is provided by the following theorem.

Theorem 1 For the polynomials $P_n(z) = z^n + c_1 z^{n-1} + \dots + c_n$ of degree $n \geqslant 1$, we have

$$\mu_n = \inf_{P_n(z)} \sup_{|z| \leqslant R} |P_n(z)| = R^n$$

and μ_n is attained only for the polynomial $\Pi_n(z) = z^n$.

PROOF Let us represent the polynomial $P_n(z)$ in the form

$$P_n(z) = z^n + c_1 z^{n-1} + \dots + c_n = z^n \left(1 + \frac{c_1}{z} + \dots + \frac{c_n}{z^n}\right) = z^n \varphi(z)$$

Remembering that the function $\varphi(z)$ is regular in the region $|z| \geqslant R$, if we twice use the maximum principle, we obtain

$$\sup_{|z| \leqslant R} |P_n(z)| = \sup_{|z| = R} |P_n(z)| = \sup_{|z| = R} R^n |\varphi(z)| = R^n \sup_{|z| \geqslant R} |\varphi(z)| \geqslant R^n |\varphi(\infty)| = R^n$$

Equality holds if and only if $\varphi(z) \equiv C$ where C is a constant, that is, if and only if $\varphi(z) \equiv \varphi(\infty) = 1$. The assertions of the theorem then follow immediately.

NOTE 1 It is easy to see that the problem of finding the polynomial that deviates the least from 0 on the set $|z| = R$ (the boundary of the region $|z| \leqslant R$) out of all the polynomials $P_n(z) = z^n + c_1 z^{n-1} + \dots + c_n$, is equivalent to Problem 1. In general, because of the maximum principle, the problem of finding the polynomial that deviates the least from zero on a closed bounded region $\bar{B}$ out of all the polynomials $P_n(z) = z^n + c_1 z^{n-1} + \dots + c_n$, is equivalent to the problem of finding the polynomial that deviates the least from zero on the boundary L of the region $\overline{B}$ out of all these same polynomials $P_n(z)$.

Let us formulate the theorem that is the dual of Theorem 1 (that is, the theorem for the solution of the problem that is the dual of Problem 1).

Theorem 1' For the polynomials $p_n(z) = c_0 z^n + c_1 z^{n-1} + \ldots + c_n$ of degree $n \geqslant 1$ satisfying the condition $\sup_{|z| \leqslant P} |p_n(z)| \leqslant 1$ the inequality $|c_0| \leqslant \frac{1}{R^n}$ holds; $|c_0| = \frac{1}{R^n}$ only for the polynomials $\pi_n(z) = \eta \left(\frac{z}{R}\right)^n$ where $|\eta| = 1$.

Let us now pose the following problem.

PROBLEM 2 Out of all the polynomials $P(z) = Az^n + c_1 z^{n-1} + \ldots + c_{n-1} z + B$ of degree $n \geqslant 2$ for fixed A and B, find the one that deviates the least from zero in the closed region $|z| \leqslant R$.

This problem is equivalent to the following problem:

PROBLEM 2' Out of all the polynomials $p(z) = c_1 z^{n-1} + \ldots + c_{n-1} z$ find the polynomial of best approximation for the function $f(z) = Az^n + B$ in the region $|z| \leqslant R$, where $R > 0$.

In view of the maximum principle, in either of these problems, we may replace the region $|z| \leqslant R$ with its boundary $|z| = R$. When we do, the problem that we obtain is equivalent to the original one.

The solution of these problems is easily obtained from the following theorem.

Theorem 2 For the polynomials

$$P(z) = Az^n + c_1 z^{n-1} + \ldots + c_{n-1} z + B$$

of degree $n \geqslant 2$ with fixed coefficients A and B,

$$\mu_n = \inf_{P(z)} \sup_{|z| \leqslant R} |P(z)| = |A| R^n + |B|, \quad R > 0. \tag{1}$$

Also, the value μ_n is attained only in the case of the polynomial $\Pi(z) = Az^n + B$.

PROOF Let us consider certain cases.

1. $A = B = 0$. Obviously, $\mu_n = 0$ if and only if $P(z) \equiv 0$, that is, if $\Pi(z) \equiv 0 = 0z^n + 0$.
2. $A = 0$, $B \neq 0$. Then,

$$\sup_{|z| \leqslant R} |P(z)| = \sup_{|z| = R} |P(z)| \geqslant |P(0)| = |B|$$

From this we easily conclude that the assertion of the theorem is valid for this case also.

3. $A \neq 0,\ B = 0$. In this case, the proof is the same as that of Theorem 1.

4. $A \neq 0,\ B \neq 0$. We note first that, for $|z| \leqslant R$

$$|\Pi(z)| \leqslant |A|\,|z^n| + |B| \leqslant |A| R^n + |B|$$

Equality holds in these inequalities if and only if $|z| = R$ and the terms Az^n and B have the same arguments, that is, if

$$z = z_k = Re^{i\frac{\varphi_0}{n}} \zeta_k, \quad k = 0, 1, \ldots, n-1, \quad \varphi_0 = \arg \frac{B}{A}, \quad \zeta_k = e^{\frac{2\pi k}{n} i}$$

Then,

$$\Pi(z_k) = AR^n e^{i\varphi_0} \zeta_k^n + B = AR^n \frac{B}{A} \left| \frac{A}{B} \right| + B = \frac{B}{|B|} (|A| R^n + |B|) \quad (2)$$

Let us suppose that

$$\sup_{|z| \leqslant R} |P(z)| \leqslant |A| R^n + |B| \quad (3)$$

for some polynomial

$$P(z) = Az^n + B + \omega(z), \quad \omega(z) = c_1 z^{n-1} + \ldots + c_{n-1} z$$

where not all the c_j are zero.

Since the n-th roots of unity $\zeta_k = e^{\frac{2\pi k}{n} i}$ (for $k = 0, 1, \ldots, n-1$) possess the following property

$$\sum_{k=0}^{n-1} \zeta_k^l = \begin{cases} 0 & \text{for } l, \text{ not a multiple of } n \\ n & \text{for } l \text{ not a multiple of } \overline{n} \end{cases}$$

it follows that

$$\frac{1}{n} \sum_{k=0}^{n-1} P(\zeta_k z) = Az^n + B = \Pi(z)$$

If we set $z = z_0 = Re^{i\frac{\varphi_0}{n}}$ we obtain

$$\Pi(z_0) = \frac{1}{n} \sum_{k=0}^{n-1} P(z_k)$$

From this, we obtain by using (2) and (3)

$$|A|R^n+|B|=|\Pi(z_0)|=\left|\frac{1}{n}\sum_{k=0}^{n-1}P(z_k)\right|\leqslant\frac{1}{n}\sum_{k=0}^{n-1}|P(z_k)|\leqslant|A|R^n+|B|$$

Consequently, the two inequalities are actually equalities, which is possible only when, first the arguments of the numbers $P(z_k)$ are equal to each other and, second, when, on the basis of (3), $|P(z_k)|=|A|R^n+|B|$ (for $k=0, 1, \ldots, n$) that is, when all the numbers

$$P(z_k)=Az_k^n+B+\omega(z_k)=\frac{B}{|B|}(|A|R^n+|B|)+\omega(z_k)$$

(for $k=0, 1, \ldots, n$) are equal. This in turn can be true only when $\omega(z_k)=C$ (for $k=0, 1, \ldots, n-1$), where C is a constant. Thus, the polynomial $\omega(z)-C$ of degree no higher than $n-1$ must have n roots: $z_0, z_1, \ldots, z_{n-1}$, which is possible only when this polynomial is identically equal to zero, that is, when, in spite of the hypothesis, all the coefficients are equal to zero. Thus, for any polynomial

$$P(z)=Az^n+c_1z^{n-1}+\ldots+c_{n-1}z+B$$

we have

$$\sup_{|z|\leqslant R}|P(z)|\geqslant|A|R^n+|B|$$

Here, equality holds only for the polynomial $\Pi(z)=Az^n+B$. This completes the proof of the theorem. Let us formulate theorems dual to Theorem 2.

Theorem 2′ Suppose that a polynomial

$$P(z)=Az^n+c_1z^{n-1}+\ldots+c_{n-1}z+B$$

of degree $n\geqslant 2$ with fixed coefficient B such that $|B|\leqslant 1$, satisfies the condition that $|P(z)|\leqslant 1$ for $|z|\leqslant R$ where $R>0$ Then,

$$|A|\leqslant\frac{1-|B|}{R^n}$$

with equality holding only for the polynomials

$$\Pi(z)=\eta\frac{1-|B|}{R^n}z^n+B$$

where $|\eta|=1$.

Theorem 2'' Suppose that a polynomial

$$P(z) = Az^n + c_1 z^{n-1} + \dots + c_{n-1} z + B$$

of degree $n \geqslant 2$ with fixed coefficient A, such that $|A| \leqslant R^{-n}$ where $R > 0$ satisfies the condition $|P(z)| \leqslant 1$ for $|z| \leqslant R$. Then, $|B| \leqslant 1 - |A| R^n$ with equality holding only for the polynomials $\Pi(z) = Az^n + (1 - |A|R^n)\eta$ where $|\eta| = 1$.

We shall mention certain corollaries to Theorem 2. For each of these corollaries, it is possible to formulate a dual assertion (sometimes more than one), but this is omitted.

COROLLARY 1 For a polynomial

$$P_n(z) = Az^n + c_1 z^{n-1} + \dots + c_{n-1} z + B$$

with fixed sum of the coefficients $A + B = C$

$$\mu_n = \inf_{P_n(z)} \sup_{|z| \leqslant R} |P_n(z)| = \begin{cases} |C| R^n & \text{for } 0 < R \leqslant 1 \\ |C| & \text{for } R \geqslant 1 \end{cases}$$

and μ_n is attained for the polynomials

$$P_n(z) = \begin{cases} Cz^n & \text{for } 0 < R < 1, \\ C(\lambda z^n + (1 - \lambda)), \ 0 \leqslant \lambda \leqslant 1, & \text{for } R = 1 \\ C & \text{for } R > 1. \end{cases}$$

To prove this corollary, we need to find the minimum of $\mu_n = |A| R^n + |B|$ with the constraint that $A + B = C$.

COROLLARY 2 For the generalised polynomials

$$Q(z) = Az^n + c_{n-1} z^{n-1} + \dots + c_{-m+1} \frac{1}{z^{m-1}} + B \frac{1}{z^m}, \quad n + m \geqslant 2$$

with fixed coefficients A and B, we have

$$\mu_n = \inf_{Q(z)} \sup_{r < |z| \leqslant R} |Q(z)| = \max_{0 < r < R} \left\{ |A| R^n + |B| \frac{1}{R^m}, \quad |A| r^n + |B| \frac{1}{r^m} \right\}$$

and μ_n is attained only for the generalised polynomial $Q(z) = Az^n + \frac{B}{z^m}$.

PROOF We set $Q(z) = \frac{1}{z^m} P(z)$. From the maximum principle, we have

$$\sup_{r < |z| \leqslant R} |Q(z)| = \max \left\{ \sup_{|z| = R} \left| \frac{1}{z^m} P(z) \right|, \quad \sup_{|z| = r} \left| \frac{1}{z^m} P(z) \right| \right\}$$

$$\geqslant \max \left\{ \frac{1}{R^m} (|A| R^{n+m} + |B|), \quad \frac{1}{r^m} (|A| r^{n+m} + |B|) \right\}$$

Equality holds only for $P(z)=Az^{n+m}+B$, that is, for the generalised polynomial $Q(z)=Az^n+\frac{B}{z^m}$.

COROLLARY 3 For the generalised polynomials

$$Q(z)=Az^n+c_{n-1}z^{n-1}+c_{-m+1}\frac{1}{z^{m-1}}+B\frac{1}{z^m}, \quad n\geqslant 1, \quad m\geqslant 1$$

with fixed sum of the coefficients $A+B=C$ we have

$$\mu_n=\inf_{Q(z)}\sup_{r\leqslant|z|\leqslant R}|Q(z)|=\begin{cases}|C|R^n & \text{for } 0<r\leqslant R\leqslant 1\\ |C|\dfrac{R^{n+m}-r^{n+m}}{(R^{n+m}-1)r^m+(1-r^{n+m})R^m} & \text{for } 0<r<1<R\\ \dfrac{|C|}{r^m} & \text{for } 1\leqslant r\leqslant R\end{cases}$$

and μ_n is attained only for

$$Q(z)=\begin{cases}Cz^n & \text{for } 0<r\leqslant R<1 \text{ and } 0<r<R\leqslant 1\\ C\dfrac{(R^m-r^m)z^n+(R^n-r^n)R^m r^m\frac{1}{z^m}}{(R^{n+m}-1)r^m+(1-r^{n+m})R^m} & \text{for } 0<r<1<R\\ C\left(\lambda z^n+\dfrac{1-\lambda}{z^m}\right), \quad 0\leqslant\lambda\leqslant 1, & \text{for } r=R=1\\ \dfrac{C}{z^m} & \text{for } 1<r\leqslant R \text{ and } 1\leqslant r<R\end{cases}$$

To prove this corollary we need to find the minimum of the quantity μ_n corresponding to Corollary 2 under the condition that $A+B=C$.

We may also consider the case in which the value of a non-constant function $f(A, B)$ is fixed, though we shall not stop to do this. This remark also applies to some of the following problems.

COROLLARY 4 For trigonometric polynomials

$$S_n(\zeta)=a_0+\sum_{k=1}^{n}(a_k\cos k\zeta+b_k\sin k\zeta), \quad \alpha\leqslant\operatorname{Im}\zeta\leqslant\beta$$

with fixed coefficients a_n and b_n, we have

$$\mu_n=\inf_{S_n(\zeta)}\sup_{\alpha\leqslant\operatorname{Im}\zeta\leqslant\beta}|S_n(\zeta)|=\frac{1}{2}\max_{\substack{\gamma=\alpha\\ \gamma=\beta}}(|a_n+b_ni|e^{n\gamma}+|a_n-ib_n|e^{-n\gamma})$$

and μ_n is attained only for the trigonometric polynomials $S_n(\zeta)=a_n\cos n\zeta+b_n\sin n\zeta$.

NOTE 2 If $\alpha=\beta=0$, then

$$\mu_n=\frac{1}{2}(|a_n+b_ni|+|a_n-b_ni|)$$

We shall obtain the proof if we shift to the variable $e^{i\zeta}$ and use Corollary 2.

We may also consider the case in which only one of the coefficients a_n or b_n is fixed.

Before proceeding to the remaining corollaries, let us recall certain properties of Zhukovskiy's function

$$w=\frac{1}{2}\left(z+\frac{1}{z}\right)$$

It maps the z plane into a two-sheeted Riemann surface which covers the entire plane with branch points at $w=\pm 1$. The circle $|z|=r$ where $r>0$ and $r\neq 1$ is mapped on to the ellipse C_r, given by the parametric equation

$$u=\frac{1}{2}\left(r+\frac{1}{r}\right)\cos\varphi,\quad v=\frac{1}{2}\left(r-\frac{1}{r}\right)\sin\varphi,\quad 0\leqslant\varphi<2\pi,\quad w=u+iv$$

the sum of the semi-axes of which is equal to r for $r>1$ and equal to $1/r$ for $0<r<1$.

For $r=1$ the ellipse degenerates into the double interval $-1\leqslant w\leqslant +1$. These ellipses have foci at the points $w=\pm 1$. We note also that Zhukovskiy's function maps the annulus $r<|z|<\frac{1}{r}$ for $0<r<1$ on to the two-sheeted closed region $\bar{B}_r$ bounded by the ellipse C_r. For $r=1$ this region obviously degenerates into the double interval $-1\leqslant w\leqslant +1$.

The following formulae are obvious:

$$z=w+\sqrt{w^2-1},\quad z^{-1}=w-\sqrt{w^2-1},\quad \sqrt{w^2-1}=\frac{1}{2}\left(z-\frac{1}{z}\right)$$

Suppose that $\omega_{2n}(z)=\sum_{k=0}^{2n}A_kz^{2n-k}$ is a polynomial of degree no higher than $2n$. It follows immediately from what has been said that

$$\frac{\omega_{2n}(z)}{z^n}=R_n(w)+\sqrt{w^2-1}\,Q_{n-1}(w) \tag{4}$$

where $R_n(w)$ and $Q_{n-1}(w)$ are polynomials of degree not exceeding n and $n-1$, respectively:

$$R_n(w)=\sum_{k=0}^{n} a_k w^{n-k}, \quad Q_{n-1}(w)=\sum_{k=0}^{n-1} b_k w^{n-1-k} \tag{5}$$

Obviously, for arbitrary polynomials (5) and a fixed value of the root $\sqrt{w^2-1}$ for arbitrary $w=w_0\neq\pm 1$ there exists a unique polynomial $\omega_{2n}(z)$ that will satisfy the identity (4). We note the following obvious relationships between the coefficients of the polynomials:

$$\begin{gathered} A_0=\frac{1}{2^n}(a_0+b_0), \quad A_{2n}=\frac{1}{2^n}(a_0-b_0) \\ a_0=2^{n-1}(A_0+A_{2n}), \quad b_0=2^{n-1}(A_0-A_{2n}) \end{gathered} \tag{6}$$

COROLLARY 5 Suppose that $R_n(w)$ and $Q_{n-1}(w)$ are of the form (4) with fixed coefficients a_0 and b_0. Then,

$$\mu_n=\inf_{q_n(w)} \sup_{w\in\overline{B}_r} |q_n(w)|=\max_{\substack{\gamma=r\\ \gamma=\frac{1}{r}}} \frac{1}{2^n}\left(|a_0+b_0|\gamma^n+|a_0-b_0|\frac{1}{\gamma^n}\right)$$

in the two-sheeted region $\overline{B}_r$, for $0<r\leqslant 1$, where

$$q_n(w)=R_n(w)+\sqrt{w^2-1}\,Q_{n-1}(w)$$

Also, μ_n is attained only for the function

$$q_n(w)=a_0T_n(w)+\sqrt{w^2-1}\,b_0U_{n-1}(w) \tag{7}$$

where

$$T_n(w)=\frac{1}{2^n}\left[\left(w+\sqrt{w^2-1}\right)^n+\left(w-\sqrt{w^2-1}\right)^n\right]$$

$$U_{n-1}(w)=\frac{\left(w+\sqrt{w^2-1}\right)^n-\left(w-\sqrt{w^2-1}\right)^n}{2^n\sqrt{w^2-1}}$$

PROOF We have

$$q_n\left(\frac{1}{2}\left(z+\frac{1}{z}\right)\right)=\frac{\omega_{2n}(z)}{z^n}, \quad \omega_{2n}(z)=\frac{1}{2^n}(a_0+b_0)z^{2n} + \ldots + \frac{1}{2^n}(a_0-b_0)$$

so that Corollary 5 follows from application of Corollary 2.

We note that the leading coefficient of each of the polynomials $T_n(w)$ and $U_{n-1}(w)$ which are of degree n and $n-1$ respectively, are equal to one.

COROLLARY 6 In the region $\overline{B}_r$, with $0<r\leqslant 1$ for the function (7) where $a_0\neq 0$ is a fixed number, we have

$$\mu_n=\inf_{q_n(w)}\sup_{w\in\overline{B}_r}|q_n(w)|\frac{1}{2^n}|a_0|\left(r^n+\frac{1}{r^n}\right)$$

and μ_n is attained only for the functions

$$q_n(w)=\begin{cases} a_0T_n(w) & \text{for} \quad 0<r<1 \\ a_0\left(T_n(w)+\lambda\sqrt{w^2-1}\,U_{n-1}(w)\right), & -1<\lambda<1 \ \text{for}\ r=1\end{cases}$$

To prove this corollary, we need to find the minimum of the quantity μ_n, corresponding to Corollary 5, as a function of b_0.

Analogously, we may consider the case in which only the coefficient b_0 or the value of some function $f(a_0, b_0)$ is fixed.

COROLLARY 7 In the region $\overline{B}_r$, with $0<r\leqslant 1$ for the function

$$r_n(w)=R_n(w)+\sqrt{w^2-1}\,Q_{n-2}(w)$$

where $R_n(w)=w^n+a_1w^{n-1}+\ldots+a_n$ is an arbitrary polynomial of degree n with leading coefficient equal to one and $Q_{n-2}(w)$ is an arbitrary polynomial of degree not exceeding $n-2$, we have

$$\mu_n=\inf_{r_n(w)}\sup_{w\in\overline{B}_r}|r_n(w)|=\frac{1}{2^n}\left(r^n+\frac{1}{r^n}\right)$$

and μ_n is attained only for the function $r_n(w)=T_n(w)$.

We derive this corollary from Corollary 5 for $a_0=1$ and $b_0=0$.

If we consider only a sub-class of the class of functions $r_n(w)$ specifically, the polynomials $R_n(w)=w^n+a_1w^{n-1}+\ldots+a_n$ the polynomial $T_n(w)$ will be all the more a polynomial that deviates least from zero; that is, the polynomial $T_n(w)$ is, of all polynomials of degree n with leading coefficient equal to unity, the one that deviates least from zero on an arbitrary ellipse C_r, for $0<r<1$ or on the interval $-1\leqslant w\leqslant 1$. Thus, $T_n(w)$ is a Chebyshev polynomial for the ellipse C_r and the interval $-1\leqslant w\leqslant 1$. This polynomial $T_n(w)$ is usually called the Chebyshev polynomial of the first kind. We note that $T_n(w)$ is a Faber polynomial for the same sets (see Chapter 2, Section 2.1.1).

If we set w cos φ, we can easily show that

$$T_n(w) = \frac{1}{2^{n-1}} \cos n \cos^{-1} w \tag{8}$$

COROLLARY 8 In the region $\bar{B}_r$, with $0 < r \leqslant 1$ for the functions

$$\rho_n(w) = R_{n-1}(w) + \sqrt{w^2 - 1}\, Q_{n-1}(w)$$

where $R_{n-1}(w)$ is an arbitrary polynomial of degree not exceeding $n - 1$ and

$$Q_{n-1}(w) = w^{n-1} + b_1 w^{n-2} + \ldots + b_{n-1}$$

is an arbitrary polynomial of degree $n - 1$ with leading coefficient equal to one, we have

$$\mu_n = \inf_{\rho_n(w)} \sup_{w \in \bar{B}_r} |\rho_n(w)| = \frac{1}{2^n}\left(r^n + \frac{1}{r^n}\right)$$

and μ_n is attained only for the function

$$\rho_n(w) = \sqrt{w^2 - 1}\, U_{n-1}(w)$$

This corollary is obtained from Corollary 5 for $a = 0$ and $b = 1$. The polynomial $U_{n-1}(w)$ is called the Chebyshev polynomial of the second kind.

In analogy with Corollary 8, we may replace the class of functions $\rho_n(w)$ with the class of functions

$$\sqrt{w^2 - 1}\, Q_{n-1}(w) = \sqrt{w^2 - 1}\,(w^{n-1} + b_1 w^{n-2} + \ldots + b_{n-1})$$

without changing μ_n the extremal function of which is $\sqrt{w^2 - 1} \times U_{n-1}(w)$.

For $U_{n-1}(w)$ we easily obtain the following formula

$$U_{n-1}(w) = \frac{1}{2^n} \cdot \frac{\sin n \cos^{-1} w}{\sqrt{1 - w^2}} \tag{9}$$

PROBLEM 3 Consider the polynomials

$$P_n(z) = c_0 z^n + c_1 z^{n-1} + \ldots + c_n$$

with fixed coefficients $c_0, c_p, c_{2p}, \ldots, c_{mp}$ where $p \geqslant 2$ and m is the greatest integer such that $mp \leqslant n$. Of these polynomials, find the one that deviates least from zero on the circle $|z| = 1$ (or on the closed disc $|z| \leqslant 1$).

Let us designate by $\mathscr{P}_n$ the set of these polynomials. We note that the polynomial

$$P_n(z) = c_0 z^n + c_p z^{n-p} + \dots + c_{mp} z^{n-mp} \in \mathscr{P}_n$$

is bounded on $|z| = 1$. Specifically

$$|P_n(z)| \leqslant |c_0| + |c_p| + \dots + |c_{mp}| = M$$

for $|z| = 1$ since the coefficients are given. Therefore, if among the polynomials $P_n(z) \in \mathscr{P}_n$ there exists a polynomial $\pi_n(z)$ that deviates least from zero on $|z| = 1$ then $|\pi_n(z)| \leqslant M$ on $|z| = 1$. This now enables us to consider the polynomials $P_n(z) \in \mathscr{P}_n$ such that $|P_n(z)| \leqslant M$ on $|z| = 1$.

Let us show that the coefficients of such polynomials are bounded. According to Cauchy's formula

$$c_k = \frac{1}{k!} P_n^{(k)}(0) = \frac{1}{2\pi i} \int\limits_{|\zeta|=1} \frac{P_n(\zeta)}{\zeta^{k+1}} \, d\zeta, \quad k = 0, 1, \dots, n$$

and, consequently,

$$|c_k| \leqslant \frac{1}{2\pi} \max_{|\zeta|=1} |P_n(\zeta)| \cdot 2\pi \leqslant M$$

Let us now show that among the polynomials $P_n(z) \in \mathscr{P}_n$ such that $|P_n(z)| \leqslant M$ on $|z| = 1$ there exists a polynomial $\pi_n(z)$ that deviates least from zero on the circle $|z| = 1$. If we set

$$\mu_n = \inf_{P_n(z) \in \mathscr{P}_n} \sup_{|z|=1} |P_n(z)| \leqslant M$$

we see that there exists a sequence of polynomials

$$P_{n,\nu}(z) = c_0^{(\nu)} z^n + \dots + c_n^{(\nu)} \in \mathscr{P}_n$$

for $\nu = 1, 2, \dots$, such that

$$\lim_{\nu \to \infty} \sup_{|z|=1} |P_{n,\nu}(z)| = \mu$$

Since the coefficients of the polynomials $P_{n,\nu}(z)$ are uniformly bounded, from the sequence of polynomials $P_{n,\nu}(z)$ for $\nu = 1, 2, \dots$, we may choose a sub-sequence $P_{n,\nu_k}(z)$ for $k = 1, 2, \dots$, such that the limits

$$\lim_{k \to \infty} c_l^{(\nu_k)} = c_l^{(0)}, \quad l = 0, 1, \dots, n \tag{10}$$

exist. Let us set

$$\pi_n(z) = c_0^{(0)}z^n + c_1^{(0)}z^{n-1} + \ldots + c_n^{(0)}$$

It is clear that $\pi_n(z) \in \mathscr{P}_n$ and

$$\sup_{|z|=1} |\pi_n(z)| \geqslant \mu_n \tag{11}$$

But

$$\pi_n(z) = \pi_n(z) - P_{n,\nu_k}(z) + P_{n,\nu_k}(z)$$

and consequently for $|z| = 1$ we have

$$|\pi_n(z)| \leqslant |\pi_n(z) - P_{n,\nu_k}(z)| + |P_{n,\nu_k}(z)| \leqslant |\pi_n(z) - P_{n,\nu_k}(z)| + \sup_{|z|=1} |P_{n,\nu_k}(z)|$$

For every fixed z, such that $|z| = 1$, on the basis of (10) we obtain in the limit as $k \to \infty$ $|\pi_n(z)| \leqslant \mu_n$. Consequently,

$$\sup_{|z|=1} |\pi_n(z)| \leqslant \mu_n$$

From this inequality and inequality (11), we conclude that

$$\sup_{|z|=1} |\pi_n(z)| = \mu_n$$

and, consequently, $\pi_n(z)$ is the polynomial that deviates least from zero on $|z| = 1$ of all the polynomials in the class $\mathscr{P}_n$.

Suppose that

$$P_n(z) = c_0z^n + c_1z^{n-1} + \ldots + c_n \in \mathscr{P}_n$$

is any polynomial that deviates least from zero on $|z| = 1$ of the polynomials in the class $\mathscr{P}_n$ and that $\zeta_k = e^{\frac{2\pi k}{p}i}$ for $k = 0, 1, \ldots, p-1$ are the p-th roots of unity. By virtue of a property of the roots ζ_k, we have

$$\sigma_n(z) = \frac{1}{p}\sum_{k=0}^{n-1} \zeta_k^{-n} P_n(\zeta_k z) = c_0z^n + c_pz^{n-p} + \ldots + c_{mp}z^{n-mp} \in \mathscr{P}_n$$

For every z, belonging to the circle $|z| = 1$, we have

$$|\sigma_n(z)| \leqslant \left|\frac{1}{p}\sum_{k=0}^{p-1} \zeta_k^{-n} P_n(\zeta_k z)\right| \leqslant \frac{1}{p}\sum_{k=0}^{n} |P_n(\zeta_k z)| \leqslant \mu_n$$

From this we conclude that $\sigma_n(z)$ of this form is a polynomial that deviates least from zero in the class $\mathscr{P}_n$ on the circle $|z|=1$ and that

$$\mu_n = \sup_{|z|=1} |\sigma_n(z)|$$

We note that the considerations that we have made do not permit us to assert that among the polynomials $P_n(z) \in \mathscr{P}_n$ there exists a unique polynomial that deviates least from zero on $|z|=1$. There may not be a unique polynomial that deviates least from zero on $|z|=1$. We shall give an example. Suppose that $\mathscr{P}_3$ is the class of polynomials of the form $P_3(z) = z^3 + c_1 z^2 + z + c_3$ (so that $(n=3,\ p=2$, and $c_0 = c_2 = 1)$. Then, on the basis of what we said above, $\pi_3(z) = z^3 + z$ is a polynomial that deviates least from zero on $|z|=1$ and $\mu_3 = 2$, since for $|z|=1$ we have $|\pi_3(z)| \leqslant |z|^3 + |z| = 2$ and $\pi_3(1) = 2$. Let us show that every polynomial of the form $P(z) = z^3 + hz^2 + z - h$ in $\mathscr{P}_3$ for $-\sqrt{2}+1 \leqslant h \leqslant \sqrt{2}+1$ is also a polynomial that deviates least from zero on $|z|=1$. If we set $z = e^{i\varphi}$, for $0 \leqslant \varphi < 2\pi$, we have

$$\begin{aligned}
|P(e^{i\varphi})| &= |e^{3i\varphi} + e^{i\varphi} + h(e^{2i\varphi} - 1)| \\
&= |e^{i\varphi}[e^{i\varphi}(e^{i\varphi} + e^{-i\varphi}) + h(e^{i\varphi} - e^{-i\varphi})]| \\
&= |e^{i\varphi} 2\cos\varphi + 2ih\sin\varphi| = 2\,|(\cos\varphi + i\sin\varphi)\cos\varphi + ih\sin\varphi| \\
&= 2\sqrt{\cos^4\varphi + (\sin\varphi\cos\varphi + h\sin\varphi)^2} \\
&= 2\sqrt{\cos^4\varphi + \sin^2\varphi\cos^2\varphi + 2h\sin^2\varphi\cos\varphi + h^2\sin^2\varphi} \\
&\leqslant 2\sqrt{\cos^2\varphi + 2\,|h|\sin^2\varphi + h^2\sin^2\varphi} \\
&= 2\sqrt{2\,|h| + h^2 + (1 - 2\,|h| - h^2)\cos^2\varphi} \leqslant 2
\end{aligned}$$

It should be noted that for the values of h indicated, the expression $1 - 2|h| - h^2$ is non-negative. Thus, in the present case, there exist infinitely many polynomials deviating least from zero on $|z|=1$.

Now, we shall present an example in which a unique polynomial deviating least from zero on $|z|=1$ does exist.

Let $\mathscr{P}_4$ denote the class of polynomials

$$P_4(z) = Az^4 + c_1 z^3 + c_3 z + B$$

(so that $n=4,\ p=2,\ c_0 = A,\ c_2 = 0,\ c_4 = B$). This class of polynomials is the sub-class of the polynomials considered in Problem 2 with $n=4$ and, consequently, $\pi_4(z) = Az^4 + B$ is the unique polynomial deviating least from zero on $|z|=1$.

We note that, in an arbitrary case, either there exists a unique polynomial that deviates least from zero on $|z|=1$ or there are infinitely many such polynomials. We shall not stop to analyse the conditions under which the first or the second case prevails.

PROBLEM 4 Consider the polynomials

$$P_n(z) = c_0 z^n + c_1 z^{n-1} + \ldots + c_n$$

with fixed coefficients c_0 and c_p, where $1 \leqslant p \leqslant n$. Find the polynomial in this class that deviates least from zero on the circle $|z| = 1$.

We shall designate by $\mathscr{P}_n$ this class of polynomials. As in Problem 3, we can show that a polynomial deviating least from zero in the class $\mathscr{P}_n$ exists. Let us show that the case in which $1 < p < n$ reduces to the case of $p = 1$. By using the considerations made in the discussion of Problem 3, we see that the polynomial $R_n(z)$ of the form

$$R_n(z) = c_0 z^n + c_p z^{n-p} + \ldots + c_{mp} z^{n-mp} \in \mathscr{P}_n$$

where m is the largest integer such that $mp \leqslant n$ is one of the polynomials that deviates least from zero on the circle $|z| = 1$. For such a polynomial, we have

$$\sup_{|z|=1} |R_n(z)| = \sup_{|z|=1} |c_0 z^{mp} + c_p z^{(m-1)p} + \ldots + c_{mp}|$$
$$= \sup_{|\zeta|=1} |c_0 \zeta^m + c_p \zeta^{m-1} + \ldots + c_{mp}|$$

on $|z| = 1$. Obviously, if we can find a polynomial that deviates least from zero on $|\zeta| = 1$, out of all the polynomials of the form $P_m(\zeta) = c_0 \zeta^m + c_p \zeta^{m-1} + \ldots c_{mp}$ with fixed coefficients c_0 and c_p, we shall also have found the polynomial $\pi_n(z) \in \mathscr{P}_n$, that deviates least from zero on $|z| = 1$. This makes it possible for us to consider only the case of $p = 1$ in solving the problem.

If $c_0 = c_1 = 0$ the polynomial $\pi_n(z) \equiv 0 \in \mathscr{P}_n$ will indeed be the polynomial that deviates least from zero on $|z| = 1$. If one of the coefficients c_0 or c_1 is equal to zero, the problem is immediately solved by invoking Theorem 1.

Therefore, we assume that $c_0 \neq 0$ and $c_1 \neq 0$. At this point, we note without proof that in this case there exists a unique polynomial that deviates least from zero on $|z| = 1$ (see Section 5.3.4).

Let us set $\theta = \arg c_0$ and $\vartheta = \arg c_1$. If $P_n(z) = c_0 z^n + c_1 z^{n-1} + \ldots + c_n \in \mathscr{P}_n$, we set $Q_n(z) = e^{i\beta} P_n(e^{i\alpha} z)$, where $\alpha = \vartheta - \theta$ and $\beta = (n-1)\theta - n\vartheta$. We then obtain

$$Q_n(z) = |c_0| z^n + |c_1| z^{n-1} + \ldots$$

and

$$\sup_{|z|=1} |P_n(z)| = \sup_{|z|=1} |Q_n(z)|$$

Here, if $\Pi_n(z)$ is a polynomial that deviates least from zero on $|z| = 1$ of all the polynomials

$$Q_n(z) = |c_0| z^n + |c_1| z^{n-1} + \ldots$$

the polynomial

$$\pi_n(z) = e^{-i\beta} \Pi_n(e^{-i\alpha} z)$$

is a polynomial that deviates least from zero on $|z| = 1$ out of all the polynomials $P_n(z) \in \mathscr{P}_n$. This now allows us to assume that $c_0 > 0$ and $c_1 > 0$. We note that all the coefficients of the polynomial $\pi_n(z) = \sum_{k=0}^{n} c_k z^{n-k}$ that deviates least from zero on $|z| = 1$ are real. If this were not the case, a polynomial

$$\overline{\pi_n(z)} = \sum_{k=0}^{n} \bar{c}_k z^{n-k} \in \mathscr{P}_n$$

distinct from $\pi_n(z)$, would also be a polynomial that deviates least from zero on $|z| = 1$ which is impossible because of the uniqueness.

We note without proof that, for the classes $\mathscr{P}_n$ (for $n = 1, 2, \ldots$) of polynomials of the form

$$P_n(z) = Az^n + Bz^{n-1} + c_2 z^{n-2} + \ldots + c_n$$

with fixed positive A and B, the quantity

$$\mu_n = \inf_{P_n(z) \in \mathscr{P}_n} \sup_{|z|=1} |P_n(z)|$$

decreases with increasing n and approaches the number

$$\mu = \frac{1}{2}\left(B + \sqrt{A^2 + B^2}\right)$$

This result is easily obtained from a theorem of Schur (see Goluzin [3]). It is also easy to show that

$$\mu_2 = \frac{1}{4\sqrt{2}} \frac{(3B+\sqrt{B^2+32A^2})^{3/2}}{(B+\sqrt{B^2+32A^2})^{1/2}}$$

and that μ_2 is attained only for the polynomial

$$P_2(z) = Az^2 + Bz + c, \qquad c = \frac{2AB}{B+\sqrt{B^2+32A^2}}$$

PROBLEM 5 Suppose that $\bar{B}$ is a closed bounded set in the z plane. In the class $\mathscr{P}_n[z_0]$ of polynomials $P_n(z)$ of degree not exceeding $n \geqslant 0$ that satisfy the condition $P_n(z_0) = 1$, where z_0 is a fixed point in the z plane, find a polynomial that deviates least from zero on $\bar{B}$. In other words, set

$$\mu_n(\bar{B}, z_0) = \inf_{P_n(z) \in \mathscr{P}_n[z_0]} \sup_{z \in \bar{B}} |P_n(z)|$$

and find a polynomial

$$\Pi_n(z) \in \mathscr{P}_n[z_0]$$

such that

$$\sup_{z \in \bar{B}} |\Pi_n(z)| = \mu_n(\bar{B}, z_0)$$

It is easy to show that such a polynomial exists. In this connection, we say that $\mu_n(\bar{B}, z_0)$ is attained for the polynomial $\Pi_n(z)$.

NOTE If $z_0 \in \bar{B}$ then $\mu_n(\bar{B}, z_0) = 1$ and $\mu_n(\bar{B}, z_0)$ is attained for example, for the polynomial $\Pi_n(z) \equiv 1$. In what follows, we assume that $z_0 \bar{\in} \bar{B}$.

In the case in which $\bar{B}$ is the closed disc $|z| \leqslant r$ where $r > 0$ the solution to Problem 5 is given by the following theorem.

Theorem 3 $\mu_n(|z| \leqslant r, z_0) = \frac{r^n}{|z_0|^n} \ (|z_0| > r)$ and $\mu_n(|z| \leqslant r, |z_0|)$ is attained only for the polynomial $\Pi_n(z) = \left(\frac{z}{z_0}\right)^n$

PROOF For every polynomial $P_n(z) \in \mathscr{P}_n[z_0]$ distinct from $\Pi_n(z)$ the function $\varphi(z) = \frac{P_n(z)}{\Pi_n(z)}$ is regular in the region

$|z| \geqslant r$, $\varphi(z_0) = 1$ and (by virtue of the maximum principle)

$$\sup_{|z|=r} |\varphi(z)| = \frac{|z_0|^n}{r^n} \sup_{|z|=r} |P_n(z)| > 1$$

consequently, $\sup\limits_{|z| \leqslant r} |P_n(z)| > \frac{r^n}{|z_0|^n}$. This completes the proof of the theorem.

This theorem is a special case of the following theorem, which can be proved in an analogous manner.

Theorem 4 Suppose that $Q_m(z)$ is a given polynomial of degree $m \geqslant 1$, that B is the set of points z satisfying the inequality $|Q_m(z)| \leqslant r$, where $r > 0$, that $z_0 \bar{\in} \bar{B}$, and that n is an arbitrary natural number. Then,

$$\mu_{nm}(\bar{B}, z_0) = \frac{r^n}{|Q_m(z_0)|^n}$$

and $\mu_{nm}(\bar{B}, z_0)$ is attained only for the polynomial

$$\Pi_{nm}(z) = \left(\frac{Q_m(z)}{Q_m(z_0)}\right)^n$$

We have already considered the Zhukovskiy function $w = \frac{1}{2}\left(z + \frac{1}{z}\right)$ and have introduced the ellipses C_r, (where $r \geqslant 1$) and the two-sheeted closed region $\bar{B}_r$. The foci of these ellipses are at the points ± 1 and the sums of their semi-axes is r; the region $\bar{B}_r$ is the image of the annulus $\frac{1}{r} \leqslant |z| \leqslant r$ under the mapping defined by Zhukovskiy's function. If we use Zhukovskiy's function, we easily obtain from Theorem 3 the following corollary.

COROLLARY Consider the two-sheeted region $\bar{B}_r$, for $r \geqslant 1$ for the functions

$$q_n(w) = P_n(w) + \sqrt{w^2 - 1}\, Q_{n-1}(w)$$

where $P_n(w)$ and $Q_{n-1}(w)$ are polynomials of degrees not exceeding n and $n-1$ respectively and satisfy the condition $q_n(w_0) = 1$ where $w_0 \in C_R$ (with $R > r$). By $q_n(w_0)$ we mean either of the two values of $q_n(w_0)$. Then,

$$\tilde{\mu}_n = \inf_{q_n(w)} \sup_{w \in \bar{B}_r} |q_n(w)| = \frac{r^n}{R^n}$$

and $\tilde{\mu}_n$ is attained only for the function

$$q_n(w) = \frac{(w + \sqrt{w^2 - 1})^n}{(w_0 + \sqrt{w_0^2 - 1})^n} = \frac{T_n(w) + \sqrt{w^2 - 1}\, U_{n-1}(w)}{T_n(w_0) + \sqrt{w_0^2 - 1}\, U_{n-1}(w)}$$

Here, the root $\sqrt{w_0^2 - 1}$ is understood to have that value (of the two possible values) that will give a greater value for $\left|w_0 + \sqrt{w_0^2 - 1}\right|$. We take $q_n(w_0) = 1$. It easily follows from this that, in particular,

$$\mu_n(\bar{B}_r, w_0) > \frac{r^n}{R^n}$$

where $w_0 \in C_R$ with $R > r$.

We denote by $\mathscr{P}_n^+[z_0]$ the class of polynomials $P_n(z)$ of degree not exceeding n that have real coefficients and that satisfy the condition $|P_n(z_0)| = 1$. We define

$$\mu_n^+(\bar{B}_r, z_0) = \inf_{P_n(z) \in \mathscr{P}_n^+[z_0]} \sup_{z \in \bar{B}_r} |P_n(z)|$$

Obviously,

$$\mu_n^+(\bar{B}_r, w_0) \geqslant \mu_n(\bar{B}_r, w_0)$$

where $w_0 \in \bar{B}_r$, and, consequently,

$$\mu_n^+(\bar{B}_r, w_0) > \frac{r^n}{R^n}$$

where $w_0 \in C_R$, with $R > r$. This result was obtained by Bernstein [1], pp. 146, 530 (see also Lemma 4, Section 1.2).

We note also that Hall [1] made a very interesting study of Problem 5 for the case in which the set $\bar{B}$ has a finite or countable number of points.

5.1.2 *Mendeleev's problem - Markov's and Bernstein's inequalities*

Suppose that $\bar{B}$ is a closed bounded set in the z plane and that $\mathscr{P}_n$ for $n \geqslant 1$ is the class of polynomials $P_n(z) = c_0 z^n + c_1 z^{n-1} + \dots + c_n$ that satisfy the condition $|P_n(z)| \leqslant M$, where $M > 0$ for $z \in B$.

PROBLEM 1. Of all the polynomials $P_n(z) \in \mathscr{P}_n$ find the one for which $\sup_{z \in \bar{B}} |P'_n(z)|$ attains its maximum value.

This problem was posed by Mendeleev for the case in which the set $\bar{B}$ is an interval on the real axis. The following problem arises in a natural manner.

PROBLEM 2 Out of all the polynomials $P_n(z) \in \mathscr{P}_n$ find the polynomial for which the quantity $|P'_n(z_0)|$, where z_0 is a fixed point, attains its maximum.

Obviously, if Problem 2 is solved for every point $z_0 \in B$, Problem 1 will also be solved. Problem 2 is a particular case of the following problem.

PROBLEM 3 From among the polynomials $P_n(z) \in \mathscr{P}_n$ find a polynomial for which the quantity

$$\lambda_n = \left| \sum_{k=1}^{n} p_k c_k \right|$$

where $p_0, \ldots, p_n$ are given numbers, attains its greatest value (see Problem 2 and its dual, Problem 1, in the introduction to this chapter).

We shall obtain Problem 2 from this problem if we set $p_k = (n-k) z_0^{n-k-1}$ for $k = 0, 1, \ldots, n$.

An extensive literature has been devoted to these three problems. The first efforts in this direction were the works of A.A. Markov [1], V.A. Markov [1], and Bernstein [2], pp. 281-286. We note also the recent works by Vidensky [1 and 3], Voronovskaya [3], and Gusev [1].

We shall give the solution of a certain problem from which the results of A.A. Markov and Bernstein can be obtained as special cases.

We recall (see introduction to this chapter) that Problem 3 has an infinite number of solutions since, if the polynomial $P_n(z) \in \mathscr{P}_n$ is a solution of Problem 3, so is the polynomial $e^{i\varphi} P_n(z) \in \mathscr{P}_n$, where $0 \leqslant \varphi < 2\pi$.

Theorem 1 Suppose that a polynomial $\omega_{2n}(z)$ of even degree $2n$, where $n \geqslant 2$, vanishes at the points $z = \pm 1$ and that $|\omega_{2n}(z)| \leqslant M$ for $M > 0$ and $|z| = 1$. Then, if we set $z = e^{i\varphi}$ where $\pi < \varphi \leqslant \pi$ we have

$$\left| \frac{\omega_{2n}(z)}{z^2 - 1} \right| \leqslant \left\{ \begin{array}{ll} \dfrac{M}{2|\sin\varphi|} & \text{for} \quad \dfrac{\pi}{2n} \leqslant \varphi \leqslant \pi - \dfrac{\pi}{2n} \\ \dfrac{M}{2} \left| \dfrac{\sin n\varphi}{\sin\varphi} \right| & \text{for} \quad |\varphi| \leqslant \dfrac{\pi}{2n}, \quad \pi - \dfrac{\pi}{2n} \leqslant |\varphi| \leqslant \pi \end{array} \right\} \leqslant \frac{M}{2} n \tag{1}$$

The first inequality reduces to an equality only for polynomials of the form

$$\omega_{2n}(z) = \frac{M}{2}\eta(z^{2n}-1), \text{ where } |\eta| = 1$$

and even then only for values of φ such that $|\varphi| \leqslant \frac{\pi}{2n}$ or $\pi - \frac{\pi}{2n} \leqslant |\varphi| \leqslant \pi$. The second inequality reduces to an equality only for $\varphi = 0$ or $\varphi = \pi$.

PROOF Consider the four points (see drawing)

$$A_1\left(\zeta_1 = e^{i\frac{\pi}{2n}}\right), \quad A_2(\zeta_2 = -\bar{\zeta}_1), \quad A_3(\zeta_3 = -\zeta_1) \text{ and } A_4(\zeta_4 = \bar{\zeta}_1)$$

on the circle $|z| = 1$.

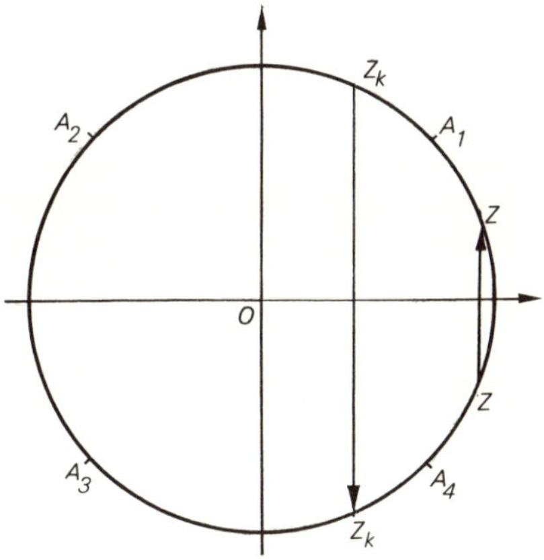

Fig. 1

If we set $z = e^{i\varphi}$ where $-\pi < \varphi \leqslant \pi$ we have on the closed arcs $\widetilde{A_1A_2}$ and $\widetilde{A_3A_4}$, i.e. for $\frac{\pi}{2n} \leqslant |\varphi| \leqslant \pi -$

$$\left|\frac{\omega_{2n}(e^{i\varphi})}{e^{2i\varphi}-1}\right| \leqslant \frac{M}{|e^{i\varphi}-e^{-i\varphi}|} = \frac{M}{2|\sin\varphi|} \leqslant \frac{M}{2\sin\frac{\pi}{2n}} < \frac{M}{2\frac{2}{\pi}\frac{\pi}{2n}} = \frac{M}{2}n$$

Thus, inequalities (1) are proven for $\frac{\pi}{2n} \leqslant |\varphi| \leqslant \pi - \frac{\pi}{2n}$ and equality does not hold in the inequality on the right.

Let us prove inequalities (1) for $|\varphi| \leqslant \frac{\pi}{2n}$ and $\pi - \frac{\pi}{2n} \leqslant |\varphi| \leqslant \pi$, that is, on the arcs $\widetilde{A_1A_4}$ and $\widetilde{A_2A_3}$. For this, we need the following representation:

$$\frac{\omega_{2n}(z)}{z^2-1} = -\frac{1}{4n}\sum_{k=1}^{2n}\left(\frac{z-\frac{1}{z}}{z_k-\frac{1}{z_k}}+1\right)\omega_{2n}(z_k)\frac{z^{2n}+1}{(zz_k-1)(z\bar{z}_k-1)} \tag{2}$$

where $z_k = e^{i\frac{2k-1}{2n}\pi}$, for $k = 1, 2, \ldots, 2n$ (these are the $(2n)$th roots of -1). Let us prove this formula. The right-hand side of Formula (2) is of the form

$$a_0 z^{2n-1} + a_1 z^{2n-2} + \ldots + a_{2n-1} + a_{2n}\frac{1}{z}$$

where

$$a_0 = -\frac{1}{4n}\sum_{k=1}^{2n}\frac{\omega_{2n}(z_k)}{z_k-\frac{1}{z_k}} = -\frac{1}{4n}\sum_{k=1}^{2n}\frac{z_k\omega_{2n}(z_k)}{z_k^2-1}$$

The polynomial $\frac{z\omega_{2n}(z)}{z^2-1}$ of degree $2n-1$ does not have a free term. Therefore, the sum of its values over all $(2n)$th roots of -1 is equal to zero. From this it follows that $a_0 = 0$. Also, $a_{2n} = 0$ since a_{2n} differs from a_0 only by sign.

Thus, in Formula (2) there are polynomials of degree $2n-2$ both on the left and on the right. It is easy to verify that their values coincide at the points z_k, where $k = 1, 2, \ldots, 2n$, which completes the proof of Formula (2).

Let us now prove inequality (1) on the open arcs A_4A_1 and A_2A_3. If we fix z arbitrarily on one of these arcs and use Formula (2), we obtain

$$\left|\frac{\omega_{2n}(z)}{z^2-1}\right| \leqslant \frac{1}{4n}\sum_{k=1}^{2n}\left|\frac{z-\frac{1}{z}}{z_k-\frac{1}{z_k}}+1\right|\cdot|\omega_{2n}(z_k)|\cdot\left|\frac{z^{2n}+1}{(zz_k-1)(z\bar{z}_k-1)}\right|$$

Remembering that $|\omega_{2n}(z_k)| \leqslant M$, we obtain

$$\left|\frac{\omega_{2n}(z)}{z^2-1}\right| \leqslant \frac{M}{4n}\sum_{k=1}^{2n}\left|\frac{z-\frac{1}{z}}{z_k-\frac{1}{z_k}}+1\right|\cdot\left|\frac{z^{2n}+1}{(zz_k-1)(z\bar{z}_k-1)}\right| \tag{3}$$

The expression on the right clearly does not depend on the choice of the polynomial $\omega_{2n}(z)$. For equality to hold in (3), it is necessary and sufficient that all the terms

$$\left(\frac{z-\dfrac{1}{z}}{z_k-\dfrac{1}{z_k}}+1\right)\omega_{2n}(z_k)\frac{z^{2n}+1}{(zz_k-1)(z\bar{z}_k-1)}$$

have identical arguments and that $|\omega_{2n}(z_k)|=M$. In these terms, the first factor is positive since the fraction

$$\xi_k=\frac{z-\dfrac{1}{z}}{z_k-\dfrac{1}{z_k}}$$

is real and $|\xi_k|<1$ (see drawing). Furthermore,

$$(zz_k-1)(z\bar{z}_k-1)=z^2-z(z_k+\bar{z}_k)+1=z[(z+\bar{z})-(z_k+\bar{z}_k)].$$

Since $|z+\bar{z}|>|z_k+\bar{z}_k|$ and $z+\bar{z}>0$ on A_4A_1 but $z+\bar{z}<0$ on A_2A_3, we have

$$\arg[(z\bar{z}_k-1)(zz_k-1)]=\begin{cases}\arg z & \text{on } A_4A_1\\ \pi-\arg z & \text{on } A_2A_3\end{cases}$$

It follows from this that the argument of the expression $\dfrac{z^{2n}+1}{(zz_k-1)(z\bar{z}_k-1)}$ for fixed z does not depend on k. Thus, all these terms will have the same argument if $\arg\omega_{2n}(z_k)$ is independent of k. Since $|\omega_{2n}(z)|=M$, we have $\omega_{2n}(z_k)=-M\eta$, where η is a fixed number whose absolute value is one.

Since we know the value of the polynomial $\omega_{2n}(z)$ at the $2n+2$ points z_k, 1, -1 the form of the polynomial for which equality holds in (3) is uniquely determined:

$$\omega_{2n}(z)=\frac{M}{2}\eta(z^{2n}-1)$$

and the absolute value of this polynomial does not exceed M for $|z|=1$. For this polynomial,

$$\frac{\omega_{2n}(z)}{z^2-1}=\frac{M}{2}\eta(1+z^2+z^4+\ldots+z^{2n-2})$$

For $z=e^{i\varphi}$ where φ is a real number satisfying either the inequality $|\varphi| \leqslant \frac{\pi}{2n}$ or the inequality $\pi - \frac{\pi}{2n} \leqslant |\varphi| \leqslant \pi$ we obtain

$$\left|\frac{\omega_{2n}(z)}{z^2-1}\right| = \frac{M}{2}\left|\frac{\sin n\varphi}{\sin\varphi}\right| \leqslant \frac{M}{2}n$$

Equality holds only for $\varphi = 0$ and $\varphi = \pi$, that is, at the points $z = \pm 1$. This proves the theorem.

COROLLARY 1 If a polynomial $\omega_n(z)$ of degree $n \geqslant 1$ vanishes at the point $z = 1$ and if $|\omega_n(z)| \leqslant M$ where $M > 0$ everywhere on the circle $|z| = 1$ then, by setting $z = e^{i\varphi}$ where $-\pi < \varphi \leqslant \pi$, we obtain

$$\left|\frac{\omega_n(z)}{z-1}\right| \leqslant \left\{\begin{array}{ll} \dfrac{M}{2\left|\sin\frac{\varphi}{2}\right|} & \text{for } \frac{\pi}{n} \leqslant |\varphi| \leqslant \pi \\ \dfrac{M}{2}\left|\dfrac{\sin\frac{n}{2}\varphi}{\sin\frac{\varphi}{2}}\right| & \text{for } |\varphi| \leqslant \frac{\pi}{n} \end{array}\right\} \leqslant \frac{M}{2}n$$

Equality holds in the inequality on the left (for $|\varphi| \leqslant \frac{\pi}{n}$) only for polynomials of the form $\omega_n(z) = \frac{M}{2}\eta(z^n - 1)$ where $|\eta| = 1$ and only for φ such that $|\varphi| \leqslant \frac{\pi}{n}$. Equality holds in the inequality on the right only when $\varphi = 0$.

To prove this corollary, we need to apply the preceding theorem to the polynomial $\omega_n(\zeta^2)$ of degree $2n$.

COROLLARY 2 Suppose that $q_n(z)$ is a polynomial of degree n such that $|q_n(z)| \leqslant M$, where $M > 0$ for $|z| = 1$. Then,

$$\left|\frac{d}{dz}\frac{q_n(z)}{z^{n/2}}\right| \leqslant \frac{M}{2}n, \qquad |z| = 1 \tag{4}$$

Equality holds only for polynomials of the form

$$q_n(z) = a_0 z^n + a_n, \text{ where } |a_0| + |a_n| = M, \text{ for } z = e^{i\theta_k}$$
$$\theta_k = \frac{1}{n}\left(\arg\frac{a_0}{a_n} + 2\pi k\right), \quad k = 0, 1, \ldots, n-1, \text{ if } a_n \neq 0$$

and for every z such that $|z| = 1$ if $a_n = 0$.

PROOF Suppose that $q_n(z) = a_0 z^n + a_1 z^{n-1} + \ldots + a_n$. We set

$$\omega_n(z) = z^n q_n\left(\frac{e^{i\theta}}{z}\right) - q_n(e^{i\theta}z) \tag{5}$$

where θ is an arbitrary real number. Since $|\omega_n(z)| \leqslant 2M$ for $|z|=1$ and since $\omega_n(1)=0$ we have by virtue of Corollary 1

$$\left|\frac{\omega_n(z)}{z-1}\right| \leqslant Mn \tag{6}$$

on the circle $|z|=1$. This yields $|\omega_n'(1)| \leqslant Mn$. On the basis of (5), we obtain

$$\omega_n'(1) = nq_n(e^{i\theta}) - 2e^{i\theta}q_n'(e^{i\theta})$$

and the preceding inequality yields

$$|nq_n(z) - 2zq_n'(z)| \leqslant Mn, \quad |z|=1$$

from which (4) follows.

If equality holds in inequality (4) for some z, for example for $z=e^{i\theta_0}$ equality must hold in (6) for $\theta=\theta_0$ and $z=1$ and, by virtue of Corollary 1,

$$\omega_n(z) = z^n q_n\left(\frac{e^{i\theta_0}}{z}\right) - q_n(e^{i\theta_0}z) = M\eta(z^n - 1)$$

where $|\eta|=1$; that is,

$$(a_0e^{in\theta_0} + a_1e^{i(n-1)\theta_0}z + \dots + a_nz^n) \\ - (a_0e^{in\theta_0}z^n + a_1e^{i(n-1)\theta_0}z^{n-1} + \dots + a_n) = M\eta(z^n - 1)$$

If we equate the coefficients of z^n, we see that $a_n - a_0e^{in\theta_0} = M\eta$ and, consequently, that the polynomial $q_n(z)$ is of the form

$$q_n(z) = a_0z^n + a_0e^{in\theta_0} + M\eta + \omega(z)$$

where $\omega(z)$ is a polynomial of degree $n-1$ without a constant term. Let us show that $\omega(z) \equiv 0$. If we set

$$z_k' = e^{i\left(\theta_0 + \frac{2k+1}{n}\pi\right)}$$

where $k=0, 1, \dots, n-1$, we have

$$q_n(z_k') = M\eta + \omega(z_k')$$

and, consequently,

$$|q_n(z_k')|^2 = M^2 + 2M\,\mathrm{Re}\,\{\bar{\eta}\omega(z_k')\} + |\omega(z_k')|^2 \leqslant M^2$$

since $|q_n(z)| \leqslant M$ for $|z|=1$. When we add these inequalities, we obtain

$$2M\,\mathrm{Re}\left\{\bar{\eta}\sum_{k=0}^{n-1}\omega(z_k')\right\}+\sum_{k=0}^{n-1}|\omega(z_k')|^2 \leqslant 0$$

Since $\sum_{k=0}^{n-1}\omega(z_k')=0$ and $\omega(z)$ is a polynomial of degree $n-1$ without a constant term, we conclude from the last inequality that $\omega(z_k')=0$, where $k=0, 1, \ldots, n-1$ and, consequently, that $\omega(z)\equiv 0$. Thus,

$$q_n(z)=a_0z^n+a_n, \quad a_n=M\eta+a_0e^{in\theta}$$

where

$$M \geqslant \sup_{|z|=1}|q_n(z)|=|a_0|+|a_n| \geqslant |a_0|+M-|a_0|=M$$

Consequently, only a polynomial of the form

$$q_n(z)=a_0z^n+a_n, \quad |a_0|+|a_n|=M$$

can be an extremal polynomial. It is easy to verify that, for such a polynomial, equality will hold in (4) at the points indicated in the corollary. This completes the proof of the corollary.

NOTE 1 Suppose that $\mathscr{P}_n$ where $n \geqslant 2$ is the class of polynomials $P_n(z)$ of degree not exceeding n that satisfy the condition

$$|P_n(z)| \leqslant M, \quad M>0, \quad \text{for} \quad |z|=1$$

If we set $q_{2n}(z)=z^nP_n(z)$ and apply Corollary 2, we obtain

$$|P_n'(z)| \leqslant Mn, \qquad |z|=1$$

Equality holds only for the polynomials $P_n(z)=M\eta z^n$ where $|\eta|=1$. Thus Problem 1 is solved for the case in which the set $\bar{B}$ is the circle $|z|=1$ and Problem 2 is solved for points z_0 belonging to the same circle. In Section 5.2 we shall give the solution of Problem 2 for every z_0 such that $|z_0|>1$.

COROLLARY 3 Suppose that $P_n(z)$ is a polynomial of degree n and that $|P_n(z)| \leqslant M$ for $M>0$ on the circle $|z|=R$ where $R>0$. Then on $|z|=R$ we have

$$|P_n'(z)| \leqslant \frac{M}{R} n. \tag{7}$$

Equality holds only for the polynomials $P_n(z) = \frac{M}{R^n} \eta z^n$ where $|\eta| = 1$.

To prove this corollary, we need to apply Corollary 2 to the polynomial $q_{2n}(z) = z^n P_n(Rz)$.

COROLLARY 4 If $|S_n(\theta)| \leqslant M$ for the trigonometric polynomial

$$S_n(\theta) = A_0 + \sum_{k=1}^{n} (A_k \cos k\theta + B_k \sin k\theta)$$

where θ assumes real values, then

$$|S_n'(\theta)| \leqslant Mn, \quad 0 \leqslant \theta < 2\pi \tag{8}$$

Equality holds only for polynomials of the form

$$S_n(\theta) = A_n \cos n\theta + B_n \sin n\theta$$

where $\frac{1}{2}(|A_n + B_n i| + |A_n - B_n i|) = M$ for

$$\theta = \theta_k = \frac{1}{2n}\left[(2k+1)\pi - \arg \frac{A_n + B_n i}{A_n - B_n i}\right], \qquad k = 0, 1, \ldots, n-1$$

if $A_n + B_n i \neq 0$ and $A_n - B_n i \neq 0$ and for all θ if $A_n = B_n i$ or $A_n = -B_n i$. Inequality (8) is usually referred to as Bernstein's first inequality.

PROOF We have

$$S_n(\theta) = e^{-in\theta}\left[A_0 e^{in\theta} + \sum_{k=1}^{n}\left(\frac{A_k - B_k i}{2} e^{i(n+k)\theta} + \frac{A_k + B_k i}{2} e^{i(n-k)\theta}\right)\right]$$

so that, if we set

$$q_{2n}(z) = A_0 z^n + \sum_{k=1}^{n}\left(\frac{A_k - B_k i}{2} z^{n+k} + \frac{A_k + B_k i}{2} z^{n-k}\right), \qquad z = e^{i\theta}$$

we obtain $|q_{2n}(z)| \leqslant M$ for $|z| = 1$.

If we apply Corollary 2, we have

$$\left|\frac{d}{dz}\left[\frac{q_{2n}(z)}{z^n}\right]\right| \leqslant Mn, \qquad |z| = 1$$

or, for $z = e^{i\theta}$

$$\left|\sum_{k=1}^{n} (kA_k \sin k\theta - kB_k \cos k\theta)\right| \leqslant Mn$$

that is, $|S_n'(\theta)| \leqslant Mn$. Equality holds only for

$$q_{2n}(z) = \frac{A_n - B_n i}{2} z^{2n} + \frac{A_n + B_n i}{2}, \quad \left|\frac{A_n - B_n i}{2}\right| + \left|\frac{A_n + B_n i}{2}\right| = M$$

when

$$z = z_k = e^{i\theta_k}, \qquad \theta_k = \frac{1}{2n}\left(2k\pi + \pi - \arg\frac{A_n + B_n i}{A_n - B_n i}\right)$$

and in this case,

$$S_n(\theta) = A_n \cos n\theta + B_n \sin n\theta$$

This proves the corollary.

COROLLARY 5 If the trigonometric polynomial

$$S_n(\theta) = A_0 + \sum_{k=1}^{n} (A_k \cos k\theta + B_k \sin k\theta)$$

vanishes at $\theta = 0$ and $\theta = \pi$ and if

$$|S_n(\theta)| \leqslant M,$$

for real values of θ, then, for real values of θ,

$$|S_n(\theta)| \leqslant \left\{\begin{matrix} M & \text{for } \frac{\pi}{2n} \leqslant |\theta| \leqslant \pi - \frac{\pi}{2n} \\ M\,|\sin n\theta| & \text{for } |\theta| \leqslant \frac{\pi}{2n} \text{ and } \pi - \frac{\pi}{2n} \leqslant |\theta| \leqslant \pi \end{matrix}\right\} \leqslant Mn\,|\sin\theta| \qquad (9)$$

Equality holds in the first inequality only when

$$S_n(\theta) = M \sin n\theta, \qquad |\eta| = 1,$$

and then only for those θ such that $|\theta| \leqslant \frac{\pi}{2n}$ or $\pi - \frac{\pi}{2n} \leqslant |\theta| \leqslant \pi$. Equality holds in the inequality on the right only for $\theta = 0$ and $\theta = \pi$.

PROOF Proceeding as in the proof of Corollary 4 and using the same notations, we have $|q_{2n}(z)| \leqslant M$ on the circle $|z| = 1$; also $q_{2n}(1) = 0$ and $q_{2n}(-1) = 0$. If we apply the preceding theorem to the function $q_{2n}(z)$ and set $z = e^{i\theta}$, where $\pi < \theta \leqslant \pi$, we establish the assertion of the corollary.

COROLLARY 6 Consider the two-valued function

$$f(x) = R_n(x) + \sqrt{x^2 - 1}\, Q_{n-1}(x), \quad -1 \leqslant x \leqslant 1,$$

where $R_n(x)$ and $Q_{n-1}(x)$ are polynomials of degrees no higher than n and $n-1$ respectively and where $R_n(1)= R_n(-1)=0$. If

$$|f(x)|\leqslant M, \quad -1\leqslant x\leqslant 1$$

for $\cos\frac{\pi}{2n}\leqslant|x|\leqslant 1$ we have

$$|f(x)|\leqslant M\,|\sin n\cos^{-1}x|\leqslant Mn\sqrt{1-x^2}$$

The inequality on the left reduces to equality only for the function $f(x)=M\eta\sin(n\cos^{-1}x)$ where $|\eta|=1$ and only for $\cos\frac{\pi}{2n}\leqslant x\leqslant 1$. The inequality on the right reduces to equality only for $x=\pm 1$.

PROOF By using the transformation $x=\frac{1}{2}\left(z+\frac{1}{z}\right)$ we construct a polynomial of degree no higher than $2n$

$$P_{2n}(z)=z^n f\left(\frac{1}{2}\left(z+\frac{1}{z}\right)\right)$$

that satisfies the conditions $|P_{2n}(z)|\leqslant M$ for $|z|=1$ and $P_{2n}(1) =P_{2n}(-1)=0$. If we set $z=e^{i\varphi}$ where $\pi<\varphi\leqslant\pi$ for $|\varphi|\leqslant\frac{\pi}{2n}$ and $\pi-\frac{\pi}{2n}\leqslant|\varphi|\leqslant\pi$ we obtain on the basis of the preceding theorem

$$\left|\frac{P_{2n}(z)}{z^2-1}\right|\leqslant\frac{M}{2}\left|\frac{\sin n\varphi}{\sin\varphi}\right|\leqslant\frac{M}{2}n$$

or

$$\left|\frac{z^n f\left(\frac{1}{2}\left(z+\frac{1}{z}\right)\right)}{z\left(z-\frac{1}{z}\right)}\right|=\left|\frac{f(\cos\varphi)}{2\sin\varphi}\right|\leqslant\frac{M}{2}\left|\frac{\sin n\varphi}{\sin\varphi}\right|\leqslant\frac{M}{2}n$$

from which we get

$$|f(x)|\leqslant M\,|\sin n\cos^{-1}x|\leqslant Mn\,|\sin\cos^{-1}x|\leqslant Mn\sqrt{1-x^2}$$

for $\cos\frac{\pi}{2n}\leqslant|x|\leqslant 1$. The inequality on the left reduces to equality only for the function

$$z^n f\left(\frac{1}{2}\left(z+\frac{1}{z}\right)\right)=\frac{M}{2}\eta(z^{2n}-1), \quad |\eta|=1$$

For $z = e^{i\varphi}$ we obtain

$$f(\cos\varphi) = \frac{M}{2}\eta \cdot 2i \sin n\varphi = M\eta i \sin n\varphi$$

or

$$f(x) = M\eta i \sin n \cos^{-1} x$$

The question of equality in the right inequality is trivial. This completes the proof of the corollary.

COROLLARY 7 Consider the two-valued function

$$f(x) = R_n(x) + \sqrt{x^2 - 1}\, Q_{n-1}(x), \quad -1 \leqslant x \leqslant 1$$

where $R_n(x)$ and $Q_{n-1}(x)$ are polynomials of degrees not exceeding n and $n-1$ respectively. Suppose that $|f(x)| \leqslant M$ for $-1 \leqslant x \leqslant 1$. Then,

$$|f'(x)| \leqslant M \frac{n}{\sqrt{1-x^2}}, \quad -1 \leqslant x \leqslant 1$$

Equality holds only for functions of the form

$$f(x) = aT_n(x) + \sqrt{x^2 - 1}\, bU_{n-1}(x)$$

under the condition that $\frac{1}{2^n}(|a+bi| + |a-bi|) = M$, at the points $x_k = \cos\theta_k$, where

$$\theta_k = \frac{1}{2n}\left((2k+1)\pi - \arg\frac{a+bi}{a-bi}\right)$$

if $a + bi \neq 0$ and $a - bi \neq 0$ and at all values of x if $a = bi$ or $a = -bi$.

PROOF We note that the function $S_n(\theta) = f(\cos\theta)$ is a trigonometric polynomial of order not exceeding n and that $|S_n(\theta)| \leqslant M$ where $-\pi < \theta \leqslant \pi$. If we apply Corollary 4 to $S_n(\theta)$ and return to the variable x, we establish the conclusion of the corollary.

COROLLARY 8 Suppose that the degree of a polynomial $R_n(x)$ does not exceed n. If $|R_n(x)| \leqslant M$ where $-1 \leqslant x \leqslant 1$, then

$$|R_n'(x)| \leqslant M \frac{n}{\sqrt{1-x^2}}, \quad -1 \leqslant x \leqslant 1 \tag{10}$$

$$|R_n'(x)| \leqslant Mn \frac{\sin n \cos^{-1} x}{\sqrt{1-x^2}} \leqslant Mn^2, \quad \cos\frac{\pi}{2n} \leqslant |x| \leqslant 1 \tag{11}$$

Equality holds in (10) only for the polynomial

$$R_n(x) = M2^{n-1}\eta T_n(x) \text{ where } |\eta| = 1, \text{ for } x = x_k = \cos\frac{(2k+1)\pi}{2n}$$
$$k = 0, 1, \ldots, 2n-1.$$

Equality holds in the first inequalities of (11) only for the same polynomial at those values of x such that $\cos(\pi/2n) \leqslant |x| \leqslant 1$. Equality holds in the second inequalities of (11) only when $x = \pm 1$

PROOF Inequality (10) follows directly from Corollary 7. From (10), we have

$$|R_n'(x)\sqrt{1-x^2}| \leqslant Mn, \quad -1 \leqslant x \leqslant 1$$

If we apply Corollary 6 to this inequality, we obtain inequality (11), in which equality can hold only if

$$R_n'(x)\sqrt{1-x^2} = Mn\eta \sin n \cos^{-1} x, \quad |\eta| = 1$$

But then $R_n(x) = -M\eta \cos n \cos^{-1} x + C$ where C is a constant.

Since $|R_n(x)| \leqslant M$ for $-1 \leqslant x \leqslant 1$ it follows that $C = 0$ and

$$R_n(x) = -M\eta \cos n \cos^{-1} x = -M2^{n-1}\eta T_n(x)$$

For this polynomial, equality does indeed hold in (11) in the cases stated in the theorem. This completes the proof of the corollary.

NOTE 2 Inequality (10) is called Bernstein's second inequality (see [1], p. 13), and the inequality $|R_n'(x)| \leqslant Mn^2$ for $-1 \leqslant x \leqslant 1$ is called Markov's inequality [1].

NOTE 3 Corollary 8, specifically the inequality $|R_n'(x)| \leqslant Mn^2$, yields the solution to Problem 1 for the case in which $\bar{B}$ is the real interval $[-1, 1]$. This corollary also yields the solution of Problem 2 for those points x for which equality holds in (10) or (11). For the remaining points of the interval $[-1, 1]$, Problem 2 has, as was noted above, been solved by V.A. Markov and Voronovsky. In Section 5.3 we shall give the solution of Problem 2 for points not belonging to the interval $[-1, 1]$.

Theorem 1 and its corollaries are due to Smirnov [3].

For certain classes of sets $\bar{B}$, estimates have been obtained for $|R_n'(x)|$ under the condition that $|R_n(z)| \leqslant M$ where $z \in \bar{B}$. In this connection, a result of Szegö [4] is of interest, and we proceed to its formulation.

Suppose that B is a simply-connected bounded region with boundary L consisting of a finite number of analytic arcs and suppose that $z_0 \in L$. Consider the least upper bound of the angles having vertex at the point z_0 such that points inferior to these angles that are sufficiently close to z_0 all lie outside L. We shall refer to this least upper bound as the external angle $\alpha\pi$, where $0 \leqslant \alpha \leqslant 2$ for L at the point z_0. Suppose that $\mathcal{P}_n$ is the class of polynomials $P_n(z)$ of degree not exceeding n that satisfy the condition

$$\sup_{z \in B} |P_n(z)| = M, \quad M > 0$$

Let us set

$$\mu_n = \mu_n(L, z_0) = \sup_{P_n \in \mathcal{P}_n} |P_n'(z_0)|$$

Theorem 2 There exist two positive numbers A and A' that depend on the properties of the curve L (but not on n) such that, for all sufficiently great n,

$$An^{\alpha} < \mu_n(L, z_0) < A'n^{\alpha}$$

Theorem 3 If, $\{b_n\}_1^{\infty}$ is an arbitrary unbounded increasing sequence of numbers, a curve L can be constructed such that, for all sufficiently great n,

$$\mu_n(L, z_0) < b_n, \quad z_0 \in L$$

We shall not stop to prove these theorems.

We also present without proof a result of Pommerenke [1].

Theorem 4 Suppose that $\bar{B}$ is a set of type $\mathfrak{M}$ (see Section 1.2) and that $\gamma(\bar{B})$ is the capacity of the set $\bar{B}$ (see Section 1.3.4). If $P_n(z)$ is a polynomial of degree n and if $|P_n(z)| \leqslant M$ for $z \in B$ then

$$|P_n'(z)| \leqslant \frac{e}{2\gamma(\bar{B})} Mn^2, \quad z \in \bar{B}$$

This inequality is analogous to Markov's inequality (see (10)). For the interval in question, Bernstein's inequality has the factor n instead of n^2 on the right, but the coefficient of this factor $\dfrac{M}{\sqrt{1-x^2}}$ increases continuously as we approach the end-points of the interval. Recently, several estimates of order n have been obtained for $|P'_n(x)|$ in the case of closed sets if we exclude from the set in question some sub-set of measure not exceeding an arbitrary given positive number ε. In this case, the coefficient of n naturally depends on ε. For rational functions, this was done in the article by Dolzhenko [1], which also contains a bibliography related to the question.

5.1.3 *A generalisation of Mendeleev's problem*

We showed earlier (see Corollary 3, Section 5.2) that if $p_n(z)$ is a polynomial of degree not exceeding n and if $|p_n(z)|\leqslant M$ where $M>0$ everywhere on the circle $|z|=1$ then $|p'_n(z)|\leqslant nM$. A more general problem is the following: suppose that a and b are constants, that $p_n(z)$ is a polynomial of degree not exceeding n, and that $f(z)$ is a function that is non-negative on $|z|=1$. Find an upper bound for $|ap_n(z)+bzp'_n(z)|$ for fixed z (such that $|z|=1$) if $|p_n(z)|\leqslant f(z)$ on the circle $|z|=1$. The idea of examining problems of this type is due to Bernstein (see, for example, [1], pp. 497-499). We shall confine our own considerations to the case in which $f(z)=|P_l(z)|$ where $P_l(z)$ is a polynomial of degree $l\leqslant n$ and we shall then obtain several corollaries. We shall first reduce this case (see Lemma 1 below) to the case in which $f(z)=|P_n(z)|$ where $P_n(z)$ is a polynomial of degree n, all zeros of which lie in the closed circle $|z|\leqslant 1$.

We shall first prove some lemmas that we shall need for what follows. We denote by K_ρ ($\overline{K}_\rho$ is the open disc and $|z|<\rho$ the closed disc $|z|\leqslant\rho$). We denote the circle $|z|=\rho$ by L_ρ.

Lemma 1 For every polynomial $Q_l(z)$ of degree l and an arbitrary natural number $n\geqslant l$ there exists a unique (up to a constant factor) polynomial $P_n(z)$ of degree n the absolute value of which is one and all the zeros of which belong to $\overline{K}_1$ such that

$$|Q_l(z)|=|P_n(z)| \quad \text{for} \quad z\in L_1 \tag{1}$$

PROOF The existence of the polynomial $P_n(z)$ follows from the obvious equality $|z-z_0|=|1-\bar{z}_0 z|$ for $|z_0|>1$

and $|z| = 1$. For $l < n$ we need an additional factor z^{n-l}. The uniqueness of the polynomial $P_n(z)$ follows immediately from the symmetry of the situation and Liouville's theorem.

Before going into the next lemma, let us consider certain regions of the complex plane. Suppose that ρ is a fixed number at least equal to one, that n is a natural number, that $D_{\rho, n}$ is the image of the disc $|\xi| < \rho$ under the mapping defined by the function $w = \frac{n\xi}{\xi - 1}$ that $B_{|z_0|}$ is the image of the disc $|\xi| < 1$ under the mapping defined by the function $w_1 = \frac{z_0}{z_0 - \xi}$ where $|z_0| \geqslant 1$ and that $B^{(1)}_{|z_0|}$ is the complement of $B_{|z_0|}$ in the extended complex plane. We denote by l_ρ and $l'_{|z_0|}$ the boundaries of $D_{\rho, n}$ and $B_{|z_0|}$, respectively. It is easy to see that $D_{\rho, n}$ for $\rho > 1$ is the region

$$\left| w - \frac{n\rho^2}{\rho^2 - 1} \right| > \frac{n\rho}{\rho^2 - 1}$$

that $D_{1, n}$ is the half-plane $\operatorname{Re} w < \frac{n}{2}$, that $B_{|z_0|}$ where $|z_0| > 1$ is the disc

$$\left| w - \frac{|z_0|^2}{|z_0|^2 - 1} \right| < \frac{|z_0|}{|z_0|^2 - 1}$$

that B_1 is the half-plane $\operatorname{Re} w_1 > \frac{1}{2}$. The function $w_1 = \frac{w}{n}$ maps the region $D_{\rho, n}$ on to the region $B^{(1)}_\rho$ and the regions $B_{|z_0|}$ are compressed as $|z_0|$ increases on the interval $1 \leqslant |z_0| < \infty$.

Lemma 2 Suppose that $Q_n(z)$ is a polynomial of degree $n \geqslant 1$ that all its zeros belong to $\overline{K}_1$, that ρ is a fixed number at least equal to one, and that w is a finite point belonging to $\overline{D}_{\rho, n}$. Then, all the zeros of the polynomial

$$L_w\{Q_n(z)\} = zQ'_n(z) - wQ_n(z) \tag{2}$$

belong to $\overline{K}_\rho$ and

$$L_w\{Q_n(z)\}_{z = z_0} = 0 \tag{3}$$

for $|z_0| = \rho$ only in the following cases: (1) if $\rho > 1$ and $w \in l_\rho$ only for the polynomials $Q_n(z) = C\left[z - \left(1 - \frac{n}{w}\right) z_0\right]^n$

where C is a constant; (2) if $\rho=1$ then (a) if z_0 is a multiple zero of the polynomial $Q_n(z)$, (b) if $Q_n(z_0)\neq 0$ and $\operatorname{Re} w=\frac{n}{2}$ only for polynomials $Q_n(z)$ all the zeros of which lie on L_1 at a point z_0, such that $w=\frac{z_0Q'_n(z_0)}{Q_n(z_0)}$.

PROOF It is obvious that Equation (3) is valid if z_0 is a multiple zero of the polynomial $Q_n(z)$ and that (3) is not valid if z_0 is a simple zero of $Q_n(z)$.

Let us now assume that $Q_n(z_0)\neq 0$. Suppose that z_ν, for $\nu=1, 2, \ldots, n$ are the zeros of $Q_n(z)$. On the right side of Equation (2), we set $z=z_0$ and we divide this equation by $nQ_n(z_0)$:

$$I=\frac{z_0Q'_n(z_0)}{nQ_n(z_0)}-\frac{w}{n}=\frac{1}{n}\sum_{1}^{n}\frac{z_0}{z_0-z_\nu}-\frac{w}{n}$$

or

$$I=\frac{1}{n}\sum_{\nu=1}^{n}w_\nu-\frac{w}{n},\qquad w_\nu=\frac{z_0}{z_0-z_\nu} \tag{4}$$

From the fact that $|z_\nu|\leqslant 1$ it follows that $w_\nu\in\bar{B}_{|z_0|}$. Therefore, the point $w^*=\frac{1}{n}\sum_{\nu=1}^{n}w_\nu$ also belongs to $\bar{B}_{|z_0|}$. If z_0 is a root of $L_w\{Q_n(z)\}$, that is, if $I=0$ or $w^*=\frac{w}{n}$ the intersection $\bar{B}_{|z_0|}\cap\bar{B}_\rho^{(1)}$ cannot be empty. Therefore, it follows that $|z_0|\leqslant\rho$ and the first assertion of the lemma is proven.

Let us now consider two cases:

1. $|z_0|=\rho>1$. In this case, the intersection just referred to is l'_ρ; that is, the point $w^*=\frac{w}{n}$ must belong to l'_ρ and $w\in l_\rho$. But $w^*\in l'_\rho$ if and only if all the points $w_\nu\in l_\rho$ coincide, and then all the $z_\nu\in L_1$ also coincide. From what has been said, it follows that $\frac{z_0}{z_0-z_\nu}=\frac{w}{n}$ (for $w\in l_\rho$) or

$$z_\nu=\left(1-\frac{w}{n}\right)z_0,\ \nu=1, 2, \ldots,\ \text{i.e. } Q_n(z)=C\left[z-\left(1-\frac{w}{n}\right)z_0\right]^n$$

2. $|z_0|=\rho=1$. In this case, reasoning as above, we can

conclude that all the z_ν must lie on L_1 and that $w \in l_1$. By setting $z_\nu = e^{i\theta_\nu}$ for $\nu = 0, 1, \ldots, n$ and $w = \frac{n}{2} + iC$ where C is a real number, we write the equation $l = 0$ in the form

$$\frac{1}{2} \sum_{\nu=1}^{n} \cot \frac{\theta_0 - \theta_\nu}{2} = -C \tag{5}$$

from which it follows that for every C, the equation $l = 0$ that is the equation $w = \frac{zQ_n'(z)}{Q_n(z)}$ has at least one root $z_0 = e^{i\theta_0}$ for $0 \leqslant \theta_0 < 2\pi$ on the circle $|z| = 1$.

This completes the proof of the lemma.

NOTE 1 If all the zeros of $Q_n(z)$ belong to K_1 and if the other conditions of the lemma are satisfied, all the zeros of $L_w\{Q_n(z)\}$ belong to K_ρ. This follows immediately from the proof that we have given for the first case.

NOTE 2 The assertion of the lemma remains valid with a corresponding change in the wording for the case in which w is a function of z.

COROLLARY 1 Suppose that $|z_0| \geqslant 1$. If $Q_n(z_0) \neq 0$, the point $\frac{z_0 Q_n'(z_0)}{Q_n(z_0)}$ belongs to the complement $D_{|z_0|, n}$. This follows from the proof of the lemma.

Let us denote by $D^*_{\rho, n}$ the region

$$\left| w + \frac{n\rho^2}{1-\rho^2} \right| > \frac{n\rho}{1-\rho^2}$$

for $0 < \rho < 1$. Let us denote by $D^*_{1, n}$ the half-plane $\operatorname{Re} w > \frac{n}{2}$. Finally, let us denote by $D^{*(1)}_{\rho, n}$ the complement of $\bar{D}^*_{\rho, n}$.

COROLLARY 2 If all the zeros of the polynomial $Q_n(z)$ lie in the region $z| \geqslant 1$ then, for $|z_0| \leqslant 1$ and $Q_n(z_0) \neq 0$, the point $\frac{z_0 Q_n(z_0)}{Q_n(z_0)}$ belongs to the complement of $D^*_{\frac{1}{|z_0|}, n}$.

To prove this, we need only apply Corollary 1 to the polynomial $z^n Q_n\left(\frac{1}{z}\right)$. The degree of $Q_n(z)$ may be less than n.

By using Note 2 above and setting $w = \frac{nz}{z - \xi}$ we obtain the following corollary.

COROLLARY 3 (Szegö (see Polya and Szegö [1], part 5, Chapter 2)). Suppose that $Q_n(z)$ is a polynomial of degree $n \geqslant 1$ all the zeros of which lie in $\bar{K}_1$ and that ξ and z_0 are arbitrary numbers such that $|\xi| \geqslant 1$ and $|z_0| \geqslant 1$. Then,

$$(\xi - z_0) Q'_n(z_0) + nQ_n(z_0) = 0 \tag{6}$$

only in the following cases:

1. $\xi = z_0$ and z_0 is a zero of the polynomial $Q_n(z)(|z_0| = |\xi| = 1)$; 2. z_0 is a multiple zero of the polynomial $Q_n(z)$ $(|z_0| = 1)$; 3. $|\xi| = 1$ and $|z_0| > 0$ only for the polynomials $Q_n(\xi) = C(z - \xi)^n$ where C is a constant and

$$(\xi - z) Q_n(z) + nQ_n(z) \equiv 0$$

4. $\xi \neq z_0$, $|z_0| = 1$, $Q_n(z_0) \neq 0$ only for polynomials all the zeros of which belong to L_1 at some point z_0 such that

$$\xi = z_0 - n\frac{Q_n(z_0)}{Q'_n(z_0)} \quad (|\xi| = 1)$$

The results given by the following theorems are connected with the investigations of Bernstein, Szegö, Videnskiy and Meyman. The first theorem is of fundamental importance. The other results are basically corollaries to it though they are presented as separate theorems. We shall give only a general outline of their proofs. The most laborious portion of the proof of the first theorem is the investigation of the cases in which the fundamental inequality reduces to equality. Proof of the inequality itself is quite easy.

Theorem 1 Suppose that $P_n(z)$ is a polynomial of degree $n \geqslant 1$ all the zeros of which belong to K_1, that $\mathscr{P}_n$ is the class of polynomials $p_n(z)$ of degree not exceeding n that satisfy the inequality $|p_n(z)| \leqslant |P_n(z)|$ on L_1, that ρ is an arbitrary fixed number at least equal to one, that a and b are arbitrary numbers such that either $b = 0$ or $\frac{a}{b} \subset \bar{D}_{\rho, n}$. Then, for $|z| \geqslant \rho$

$$|bzp'_n(z) - ap_n(z)| \leqslant |bzP'_n(z) - aP_n(z)|, \quad p_n(z) \in \mathscr{P}_n \tag{7}$$

Equality holds for $z = z_0$ where $|z_0| \geqslant \rho$, only in the following cases: (1) if $a = b = 0$ equality holds for every polynomial $p_n(z) \in \mathscr{P}_n$; (2) if $a \neq 0$, $b = 0$, and $|z_0| = 1$ equality holds only for the polynomials $p_n(z) \in \mathscr{P}_n$ that satisfy the condition $|p_n(z_0)| = |P_n(z_0)|$; (3) if $|a| + |b| \neq 0$, and $|z_0| > 1$ equality holds only for the polynomials $p_n(z) = \eta P_n(z)$ where $|\eta| = 1$; (4) if $b \neq 0$, and $|z_0| = 1$ then (a) if $w = \frac{a}{b}$ is an interior point of $\bar{D}_{1,n}$, equality holds only for the polynomials $p_n(z) = \eta P_n(z)$ where $|\eta| = 1$; but (b) if $w = \frac{a}{b} \in l_1$, equality holds only for the polynomials

$$p_n(z) = \eta\left[(1-\lambda) P_n(z) - \lambda e^{i\alpha} z^n \overline{P_n\left(\frac{1}{\bar{z}}\right)}\right], \quad 0 \leqslant \lambda \leqslant 1, \ |\eta| = 1$$

$$\alpha = \arg\left\{\frac{\bar{z}_0^n}{b^2}\left(bz_0 P_n'(z_0) - aP_n(z_0)\right)^2\right\} \tag{8}$$

PROOF The function $\frac{p_n(z)}{P_n(z)}$ is regular in the region $|z| \geqslant 1$ and $\left|\frac{p_n(z)}{P_n(z)}\right| \leqslant 1$ on L_1. Consequently, either this function is a constant not exceeding one or

$$|p_n(z)| < |P_n(z)| \quad \text{for} \quad |z| > 1; \quad p_n(z) \in \mathscr{P}_n \tag{9}$$

In the first case, inequality (7) is obvious, and therefore we shall consider the case (9). Suppose that ξ, is an arbitrary fixed number such that $|\xi| < 1$. All the zeros of the polynomial $\xi p_n(z) + P_n(z)$ belong to K_1 . By virtue of Note 1 to Lemma 2, for an arbitrary fixed z such that $|z| \geqslant \rho$ and for arbitrary ξ, such that $|\xi| < 1$, the expression

$$\xi\left[bzp_n'(z) - ap_n(z)\right] + \left[bzP_n'(z) - aP_n(z)\right]$$

is non-zero. From this it follows that

$$|\xi|\,|bzp_n'(z) - ap_n(z)| < |bzP_n'(z) - aP_n(z)|, \ |z| \geqslant \rho, \ |\xi| < 1$$

Taking the limit as $|\xi| \to 1$ we obtain (7).

Let us now investigate the cases in which equality holds in (7). We note first of all that this cannot happen if $|p_n(z)| < |P_n(z)|$ on L_1 since in this case we can replace $p_n(z)$ in (7) with $(1+\mu)p_n(z)$ where μ is a sufficiently small positive number.

Assertions (1) and (2) are obvious; so is assertion (3) for $b = 0$. Let us suppose that equality holds in (7) for $p_n(z) = p_n^*(z) \in \mathscr{P}_n$ when $z = z_0$ with $|z_0| \geqslant \rho$. If we replace $p_n^*(z)$ with $e^{i\alpha} p_n^*(z)$ where α is chosen suitably, we assume

$$bz_0 p_n^{*\prime}(z_0) - ap_n^*(z_0) = bz_0 P_n'(z_0) - aP_n(z_0), \quad |z_0| \geqslant \rho. \qquad (10)$$

The case in which $p_n^*(z) \equiv P_n(z)$ corresponds to the fact that, when equality holds in (7), we have

$$p_n(z) = \eta P_n(z), \quad |\eta| = 1$$

This last case includes assertions (3), (4a), and (4b) for $\lambda = 0$. Thus, we need to show that for $p_n^*(z) \equiv P_n(z)$ the case $|z_0| > 1$ is impossible and that the case $|z_0| = 1$ leads to the $p_n(z)$ mentioned in (4b).

Consider the polynomial

$$q_n(z) = (1 - \lambda)\, p_n^*(z) + \lambda P_n(z)$$

where λ is a fixed number in (0, 1). On L_1 we have

$$|q_n(z)| \leqslant (1 - \lambda)\,|p_n^*(z)| + \lambda\,|P_n(z)| \\ \leqslant (1 - \lambda)\,|P_n(z)| + \lambda\,|P_n(z)| = |P_n(z)| \qquad (11)$$

Equation (10) remains valid if we replace $p_n^*(z)$ with $q_n(z)$. On L_1 there is at least one point z' at which $|q_n(z')| = |P_n(z')|$. It follows from this equation by virtue of (11) that $q_n(z') = P_n(z')$. Replacing $p_n^*(z)$ with $q_n(z)$ if it is necessary, we may assume that $|p_n^*(z')| = |P_n(z')|$ implies $p_n^*(z') = P_n(z')$. The condition $p_n^*(z) \not\equiv P_n(z)$ remains in effect when we make this substitution. The polynomial $X(z) = P_n(z) - p_n^*(z)$ has at least one zero on L_1. Let us show that it has n simple zeros on L_1. In this connection, we note that the substitution of $q_n(z)$ for $p_n^*(z)$ can only change $X(z)$ by a constant non-zero factor $(1 - \lambda)$. We shall prove this by contradiction. Let us suppose that $X(z)$ has only $r < n$ distinct roots $z_1, z_2, \ldots, z_r$ on L_1. We introduce the polynomial of degree not exceeding n:

$$R_n(z) = p_n^*(z) - uq(z)$$

$$q(z) = \begin{cases} \prod_{\nu=0}^{r} (z - z_\nu), & \text{if } z_0 \neq z_\nu, \ \nu = 1, 2, \ldots, r \\ \prod_{\nu=1}^{r} (z - z_\nu) & \text{otherwise} \end{cases}$$

where the number u is defined by the equation

$$[bz_0p_n^{*'}(z_0) - ap_n^*(z_0)] - ubz_0q'(z_0) = 0.$$

Here we note that $q_n'(z_0) \neq 0$ and that the expression in the square brackets is equal to $bz_0P_n'(z_0) - aP_n(z_0)$ which is non-zero by virtue of Note 1 to Lemma 2 (with $|z_0| \geqslant \rho$) so that $u \neq 0$. Consider the polynomial of degree not exceeding n

$$p_n(z) = p_n^*(z) - \mu R(z) = (1-\mu)\, p_n^*(z) + \mu u q(z)$$

where $0 < \mu < 1$. It is easy to see that equality holds in (7) for this polynomial with $z = z_0$. If we can show that $|p_n(z)| < |P_n(z)|$ on L_1 for sufficiently small μ, this contradiction will enable us to conclude that $X(z)$ has n simple zeros on L_1.

Let us set $H = \min\limits_{|z|=1} |P_n(z)| > 0$. We denote by E the set of points on the circle $|z| = 1$ that lie at a distance less than

$$\frac{H}{2^{r-1}|u|(|z_0|+1)}$$

from the set of points $z_1, z_2, \ldots, z_r$ and we denote by E' the set consisting of all other points on the circle $|z| = 1$. For $z \in E$ we have

$$|q(z)| \leqslant \left| \prod_{\nu=1}^{r} (z - z_\nu)(|z_0| + 1) \right| < \frac{H}{2^{r-1}|u|(|z_0|+1)} 2^{r-1}(|z_0|+1) = \frac{H}{|u|}$$

so that, for $z \in E$

$$\begin{aligned} |p_n(z)| = |(1-\mu)\, p_n^*(z) + \mu u q(z)| &\leqslant (1-\mu)|p_n^*(z)| + \mu|u||q(z)| \\ &< (1-\mu)|P_n(z)| + \mu|u|\frac{H}{|u|} \\ &\leqslant (1-\mu)|P_n(z)| + \mu|P_n(z)| = |P_n(z)| \end{aligned}$$

We now set $h = \min\limits_{z \in E'} (|P_n(z)| - |p_n^*(z)|) > 0$, and $K = \max\limits_{z \in E'} |q(z)|$ Then, for $z \in E'$ we have

$$\begin{aligned} |p_n(z)| \leqslant (1-\mu)|p_n^*(z)| + \mu|u|K &\leqslant (1-\mu)(|P_n(z)| - h) \\ &+ \mu|u|K = |P_n(z)| + \mu(|u|K + h - |P_n(z)|) - h \end{aligned}$$

Obviously, for sufficiently small μ in [0, 1], the last expression is less than $|P_n(z)|$ for $z \in E'$. Finally, $|p_n(z)| < |P_n(z)|$

on L_1 for sufficiently small μ. This proves that the polynomial $X(z)$ has n simple zeros on L_1. Thus, we may assert that if, for $b \neq 0$, the polynomial $p_n^*(z)$ satisfies condition (10) and $p_n^*(z) \not\equiv P_n(z)$, the polynomial $X(z) = P_n(z) - p_n^*(z)$ has n simple zeros on L_1.

If we apply Lemma 2 to the polynomial $X(z)$ we see that the case $|z_0| > 1$ is impossible and that, for $|z_0| = 1$, we must have $w \in l_1$, that is, $\operatorname{Re} w = \frac{n}{2}$. Let us examine this case. From the fact that all the zeros of $X(z)$ lie on L_1, it follows immediately that

$$X(z) = e^{i\alpha} z^n \overline{X\left(\frac{1}{\bar{z}}\right)}$$

where α is some constant in $[0, 2\pi]$. We form the two nonnegative trigonometric polynomials:

$$|P_n(z)|^2 - |p_n^*(z)|^2, \quad |P_n(z) - p_n^*(z)|^2, \quad z = e^{i\theta}, \quad 0 \leqslant \theta < 2\pi$$

The second of these has n distinct (double) zeros in the interval $0 \leqslant \theta < 2\pi$. Furthermore, these zeros are multiple zeros of the first polynomial, the order of which does not exceed $2n$. From this it follows that

$$|P_n(z)|^2 - |p_n^*(z)|^2 = \tau\, |P_n(z) - p_n^*(z)|^2, \quad \tau \geqslant 0$$

The case in which $\tau = 0$ corresponds to the case in which the trigonometric polynomial on the left is identically equal to zero. This identity on L_1 may be successively transformed (by use of the fact that $X(z) = P_n(z) - p_n^*(z)$) as follows:

$$\overline{X(z)}\, P_n(z) + X(z)\, \overline{P_n(z)} = (\tau + 1)\, X(z)\, \overline{X(z)},$$

$$z^n\, \overline{X\left(\frac{1}{\bar{z}}\right)} P_n(z) + X(z)\, z^n\, \overline{P_n\left(\frac{1}{\bar{z}}\right)} = (\tau + 1)\, X(z)\, z^n\, \overline{X\left(\frac{1}{\bar{z}}\right)}$$

$$\left(e^{-i\alpha} P_n(z) + z^n \overline{P_n\left(\frac{1}{\bar{z}}\right)}\right) X(z) = (\tau + 1)\, e^{-i\alpha} X(z)^2$$

Obviously, the last of these equations is valid on the entire z plane. Since $X(z) \not\equiv 0$ we obtain from this equation

$$X(z) = \frac{1}{\tau + 1}\left(P_n(z) + e^{i\alpha} z^n \overline{P_n\left(\frac{1}{\bar{z}}\right)}\right)$$

and

$$p_n^*(z) = P_n(z) - X(z) = \frac{\tau}{\tau+1} P_n(z) - \frac{1}{\tau+1} e^{i\alpha} z^n \overline{P_n\left(\frac{1}{\bar{z}}\right)}$$

Thus, the polynomial $p_n^*(z)$ must be of the form

$$p_n^*(z) = (1-\lambda) P_n(z) - \lambda e^{i\alpha} z^n \overline{P_n\left(\frac{1}{\bar{z}}\right)}, \quad 0 < \lambda \leqslant 1, \quad 0 \leqslant \alpha < 2\pi$$

On the circle $|z| = 1$ we have

$$|p_n^*(z)| \leqslant (1-\lambda)|P_n(z)| + \lambda |P_n(z)| = |P_n(z)|$$

Consequently, for arbitrary λ in [0, 1] and arbitrary α in [0, 2π], the polynomial $p_n^*(z) \in \mathscr{P}_n$. For $p_n^*(z)$ Equation (10) must be valid. In the present case, for $0 < \lambda \leqslant 1$ this equation reduces to the equation

$$e^{i\alpha} z_0^n [(a - nb)\overline{P_n(z_0)} + b(\overline{z_0 P_n'(z_0)})] = b z_0 P_n'(z_0) - a P_n(z_0)$$

which is independent of λ, or

$$e^{i\alpha} z_0^n \overline{P_n(z_0)} \left[\overline{\left(\frac{z_0 P_n'(z_0)}{P_n(z_0)} \right)} + \frac{a}{b} - n \right] = P_n(z_0) \left[\frac{z_0 P_n'(z_0)}{P_n(z_0)} - \frac{a}{b} \right]$$

It is easy to see that the real parts of the expressions in the square brackets are equal and that the imaginary parts differ in sign. Consequently,

$$e^{i\alpha} z_0^n \overline{P_n(z_0)} \overline{\left[\frac{z_0 P_n'(z_0)}{P_n(z_0)} - \frac{a}{b} \right]} = P_n(z_0) \left[\frac{z_0 P_n'(z_0)}{P_n(z_0)} - \frac{a}{b} \right]$$

From this, we obtain Formula (8) for α with arbitrary fixed z_0 such that $|z_0| = 1$. Assertion (4b) regarding the equality sign in (7), and hence the entire theorem, is proven.

NOTE 3 The value $\lambda = 0$ leads to the case $p_n^*(z) \equiv P_n(z)$ that we referred to above. For $\lambda = 1$ we have

$$p_n(z) = \eta e^{i\alpha} z^n P_n \overline{\left(\frac{1}{\bar{z}}\right)}$$

In this expression, we may eliminate the factor $e^{i\alpha}$. If we set $\eta = 1$ and remember that α is determined by Formula (8), we see that $p_n(z)$ satisfies Equation (10).

NOTE 4 It is easy to show that

$$P_n(z_0) - p_n^*(z_0) \neq 0.$$

NOTE 5 It follows from the proof of the theorem that all the zeros of the polynomial

$$(\tau + 1) X(z) = P_n(z) + e^{i\alpha} z^n \overline{P_n\left(\frac{1}{\bar{z}}\right)}$$

where α is given by Formula (8), lie on the circle $|z| = 1$ and that they are all simple. By a suitable choice of z_0 and $\operatorname{Im} \frac{a}{b}$ we can obtain an arbitrary value of α such that $0 \leqslant \alpha < 2\pi$ in Formula (8).

NOTE 6 Suppose that all the zeros of the polynomial $P_n(z)$ belong to K_1 and that all the other conditions of the theorem are satisfied. Then, inequality (7) holds for $|z| \geqslant \rho$. To prove this, we replace $p_n(z)$ and $P_n(z)$ with $p_n(rz)$ and $P_n(rz)$ where $r > 1$ in (7) and then let r approach one. Conditions can be found under which equality holds in the inequality thus obtained for $z = z_0$, where $|z_0| \geqslant \rho$.

NOTE 7 In the formulation of the theorem, we may treat the numbers a and b as functions of z. Here, it is obvious that inequality (7) will hold at those points z such that $|z| \geqslant 1$ at which either $b = 0$ or $a/b \in \bar{D}_{|z|}$. This assertion remains valid in the case in which all the zeros of the polynomial $P_n(z)$ belong to $\bar{K}_1$.

Let us note one particular case of the theorem. If we set $= n$, $b = 2$ and $P_n(z) = Mz^n$ where $M > 0$ we obtain the following corollary.

COROLLARY 1 If $p_n(z)$ is a polynomial of degree not exceeding n and if $|p_n(z)| \leqslant M$ where $M > 0$ on L_1, then

$$|2zp_n'(z) - np_n(z)| \leqslant Mn|z|^n, \quad |z| \geqslant 1$$

Equality holds for $z = z_0$, with $|z_0| \geqslant 1$ as follows: (1) for $|z_0| > 1$ only in the case of the polynomials $p_n(z) = M\eta z^n$, where $|\eta| = 1$; (2) for $|z_0| = 1$ only in the case of the polynomials $p_n(z) = M\eta[(1 - \lambda) z^n - \lambda e^{i\alpha}]$ where $0 \leqslant \lambda \leqslant 1$, $|\eta| = 1$ and $\alpha = n \arg z_0$.

We have already obtained this result on the circle $|z| = 1$ (see Corollary 2 of Theorem 1, Section 5.1.2).

COROLLARY 2 Suppose that $P_n(z)$ is a polynomial of degree $n \geqslant 1$ all the zeros of which belong to $\bar{K}_1$, that $p_n(z)$ is an arbitrary polynomial of degree not exceeding n, that $\bar{S}$ is the image in the w plane of the closed region

$|z| \geqslant 1$ under the mapping defined by the function $w = \frac{p_n(z)}{P_n(z)}$, that ρ is a constant at least equal to one, and that a and b are arbitrary numbers such that either $b=0$ or $b \neq 0$ and $\frac{a}{b} \in \bar{D}_{\rho, n}$. Then, for every z such that $|z| \geqslant \rho$

$$\frac{ap_n(z) - bzp'_n(z)}{aP_n(z) - bzP'_n(z)} = w \in \bar{S}$$

PROOF Let us set

$$L\{f(z)\} = af(z) - bzf'(z)$$

where $f(z)$ is a regular function. Let K denote an arbitrary region bounded by a circle or a straight line such that the intersection $\bar{K} \cap \bar{S}$ is empty and $\xi = \frac{\alpha w + \beta}{\gamma w + \delta}$ is a function mapping the region K on to the open disc $|z| \geqslant \rho$. Then, for $|\xi| < 1$

$$\left| \frac{\alpha \frac{p_n(z)}{P_n(z)} + \beta}{\gamma \frac{p_n(z)}{P_n(z)} + \delta} \right| > 1$$

Consequently, in the region $|z| \geqslant \rho$

$$|\gamma p_n(z) + \delta P_n(z)| < |\alpha p_n(z) + \beta P_n(z)|$$

and all the zeros of the polynomial $\alpha p_n(z) + \beta P_n(z)$ lie in the disc $|z| < 1$. Therefore,

$$|\gamma L\{p_n(z)\} + \delta L\{P_n(z)\}| < |\alpha L\{p_n(z)\} + \beta L\{P_n(z)\}|, \quad |z| \geqslant \rho,$$

or

$$\left| \frac{\alpha \frac{L\{p_n(z)\}}{L\{P_n(z)\}} + \beta}{\gamma \frac{L\{p_n(z)\}}{L\{P_n(z)\}} + \delta} \right| > 1, \quad |z| \geqslant \rho$$

From this we conclude that the point $w = \frac{L\{p_n(z)\}}{L\{P_n(z)\}} \overline{\in} K$. Since K is arbitrary, this proves the corollary. The special case of this corollary when $a=0$ (here we may take $\rho = 1$) was proven by de Bruijn [1].

Let n and m denote natural numbers. Suppose that R and

ρ are given numbers such that $R>1$ and $1\leqslant\rho\leqslant R$. We denote by $D_{\rho(n,m,R)}$ the region in the ξ plane defined by the inequalities

$$\left|\xi-\left(\frac{n\rho^2}{\rho^2-1}+\frac{m\rho^2}{R^2-\rho^2}\right)\right|>\frac{n\rho}{\rho^2-1}+\frac{mR\rho}{R^2-\rho^2}\quad\text{for } 1<\rho<R$$

$$\operatorname{Re}\xi\leqslant\frac{n}{2}-\frac{m}{R+1}\quad\text{for } \rho=1$$

$$\operatorname{Re}\xi\leqslant\frac{nR}{R+1}-\frac{m}{2}\quad\text{for } \rho=R$$

By using Note 5 to the preceding theorem, we can easily prove the following corollary.

COROLLARY 3 Suppose that $P_n(z)$ is a polynomial of degree $n\geqslant 1$ all the zeros of which belong to $\bar{K}_1$, that $\mathscr{P}_n$ is the class of polynomials $p_n(z)$ of degree not exceeding n which satisfy the inequality $|p_n(z)|\leqslant|P_n(z)|$ everywhere on L_1, that $Q_m(z)$ is a polynomial of degree $m\geqslant 1$ all the zeros of which lie in the region $|z|\geqslant R$ where $R>1$, that ρ is a fixed number in 1 R and that a and b are arbitrary numbers such that either $b=0$ or $\frac{a}{b}\in D_{\rho(n,m,R)}$. Then, for $|z|=\rho$ and $(Q_m(z)\neq 0)$

$$\left|a\frac{p_n(z)}{Q_m(z)}-bz\frac{d}{dz}\frac{p_n(z)}{Q_m(z)}\right|\leqslant\left|a\frac{P_n(z)}{Q_m(z)}-bz\frac{d}{dz}\frac{P_n(z)}{Q_m(z)}\right|$$

Theorem 2 Suppose that B is the disc $|z-w|<R$ where $R>0$ in the z plane (or that B is the region $|z-w|>R$), that $P_n(z)$ is a polynomial of degree $n\geqslant 1$ all the zeros of which lie in the region B, that $\mathscr{P}_n$ is the class of polynomials $p_n(z)$ of degree not exceeding n which satisfy the inequality $|p_n(z)|\leqslant|P_n(z)|$ everywhere on the circle $|z-w|=R$, that ρ is a constant such that $\rho\geqslant 1$ (respectively $0<\rho\leqslant 1$), and that a and b are arbitrary numbers such that either $b=0$ or $\frac{a}{b}\in\bar{D}_{\rho,n}$ (respectively $\frac{a}{b}\in\overline{D^*_{\rho,n}}$). Then, for $|z-w|\geqslant R\rho$ (respectively for $|z-w|\leqslant R\rho$)

$$|b(z-w)p_n'(z)-ap_n(z)|\leqslant|b(z-w)P_n'(z)-aP_n(z)|\qquad p_n(z)\in\mathscr{P}_n\tag{12}$$

Equality holds in (12) for $z=z_0$ where $|z_0-w|\geqslant R\rho$ (respectively $|z_0-w|\leqslant R\rho$) only in the following cases: (1) if $a=b=0$ equality holds only for every polynomial $p_n(z)\in\mathscr{P}_n$; (2) if

$a \neq 0$, $b=0$ and $|z_0-w|=R$, equality holds only for the polynomials $p_n(z)\in\mathscr{P}_n$ that satisfy the condition $|p_n(z_0)|=|P_n(z_0)|$; (3) if $|a|+|b|\neq 0$ and $z_0\overline{\in} \overline{B}$, equality holds only for the polynomials $p_n(z)=\eta P_n(z)$ where $|\eta|=1$; (4) if $b\neq 0$ and $|z_0-w|=R$ equality holds only for the polynomials $p_n(z)=\eta P_n(z)$ (with $|\eta|=1$) if $\frac{a}{b}\overline{\in} l_1$, and only for the polynomials

$$p_n(z)=\eta\left[(1-\lambda)P_n(z)-\lambda e^{i\alpha}\left(\frac{z-w}{R}\right)^n\overline{P_n\left(\frac{R^2}{\overline{z-w}}+w\right)}\right], \tag{13}$$

$$0\leqslant\lambda\leqslant 1,\quad |\eta|=1,$$

$$\alpha=\arg\left\{\frac{\overline{(z_0-w)^n}}{b^2}\left(b(z_0-w)P_n'(z_0)-aP_n(z_0)\right)^2\right\}.$$

if $\frac{a}{b}\in l_1$.

PROOF For the case in which $|z-w|<R$ the proof is carried out by means of the transformation $z=R\zeta+w$ and application of Theorem 1. For the region $|z-w|>R$ we use the transformation $z=\frac{R}{\zeta}+w$ and we replace a, b $p_n(z)$ and $P_n(z)$ in Theorem 1 with $-b$, $a-nb$, $\zeta^n p_n\left(\frac{R}{\zeta}+w\right)$ and $\zeta^n P_n\left(\frac{R}{\zeta}+w\right)$ respectively.

NOTE 8 For the case in which the region is $|z-w|>R$ we may use a polynomial of degree no higher than n as $P_n(z)$.

COROLLARY If $p_n(z)$ is a polynomial of degree $n\geqslant 1$ all the zeros of which lie in the region $|z|\geqslant 1$ and if $|p_n(z)|\leqslant M$ everywhere on L_1, then $|p_n'(z)|\leqslant\frac{M}{2}n$ for $z\in\overline{K}_1$.

PROOF Let us set $a=n$, $b=1-\frac{\xi}{z}$ where $|\xi|=1$. Then if $\xi\neq z$, we have $\frac{a}{b}\in D^*_{\rho,n}$. If we set $P_n(z)=M$ in (12), we have

$$|(\xi-z)p_n'(z)+np_n(z)|\leqslant Mn,\quad |z|\leqslant 1,$$

or, since $\arg\xi$ is arbitrary,

$$|p_n'(z)|+|np_n(z)-zp_n'(z)|\leqslant Mn \tag{14}$$

We now introduce the polynomial $q_n(z)=z^n\overline{p_n\left(\frac{1}{\bar{z}}\right)}$, all zeros of which belong to $\overline{K}_1$. On L_1, we have $|p_n(z)|\leqslant|q_n(z)|$.

Therefore, $|p_n'(z)| \leqslant |q_n'(z)|$ for $|z| \geqslant 1$ or

$$|p_n'(z)| \leqslant \left| nz^{n-1}\overline{p_n\left(\frac{1}{\bar{z}}\right)} - z^{n-1}\frac{1}{\bar{z}}\,\overline{p_n'\left(\frac{1}{\bar{z}}\right)}\right|$$

On L_1, this yields

$$|p_n'(z)| \leqslant |np_n(z) - zp_n'(z)| \tag{15}$$

From this inequality and inequality (14), we obtain

$$2|p_n'(z)| \leqslant Mn \quad \text{for} \quad z \in L_1 \tag{16}$$

Obviously, this inequality also holds in $\overline{K}_1$ This completes the proof of the corollary.

Theorem 3 Suppose that $P_n(z)$ is a polynomial of degree $n \geqslant 1$ all the zeros of which lie in the half-plane $\mathrm{Re}(e^{-i\theta}z) > R$, where $R \geqslant 0$ and $0 \leqslant \theta < 2\pi$, that $\mathscr{P}_n$ is the class of polynomials $p_n(z)$ of degree not exceeding n, which satisfy the inequality $|p_n(z)| \leqslant |P_n(z)|$ everywhere on the straight line $\mathrm{Re}(e^{-i\theta}z) = R$, that u_0 is an arbitrary non-negative constant, that $\overline{G}_{u_0}$ is the closed region

$$\left|\xi - \frac{n}{2u_0}e^{-i\theta}\right| \geqslant \frac{n}{2u_0}$$

of the ξ plane if $u_0 \geqslant 0$ and is the closed region $\mathrm{Re}(e^{i\theta}\xi) \leqslant 0$ if $u_0 = 0$, that a and b are arbitrary numbers such that either $b = 0$ or $\frac{a}{b} \in \overline{G}_{u_0}$. Then, where z is such that $\mathrm{Re}(e^{-i\theta}z) \leqslant R - u_0$, we have

$$|ap_n(z) + bp_n'(z)| \leqslant |aP_n(z) + bP_n'(z)|, \quad p_n(z) \in \mathscr{P}_n. \tag{17}$$

Equality holds in (17) for $z = z_0$ and $\mathrm{Re}(e^{-i\theta}z_0) \leqslant R - u_0$ only in the following cases: (1) if $a = b = 0$, equality holds for every polynomial $p_n(z) \in \mathscr{P}_n$; (2) if $a \neq 0$, $b = 0$ and $\mathrm{Re}(e^{-i\theta}z_0) = R$, equality holds only for polynomials $p_n(z) \in \mathscr{P}_n$ that satisfy the condition $|p_n(z_0)| = |P_n(z_0)|$; (3) if $|a| + |b| \neq 0$ and $\mathrm{Re}(e^{-i\theta}z_0) < R$, equality holds only for the polynomials $p_n(z) = \eta P_n(z)$ where $|\eta| = 1$; (4) if $b \neq 0$ and $\mathrm{Re}(e^{-i\theta}z_0) = R$, equality holds only for the polynomials $p_n(z) = \eta P_n(z)$ where $|\eta| = 1$ if

$$\mathrm{Re}\left(e^{i\theta}\frac{a}{b}\right) < 0$$

and only for the polynomials

$$p_n(z) = \eta\,[(1-\lambda)P_n(z) - \lambda e^{-i\alpha}\overline{P_n(2\,\mathrm{Re}^{i\theta} - e^{2i\theta}\bar{z})}], \quad 0 \leqslant \lambda \leqslant 1, |\eta| = 1$$
$$\alpha = 2\left(\theta + \arg\{aP_n(z_0) + bP'_n(z_0)\}\right) \tag{18}$$

if $\mathrm{Re}\left(e^{i\theta}\frac{a}{b}\right) = 0$.

PROOF Let us prove inequality (17) for $z = z_0$, and $\mathrm{Re}(e^{-i\theta}z_0) \leqslant R - u_0$. We set $u = R - \mathrm{Re}(e^{-i\theta}z_0) \geqslant u_0$. Let us suppose first that $u > 0$. The function

$$\zeta = \frac{4u - e^{-i\theta}(z - z_0)}{2u + e^{-i\theta}(z - z_0)} \qquad \left(z = 2e^{i\theta}u\,\frac{2-\zeta}{1+\zeta} + z_0\right)$$

maps the half-plane $\mathrm{Re}(e^{-i\theta}z) > R$ on to the open disc $|\zeta| < 1$ Consequently, all zeros of the polynomial

$$P^*_n(\zeta) = (1+\zeta)^n P_n\left(2e^{i\theta}u\,\frac{2-\zeta}{1+\zeta} + z_0\right)$$

of degree n belong to the disc $|\zeta| < 1$. If we replace z, a, b, $p_n(z)$ and $P_n(z)$ with 2,

$$a_1 = a + \frac{nb}{2u_0}e^{-i\theta}, \quad b_1 = \frac{3}{4}\frac{b}{u_0}e^{-i\theta}$$
$$p^*_n(\zeta) = (1+\zeta)^n p_n\left(2e^{i\theta}\frac{2-\zeta}{1+\zeta} + z_0\right) \qquad P^*_n(\zeta)$$

respectively and keep in mind that $\frac{a_1}{b_1} \in \bar{D}_{2,n}$, we obtain inequality (17) for $z = z_0$.

If $u = 0$ (in which case $u_0 = 0$), we set

$$\zeta = \frac{1 - 2e^{-i\theta}(z - z_0)}{1 + 2e^{-i\theta}(z - z_0)}$$

and then reason analogously to what we did above. It is not difficult to find the polynomials $p_n(z) \in \mathscr{P}_n$ in which the equality sign should be taken in (17), and we shall not stop for this. This completes the proof of the theorem.

It is possible to make notes and derive corollaries in connection with Theorems 2 and 3 corresponding to the notes and corollaries following Theorem 1.

We mention now that it is possible to carry the results that we have obtained above over to regions other than discs and half-planes. Suppose that B is some region in the z

plane, that $P_n(z)$ is a polynomial of degree n, all zeros of which lie in B, and that $\mathscr{P}_n$ is the class of polynomials $p_n(z)$ of degree not exceeding n that satisfy the condition $|p_n(z)| \leqslant |P_n(z)|$ everywhere on the boundary L of the region B. Suppose that a circle (or straight line) C exists which partitions the z plane into two regions G_1 and G_2 in such a way that $B \subset G_1$. Since the function $\frac{p_n(z)}{P_n(z)}$ is regular outside B and since $\left|\frac{p_n(z)}{P_n(z)}\right| < 1$ on L, it follows that $|p_n(z)| \leqslant |P_n(z)|$ outside B (in particular, on C) and that all zeros of the polynomial $P_n(z)$ lie in G_1. Therefore we may apply Theorem 2 (respectively Theorem 3) to finding an estimate for the expression mentioned in its formulation for $z \in G_2$. These considerations enable us to obtain estimates analogous to the preceding ones for regions B possessing certain special properties. Let us look at some of these.

Suppose that B is a convex bounded region with boundary L, that $w_0 \in B$, that C_z is the circle of minimum radius passing through the point $z \in L$ that contains the region B, and that w_z is the centre of this circle. If

$$\rho = \min_{z \in L}\left|\frac{z - w_0}{z - w_z}\right| \qquad \theta_1 \leqslant \arg\frac{z - w_0}{z - w_z} \leqslant \theta_2, \quad z \in L$$

the inequality

$$|ap_n(z) + b(z - w_0)\, p'_n(z)| \leqslant |aP_n(z) + b(z - w_0)\, P'_n(z)| \qquad p_n(z) \in \mathscr{P}_n \tag{19}$$

holds for every point $z \bar{\in} B$ and all numbers a and b such that either $b = 0$ or the point $\xi = a/b$ belongs to the closed region which contains the point $\xi = 0$ and is bounded by the smooth curve consisting of the arc of the circle

$$\xi = \frac{n}{2}\rho e^{i\theta}, \quad \pi + \theta_1 \leqslant \theta \leqslant \pi + \theta_2$$

and the two rays issuing from the end points of that arc. Also, the inequality

$$|ap_n(z) + bp'_n(z)| \leqslant |aP_n(z) + bP'_n(z)|, \quad p_n(z) \in \mathscr{P}_n \tag{20}$$

holds for every point $z \bar{\in} B$ if either $b = 0$ or

$$b \neq 0 \text{ and } \left|\frac{a}{b}\right| \leqslant \frac{1}{r}, \quad r = \max_{z \in L}|z - w_z|$$

Suppose that B is an unbounded convex region, and that θ_z is the argument of the outer normal (with respect to B) to the supporting straight line and to the region B at the point $z \in L$. If $\alpha \leqslant \theta_z \leqslant \beta$, $z \in L$ and $\beta - \alpha \leqslant \pi$ then for every point $z \bar{\in} B$ inequality (2) holds for all numbers a and b such that $a \neq 0$ and either $b = 0$ or

$$-\alpha - \frac{\pi}{2} \leqslant \arg \frac{a}{b} \leqslant -\beta + \frac{\pi}{2}$$

If B is an arbitrary convex region, then $|p_n'(z)| \leqslant |P_n'(z)|$ for every point $z \bar{\in} B$.

NOTE 9 The conditions imposed on the numbers a and b in (19) and (20) for fixed $z = z_0 \bar{\in} B$ can be weakened somewhat, especially if $z_0 \bar{\in} \bar{B}$.

NOTE 10 Assertions analogous to those that we have expounded above remain valid for the case in which all the zeros of the polynomial $P_n(z)$ lie in a region B whose complement is convex.

For what follows, we shall need the following lemma.

Lemma 3 Suppose that $P_{2n}(z)$ is a polynomial of degree $2n$ and that

$$F_n(x) = \frac{P_{2n}(x + \sqrt{x^2 - 1})}{(x + \sqrt{x^2 - 1})^n} = M_n(x) + i\sqrt{1 - x^2}\, R_{n-1}(x)$$

where $M_n(x)$ and $R_{n-1}(x)$ are polynomials of degrees not exceeding n and $n-1$ respectively. For all zeros of the polynomial $P_{2n}(z)$ to lie in the disc $|z| < 1$ it is necessary and sufficient that the real and imaginary parts of the function $F_n(x)$ have each $2n$ zeros as a complete circuit is made of the slit $-1 < x < 1$, that the zeros of the real and imaginary parts alternate, and that

$$\operatorname{Im} \frac{\sqrt{1 - x^2}\, F_n'(x)}{F_n(x)} < 0$$

at at least one point x in the interval $-1 < x < 1$.

PROOF If we set $w = g(z) = \frac{P_{2n}(z)}{z^n}$ and $x = \frac{1}{2}\left(z + \frac{1}{z}\right)$ we have

$$F_n(x) = g\left(x + \sqrt{x^2 - 1}\right)$$

We obtain proof of the lemma by examining the motion of the point w as we let the point z describe the circle L_1.

Theorem 4 Suppose that $F_n(x) = M_n(x) + i\sqrt{1-x^2}\,R_{n-1}(x)$ where $M_n(x)$ and $R_{n-1}(x)$ are polynomials of degrees n and $n-1$ respectively, that $\operatorname{Re} F_n(x)$ and $\operatorname{Im} F_n(x)$ have each $2n$ zeros on encircling the slit $-1 < x < 1$, that the zeros alternate and that

$$\operatorname{Im}\frac{\sqrt{1-x^2}\,F_n'(x)}{F_n(x)} < 0$$

at at least one point x such that $-1 < x < 1$, that F_n is the class of functions $f_n(x) = m_n(x) + i\sqrt{1-x^2}\,r_{n-1}(x)$ where $m_n(x)$ and $r_{n-1}(x)$ are polynomials of degrees not exceeding n and $n-1$ respectively which satisfy the condition $|f_n(x)| \leqslant |F_n(x)|$ on both edges of the cut $[-1; 1]$, that x is an arbitrary point in the x plane with a cut along the segment $-1 \leqslant x \leqslant 1$, that ρ is the sum of the semi-axes of the ellipse with foci at the points -1 and 1 that passes through the point x_0, that D_ρ is the closed region of the ξ plane $\left|\xi - i\frac{\rho^2+1}{\rho^2-1}n\right| \geqslant \frac{2\rho n}{\rho^2-1}$ if $\rho > 1$ and is the closed region $\operatorname{Im}\xi \geqslant 0$ if $\rho = 1$, and that a and b are arbitrary numbers such that either $b = 0$ or $\frac{a}{b} \in \overline{D}_\rho$. Then,

$$|af_n(x) + b\sqrt{1-x^2}\,f_n'(x)| \leqslant |aF_n(x) + b\sqrt{1-x^2}\,F_n'(x)| \qquad f(x) \in F_n. \tag{21}$$

For $x = x_0$ equality in (21) holds only when: (1) if $a = b = 0$, equality holds for every function $f_n(x) \in F_n$; (2) if $a \neq 0$, $b = 0$ and $-1 \leqslant x_0 \leqslant 1$, equality holds only for functions $f_n(x) \in F_n$ that satisfy the condition $|f_n(x_0)| = |F_n(x_0)|$; (3) if $|a| + |b| \neq 0$ and $x_0 \overline{\in} [-1, 1]$, equality holds only for functions $f_n(x) = \eta F_n(x)$ where $|\eta| = 1$; (4) if $b \neq 0$, $-1 \leqslant x_0 \leqslant 1$ and also $\operatorname{Im}\frac{a}{b} > 0$, equality holds only for the functions $f_n(x) = \eta F_n(x)$ where $|\eta| = 1$; (5) if $b \neq 0, -1 \leqslant x_0 \leqslant 1$; and $\operatorname{Im}\frac{a}{b} = 0$, equality holds only for the functions

$$f_n(x) = \eta[(1-\lambda)F_n(x) + \lambda e^{i\alpha}\overline{F_n(\overline{x})}], \quad 0 \leqslant \lambda \leqslant 1, \quad |\eta| = 1$$
$$\alpha = 2\arg\left[\frac{a}{b}F_n(x_0) + \sqrt{1-x_0^2}\,F_n'(x_0)\right] \tag{22}$$

To prove this theorem, we need only use Theorem 1 and make the transformation $x = \frac{1}{2}\left(z + \frac{1}{z}\right)$.

Notes might be made in connection with Theorem 4 analogous to the remarks made following Theorem 1.

COROLLARY Suppose that $0 < \omega < \pi$, and that

$$\Phi_n(\theta) = \mu_n(\theta) + i\sqrt{1 - \left(\frac{\tan\theta/2}{\tan\omega/2}\right)^2}\,\nu_n(\theta)$$

where

$$\mu_n(\theta) = \sum_{k=0}^{n}(a_k\cos k\theta + b_k\sin k\theta),$$

$$\nu_n(\theta) = \sum_{k=0}^{n-1}(c_k\cos k\theta + d_k\sin k\theta) + d_n\sin n\theta$$

are trigonometric polynomials of degree n, with real coefficients, with $2n$ and $2n-1$ zeros respectively in the interval $(-\omega, \omega)$. Suppose that these zeros alternate. Let $\mathscr{P}_n$ denote the class of trigonometric polynomials $S_n(\theta)$ of degree not exceeding n that satisfy the inequality $|S_n(\theta)| \leqslant |\Phi_n(\theta)|$ everywhere on the interval $-\omega \leqslant \theta \leqslant \omega$. Then, for every real A,

$$\begin{gathered} |S_n'(\theta) + AS_n(\theta)| \leqslant |\Phi_n'(\theta) + A\Phi_n(\theta)| \\ -\omega < \theta < \omega, \quad S_n(\theta) \in \mathscr{P}_n. \end{gathered} \tag{23}$$

Equality holds in (23) only for trigonometric polynomials $S_n(\theta) = \eta\mu_n(\theta)$ where $|\eta| = 1$ at the $2n$ points θ_ν that are zeros of the function

$$\frac{d}{d\theta}\left[\sqrt{1 - \left(\frac{\tan\theta/2}{\tan\omega/2}\right)^2}\,\nu_n(\theta)\right]$$

PROOF Let us make a transformation of the functions $S_n(\theta)$, $\mu_n(\theta)$ and $\nu_n(\theta)$ by setting $x = \frac{1}{\alpha}\tan\frac{\theta}{2}$ where $\alpha = \tan\frac{\omega}{2}$

$$S_n(\theta) = \frac{P_{2n}(x)}{(1+\alpha^2x^2)^n}; \quad \mu_n(\theta) = \frac{M_{2n}(x)}{(1+\alpha^2x^2)^n}; \quad \nu_n(\theta) = \frac{R_{2n-1}(x)}{(1+\alpha^2x^2)^n}$$

Then,

$$\begin{gathered} |P_{2n}(x)| \leqslant |F_{2n}(x)|, \quad F_{2n}(x) = M_{2n}(x) + i\sqrt{1-x^2}\,R_{2n-1}(x) \\ -1 \leqslant x \leqslant 1 \end{gathered}$$

If we apply Theorem 4 and set

$$a = -\frac{nx}{\alpha(1+\alpha^2x^2)^n} + \frac{A}{(1+\alpha^2x^2)^n}, \quad b = \frac{1}{2\alpha(1+\alpha^2x^2)^{n-1}\sqrt{1-x^2}}$$

we obtain

$$\left|\frac{d}{dx}\left[\frac{P_{2n}(x)}{(1+\alpha^2x^2)^n}\right]\frac{dx}{d\theta} + A\frac{P_{2n}(x)}{(1+\alpha^2x^2)^n}\right| \leqslant \left|\frac{d}{dx}\left[\frac{F_{2n}(x)}{(1+\alpha^2x^2)^n}\right]\frac{dx}{d\theta} + A\frac{F_{2n}(x)}{(1+\alpha^2x^2)^n}\right| \tag{24}$$

If we return to the variable θ, we obtain the desired inequality. The assertion regarding equality follows easily from Theorem 4. This completes the proof of the corollary.

NOTE 11 Inequality (23) remains valid if $\operatorname{Im} A > 0$ but equality in (23) then does not hold for any θ, in the interval $(-\omega < \theta < \omega)$.

Let us note a special case of this corollary. If we set

$$F_{2n}(x) = M\left(\sqrt{1+\alpha^2}\,x - i\sqrt{1-x^2}\right)^{2n}$$

where $M > 0$, we see that if $|S_n(\theta)| \leqslant M$, for $-\omega < \theta < \omega$, then

$$|S_n'(\theta) + AS_n(\theta)| \leqslant M\sqrt{\frac{n^2\cos^2\frac{\theta}{2}}{\sin^2\frac{\omega}{2} - \sin^2\frac{\theta}{2}} + A^2}$$

This corollary and this special case of it were proven by Vidensky [5] for $A = 0$.

The inequalities obtained in the preceding theorems contain only first-order derivatives. We can obtain inequalities containing derivatives of higher orders.

Let us introduce some notations that we shall use in the formulation of the following theorem. Suppose that $f(z)$ is an arbitrary function that is regular at some point. We define

$$L^0\{f(z)\} = f(z),\ L^1\{f(z)\} = zf'(z),\ \ldots$$
$$L^k\{f(z)\} = L^1\{L^{k-1}\{f(z)\}\},\ \ldots$$

If $Q(w) = \sum_{l=0}^{k} a_l w^{k-l}$ is a polynomial, we define

$$Q_L\{f(z)\} = \sum_{l=0}^{k} a_l L^{k-l}\{f(z)\}$$

Theorem 5 Suppose that $P_n(z)$ is a polynomial of degree $n \geqslant 1$ all the zeros of which belong to K_1, that $\mathcal{P}_n$ is the class of polynomials $p_n(z)$ of degree not exceeding n, that satisfy the inequality $|p_n(z)| \leqslant |P_n(z)|$ everywhere on L_1, and that

$$Q(w) = \sum_{l=0}^{k} a_l w^{k-l}$$

is a polynomial of degree $k \geqslant 1$ all the zeros of which lie in the closed half-plane $\operatorname{Re} w \leqslant 0$. Then, in the region $|z| \geqslant 1$

$$|Q_L\{z^{-n/2} p_n(z)\}| \leqslant |Q_L\{z^{-n/2} P_n(z)\}|, \quad p_n(z) \in \mathcal{P}_n \tag{25}$$

Equality holds in (25) for arbitrarily given $z = z_0$, where $|z_0| \geqslant 1$, only in the following cases: (1) if $|z_0| > 1$ (or if $|z_0| = 1$ and at least one zero of the polynomial $Q(w)$ lies in the half-plane $\operatorname{Re} w < 0$), equality holds only for polynomials $p_n(z) = \eta P_n(z)$ where $|\eta| = 1$; (2) if $|z_0| = 1$ and all zeros of the polynomial $Q(w)$ lie on the imaginary axis, equality holds only for the polynomials

$$\begin{gathered} p_n(z) = \eta\left[(1-\lambda)P_n(z) + (-1)^k \lambda e^{i\alpha} z^n \overline{P_n\left(\frac{1}{\bar{z}}\right)}\right] \\ 0 \leqslant \lambda \leqslant 1, \quad |\eta| = 1 \\ \alpha = \arg\left\{\frac{\bar{z}_0^n}{a_0^2}\left[Q_L\left\{z^{-\frac{n}{2}} P_n(z)\right\}\right]^2_{z=z_0}\right\} \end{gathered} \tag{26}$$

PROOF For $k = 1$ we have a special case of Theorem 1. For $k = 2, 3, \ldots$ inequality (25) and the assertions regarding the equality sign in it are proven by induction.

NOTE 12 We have also proven that if all zeros of polynomial $P_n(z)$ lie within the disc $|z| < 1$ so do all the zeros of a polynomial

$$z^{\frac{n}{2}} Q_L\left\{z^{-\frac{n}{2}} P_n(z)\right\}$$

where ($Q(w)$ is a polynomial all the zeros of which lie in the region $\operatorname{Re} w \leqslant 0$).

NOTE 13 The theorem may be extended with a slight change in the wording to the case when all zeros of the polynomial $P_n(z)$ lie in the disc $|z| \leqslant 1$.

NOTE 14 For $|z| \geqslant R > 1$ inequality (25) remains valid under weaker hypotheses concerning the zeros of the polynomial $Q(w)$.

NOTE 15 Obviously, the theorem remains valid if the coefficients of the polynomial $Q(w)$ depend on z provided all the zeros of this polynomial lie in the region $\mathrm{Re}\, w \leqslant 0$ for all z such that $|z| \geqslant 1$. If this requirement is satisfied for certain values of z, such that $|z| \geqslant 1$, the assertions of the theorem can be extended only to such z.

NOTE 16 If

$$P_n(z) = \sum_{\nu=0}^{n} A_\nu z^\nu$$

and

$$p_n(z) = \sum_{\nu=0}^{n} c_\nu z^\nu$$

inequality (25) can be written as follows:

$$\left| \sum_{\nu=0}^{n} c_\nu Q\left(\nu - \frac{n}{2}\right) z^\nu \right| \leqslant \left| \sum_{\nu=0}^{n} A_\nu Q\left(\nu - \frac{n}{2}\right) z^\nu \right|, \quad |z| \geqslant 1 \tag{27}$$

It is natural to ask the following question. What necessary and sufficient conditions must the numbers α_ν, where $\nu = 0, 1, \ldots, n$, satisfy for the inequality

$$\left| \sum_{\nu=0}^{n} \alpha_\nu c_\nu \right| \leqslant \left| \sum_{\nu=0}^{n} \alpha_\nu A_\nu \right|$$

to hold for every polynomial $p_n(z) \in \mathscr{P}_n$ and for equality to hold in it only for polynomials of the form (26) for given α, such that $0 \leqslant \alpha < 2\pi$? We shall answer this question in Section 5.4 (see the remark following Corollary 1 to Theorem 3').

Theorem 6 Suppose that

$$S_n(\xi) = \sum_{k=0}^{n} (\alpha_k \cos k\xi + \beta_k \sin k\xi)$$

where $\alpha_n + \beta_n i \neq 0$ is a trigonometric polynomial of degree n all the zeros of which lie in the half-plane $\mathrm{Im}\, \xi < 0$, that Σ_n is the class of trigonometric polynomials $\sigma_n(\xi)$ of degree not exceeding n that satisfy the inequality $|\sigma_n(\xi)| \leqslant |S_n(\xi)|$ everywhere on the real axis, and that

$$Q(w) = \sum_{l=0}^{k} a_l w^{k-l}$$

is a polynomial of degree $k \geqslant 1$, all the zeros of which lie in the half-plane $\operatorname{Im} w \geqslant 0$. Then, everywhere in the region $\operatorname{Im} \xi \geqslant 0$

$$\left| \sum_{l=0}^{k} a_l \sigma^{(k-l)}(\xi) \right| \leqslant \left| \sum_{l=0}^{k} a_l S_n^{(k-l)}(\xi) \right|, \quad \sigma_n(\xi) \in \Sigma_n \tag{28}$$

Equality holds in (28) for arbitrarily given $\xi = \xi_0$, where $\operatorname{Im} \xi_0 \geqslant 0$ only in the following cases: (1) if $\operatorname{Im} \xi_0 > 0$ or if at least one zero of the polynomial $Q(w)$ lies in the half-plane $\operatorname{Im} w > 0$, equality holds only for the trigonometric polynomials $\sigma_n(\xi) = \eta S_n(\xi)$, where $|\eta| = 1$; (2) if $\operatorname{Im} \xi_0 = 0$ and all the zeros of the polynomial $Q(w)$ lie on the real axis, equality holds only for the trigonometric polynomials

$$\sigma_n(\xi) = \eta [(1-\lambda) S_n(\xi) + (-1)^k \lambda e^{i\alpha} \overline{S_n(\bar{\xi})}], \quad 0 \leqslant \lambda \leqslant 1, \quad |\eta| = 1$$
$$\alpha = \arg \left\{ \frac{1}{a_0^2} \left(\sum_{l=0}^{k} a_k S_n^{(k-l)}(\xi_0)^2 \right\} \right. \tag{29}$$

This theorem is easily proven by replacing n with $2n$ in Theorem 5, setting

$$p_{2n}(z) = z^n \sigma_n(i \ln z), \quad P_{2n}(z) = z^n S_n(i \ln z)$$

and returning to the variable ξ by taking $z = e^{-i\xi}$.

We note that all the notes made in connection with Theorem 5 remain in force for this theorem with appropriate changes in the wording.

If we set $S_n(\xi) = M(\cos n\xi - i \sin n\xi)$, in the theorem, where $M > 0$, we obtain the following corollary.

COROLLARY 1 (Bernstein) If $\sigma_n(\xi)$ is a trigonometric polynomial of degree not exceeding n and if $|\sigma_n(\xi)| \leqslant M$ everywhere on the real axis, then, throughout the region $\operatorname{Im} \xi \geqslant 0$

$$|\sigma_n^{(m)}(\xi)| \leqslant M n^m e^{n \operatorname{Im} \xi}, \quad m = 0, 1, \ldots$$

COROLLARY 2 (Bernstein [1], p. 337) If $\sigma_n(\theta)$ is a trigonometric polynomial of degree not exceeding n such that $|\sigma_n(\theta)| \leqslant M$, where $0 \leqslant \theta < 2\pi$ and if

$$Q(w)=\sum_{l=0}^{k} a_l w^{k-l}$$

is a real polynomial of degree k all the zeros of which are real, then

$$\left|\sum_{l=0}^{k} a_l \sigma_n^{(k-l)}(\theta)\right| \leqslant M\,|Q(in)|, \quad 0 \leqslant \theta < 2\pi$$

Here, equality holds only for the trigonometric polynomials

$$\sigma_n(\theta)=M\eta\,[(1-\lambda)\,e^{in\theta}+\lambda e^{i\alpha}e^{-in\theta}]$$

$$0 \leqslant \lambda \leqslant 1, \; |\eta|=1, \; 0 \leqslant \alpha < 2\pi$$

Dochev [1] has shown that the assertion of Corollary 2 (without the statement regarding equality) is also valid when the zeros of the polynomial $Q(w)$ lie in the region

$$y^2-\frac{x^2}{2n-1} \leqslant \frac{n}{2}, \quad w=x+yi$$

Theorem 7 Suppose that $M_n(x)$ is a polynomial of degree $n \geqslant 1$ all the zeros of which lie in the half-plane $\operatorname{Im} x \leqslant 0$, that $\mathfrak{M}_n$ is the class of polynomials $m_n(x)$ of degree not exceeding n which satisfy the inequality $|m_n(x)| \leqslant |M_n(x)|$ everywhere on the real axis, and that

$$Q_k(w)=\sum_{l=0}^{k} a_l w^{k-l}$$

is a polynomial of degree $k \geqslant 1$ all the zeros of which lie in the half-plane $\operatorname{Re} w \geqslant 0$. Then, in the region $\operatorname{Im} x \geqslant 0$

$$\left|\sum_{l=0}^{k} a_l m_n^{(k-l)}(x)\right| \leqslant \left|\sum_{l=0}^{k} a_l M_n^{(k-l)}(x)\right|, \quad m_n(x) \in \mathfrak{M}_n \tag{30}$$

Equality holds in (30) for arbitrary given $x=x_0$ where $\operatorname{Im} x_0 \geqslant 0$, only in the following cases: (1) if $\operatorname{Im} x_0 > 0$ and if at least one zero of the polynomial $Q_k(w)$ lies in the half-plane $\operatorname{Im} w > 0$, equality holds only for the polynomials $m_n(x)=\eta\, M_n(x)$, where $|\eta|=1$; (2) if $\operatorname{Im} x_0=0$ and all the zeros of the polynomial $Q_k(w)$ lie on the real axis, equality holds only for the polynomial.

$$m_n(x) = \eta[(1-\lambda)M_n(x) + \lambda e^{i\alpha}\overline{M_n(\bar{x})}], \quad 0 \leqslant \lambda \leqslant 1, \quad |\eta| = 1$$

$$\alpha = \arg\left\{\frac{1}{a_0^2}\left(\sum_{l=0}^{k} a_l M_n^{(k-l)}(x_0)\right)^2\right\} \tag{31}$$

Proof of this theorem does not differ materially from Theorem 5 and it rests on Theorem 3. Theorem 7 was obtained in a different manner by Meiman [1] without a clarification of the cases under which equality holds.

Before giving the next theorem, we introduce some notations. Suppose that $f(z)$ is an arbitrary function that is regular at a point $z_0 \neq 0$. We define

$$\Psi^0\{f(z)\} = f(z), \; \Psi^1\{f(z)\} = \sqrt{z}\, f'(z), \; \ldots$$
$$\ldots, \; \Psi^k\{f(z)\} = \Psi^1\{\Psi^{k-1}\{f(z)\}\}$$

If $Q(w) = \sum_{l=0}^{k} a_l w^{k-l}$ is a polynomial, we set

$$Q_\Psi\{f(z)\} = \sum_{l=0}^{k} a_l \Psi^{k-l}\{f(z)\}$$

By using the mapping $z = x^2$ of the half-plane $\operatorname{Im} x > 0$ on to the z plane with cut $[0, \infty]$ and by using Theorem 7, we may prove the following.

Theorem 8 Suppose that $F_n(z) = A_n(z) + \sqrt{z}\, B_n(z)$, where $A_n(z)$ is a polynomial of degree n, and that $B_n(z)$ is a polynomial of degree n (respectively $n-1$), that $\operatorname{Re} F_n(z)$ and $\operatorname{Im} F_n(z)$ have each $2n+1$ (respectively $2n$) zeros as a complete circuit is made of the cut $[0, +\infty]$, that these zeros alternate, that $\operatorname{Im} \frac{\sqrt{z}\, F'(z)}{F(z)} < 0$ at at least one point z such that $0 < z < +\infty$, that $\mathcal{F}_n^*$ is the class of functions $f_n(z) = a_n(z) + \sqrt{z}\, b_n(z)$ where $a_n(z)$ is a polynomial of degree not exceeding n and $b_n(z)$ is a polynomial of degree not exceeding n (respectively not exceeding $n-1$), that both these last two polynomials satisfy the condition $|f_n(z)| \leqslant |F_n(z)|$ on both edges of the cut $[0, \infty]$, and that

$$Q_k(w) = \sum_{l=0}^{k} a_l w^{k-l}$$

is a polynomial of degree $k \geqslant 1$ all the zeros of which lie in the region $\operatorname{Im} w \geqslant 0$.. Then, outside the cut $[0, \infty]$ and on its edges,

$$|Q_k\{f(z)\}| \leqslant |Q_k\{F_n(z)\}|, \quad f_n(z) \in \mathscr{F}_n^* \tag{32}$$

Equality holds for arbitrary $z = z_0$ only in the following cases: (1) if z_0 lies outside the cut $[0, \infty]$ or if z_0 lies on the cut $[0, \infty]$ and at least one zero of the polynomial $Q_k(w)$ lies in the half-plane $\operatorname{Im} w > 0$, equality holds only for the functions $f_n(z) = \eta F_n(z)$ where $|\eta| = 1$; (2) if z_0 lies on the cut $[0, +\infty]$ and all zeros of the polynomial $Q_k(w)$ lie on the real axis, equality holds for the functions

$$f_n(z) = \eta[(1-\lambda)F_n(z) + \lambda e^{i\alpha}\overline{F_n(\bar{z})}], \quad 0 \leqslant \lambda \leqslant 1, \quad |\eta| = 1$$
$$\alpha = \arg\left\{\frac{1}{a_0^2}[Q_k\{F_n(z)\}_{z=z_0}]^2\right\} \tag{33}$$

We introduce notations analogous to those used above. Suppose that $f(z)$ is an arbitrary function that is regular at some point. We define

$$\Phi^0\{f(z)\} = f(z); \quad \Phi^1\{f(z)\} = \sqrt{1-z^2}\, f'(z)$$
$$\Phi^k\{f(z)\} = \Phi^1\{\Phi^{k-1}\{f(z)\}\}$$

If $Q(w) = \sum_{l=0}^{k} a_l w^{k-l}$ is a polynomial, we define

$$Q_\Phi\{f(z)\} = \sum_{l=0}^{k} a_l \Phi^{k-l}\{f(z)\}$$

By using Theorem 4, we can easily prove the following theorem.

Theorem 9 Suppose that $F_n(x) = M_n(x) + i\sqrt{1-x^2}\, R_{n-1}(x)$ where $M_n(x)$ and $R_{n-1}(x)$ are polynomials of degrees n and $n-1$, respectively, is a function such that $\operatorname{Re} F_n(x)$ and $\operatorname{Im} F_n(x)$ each have $2n$ zeros as a complete circuit is made of the cut $[-1, 1]$, that these zeros alternate, that

$$\operatorname{Im} \frac{\sqrt{1-x^2}\, F'(x)}{F(x)} < 0$$

at at least one point x such that $-1 < x < 1$, that $\mathscr{F}_n$ is the class of functions

$$f_n(x) = m_n(x) + i\sqrt{1-x^2}\, r_{n-1}(x)$$

where $m_n(x)$ and $r_{n-1}(x)$ are polynomials of degrees not exceeding n and $n-1$ respectively which satisfy the condition $|f_n(x)| \leqslant |F_n(x)|$ on both edges of the cut [–1, 1], that $Q(w)=\sum_{l=0}^{k} a_l w^{k-l}$ is a polynomial of degree $k \geqslant 1$ all the zeros of which lie in the region $\operatorname{Im} w \leqslant 0$.. Then, outside the cut [–1, 1] and on its edges,

$$|Q_\Phi\{f_n(x)\}| \leqslant |Q_\Phi\{F_n(x)\}| \tag{34}$$

for every function $f_n(x) \in \mathcal{F}_n$. Equality holds for arbitrarily given $x = x_0$ only in the following cases: (1) if x_0 lies outside the cut [–1, 1] and if x_0 lies on the cut [–1, 1] and at least one zero of the polynomial $Q(w)$ lies in the plane $\operatorname{Im} w < 0$, equality holds only for the functions $f_n(x) = \eta F_n(x)$ where $|\eta| = 1$; (2) if x_0 lies on the cut [–1, 1] and all zeros of the polynomial $Q_k(w)$ lie on the real axis, equality holds only for the functions

$$f_n(x) = \eta[(1-\lambda)F_n(x) + \lambda(-1)^k e^{i\alpha}\overline{F_n(\bar{x})}],\ 0 \leqslant \lambda \leqslant 1,\ |\eta| = 1$$
$$\alpha = \arg\left\{\frac{1}{a_0^2}[Q_\Phi\{F_n(x)\}]^2_{x=x_0}\right\} \tag{35}$$

All the notes made in connection with Theorem 5 remain in effect with a corresponding change in wording for this theorem. Of particular interest is the special case of the corollary in which $M_n(x)$ and $R_{n-1}(x)$ are polynomials with real coefficients and $f_n(x) = p_n(x)$ is a polynomial of degree not exceeding n or else $f_n(x) = p_{n-1}(x)\sqrt{1-x^2}$ where $p_{n-1}(x)$ is a polynomial of degree not exceeding $n-1$. In the first case, equality holds in (34) only when all zeros of the polynomial $Q(w)$ are real and k is an odd number and only for the polynomial $p_n(x) = \eta M_n(x)$, where $|\eta| = 1$ at those points of the interval (–1, 1) at which $R_{n-1}(x)$ vanishes. In the second case, equality holds in (34) only when all zeros of the polynomial $Q(w)$ are real and k is an even number and only for the polynomial

$$p_{n-1}(x) = \eta R_{n-1}(x), \quad |\eta| = 1$$

at those points of the interval (–1, 1) at which $M_n(x) = 0$.

We close this section with two theorems due to Videnskiy for polynomials with real coefficients.

Theorem 10 (Vidensky [1] *)* Let us suppose $F_n(z) = A_n(z) + i\sqrt{z}\,B_n(z)$ where $A_n(z)$ and $B_n(z)$ are polynomials of degree n with real coefficients $B_n(z)$ may be of degree $n-1$. Suppose also that all zeros of these polynomials lie on the interval $0 \leqslant z < \infty$, that they alternate and that $\mathcal{F}_n$ is the class of polynomials $p_n(z)$ of degree not exceeding n that satisfy the condition $|p_n(z)| \leqslant |F_n(z)|$, where $0 \leqslant z < \infty$.

Then, for every $k = 1, 2, \ldots, n$,

$$|p_n^{(k)}(z)| \leqslant |F_n^{(k)}(z)|, \quad 0 \leqslant z < +\infty$$

Equality holds only for the polynomials $p_n(z) = \eta M_n(z)$ where $|\eta| = 1$, at points at which the function

$$\frac{d^k}{dz^k}(\sqrt{z}\,B_n(z))$$

vanishes.

Theorem 11 (Vidensky [1] *)* Suppose that

$$F_n(x) = M_n(x) + i\sqrt{1-x^2}\,R_{n-1}(x)$$

where $M_n(x)$ and $R_{n-1}(x)$ are polynomials of degrees n and $n-1$ respectively with real coefficients, that all their zeros lie in the interval (–1, 1) and that these zeros alternate. Let $\mathcal{P}_n$ be the class of polynomials $p_n(x)$ of degree not exceeding n, that satisfy the condition

$$|p_n(x)| \leqslant |F_n(x)|, \quad -1 \leqslant x \leqslant 1$$

Then, for arbitrary $k = 1, 2, \ldots, n$

$$|p_n^{(k)}(x)| \leqslant |F_n^{(k)}(x)|, \quad -1 \leqslant x \leqslant 1$$

Equality holds only for the polynomials $p_n(x) = \eta M_n(x)$ where $|\eta| = 1$ at points at which the function

$$\frac{d^k}{dx^k}(\sqrt{1-x^2}\,R_{n-1}(x))$$

vanishes.

From this, we may, in particular, obtain the following result of Schaeffer and Duffin [1]: if $|p_n(x)| \leqslant M$ for $-1 \leqslant x \leqslant 1$, where $M > 0$, then

$$|p_n^{(k)}(x)| \leqslant \left|\frac{d^k}{dx^k}(x+i\sqrt{1-x^2})^n\right|, \quad -1 \leqslant x \leqslant 1, \ k=1, 2, \ldots, n$$

By using this result, these same authors obtained a simple proof of a familiar result of V.A. Markov [1]: if

$$|p_n(x)| \leqslant M, \quad -1 \leqslant x \leqslant 1, \quad M > 0$$

then

$$|p_n^{(k)}(x)| \leqslant M \frac{n^2(n^2-1^2)\ldots(n^2-(k-1)^2)}{1\cdot 3 \ldots (2k-1)}$$
$$-1 \leqslant x \leqslant 1, \quad k=1, 2, \ldots, n.$$

5.2 THE SMALLEST DEVIATION FROM ZERO OF FUNCTIONS DEPENDING ON PARAMETERS

5.2.1 Definition and basic theorem

In the preceding section, we gave the formulation and solution of a number of specific problems. In the present and subsequent sections, we shall state certain general problems regarding the least deviation from zero and the best approximation and we shall also prove certain general theorems that deal with these problems. In what follows, we shall often have occasion to deal with a real n-dimensional space $\mathfrak{R}_n$. Let us recall the basic definitions and features of the geometry of the space $\mathfrak{R}_n$. An n-tuple of real numbers $(c_1, c_2, \ldots, c_n)$ is called a point X in the space $\mathfrak{R}_n$. The distance between two points

$$X_0(c_1^{(0)}, c_2^{(0)}, \ldots, c_n^{(0)})$$

and $X_1(c_1^{(1)}, c_2^{(1)}, \ldots, c_n^{(1)})$ in the space $\mathfrak{R}_n$ is defined by the formula

$$\rho = \rho(X_0, X_1) = \sqrt{\sum_{j=1}^{n}(c_j^{(1)} - c_j^{(0)})^2}$$

The set of all points X such that $\rho(X_0, X) < \varepsilon$ for $\varepsilon > 0$, is called an ε-neighbourhood of the point X_0. A point $X_0 \in \mathfrak{R}_n$ is called a cluster point of a set K of points $X \in \mathfrak{R}_n$ if every ε-neighbourhood of the point X_0 holds points of the set K distinct from X_0. This is equivalent to saying that an arbitrary ε-neighbourhood of the point X_0 contains an infinite set of points $X \in K$. A set of K of points $X \in \mathfrak{R}_n$ is said to be closed if all the cluster points of the set K belong to K. The closure of a set K is the set $\bar{K}$ consisting of all the points of K and all its cluster points. The set $\bar{K}$ is closed.

Let K' be a collection of non-empty sets K of points of $X \in \mathfrak{R}_n$. If each of the sets $K \in K'$ is closed, their intersection K_0 will be a closed set. If K_0 is the empty set, it is by convention considered closed. A set K of points $X \in \mathfrak{R}_n$ is said to be bounded if there exists a positive integer M such that $\rho(X, O) \leqslant M$ for an arbitrary point $X \in K$ and O (0, 0,...,0). If we have a sequence of closed non-empty sets of points $X \in \mathfrak{R}_n$, each containing the next

$$K_1 \supset K_2 \supset \ldots \supset K_j \supset \ldots$$

and if some set K_j is bounded, the intersection

$$K_0 = \bigcap_{j=1}^{\infty} K_j$$

of the sets K_j is a non-empty (bounded closed) set. The set K_0 is the set of points belonging to all K_j, for $j = 1$ 2, ...

We sometimes refer to the point $X(c_1, c_2, \ldots, c_n)$ as a vector and the numbers c_j, for $j = 1, 2, \ldots, n$, as the coordinates of the point or of the vector. Let us define the concept of the sum $X_0 + X_1$ of two vectors and the product τX_0 of the vector X_0 and the real number τ, where

$$X_0 (c_1^{(0)}, c_2^{(0)}, \ldots, c_n^{(0)})$$

and

$$X_1 (c_1^{(1)}, c_2^{(1)}, \ldots, c_n^{(1)})$$

are two vectors in $\mathfrak{R}_n$. The coordinates of the vector $X_0 + X_1$ are taken equal to $(c_1^{(0)} + c_1^{(1)}, c_2^{(0)} + c_2^{(1)}, \ldots, c_n^{(0)} + c_n^{(1)})$ and the coordinates of the vector τX_0 are taken equal to

$$(\tau c_1^{(0)}, \tau c_2^{(0)}, \ldots, \tau c_n^{(0)})$$

The concept of the sum is naturally generalised to an arbitrary number of vectors added together. A set K of points $X \in \mathfrak{R}_n$ is said to be convex if, for any two points X_0 and X_1 belonging to K, all points of the form $X_\tau = (1 - \tau) X_0 + \tau X_1$ for $0 < \tau < 1$ also belong to K. We say that these points

X_τ constitute the line segment between the points X_0 and X_1 The intersection of convex sets is a convex set - though it may possibly be empty.

What we have just said leads us in a natural way to the definition of a complex n-dimensional set $\tilde{\mathfrak{R}}_n$. A point $\tilde{X}$ in such a space is defined as an n-tuple of complex numbers $(c_1, c_2, \ldots, c_n)$. In the expression for the distance between two points, the squares of the distances between corresponding components of the n-tuples are replaced with the squares of the absolute values of these distances:

$$\rho=\rho(\tilde{X}_0, \tilde{X}_1)=\sqrt{\sum_{j=1}^{n}|c_j^{(1)}-c_j^{(0)}|^2}$$

We define ε-neighbourhood, cluster point and closed set just as before. All that was said with regard to the intersection of two or more sets still holds. Separating the real and imaginary parts of each components of an n-tuple, we set $c_k=a_k+b_k i$. Then, instead of a complex space $\tilde{\mathfrak{R}}_n$ the coordinates of the elements of which are $(c_1, \ldots, c_n)$ we may consider the real space $\mathfrak{R}_{2n}$ the coordinates of the elements of which are $(a_1,b_1, a_2,b_2, \ldots, a_n,b_n)$. Then, the expressions for the distance between corresponding pairs of points in $\tilde{\mathfrak{R}}_n$ and $\mathfrak{R}_{2n}$ coincide.

Let us turn to the problem mentioned in the title to this section. Suppose that D is a non-empty set of elements ξ and that $F(\xi, X)=F(\xi, c_1, \ldots, c_n)$ where $X=(c_1, \ldots, c_n)$ is an arbitrary real non-negative function defined for $\xi\in D$ and $X\in\mathfrak{R}_n$. This function may assume either finite or infinite values. In what follows, we shall consider the following cases: suppose that D is a bounded sub-set of the complex plane, that ξ denotes a point in this set, and that $f(\xi)$ is a function defined on the set D. We define

$$F(\xi, X)=F(\xi, c_1, \ldots, c_{2n})=\left|\xi^n+\sum_{k=0}^{n-1}a_k\xi^k\right|, \quad a_k=c_{2k+1}+ic_{2k+2}$$
$$X(c_1, c_2, \ldots, c_{2n})\in\mathfrak{R}_{2n},$$

or

$$F(\xi, X)=F(\xi, c_1, \ldots, c_{2n+2})=\left|f(\xi)-\sum_{k=0}^{n}a_k\xi^k\right|$$
$$a_k=c_{2k+1}+ic_{2k+2},$$
$$X(c_1, c_2, \ldots, c_{2n+2})\in\mathfrak{R}_{2n+2}$$

Polynomials in ξ may be replaced, for example, with rational functions of ξ. We shall also have occasion to consider the functions $F(\xi, \tilde{X})$ where $\xi \in D$ and $\tilde{X} \in \tilde{\mathfrak{R}}_n$.

DEFINITION The non-negative number

$$\mu = \mu(D) = \inf_{X \in \mathfrak{R}_n} \sup_{\xi \in D} F(\xi, X)$$

is called the least deviation from zero of the function $F(\xi, X)$, where $\xi \in D$ and $X \in \mathfrak{R}_n$.

In what follows, we shall assume without explicit mention that μ is a finite number, that is, that there exists at least one point $X \in \mathfrak{R}_n$ such that the function $F(\xi, X)$ is bounded on D. At the same time it is possible that for some X $\sup_{\xi \in D} F(\xi, X) = +\infty$ or even $F(\xi, X) = +\infty$ for $\xi \in D$. The question arises: for what functions $F(\xi, X)$ does a point $X_0 \in \mathfrak{R}_n$, exist such that

$$\sup_{\xi \in D} F(\xi, X_0) = \mu \tag{1}$$

Before proving the theorem dealing with this problem, let us introduce some notations that we shall be using quite frequently in what follows. We denote by $K(\xi, \lambda)$, where $\lambda \geqslant 0$ and $\xi \in D$, the set of all points $X \in \mathfrak{R}_n$ such that

$$F(\xi, X) < \lambda, \quad K(D', \lambda) = \bigcap_{\xi \in D'} K(\xi, \lambda)$$

$D' \subset D$, and $K(\lambda) = K(D, \lambda)$. Obviously,

$$K(\xi, \lambda') \subset K(\xi, \lambda'') \qquad K(D', \lambda') \subset K(D', \lambda'') \qquad 0 \leqslant \lambda' < \lambda'' \tag{2}$$

and, in particular,

$$K(\lambda') \subset K(\lambda'') \quad \text{for} \quad 0 \leqslant \lambda' < \lambda'' \tag{3}$$

Also,

$$K(D'', \lambda) \subset K(D', \lambda) \quad \text{for} \quad D' \subset D'' \tag{4}$$

We note that $K(D', \lambda)$ is the set of points $X \in \mathfrak{R}_n$ such that $F(\xi, X) \leqslant \lambda$ for every point $\xi \in D'$. In other words, $K(D', \lambda)$ is the set of points $X \in \mathfrak{R}_n$ such that

$$\sup_{\xi \in D'} F(\xi, X) \leqslant \lambda \tag{5}$$

In particular, $K(\lambda)$ is the set of points $X \in \mathfrak{R}_n$ such that

$$\sup_{\xi \in D} F(\xi, X) \leqslant \lambda. \tag{6}$$

Let us show that the set $K(\lambda)$ and hence $K(\xi, \lambda)$ is non-empty for $\lambda > \mu$. It follows from the definition of μ that a point $X_0 \in \Re_n$ exists such that

$$\sup_{\xi \in D} F(\xi, X_0) \leqslant \lambda$$

From this it follows that $F(\xi, X_0) \leqslant \lambda$ for all $\xi \in D$; that is, $X_0 \in K(\xi, \lambda)$ for all $\xi \in D$ and $X_0 \in K(\lambda)$.

Theorem 1 If there exists a $\lambda_0 > \mu$ such that for every λ, $\mu < \lambda < \lambda_0$ and for $\xi \in D$ the set $K(\xi, \lambda)$ is closed and the set $K(\lambda_0)$ is bounded, there exists a point $X_0 \in \Re_n$ such that Equation (1) holds.

PROOF Let λ_j (for $j = 1, 2, \ldots$), where $\mu < \lambda_j < \lambda_0$, be an arbitrary decreasing sequence of numbers that approaches μ. It follows from the first condition of the theorem that the non-empty sets $K(\lambda_j)$ are closed. From (3),

$$K(\lambda_1) \supset K(\lambda_2) \supset \ldots$$

and, from the second condition of the theorem, all the sets $K(\lambda_j)$ are bounded. It follows from this that the intersection

$$K = \bigcap_{j=1}^{\infty} K(\lambda_j)$$

is non-empty. Suppose that $X_0 \in K$. Then, $X_0 \in K(\lambda_j)$ (for $j = 1, 2, \ldots$), and consequently $X_0 \in K(\xi, \lambda_j)$ (for $j = 1, 2, \ldots$), where $\xi \in D$; that is, $F(\xi, X_0) \leqslant \lambda_j$ (for $j = 1, 2, \ldots$). Since

$$\lim_{j \to \infty} \lambda_j = \mu$$

it follows from what has been said that

$$\mu \leqslant \sup_{\xi \in D} F(\xi, X_0) \leqslant \mu$$

that is, Equation (1) holds and the theorem is proven.

NOTE 1 We shall exhibit a case in which all the sets $K(\xi, \lambda)$, for $\lambda \geqslant 0$ and $\xi \in D$, are closed. Let us suppose that the function $F(\xi, X)$ is continuous with respect to X for $X \in \Re_n$ and arbitrary fixed ξ; that is, for arbitrary $X_1 \in \Re_n$

and arbitrary $\varepsilon > 0$ there exists a $\delta > 0$ such that

$$|F(\xi, X) - F(\xi, X_1)| < \varepsilon \quad \text{for} \quad \rho(X, X_1) < \delta \tag{7}$$

It is easy to see that under these conditions the sets $K(\xi, \lambda)$ are closed for $\lambda \geqslant 0$ and $\xi \in D$. Specifically, if X_1 is a cluster point of the set $K(\xi, \lambda)$, a δ-neighbourhood of the point X_1 contains points $X \in K(\xi, \lambda)$ distinct from X_1. Equation (7) then holds and

$$F(\xi, X_1) \leqslant |F(\xi, X_1) - F(\xi, X)| + F(\xi, X) < \varepsilon + \lambda$$

for $\rho(X, X_1) < \delta$, $X \in K(\xi, \lambda)$, and $\xi \in D$. Since ε can be arbitrarily small, it follows that $F(\xi, X_1) \leqslant \lambda$, and this means that the cluster point X_1 of the set $K(\xi, \lambda)$ belongs to this set; that is, $K(\xi, \lambda)$ is a closed set. We note that all that has been said applied to an arbitrary fixed element $\xi \in D$.

5.2.2 *The least deviation from zero of a parameter-dependent function that is defined on sub-sets of a basic set*

We note first of all that if D' is a non-empty sub-set of a set D, then

$$\mu(D') \leqslant \mu(D), \quad D' \subset D \tag{1}$$

This follows immediately from the definition of μ and the inequality

$$\sup_{\xi \in D'} F(\xi, X) \leqslant \sup_{\xi \in D} F(\xi, X), \quad X \in \mathfrak{R}_n$$

We denote by $\mathfrak{M}_k$, for $k = 1, 2, \ldots$, the collection of all possible sets D_k consisting of k elements $\xi_1, \ldots, \xi_k$ (some of which may be identical) belonging to the set D. We denote by $\mu_k = \mu_k(D)$ the least upper bound of the set of least deviations from zero of the function $F(\xi, X)$ defined on all possible sets $D_k \in \mathfrak{M}_k$:

$$\mu_k = \mu_k(D) = \sup_{D_k \in \mathfrak{M}_k} \mu(D_k) = \sup_{D_k \in \mathfrak{M}_k} \inf_{X \in \mathfrak{R}_n} \sup_{\xi \in D_k} F(\xi, X)$$

Theorem 1 $\mu_k \leqslant \mu_{k+1} \leqslant \mu$, for $k = 1, 2, \ldots$

PROOF For arbitrary $D_k \in \mathfrak{M}_k$, we have $\mu(D_k) \leqslant \mu(D)$ and consequently, $\mu_k(D) \leqslant \mu(D)$. Let D_k be an arbitrary set in $\mathfrak{M}_k$. Since there exists a set $D_{k+1}^{(0)}$ in $\mathfrak{M}_{k+1}$ such that $D_k \subset D_{k+1}^{(0)}$.

we have

$$\mu(D_k) \leqslant \sup_{D_{k+1} \in \mathfrak{M}_{k+1}} \mu(D_{k+1}) = \mu_{k+1}(D) \leqslant \mu(D)$$

and, consequently,

$$\mu_k(D) = \sup_{D_k \in \mathfrak{M}_k} \mu(D_k) \leqslant \mu_{k+1}(D) \leqslant \mu(D)$$

which completes the proof of the theorem.

NOTE 1 It follows from the definition of μ_k that, for arbitrary $D_k \in \mathfrak{M}_k$

$$\inf_{X \in \mathfrak{R}_n} \sup_{\xi \in D_k} F(\xi, X) \leqslant \mu_k$$

and, consequently, for arbitrary $\lambda > \mu_k$, there exists a point $X_0 \in \mathfrak{R}_n$ such that

$$\sup_{\xi \in D_k} F(\xi, X_0) \leqslant \lambda$$

that is, $X_0 \in K(\xi, \lambda)$ for $\xi \in D_k$. Therefore,

$$X_0 \in K(D_k, \lambda) = \bigcap_{\xi \in D_k} K(\xi, \lambda)$$

It follows from this that for an arbitrary choice of $D_k \in \mathfrak{M}_k$ the sets $K(D_k, \lambda)$ are non-empty for $\lambda > \mu_k$. For $k=1$, it follows from the considerations made above that the sets $K(\xi, \lambda)$ are non-empty for $\lambda > \mu_1$.

In Theorem 3 we shall give sufficient conditions to impose on $F(\xi, X)$ for $\mu_{n+1}(D)$ to be equal to $\mu(D)$.

First we shall state without proof a theorem regarding the intersection of convex sets in $\mathfrak{R}_n$.

Theorem 2 (Khelli [1]) If the collection K of closed convex sets of points $X \in \mathfrak{R}_n$ contains at least $n+1$ sets (some of which may be the same), if the intersection of $n+1$ arbitrarily chosen sets in K is non-empty, and if the intersection of some finite number of sets in K is bounded, then the intersection of all sets belonging to K is non-empty.

By using what was said above, we shall prove the following theorem.

Theorem 3 (Shnirel'man [1]*)* If there exists a $\lambda_0 > \mu_{n+1}$ such that, for all $\xi \in D$ and arbitrary λ such that $\mu_{n+1} < \lambda < \lambda_0$, the set $K(\xi, \lambda)$ is closed and convex and the intersection of some finite number of sets $K(\xi, \lambda)$ is bounded, then $\mu_{n+1}(D) = \mu(D)$.

PROOF According to Note 1, the sets

$$K(D_{n+1}, \lambda) = \bigcap_{\xi \in D_{n+1}} K(\xi, \lambda)$$

are non-empty and convex for arbitrary $D_{n+1} \in \mathfrak{M}_{n+1}$, and the intersection of some finite number of sets $K(\xi, \lambda)$ is, from the hypothesis of the theorem, bounded (since $K(\xi, \lambda) \subset K(\xi, \lambda_0)$). Therefore, on the basis of Theorem 2, the intersection $K(\lambda)$ of all sets $K(\xi, \lambda)$ for $\xi \in D$ is non-empty and $K(\lambda)$ is a closed convex bounded set.

Let λ_j (for $j = 1, 2, \ldots$), where $\mu_{n+1} < \lambda_j < \lambda_0$, be a decreasing sequence of numbers that approaches μ_{n+1}. Then, $K(\lambda_{j+1}) \subset K(\lambda_j)$, and the intersection

$$K = \bigcap_{j=1}^{\infty} K(\lambda_j)$$

is a non-empty closed convex set. Choose $X_0 \in K$. Then, $X_0 \in K(\lambda_j)$ (for $j = 1, 2, \ldots$) and, consequently, $X_0 \in K(\xi, \lambda_j)$ that is, $F(\xi, X_0) \leqslant \lambda_j$ for $\xi \in D$. If we take the limit as $j \to \infty$ we obtain $F(\xi, X_0) \leqslant \mu_{n+1}$ for $\xi \in D$. From this, by using the definition of μ, we arrive at the inequality

$$\mu \leqslant \sup_{\xi \in D} F(\xi, X_0) \leqslant \mu_{n+1} \tag{2}$$

But, according to Theorem 1, $\mu_{n+1} \leqslant \mu$; consequently, $\mu_{n+1} = \mu$. This completes the proof of the theorem.

NOTE 2 It follows from Formula (2) and the inequality $\mu_{n+1} = \mu$ that, for every point $X' \in K$

$$\sup_{\xi \in D} F(\xi, X') = \mu \tag{3}$$

that is, $X' \in K(\xi, \mu)$ for $\xi \in D$; consequently, $X' \in K(\mu)$. Thus, $K(\mu)$ is non-empty and $K \subset K(\mu)$. It then follows from the definition of μ that $K(\mu)$ is the set of all points X' for which Equation (3) is satisfied. Therefore, $K(\mu) \subset K$, which together with what was said above, yields $K = K(\mu)$.

NOTE 3 It is easily proven that

$$K(\xi, \mu) = \bigcap_{j=1}^{\infty} K(\xi, \lambda_j)$$

and from this it follows that $K(\xi, \mu)$ for $\xi \in D$ is a closed convex set. The same is obviously true of

$$K(D', \mu) = \bigcap_{\xi \in D'} K(\xi, \mu)$$

for $D' \subset D$. Here,

$$K(D, \mu) = K(\mu) \subset K(D', \mu) \subset K(\xi, \mu), \quad \xi \in D' \subset D$$

We shall prove one more theorem, making certain hypotheses with regard to D and the function $F(\xi, X)$.

Let us suppose that D is a non-empty compact sub-set of some complete metric space or, what amounts to the same thing, that D is a compact metric space. We shall say that a function $f(\xi)$, for $\xi \in D$ is continuous on D if, for every element $\xi_0 \in D$ and arbitrary $\varepsilon > 0$, there exists a $\delta > 0$ such that $|f(\xi) - f(\xi_0)| < \varepsilon$ when $\rho(\xi, \xi_0) < \delta$, where $\rho(\xi_1, \xi_2)$ is the distance between two elements ξ_1 and ξ_2 of the metric space in question.

We note that the real metric space $\mathfrak{R}_n$ and the complex metric space $\tilde{\mathfrak{R}}_n$ are complete metric spaces with respect to the definition given above for $\rho(X_1, X_2)$ and that all bounded closed sets of points in $\tilde{\mathfrak{R}}_n$ or $\mathfrak{R}_n$ are compact sub-sets of $\mathfrak{R}_n$ or $\tilde{\mathfrak{R}}_n$.

Theorem 4 If D is a set of the type mentioned above, if $F(\xi, X)$, where $\xi \in D$ and $X \in \mathfrak{R}_n$, is, for every fixed X, a continuous function of ξ, and if p is an arbitrary natural number, then there exists a set $D_m = \{\xi_l\}_1^m$, where $1 \leqslant m \leqslant p$, of distinct elements $\xi_l \in D$ such that $\mu(D_m) = \mu_p(D)$.

PROOF If the set D has no more than p elements, the assertion of the theorem is obvious, so that we do not need to consider this case further. It follows from the definition of $\mu_p(D)$ that there exists a sequence of sets $D_{p,\nu} = \{\xi_{l,\nu}\}_{l=1}^p$ of elements in D such that

$$\mu_{p,\nu} = \inf_{X \in \mathfrak{R}_n} \sup_{\xi \in D_{p,\nu}} F(\xi, X) \to \mu_p(D) \tag{4}$$

as $\nu \to \infty$

We choose a sequence ν_j of indices ν such that the limits $\lim\limits_{j\to\infty} \xi_{l,\nu_j} = \xi_l$, exist for $l = 1, 2, \ldots, p$. Since D is compact

$\xi_l \in D$ for $l=1, 2, \ldots, p$. Let us set $D_m = \{\xi_l\}_1^p$, where m is the number of distinct elements ξ_l, for $l=1, 2, \ldots, p$. Clearly, $1 \leqslant m \leqslant p$. It follows from the definition

$$\mu(D_m) = \inf_{X \in \mathfrak{R}_n} \sup_{\xi \in D_m} F(\xi, X)$$

that $\mu(D_m) \leqslant \mu_p(D)$. Therefore, to prove the theorem, we need only prove the reverse inequality $\mu_p(D) \leqslant \mu(D_m)$. For arbitrary $\varepsilon > 0$, there exists a point $X_0 \in \mathfrak{R}_n$ such that

$$\sup_{\xi \in D_m} F(\xi, X_0) \leqslant \mu(D_m) + \frac{\varepsilon}{2}$$

For sufficiently large j and for $l=1, 2, \ldots, p$, we have, by virtue of the continuity of $F(\xi, X_0)$ on D

$$F(\xi_{l, \nu_j}, X_0) \leqslant |F(\xi_{l, \nu_j}, X_0) - F(\xi_l, X_0)| + F(\xi_l, X_0) \leqslant \frac{\varepsilon}{2} + \mu(D_m) + \frac{\varepsilon}{2}$$

and for such values of j,

$$\mu_{p, \nu_j} \leqslant \sup_{\xi \in D_{p, \nu_j}} F(\xi, X_0) \leqslant \mu(D_m) + \varepsilon$$

that is,

$$\mu_{p, \nu_j} \leqslant \mu(D_m) + \varepsilon$$

By virtue of (4) we then have in the limit

$$\mu_p(D) \leqslant \mu(D_m)$$

This completes the proof of the theorem.

Obviously, this theorem remains valid in the case of a complex space $\tilde{\mathfrak{R}}_n$, which should be considered as the real space $\mathfrak{R}_{2n}$, of which we spoke earlier.

NOTE 4 This theorem will be used later for $p = n+1$.

5.3 THE BEST UNIFORM APPROXIMATION OF A FUNCTION BY MEANS OF GENERALISED POLYNOMIALS

5.3.1 *Statement of the problem*

The first statement of the problem of the best uniform approximation of a function by means of generalised polynomials consists in the following: suppose that B is a non-empty set of arbitrary elements z and that $\zeta_j = \varphi_j(z)$ (for $j = 0, 1, \ldots, n$) are complex functions defined on B that assume only finite values on B. We shall say that the functions of

$$p(z, \widetilde{X}) = \sum_{j=1}^{n} c_j \varphi_j(z), \quad z \in B, \quad \widetilde{X}(c_1, \ldots, c_n) \in \widetilde{\mathfrak{R}}_n$$

is a generalised polynomial or simply a polynomial when this is not likely to lead to confusion. We shall refer to the set of all such polynomials as the class $\mathscr{P}_n$. We set

$$F_*(z, \widetilde{X}) = |\varphi_0(z) - p(z, \widetilde{X})|, \quad z \in B, \quad \widetilde{X} \in \widetilde{\mathfrak{R}}_n$$

We shall refer to the least deviation of the function $F_*(z, \widetilde{X})$ from zero on B

$$\mu = \mu(B) = \inf_{\widetilde{X} \in \widetilde{\mathfrak{R}}_n} \sup_{z \in B} |\varphi_0(z) - p(z, \widetilde{X})| \tag{1}$$

as the measure of best approximation of the function $\varphi_0(z)$ by means of the generalised polynomials $p(z, \widetilde{X}) \in \mathscr{P}_n$ on B.

We now consider the functions $F_*(z, \widetilde{X})$, for which $\mu < +\infty$ and for which, consequently, there exists a point $\widetilde{X}_0 \in \widetilde{\mathfrak{R}}_n$ such that the function $|\varphi_0(z) - p(z, \widetilde{X}_0)|$ is bounded on B. We shall say that the polynomial $\pi(z) = p(z, \widetilde{X}_0)$ that satisfies the condition

$$\sup_{z \in B} |\varphi_0(z) - \pi(z)| = \mu \tag{2}$$

is a polynomial of best approximation of the function $\varphi_0(z)$ on B by means of the generalised polynomials $p(z, \widetilde{X}) \in \mathscr{P}_n$.

In what follows, we shall investigate the properties of $\pi(z)$ and, in particular, the existence and uniqueness theorems. We shall also seek sufficient conditions for a given polynomial to be a polynomial of best approximation.

We now give a second statement of the problem. In addition to the space $\widetilde{\mathfrak{R}}_n$ of n-tuples of complex numbers $\widetilde{X}(c_1, \ldots, c_n)$, we introduce the space $\widetilde{\mathfrak{R}}_{n+1}$ of n $(+1)$ tuples $\xi(\zeta_0, \zeta_1, \ldots, \zeta_n)$. We shall refer to the linear form of the coordinates $\zeta_1, \ldots, \zeta_n$ of points $\xi \in \widetilde{\mathfrak{R}}_{n+1}$

$$P(\xi, \widetilde{X}) = \sum_{j=1}^{n} c_j \zeta_j; \quad \zeta \in \widetilde{\mathfrak{R}}_{n+1}, \quad \widetilde{X} \in \widetilde{\mathfrak{R}}_n$$

as a generalised polynomial or simply as a polynomial and we shall refer to the numbers c_j as its coefficients. Suppose that D is a non-empty set of points $\xi(\zeta_0, \zeta_1, \ldots, \zeta_n) \in \widetilde{\mathfrak{R}}_{n+1}$. We introduce the function

$$F(\xi, \widetilde{X}) = |\zeta_0 - P(\xi, \widetilde{X})|, \quad \xi \in D, \quad \widetilde{X} \in \widetilde{\mathfrak{R}}_n$$

We shall call the quantity

$$\mu = \mu(D) = \inf_{\widetilde{X} \in \widetilde{\mathfrak{R}}_n} \sup_{\xi \in D} F(\xi, \widetilde{X}) \tag{3}$$

the measure of best approximation on D and we shall refer to the polynomials $\pi(\xi) = P(\xi, \widetilde{X}_0)$, where $\widetilde{X}_0 \in \widetilde{\mathfrak{R}}_n$ that satisfies the condition

$$\sup_{\xi \in D} F(\xi, \widetilde{X}_0) = \sup_{\xi \in D} |\zeta_0 - \pi(\xi)| = \mu. \tag{4}$$

as the polynomials of best approximation on D.

The problem of best approximation on the set D of points $\xi(\zeta_0, \zeta_1, \ldots, \zeta_n) \in \widetilde{\mathfrak{R}}_{n+1}$ reduces to the original statement if the coordinates $\zeta_0, \zeta_1, \ldots, \zeta_n$ of points $\xi \in D$ are regarded as functions of ξ: $\zeta_j = \zeta_j(\xi)$, for $j = 0, 1, \ldots n$ defined on the set of points $\xi \in D$. Then,

$$P(\xi, \widetilde{X}) = \sum_{j=1}^{n} c_j \zeta_j(\xi), \quad \xi \in D, \quad \widetilde{X}(c_1, \ldots, c_n) \in \widetilde{\mathfrak{R}}_n$$

and

$$F(\xi, X) = |\zeta_0(\xi) - P(\xi, \widetilde{X})|, \quad \xi \in D, \quad \widetilde{X} \in \widetilde{\mathfrak{R}}_n$$

The original statement of the problem of best approximation of the function $\varphi_0(z)$ on B by means of the generalised polynomials $p(z, \widetilde{X}) \in \mathscr{P}_n$ reduces to the problem of best approximation on D if we take as the set D the set of points $\xi(\varphi_0(z), \ldots, \varphi_n(z)) \in \widetilde{\mathfrak{R}}_{n+1}$ corresponding to all possible $z \in B$. The generalisation of the original statement of the problem reduces to the problem of best approximation on D even in the case when the function $\varphi_0(z)$ is multiple-valued on B, that is, when to each $z \in B$ there may correspond not one but several or even infinitely many values of the function, in which case the approximation is made by means of single-valued functions.

We shall also consider the problem of best approximation

on the set D of points $\xi \in \widetilde{\Re}_{n+1}$, and we shall make certain remarks related to the problem of best approximation in its original formulation.

5.3.2 *The fundamental theorem*

In formulating the problem of best approximation on D, we have not required that the set D be bounded; that is, we have not required that a number $M > 0$ exist such that $|\zeta_j| \leqslant M$, where $j = 0, 1, \ldots, n$, for an arbitrary point

$$\xi(\zeta_0, \ldots, \zeta_n) \in D$$

It can be shown that the problem of best approximation in the case in which D is unbounded can be reduced to the problem of best approximation on some other set D^* that is bounded. In what follows, we assume that the set D is bounded.

We shall say that points $\xi \in D$ are linearly independent on D if the identity

$$c_0^{(0)}\zeta_0 + c_1^{(0)}\zeta_1 + \ldots + c_n^{(0)}\zeta_n \equiv 0, \quad \xi(\zeta_0, \ldots, \zeta_n) \in D \tag{1}$$

where $c_j^{(0)}$, for $j = 0, 1, \ldots, n$, are non-negative integers, is possible if and only if all the $c_j^{(0)} = 0$.

Suppose that the identity (1) holds and that $c_0^{(0)} \neq 0$, so that, consequently,

$$\zeta_0 \equiv -\frac{c_1^{(0)}}{c_0^{(0)}}\zeta_1 - \frac{c_2^{(0)}}{c_0^{(0)}}\zeta_2 - \ldots - \frac{c_n^{(0)}}{c_0^{(0)}}\zeta_n, \quad \xi(\zeta_0, \ldots, \zeta_n) \in D$$

Then, $\mu(D) = 0$ and $\pi(\xi) = P(\xi, \widetilde{X}_0)$, where

$$\widetilde{X}_0\left(-\frac{c_1^{(0)}}{c_0^{(0)}}, \ldots, -\frac{c_n^{(0)}}{c_0^{(0)}}\right)$$

is a polynomial of best approximation on D. We also assume that there exist numbers $d_1, d_2, \ldots, d_n$, such that

$$\zeta_0 \equiv d_1\zeta_1 + d_2\zeta_2 + \ldots + d_n\zeta_n, \quad \xi(\zeta_0, \zeta_1, \ldots, \zeta_n) \in D$$

Let us introduce the concept of the rank of the set D. Let $D_n = \{\xi_j\}_1^n$ an arbitrary set of points (not necessarily distinct), so $\xi_j(\zeta_0^{(j)}, \zeta_1^{(j)}, \ldots, \zeta_n^{(j)}) \in D$, where $j = 1, 2, \ldots, n$

and $D_n \in \mathfrak{M}_n$. We denote by $r(D_n)$ the rank of the matrix

$$A(D_n) = \left\| \begin{matrix} \zeta_1^{(1)}, & \zeta_2^{(1)}, & \ldots, & \zeta_n^{(1)} \\ \zeta_1^{(2)}, & \zeta_2^{(2)}, & \ldots, & \zeta_n^{(2)} \\ \cdot & \cdot & \cdot & \cdot \\ \zeta_1^{(n)}, & \zeta_2^{(n)}, & \ldots, & \zeta_n^{(n)} \end{matrix} \right\|$$

We shall call the number $r = \sup_{D_n \in \mathfrak{M}_n} r(D_n)$ the rank of the set D. Obviously, $0 \leqslant r \leqslant n$ and, if $r = 0$, the coordinates $\zeta_1, \ldots, \zeta_n$ of all the points $\xi \in D$ are equal to zero and the problem of best approximation is of no interest. Let us suppose that $r \geqslant 1$. Then, there exists a system of points $\xi_j \in D$, for $j = 1, 2, \ldots, r$, such that, for example, the determinant

$$\Delta_r = \begin{vmatrix} \zeta_1^{(1)}, & \zeta_2^{(1)}, & \ldots, & \zeta_r^{(1)} \\ \zeta_1^{(2)}, & \zeta_2^{(2)}, & \ldots, & \zeta_r^{(2)} \\ \cdot & \cdot & \cdot & \cdot \\ \zeta_1^{(r)}, & \zeta_2^{(r)}, & \ldots, & \zeta_r^{(r)} \end{vmatrix} \neq 0$$

but, for every point $\xi(\zeta_0, \zeta_1, \ldots, \zeta_n) \in D$, the determinants

$$\Delta_r^{(\nu)} = \begin{vmatrix} \zeta_1^{(1)}, & \zeta_2^{(1)}, & \ldots, & \zeta_r^{(1)}, & \zeta_\nu^{(1)} \\ \cdot & \cdot & \cdot & \cdot & \cdot \\ \zeta_1^{(r)}, & \zeta_2^{(r)}, & \ldots, & \zeta_r^{(r)}, & \zeta_\nu^{(r)} \\ \zeta_1, & \zeta_2, & \ldots, & \zeta_r, & \zeta_\nu \end{vmatrix} = 0, \quad \nu = r+1, r+2, \ldots, n$$

If we expand these determinants according to the elements of the last row, we see that the identities

$$\zeta_\nu = d_1^{(\nu)}\zeta_1 + d_2^{(\nu)}\zeta_2 + \ldots + d_r^{(\nu)}\zeta_r, \quad \nu = r+1, \ldots, n, \quad \xi(\zeta_0, \ldots, \zeta_n) \in D \qquad (2)$$

are valid for all points $\xi \in D$. It is easy to see that, if these identities hold, the set D is of rank not exceeding r. Here, we note that there exists a point $\xi_0 = \xi(\zeta_0^{(0)}, \ldots, \zeta_n^{(0)}) \in D$ such that the determinant

$$\Delta_r^{(0)} = \begin{vmatrix} \zeta_0^{(0)}, & \zeta_1^{(0)}, & \ldots, & \zeta_r^{(0)} \\ \zeta_0^{(1)}, & \zeta_1^{(1)}; & \ldots, & \zeta_r^{(1)} \\ \cdot & \cdot & \cdot & \cdot \\ \zeta_0^{(r)}, & \zeta_1^{(r)}, & \ldots, & \zeta_r^{(r)} \end{vmatrix} \neq 0 \qquad (3)$$

In the opposite case, it is easy to show that the identity

$$\zeta_0 = d_1^{(0)}\zeta_1 + d_2^{(0)}\zeta_2 + \ldots + d_r^{(0)}\zeta_r, \quad \xi(\zeta_0, \ldots, \zeta_n) \in D$$

is valid.

It follows from what has been said that if a set is of rank r, it must contain at least $r+1$ points. Suppose that a set D is of rank r and that the identities (2) hold. We introduce the set D^* of points $\xi^*(\zeta_0, \zeta_1, \ldots, \zeta_r) \in \widetilde{\mathfrak{R}}_{r+1}$ such that a point $\xi(\zeta_0, \zeta_1, \ldots, \zeta_n) \in D$ exists and we assign to each point $\widetilde{X}(c_1, \ldots, c_n) \in \widetilde{R}_n$ the point $\widetilde{X}^*(c_1^*, \ldots, c_r^*) \in \widetilde{\mathfrak{R}}_r$, where

$$c_j^* = c_j + \sum_{\nu=r+1}^{n} c_\nu d_j^{(\nu)}$$

(for $j = 1, 2, \ldots, r$). Then,

$$F(\xi, \widetilde{X}) = \left|\zeta_0 - \sum_{j=1}^{n} c_j\zeta_j\right| = \left|\zeta_0 - \sum_{j=1}^{r} c_j\zeta_j - \sum_{\nu=r+1}^{n} c_\nu \sum_{j=1}^{r} d_j^{(\nu)}\zeta_j\right|$$

$$= \left|\zeta_0 - \sum_{j=1}^{r} c_j^*\zeta_j\right| = F(\xi^*, \widetilde{X}^*), \quad \xi^* \in D^*, \quad \widetilde{X}^* \in \widetilde{\mathfrak{R}}_r$$

It follows from this that $\mu(D) = \mu(D^*)$ and, if $\pi^*(\xi^*)$ is a polynomial of best approximation on D^*, then every polynomial

$$\pi(\xi) = \pi^*(\xi^*) + \sum_{\nu=r+1}^{n} A_\nu \left(\sum_{j=1}^{r} d_j^{(\nu)}\zeta_j - \zeta_\nu\right)$$

where the A_ν (for $\nu = r+1, \ldots, n$), are arbitrary constants, is a polynomial of best approximation on D. By using all polynomials of best approximation on D^*, we obtain all polynomials of best approximation on D. As in Section 5.2, we denote by $\mathfrak{M}_k$, for $k = 1, 2, \ldots$, the set of all possible D_k, consisting of k (not necessarily distinct) points belonging to the set D of points $\xi \in \widetilde{\mathfrak{R}}_{n+1}$, and we set

$$\mu_k = \mu_k(D) = \sup_{D_k \in \mathfrak{M}_k} \mu(D_k), \quad \mu(D') = \inf_{\widetilde{X} \in \widetilde{\mathfrak{R}}_n} \sup_{\xi \in D'} F(\xi, X), \quad D' \subset D$$

$$\mu'(D') = \inf_{X \in \mathfrak{R}_n} \sup_{\xi \in D'} F(\xi, X), \quad D' \subset D \tag{4}$$

$$\mu'_k = \mu'_k(D) = \sup_{D_k \in \mathfrak{M}_k} \mu'(D_k) \tag{5}$$

Obviously, $\mu_k \leqslant \mu'_k$. However, if all the coordinates of the points $\xi \in D$ are real, $\mu_k = \mu'_k$ since, by assigning to an arbitrary point $X(a_1+b_1i, \ldots, a_n+b_ni) \in \tilde{\mathfrak{R}}_n$ the point $X(a_1, \ldots, a_n) \in \mathfrak{R}_n$, we have

$$F(\xi, \tilde{X}) = \left| \zeta_0 - \sum_{j=1}^{n} (a_j + b_j i) \zeta_j \right|$$

$$= \sqrt{\left(\zeta_0 - \sum_{j=1}^{n} a_j\zeta_j\right)^2 + \left(\sum_{j=1}^{n} b_j\zeta_j\right)^2} \geqslant \left|\zeta_0 - \sum_{j=1}^{n} a_j\zeta_j\right| = F(\xi, X)$$

Theorem 1 (the fundamental theorem on best approximation). Suppose that D is a non-empty bounded set of points $\xi \in \mathfrak{R}_{n+1}$ of rank r, where $1 \leqslant r \leqslant n$. Then: (1) there exists a polynomial $\pi(\xi) = P(\xi, \tilde{X}_0)$ where $\tilde{X}_0 \in \tilde{\mathfrak{R}}_n$, of best approximation on D (see Formula (4) in Section 5.3.1); (2) $\mu = \mu_{2r+1}$ (see Formula (3) of Section 5.3.1 and Formula (4)); (3) if D is a closed bounded set, there exist a set $D_m = \{\xi_l\}_1^m$, where $1 \leqslant m \leqslant 2r+1$, of distinct points $\xi_l \in D$ such that $\mu(D_m) = \mu$; (4) if all coordinates $\zeta_0, \zeta_1, \ldots, \zeta_n$ of the points $\xi(\zeta_0, \zeta_1, \ldots, \zeta_n) \in D$ are real, then (a) a polynomial of best approximation on D has only real coefficients: $\pi(\xi) = P(\xi, X_0)$, where $X_0 \in \mathfrak{R}_n$; (b) $\mu = \mu_{r+1} = \mu'_{r+1}$ (see Formula (5)); (c) if D is a closed bounded set, there exists a set $D_m = \{\xi_l\}_1^m$, where $1 \leqslant m \leqslant r+1$, such that $\mu(D_m) = \mu$.

PROOF Suppose that the identities (2) hold. As above, we introduce the set D^* of points $\xi^*(\zeta_0, \zeta_1, \ldots, \zeta_r) \in \tilde{\mathfrak{R}}_{r+1}$ such that the point $\xi(\zeta_0, \zeta_1, \ldots, \zeta_n) \in D$, and we consider the function $F^*(\xi^*, \tilde{X}) = |\zeta_0 - P(\xi^*, X)|$, where $\xi^*(\zeta_0, \zeta_1, \ldots, \zeta_r) \in D^*$ and $\tilde{X} \in \tilde{\mathfrak{R}}_r$. In the general case, this function depends on $2r$ real parameters. If the coordinates $\zeta_0, \zeta_1, \ldots, \zeta_r$ of the point $\xi^* \in D^*$ are all real, let us consider the function $F^*(\xi^*, X)$, where $\xi^* \in D^*$ and $X \in \mathfrak{R}_r$, which depends on r real parameters. Now we consider only the general case (the case in which the coordinates are real is treated analogously). Let us show that in the present case all the conditions of Theorem 1 of Section 5.2.1 and Theorems 3 and 4 of Section 5.2.2 are satisfied. Obviously, the function $F^*(\xi^*, \tilde{X})$ is continuous with respect to $\tilde{X} \in \tilde{\mathfrak{R}}_r$ for every fixed $\xi^* \in D^*$ and, consequently, the sets $K(\xi^*, \lambda)$, for $\lambda \geqslant 0$, are closed (see Section 5.2.1). Let us show that these sets are bounded (they may in fact be empty). Suppose that $\tilde{X}(c_1, \ldots, c_n) \in K(\xi^*, \lambda)$. Since we assume that the relation

$$\zeta_0 = d_1\zeta_1 + d_2\zeta_2 + \ldots + d_n\zeta_n, \quad \xi(\zeta_0, \zeta_1, \ldots, \zeta_n) \in D$$

does not hold, there exist points

$$\xi_\nu^*(\zeta_0^{(\nu)}, \zeta_1^{(\nu)}, \ldots, \zeta_r^{(\nu)}) \in D^* \qquad \nu = 0, 1, \ldots, r$$

such that the determinant $\Delta_r^{(0)}$ (3) is non-zero. If we treat the equations

$$\begin{aligned}
\zeta_0^{(0)} - c_1\zeta_1^{(0)} - \ldots - c_r\zeta_r^{(0)} &= \varphi(\xi_0^*, \tilde{X}) \\
\zeta_0^{(1)} - c_1\zeta_1^{(1)} - \ldots - c_r\zeta_r^{(1)} &= \varphi(\xi_1^*, \tilde{X}) \\
\cdots\cdots\cdots\cdots\cdots\cdots \\
\zeta_0^{(r)} - c_1\zeta_1^{(r)} - \ldots - c_r\zeta_r^{(r)} &= \varphi(\xi_r^*, \tilde{X})
\end{aligned}$$

where $\varphi(\xi^*, X) = \zeta_0 - P(\xi^*, \tilde{X})$, as a system of equations in $1, c_1, c_2, \ldots, c_r$ and apply Cramer's rule, we obtain the estimate

$$|c_j| \leqslant (r+1)\frac{\delta\lambda}{|\Delta_r^{(0)}|} \qquad j = 1, 2, \ldots, r$$

where δ is the largest of the absolute values of the minors of the r th-order determinant $\Delta_r^{(0)}$. It follows from this estimate that the set $K(\xi^*, \lambda)$ is bounded. Let us show, for example, that the set $K(\xi^* \lambda)$ is convex if it is not empty. We need to show that if the points $\tilde{X}_1$ and $\tilde{X}_2$ belong to $K(\xi^*, \lambda)$, that is, if $F^*(\xi^*, \tilde{X}_1) \leqslant \lambda$ and $F^*(\xi^*, \tilde{X}_2) \leqslant \lambda$ for all $\xi^* \in D^*$, the point $\tau\tilde{X}_1 + (1-\tau)\tilde{X}_2 \in K(\xi^*, \lambda)$ for every τ such that $0 < \tau < 1$. We have

$$\begin{aligned}
F^*(\xi^*, \tau\tilde{X}_1 + (1-\tau)\tilde{X}_2) &= |\zeta_0 - P(\xi^*, \tau\tilde{X}_1 + (1-\tau)\tilde{X}_2)| \\
&= |\tau[\zeta_0 - P(\xi^*, \tilde{X}_1)] + (1-\tau)[\zeta_0 - P(\xi^*, \tilde{X}_2)]| \\
&\leqslant \tau F^*(\xi^*, \tilde{X}_1) + (1-\tau)F^*(\xi^*, \tilde{X}_2) \leqslant \tau\lambda + (1-\tau)\lambda = \lambda
\end{aligned}$$

Thus, all the conditions of Theorem 1, Section 5.2.1, and Theorem 3, Section 5.2.2, are satisfied for a function $F^*(\xi^*, \tilde{X})$, where $\xi^* \in D^*$ and $\tilde{X} \in \mathfrak{R}_r$. On the basis of Theorem 1, there exists a point $\tilde{X}_0^*(c_1^{(0)}, \ldots, c_r^{(0)}) \subset \mathfrak{R}_r$ such that

$$\sup_{\xi^* \in D^*} F^*(\xi^*, \tilde{X}_0^*) = \sup_{\xi^* \in D^*} |\zeta_0 - P(\xi^*, \tilde{X}_0^*)| = \mu(D^*)$$

and the polynomial

$$\pi^*(\xi^*) = P(\xi^*, \tilde{X}_0^*) = \sum_{j=1}^{r} c_j^{(0)}\zeta_j$$

is a polynomial of best approximation on the set D^*.

It follows from what was said above that $\mu = \mu(D) = \mu(D^*)$ and that the polynomial $\pi^*(\xi)$ for $\xi \in D$ is a polynomial of best approximation on D. On the basis of Theorem 3, Section 5.2.2, $\mu_{2r+1}(D^*)=\mu(D^*)$ and, from the above, $\mu_{2r+1}(D)=\mu_{2r+1}(D^*)$; that is, $\mu_{2r+1}=\mu$. If D is a closed and bounded set, the set D^* must also be closed and bounded and, according to Theorem 4, Section 5.2.2, there exists a set $D_m^*=\{\xi_l^*\}_1^m$ where $1\leqslant\mu\leqslant 2r+1$, of points $\xi_l^*\in D^*$ such that

$$\mu(D_m^*)=\mu(D^*)=\mu,$$

also every set $D_m=\{\xi_l\}_1^m$ of points $\xi_l(\zeta_0^{(l)},\zeta_1^{(l)},\ldots,\zeta_n^{(l)})\in D$ such that $\zeta_k^{(l)}=\zeta_k^*$ (for $k=0,1,\ldots,r$) possesses the same property as does $\mu(D_m)=\mu(D_m^*)=\mu$. This completes the proof of the theorem.

NOTE 1 The set $K(\mu)=K(D,\mu)$ of points $\tilde{X}\in\tilde{\Re}_n$ such that $P(\xi,\tilde{X})$ is a polynomial of best approximation on D is a closed convex set in $\Re_{2n}$ (or in $\tilde{\Re}_n$) (see Note 2 to Theorem 3, Section 5.2.2). This remains valid for the set $K(D_m,\mu)$ of points $\tilde{X}\in\tilde{\Re}_n$ such that $P(\xi,\tilde{X})$ is a polynomial of best approximation on D_m; also, $K(\mu)\subset K(D_m,\mu)$. It follows from this last statement that, if there exists a unique polynomial of best approximation on D_m, this polynomial is the unique polynomial of best approximation on D. An analogous assertion holds for the case in which all the coordinates of the points $\xi\in D$ are real numbers.

NOTE 2 Assertion (3) of the theorem stipulates $1\leqslant m\leqslant 2r+1$. For arbitrary m such that $1\leqslant m\leqslant 2r+1$, we can exhibit a set D of rank r for which assertion (3) of the theorem will be satisfied with that m. Analogous examples can also be exhibited for assertion (4c) of the theorem.

NOTE 3 If we do not specify the rank of the set D in the formulation of the theorem, the inequality $1\leqslant m\leqslant 2n+1$ will be satisfied for m in assertion (3) of the theorem and the inequality $1\leqslant m\leqslant n+1$ will be satisfied in assertion (4c).

Let us turn to the first statement of the problem of best uniform approximation of a function by means of generalised polynomials. We keep the notations used at the beginning of Section 5.3.1 and, as before, we set

$$\mu_k=\mu_k(B)=\sup_{B_k\in\mathfrak{M}_k}\mu(B_k),\quad \mu(B')=\inf_{\tilde{X}\in\tilde{\Re}_n}\sup_{z\in B'}F_*(z,\tilde{X}),\quad B'\subset B,$$

$$\mu'(B')=\inf_{X\in\Re_n}\sup_{z\in B'}F_*(z,X),\quad B'\subset B,\quad \mu_k'=\mu_k'(B)=\sup_{B_k\in\mathfrak{M}_k}\mu'(B_k),$$

where $\mathfrak{M}_k$, for $k=1, 2, \ldots$, is the collection of all possible sets B_k consisting of k (not necessarily distinct) elements in B. Obviously, $\mu_k \leqslant \mu_k'$, and if all the functions $\varphi_j(z)$, for $j=0, 1, \ldots, n$, are real, then $\mu_k = \mu_k'$. We note also that the set D of points $\xi(\varphi_0(z), \ldots, \varphi_n(z)) \in \mathfrak{R}_{n+1}$, for $z \in B$, will be a closed bounded set - for example, in the case in which B is compact and the functions $\varphi_j(z)$, for $j=0, 1, \ldots, n$, are continuous on B.

COROLLARY 1 Suppose that B is a non-empty set of arbitrary elements z, that $\zeta_i = \varphi_i(z)$, for $j=0, 1, \ldots, n$, are functions bounded on B, and that $\mathscr{P}_n$ is the class of generalised polynomials

$$p(z, \tilde{X}) = \sum_{j=1}^{n} c_j \varphi_j(z), \quad z \in B, \quad \tilde{X}(c_1, \ldots, c_n) \in \tilde{\mathfrak{R}}_n$$

Then: (1) there exists a polynomial $\pi(z) = p(z, \tilde{X}_0)$, where $\tilde{X}_0 \in \tilde{\mathfrak{R}}_n$ of best approximation for the function $\varphi_0(z)$ on B by means of generalised polynomials of the class $\mathscr{P}_n$ (see Formula (2)); (2) $\mu = \mu_{2n+1}$; (3) if B is compact and the functions $\varphi_j(z)$, for $j=0, 1, \ldots, n$, are continuous on B, there exists a set $B_m = \{z_l\}_1^m$ (where $1 \leqslant m \leqslant 2n+1$) of distinct elements $z_l \in B$ such that $\mu(B_m) = \mu$; (4) if all the functions $\varphi_j(z)$, for $j=0, 1, \ldots, n$ are real, then (a) a polynomial of best approximation $\varphi_0(z)$ on B has only real coefficients: $\pi(z) = p(z, X_0)$ for $X_0 \in \mathfrak{R}_n$; (b) $\mu = \mu_{n+1} = \mu_{n+1}'$; (c) if conditions (3) are satisfied, there exists a set $B_m = \{z_l\}_1^m$ (where $1 \leqslant m \leqslant n+1$), of distinct elements $z_l \in B$ such that $\mu(B_m) = \mu$.

NOTE 4 In Theorem 1, we may remove the requirement that the set D be bounded and, in Corollary 1, we may remove the requirement that the functions $\varphi_j(z)$ be bounded.

5.3.3 *Necessary and sufficient conditions for a given polynomial to be a polynomial of best approximation*

Suppose that D is a bounded set of points $\xi(\zeta_0, \zeta_1, \ldots, \zeta_n) \in \tilde{\mathfrak{R}}_{n+1}$ and that $\pi(\xi) = P(\xi, \tilde{X}_0)$, where $\tilde{X}_0(c_1^{(0)}, \ldots, c_n^{(0)}) \in \tilde{\mathfrak{R}}_n$ is a generalised polynomial. We set

$$\rho = \sup_{\xi \in D} F(\xi, \tilde{X}_0) = \sup_{\xi \in D} |\zeta_0 - \pi(\xi)| = \sup_{\xi \in \overline{D}} |\zeta_0 - \pi(\xi)|$$

and denote by $D(\rho)=D(\rho, \pi)$ the set of all points $\xi \in \bar{D}$ such that $|\zeta_0-\pi(\xi)|=\rho$. Obviously, $D(\rho)$ is a closed set.

Theorem 1 (Kolmogorov [1]*)* For a generalised polynomial $\pi(\xi)=P(\xi, \tilde{X}_0)$ to be a polynomial of best approximation on a bounded set D, it is necessary and sufficient that, for an arbitrary polynomial $P(\xi, \tilde{X})$, where $\tilde{X} \in \tilde{\mathfrak{R}}_n$, the inequality

$$\inf_{\xi \in D(\rho)} \operatorname{Re}\{\overline{P(\xi, \tilde{X})}(\zeta_0-\pi(\xi))\} \leqslant 0 \tag{1}$$

be satisfied.

PROOF: NECESSITY By hypothesis, $\pi(\xi)$ is a polynomial of best approximation on D. Let us suppose that Kolmogorov's condition is not satisfied, that is, that there exists a polynomial $P(\xi, \tilde{X})$ such that

$$\inf_{\xi \in D(\rho)} \operatorname{Re}\{\overline{P(\xi, \tilde{X})}(\zeta_0-\pi(\xi))\}=\varepsilon>0 \tag{2}$$

We denote by D' the set of all points $\xi \in \bar{D}$ such that

$$\operatorname{Re}\{\overline{P(\xi, \tilde{X})}(\zeta_0-\pi(\xi))\} \leqslant \frac{\varepsilon}{2}$$

Obviously, D' is a closed (possibly empty) set. By virtue of (2), D and $D(\rho)$ are disjoint; that is,

$$\rho'=\sup_{\xi \in D'} |\zeta_0-\pi(\xi)|<\rho$$

We define

$$h=\rho-\rho', \quad H=\sup_{\xi \in D} |P(\xi, \tilde{X})| \quad \lambda=\min\left\{\frac{\varepsilon}{2H^2}, \frac{h}{2H}\right\}$$
$$Q(\xi)=\pi(\xi)+\lambda P(\xi, \tilde{X})=P(\xi, \tilde{X}_0+\lambda \tilde{X})$$

On $\bar{D}-D'$, we have

$$\begin{aligned}
|\zeta_0-Q(\xi)|^2 &=|\zeta_0-\pi(\xi)-\lambda P(\xi, \tilde{X})|^2 \\
&=|\zeta_0-\pi(\xi)|^2+\lambda^2|P(\xi, \tilde{X})|^2-2\lambda \operatorname{Re}\{\overline{P(\xi, \tilde{X})}(\zeta_0-\pi(\xi))\}<\rho^2 \\
&\quad+\lambda^2 H^2-2\lambda \cdot \frac{\varepsilon}{2} \leqslant \rho^2+\lambda \frac{\varepsilon}{2H^2} H^2-\lambda \varepsilon=\rho^2-\lambda \frac{\varepsilon}{2}<\rho^2
\end{aligned}$$

Now if $\xi \in D'$, we have

$$\begin{aligned}
|\zeta_0-Q(\xi)| &=|\zeta_0-\pi(\xi)-\lambda P(\xi, \tilde{X})| \leqslant|\zeta_0-\pi(\xi)|+\lambda|P(\xi, \tilde{X})| \\
&\leqslant \rho'+\lambda H \leqslant \rho-h+\frac{h}{2H} H=\rho-\frac{1}{2} h<\rho
\end{aligned}$$

that is, on the set D, we have

$$F(\xi, \tilde{X}_0+\lambda\tilde{X})=|\zeta_0-Q(\xi)|\leqslant\rho''<\rho$$

which contradicts the hypothesis that $\pi(\xi)$ is a polynomial of best approximation. This proves the necessity.

SUFFICIENCY Let $Q(\xi)=P(\xi, \tilde{X})$, where $\tilde{X}\in\tilde{\mathfrak{N}}_n$ be a polynomial in the class $\mathscr{P}_n$. By virtue of Kolmogorov's condition, there exists a point $\xi\in D(\rho)$ such that

$$\operatorname{Re}\{\overline{P(\xi, \tilde{X}-\tilde{X}_0)}(\zeta_0-\pi(\xi))\}\leqslant 0$$

For this point ξ,

$$|\zeta_0-P(\xi, \tilde{X})|^2=|\zeta_0-P(\xi, \tilde{X}_0)-P(\xi, \tilde{X}-\tilde{X}_0)|^2$$

$$=|\zeta_0-\pi(\xi)|^2+|P(\xi, \tilde{X}-\tilde{X}_0)|^2$$

$$-2\operatorname{Re}\{\overline{P(\xi, \tilde{X}-\tilde{X}_0)}(\zeta_0-\pi(\xi))\}\geqslant|\zeta_0-\pi(\xi)|^2=\rho^2$$

Consequently, $\sup_{\xi\in D}|\zeta_0-P(\xi, \tilde{X})|\geqslant\rho$ and $\pi(\xi)$ is a polynomial of best approximation on D. This completes the proof of the theorem.

We shall call condition (1) Kolmogorov's condition. A special case of this theorem was proven by Tonelli [1].

COROLLARY 1 If $\pi(\xi)$ is a polynomial of best approximation on a bounded set D, it is a polynomial of best approximation on every set $D'\subset\bar{D}$ such that $D(\mu)\subset\bar{D}'$, in particular, on $D(\mu)$.

COROLLARY 2 If D is a closed bounded set, there exists in accordance with Theorem 1 a set $D_m=\{\xi_l\}_1^m$ where $1\leqslant m\leqslant 2n+1$, of points $\xi_l\in D$ such that $\mu(D_m)=\mu(D)=\mu$. Corollary 1 also permits us to assume that $D_m\subset D(\mu)$.

Let us return to the first statement of the problem of best uniform approximation of a function by means of generalised polynomials.

Suppose that B is compact and that the functions $\varphi_j(z)$, for $j=0, 1, \dots n$, are continuous on B. Let

$$\pi(z)=\sum_{j=1}^{n}c_j^{(0)}\varphi_j(z)$$

be a polynomial in the class $B(\rho)$. Let us set

$$\rho = \sup_{z \in B} |\varphi_0(z) - \pi(z)|$$

and let us denote by $B(\rho)$ the set of points $z \in B$ for which $|\varphi_0(z) - \pi(z)| = \rho$.

COROLLARY 3 For a polynomial $\pi(z)$ to be a polynomial of best approximation on B for $\varphi_0(z)$, it is necessary and sufficient that, for an arbitrary polynomial

$$p(z) = \sum_{j=1}^{n} c_j \varphi_j(z)$$

the inequality

$$\inf_{z \in B(\rho)} \operatorname{Re} \{ \overline{p(z)} (\varphi_0(z) - \pi(z)) \} \leqslant 0$$

be satisfied.

To prove this corollary, we introduce the set D of points $\xi(\varphi_0(z), \ldots, \varphi_n(z)) \in \widetilde{\mathfrak{R}}_{n+1}$, where $z \in B$, and the set $D(\rho)$ of points $\xi(\varphi_0(z), \ldots, \varphi_n(z)) \in \widetilde{\mathfrak{R}}_{n+1}$, where $z \in B(\rho)$. We then apply Theorem 1, which is possible in view of the conditions imposed on B and the functions $\varphi_j(z)$.

We note that the assertion of Corollary 3 remains in force if we impose less stringent conditions on the set B and the functions $\varphi_j(z)$, but we shall not do this.

EXAMPLE Let B be a disc $|z| \leqslant R$, where $R > 0$, in the complex plane. Find the measure of best approximation of the function z^n on B by means of the polynomials

$$P(z) = c_1 z^{n-1} + c_2 z^{n-2} + \ldots + c_n$$

of degree not exceeding $n-1$:

$$\mu = \inf_{P(z)} \sup_{|z| \leqslant R} |z^n - P(z)|$$

and find the polynomial $\pi(z)$ of best approximation.

By using Corollary 3, let us show that $\pi(z) \equiv 0$ is a polynomial of best approximation of z^n on B. We obtained this result in Section 5.1. In the case at hand, $\rho = R^n$ and the set $B(\rho)$ is the circle $|z| = R$. We need to show that

$$I(P(z)) = \inf_{|z| = R} \operatorname{Re} \{ z^n \overline{P(z)} \} \leqslant 0$$

for every polynomial $P(z)$ of degree not exceeding $n-1$.

If we set

$$P(z) = \sum_{k=0}^{n-1} (a_k + ib_k) z^k$$

and $z = e^{i\psi}$, where $0 \leqslant \psi < 2\pi$, we have

$$I(P(z)) = \inf_{0 \leqslant \psi < 2\pi} \sum_{k=0}^{n-1} (a_k \cos(n-k)\psi + b_k \sin(n-k)\varphi)$$

Since every trigonometric polynomial without a constant term has zeros on the interval $0 \leqslant \psi < 2\pi$ (since the integral of such a trigonometric polynomial over the interval $[0, 2\pi]$ is equal to zero), we have $I(P(z)) \leqslant 0$ and, consequently, $\pi(z) \equiv 0$ is a polynomial of best approximation of the function z^n on $|z| \leqslant R$ and $\mu = R^n$.

To prove the next theorem, we need the following lemma.

Lemma 1 Suppose that $u_{l,j}$, (where $l = 1, 2, \ldots, m$ and $j = 1, 2, \ldots, n$) are given numbers. For numbers $\delta_l \geqslant 0$, where $l = 1, 2, \ldots, m$, not all zero, such that

$$\sum_{l=1}^{m} \delta_l u_{l,j} = 0, \quad j = 1, 2, \ldots, n \tag{3}$$

to exist, it is necessary and sufficient that for an arbitrary system of numbers c_j, for $j = 1, 2, \ldots, n$, the inequalities

$$\operatorname{Re} \sum_{j=1}^{n} c_j u_{l,j} > 0, \quad l = 1, 2, \ldots, m \tag{4}$$

should not be satisfied for all l.

PROOF: NECESSITY If (3) holds for $\delta_l \geqslant 0$, it follows from the obvious equality

$$\sum_{l=1}^{m} \delta_l \operatorname{Re} \sum_{j=1}^{n} c_j u_{l,j} = \operatorname{Re} \sum_{j=1}^{n} c_j \sum_{l=1}^{m} \delta_l u_{l,j} = 0$$

that inequalities (4) cannot be satisfied for all l and in an arbitrary system of numbers c_j.

SUFFICIENCY We introduce the function

$$v = v(\delta_1, \delta_2, \ldots, \delta_m) = \sum_{j=1}^{n} \left| \sum_{l=1}^{m} \delta_l u_{l,j} \right|^2, \quad \delta_l \geqslant 0, \quad \sum_{l=1}^{m} \delta_l = 1$$

Since the regions of variation of the independent variables are closed and bounded, this function attains a minimum value $v_0 \geqslant 0$, let us say at $\delta_l = \delta_l^{(0)}$. The equation $v_0 = 0$ is equivalent to the assertion of the lemma, so that we need to show that if inequalities (4) cannot be satisfied for all l with any choice of c_j, then v_0 must be equal to zero. Let us assume the opposite. Suppose that inequalities (4) cannot be satisfied for all l with any choice of c_j and $v_0 > 0$. We set

$$c_j = \sum_{l=1}^{m} \delta_l^{(0)} \bar{u}_{l,j}$$

Let us suppose that, for example, inequality (4) does not hold for $l = m$; that is,

$$\operatorname{Re} \sum_{j=1}^{n} \left(\sum_{l=1}^{m} \delta_l^{(0)} \bar{u}_{l,j} \right) u_{m,j} \leqslant 0$$

If we set

$$v_* = \sum_{j=1}^{n} |u_{m,j}|^2, \qquad \lambda = \frac{v_*}{v_* + v_0}$$

$$\delta_l = \begin{cases} \lambda \delta_l^{(0)}, & l = 1, 2, \ldots, m-1 \\ (1-\lambda) + \lambda \delta_m^{(0)}, & l = m, \end{cases}$$

we obtain

$$v = \sum_{j=1}^{n} \left| \sum_{l=1}^{m} \delta_l u_{l,j} \right|^2 = \sum_{j=1}^{n} \left| (1-\lambda) u_{m,j} + \lambda \sum_{l=1}^{m} \delta_l^{(0)} u_{l,j} \right|^2$$

$$= (1-\lambda)^2 \sum_{j=1}^{n} |u_{m,j}|^2 + 2\lambda(1-\lambda) \operatorname{Re} \left\{ \sum_{j=1}^{n} \sum_{l=1}^{m} \delta_l^{(0)} \bar{u}_{l,j} u_{m,j} \right\}$$

$$+ \lambda^2 \sum_{j=1}^{n} \left| \sum_{l=1}^{m} \delta_l^{(0)} u_{l,j} \right|^2 \leqslant (1-\lambda)^2 v_* + \lambda^2 v_0 = \frac{v_* v_0}{v_* + v_0} < v_0$$

But this contradicts the definition of the number v_0. This completes the proof of the lemma.

NOTE 1 If the numbers $u_{l,j}$ are real, the lemma remains in force even when the numbers c_j are assumed to be real.

We shall keep the notations introduced above.

DEFINITION Suppose that D is a bounded set of points $\xi \in \tilde{\mathfrak{R}}_n$. The set $D_m^{(0)} = \{\xi_l\}_1^m$ of points $\xi_l \in \bar{D}$ is said to be characteristic‡ for the set D, if $\mu(D_m^{(0)})=\mu(D)$, and if, for an arbitrary sub-set $D^*\subset D_m^{(0)}$, where $D^*\neq D_m^{(0)}$, we have $\mu(D^*)<\mu(D_m^{(0)})$.

NOTE 2 It follows from Theorem 1, Section 5.3.2, that there exists at least one characteristic set $D_m^{(0)}$ for D and $1\leqslant m\leqslant 2n+1$.

NOTE 3 The set $D_m^{(0)}$ is characteristic for every set D' such that $D_m^{(0)}\subset D'\subset \bar{D}$, and every polynomial of best approximation on D' is a polynomial of best approximation on $D_m^{(0)}$. The last assertion follows from Note 2 to Theorem 3, Section 5.2.2.

NOTE 4 If $\pi(\xi)$ is a polynomial of best approximation on $D_m^{(0)}$, then, for every element

$$\xi_l=\xi(\zeta_0^{(l)},\ \ldots,\ \zeta_n^{(l)})\in D_m^{(0)},\quad l=1,\ 2,\ \ldots,\ m,$$

we have $|\zeta_0^{(l)}-\pi(\xi_l)|=\mu$.

To see this, let us suppose that, for example, $|\zeta_0^{(m)}-\pi(\xi_m)|<\mu$, for an element $\xi_m\in D_m^{(0)}$. Let us find the polynomial $\pi^*(\xi)$ of best approximation on the set $D_{m-1}=\{\xi_l\}_1^{m-1}$. Here, $|\zeta_0^{(l)}-\pi^*(\xi_l)|<\mu$, for $l=1,\ 2,\ \ldots,\ m-1$. From this, we easily see that

$$|\zeta_0^{(l)}-\pi_\lambda(\xi_l)|<\mu,\quad l=1,\ 2,\ \ldots,\ m.$$

for the polynomial

$$\pi_\lambda(\xi)=\lambda\pi^*(\xi)+(1-\lambda)\pi(\xi),\quad 0<\lambda<1,$$

if λ is sufficiently small. But this contradicts the condition that $\pi(\xi)$ is a polynomial of best approximation on $D_m^{(0)}$.

NOTE 5 It is possible to exhibit an example of a set D, for which several or even infinitely many characteristic sets exist; also, these may have a different number of points. There may be distinct polynomials of best approximation on distinct characteristic sets.

Suppose that B is compact and that the functions $\varphi_j(z)$, where $j=0,\ 1,\ \ldots,\ n$, are continuous on B. We shall say that the set $B_m^{(0)}=\{z_l\}_1^m$ of points $z_l\in B$ is characteristic for the set B and the functions $\varphi_j(z)$, for $j=0,\ 1,\ \ldots,\ n$, if $\mu(B_m^{(0)})=\mu(B)$ and if $\mu(B^*)<\mu(B_m^{(0)})$ for an arbitrary sub-set $B^*\subset B_m^{(0)}$ where $B^*\neq B_m^{(0)}$

‡ This concept and terminology was introduced by Vidensky [2].

Obviously, notes analogous to those above with regard to the set $D_m^{(0)}$ are applicable here.

Theorem 2 (Remez [1] *)**For a generalised polynomial $\pi(\xi) = P(\xi, \tilde{X}_0)$, where $\tilde{X}_0 \in \tilde{\mathfrak{R}}_n$ to be a polynomial of best approximation on a bounded set D, it is necessary and sufficient that there exist a set $D_m = \{\xi_l\}_1^m$ (where $1 \leqslant m \leqslant 2n+1$) of points $\xi_l = (\zeta_0(\xi_l), \ldots, \zeta_n(\xi_l)) \in D(\rho)$ and numbers $\delta_l > 0$ (where $l = 1, 2, \ldots, m$), such that

$$\sum_{l=1}^{m} \delta_l \overline{(\zeta_0(\xi_l) - \pi(\xi_l))}\, \zeta_j(\xi_l) = 0, \quad j = 1, 2, \ldots, n \tag{5}$$

PROOF: SUFFICIENCY Since conditions (5) are satisfied, according to the preceding lemma, the inequalities

$$\operatorname{Re} \sum_{j=1}^{n} c_j \zeta_j(\xi_l) \overline{(\zeta_0(\xi_l) - \pi(\xi_l))} > 0, \quad l = 1, 2, \ldots, m \tag{6}$$

are not simultaneously satisfied for arbitrary $c_1, c_2, \ldots, c_n$; that is, for an arbitrary polynomial

$$P(\xi, \tilde{X}) = \sum_{j=1}^{n} c_j \zeta_j$$

where $\xi(\zeta_1, \ldots, \zeta_n) \in \tilde{\mathfrak{R}}_n$, we have

$$\inf_{l=1, 2, \ldots, n} \operatorname{Re} \{P(\xi_l, \tilde{X}) \overline{(\zeta_0(\xi_l) - \pi(\xi_l))}\} \leqslant 0 \tag{7}$$

and hence a fortiori we have Kolmogorov's inequality

$$\inf_{\xi \in D(\rho)} \operatorname{Re} \{P(\xi, \tilde{X}) \overline{(\zeta_0 - \pi(\xi))}\} \leqslant 0$$

NECESSITY Suppose that $D_m = \{\xi_l\}_1^m$ is the characteristic set for D. Since $\pi(\xi)$ is a polynomial of best approximation on D and hence on D_m, it follows from Kolmogorov's condition that (7) is satisfied for an arbitrary polynomial $P(\xi_l, \tilde{X})$. Consequently, inequalities (6) are not satisfied simultaneously for arbitrary $c_1, \ldots c_n$. Therefore, in view of the lemma, there exist numbers $\delta_l \geqslant 0$, where $\sum_{l=1}^{m} \delta_l \neq 0$, such that conditions (5) are satisfied. All the δ_l are positive because if some δ_l were equal to zero we could

* See also the works by Remez [2,3], Ivanov [1,2,3], and Vidensky [2,4].

remove the corresponding points ξ_l from the set D_m and, because of the sufficiency of Remez's conditions, conclude that the polynomial $\pi(\xi)$ would also be a polynomial of best approximation on the resulting sub-set of D_m. This contradicts the definition of a characteristic set and completes the proof of the theorem.

We shall call conditions (5) Remez's conditions.

NOTE 6 We may always assume that $\sum_{l=1}^{m} \delta_l = 1$ in Remez's conditions.

NOTE 7 In the proof of the necessity of Remez's conditions, we show that they must be satisfied on every characteristic set.

NOTE 8 If we set

$$\zeta_0(\xi_l) - \pi(\xi_l) = \mu e^{i\theta_l}, \quad \mu = \rho, \quad 0 \leqslant \theta_l < 2\pi \tag{8}$$

Remez's conditions (5) take the form

$$\sum_{l=1}^{m} \delta_l e^{-i\theta_l} \zeta_j(\xi_l) = 0, \quad j = 1, 2, \ldots, n \tag{9}$$

NOTE 9 Obviously, in the statement of the theorem, the inequality $1 \leqslant m \leqslant 2n+1$ may be replaced with the inequality $1 \leqslant m < +\infty$, and the conditions $\delta_l > 0$, for $l = 1, 2, \ldots, m$, may be replaced with the conditions $\delta_l \geqslant 0$, for $l = 1, 2, \ldots, m$, where

$$\sum_{l=1}^{m} \delta_l > 0$$

COROLLARY 1 For Remez's conditions to be satisfied, it is necessary and sufficient, that, for every polynomial

$$P(\xi, \tilde{X}) = \sum_{j=1}^{n} c_j \zeta_j(\xi) \quad \tilde{X}(c_1, \ldots, c_n) \in \tilde{\mathfrak{R}}_n$$

the condition

$$\sum_{l=1}^{m} \delta_l \overline{(\zeta_0(\xi_l) - \pi(\xi_l))} P(\xi_l, \tilde{X}) = 0 \quad \tilde{X} \in \tilde{\mathfrak{R}}_n \tag{10}$$

or

$$\sum_{l=1}^{m} \delta_l e^{-i\theta_l} P(\xi_l, \tilde{X}) = 0, \quad \tilde{X} \in \tilde{\mathfrak{R}}_n \tag{11}$$

be satisfied.

COROLLARY 2 If $\sum_{l=1}^{m} \delta_l = 1$ in Remez's conditions, then

$$\mu = \sum_{l=1}^{m} \delta_l e^{-i\theta_l} \zeta_0(\xi_l) \tag{12}$$

To see this, set $P(\xi, \tilde{X}) = \pi(\xi)$ in (11). Then, in view of (8),

$$\sum_{l=1}^{m} \delta_l e^{-i\theta_l} \zeta_0(\xi_l) - \sum_{l=1}^{m} \delta_l \mu = 0$$

from which (12) follows.

COROLLARY 3 If we set $\sum_{l=1}^{m} \delta_l = 1$ in Remez's conditions and use the notations

$$\zeta_0(\xi_l) - Q(\xi_l) = \lambda_l e^{i\varphi_l}, \quad l = 1, 2, \ldots, m, \quad \lambda_l \geqslant 0, \ 0 \leqslant \varphi_l < 2\pi,$$

for an arbitrary polynomial $Q(\xi) = P(\xi, \tilde{X})$, then

$$\mu \leqslant \sum_{l=1}^{m} \delta_l \lambda_l. \tag{13}$$

Equality holds in (13) if and only if $\varphi_l = \theta_l$ for all l such that $\lambda_l \neq 0$. If $\varphi_l = \theta_l + \gamma$ for all l such that $\lambda_l \neq 0$, where γ is a constant, γ must be zero. If equality holds in (13), then

$$\min_{l=1, 2, \ldots, m} \lambda_l \leqslant \mu \leqslant \max_{l=1, 2, \ldots, m} \lambda_l \tag{14}$$

and, for an arbitrary polynomial $P(\xi)$, the inequalities

$$|\zeta_0(\xi_l) - P(\xi_l)| < \lambda_l, \quad l = 1, 2, \ldots, m \tag{15}$$

are not satisfied simultaneously.

PROOF Since the measures of best approximation of the functions $\zeta_0(\xi)$ and $\zeta_0(\xi) - Q(\xi)$ by means of the polynomials $P(\xi, \tilde{X})$ on D are the same, it follows on the basis of (12) that

$$\mu = \sum_{l=1}^{m} \delta_l e^{-i\theta_l} (\zeta_0(\xi_l) - Q(\xi_l)) = \sum_{l=1}^{m} \delta_l \lambda_l e^{i(\varphi_l - \theta_l)}$$
$$= \left| \sum_{l=1}^{m} \delta_l \lambda_l e^{i(\varphi_l - \theta_l)} \right| \leqslant \sum_{l=1}^{m} \delta_l \lambda_l$$

Equality holds only when the arguments of all the non-zero

terms $\delta_l \lambda_l e^{i(\varphi_l - \theta_l)}$ are the same, that is, if $\varphi_l = \theta_l + \gamma$. Also,

$$\mu = \sum_{l=1}^{m} \delta_l \lambda_l e^{i\gamma} = e^{i\gamma} \sum_{l=1}^{m} \delta_l \lambda_l$$

and consequently $\gamma = 0$ and $\varphi_l = \theta_l$ for $l = 1, 2, \ldots, m$.

If equality holds in (13), inequalities (14) are obvious.

Let us now show that if equality holds in (13), inequalities (15) are satisfied simultaneously.

If we set

$$\zeta_0(\xi_l) - P(\xi_l) = \lambda_l' e^{i\varphi_l'}, \ \lambda_l' \geqslant 0, \ 0 \leqslant \varphi_l' < 2\pi, \ l = 1, 2, \ldots, m$$

and assume that $\lambda_l' < \lambda_l$, for $l = 1, 2, \ldots, m$, we have

$$\mu = \sum_{l=1}^{m} \delta_l e^{-i\theta_l} \lambda_l' e^{i\varphi_l'} = \left| \sum_{l=1}^{m} \delta_l \lambda_l' e^{i(\varphi_l' - \theta_l)} \right| \leqslant \sum_{l=1}^{m} \delta_l \lambda_l' < \sum_{l=1}^{m} \delta_l \lambda_l = \mu$$

This contradiction proves the validity of the last assertion and completes the proof of the corollary.

COROLLARY 4 All polynomials of best approximation on a characteristic set $D_m^{(0)} = \{\xi_l\}_1^m$ have identical values at an arbitrary point $\xi_l \in D_m^{(0)}$.

PROOF If $Q(\xi)$ is an arbitrary polynomial of best approximation on $D_m^{(0)}$, then $\lambda_l = \mu$ and, consequently, equality holds in (13) and $\varphi_l = \theta_l$, for $l = 1, 2, \ldots, m$.

An immediate consequence of Theorem 2 is the following theorem.

Theorem 2′ Suppose that B is compact and that the functions $\varphi_j(z)$ for $j = 0, 1, \ldots, n$, are continuous on B. For a generalised polynomial

$$\pi(z) = \sum_{j=1}^{n} c_j^{(0)} \varphi_j(z)$$

to be the polynomial of best approximation for $\varphi_0(z)$ on B out of all the generalised polynomials

$$p(z) = \sum_{j=1}^{n} c_j \varphi_j(z)$$

it is necessary and sufficient that there exist a set $B_m = \{z_l\}_1^m$, where $1 \leqslant m \leqslant 2n + 1$, of elements $z_l \in B(\rho)$ and numbers $\delta_l > 0$, where $l = 1, 2, \ldots, m$, such that

$$\sum_{l=1}^{m} \delta_l \overline{(\varphi_0(z_l) - \pi(z_l))} \varphi_j(z_l) = 0, \quad j = 1, 2, \ldots, n \tag{16}$$

Notes and corollaries analogous to those following Theorem 2 are applicable to this theorem. In particular, in analogy with Note 8, we have the following: if we write $\varphi_0(z) - \pi(z_l) = \mu e^{i\theta_l}$, where $\mu > 0$, and $0 \leqslant \theta_l < 2\pi$, then Remez's conditions (16) take the form

$$\sum_{l=1}^{m} \delta_l e^{-i\theta_l} \varphi_j(z_l) = 0, \quad j = 1, 2, \ldots, n \tag{17}$$

Let us now examine in somewhat greater detail the case in which B is a closed bounded set of points in the complex z plane (it is a compact sub-set of the complete metric space $\mathfrak{R}_1$) and $\varphi_j(z) = z^{n+1-j}$ for $j = 1, 2, \ldots, n+1$ and in which $\varphi_0(z)$ is an arbitrary function continuous on B. In other words, let us examine the uniform approximation of the function $\varphi_0(z)$ on B by means of the algebraic polynomials $p_n(z) = c_0 z^n + c_1 z^{n-1} + \ldots + c_n$ of degree not exceeding n.

If the set B holds no more than $n+1$ points, every function $\varphi_0(z)$ defined on B can be represented by a polynomial of degree not exceeding n (this follows easily from Lagrange's formula), and, consequently, the measure μ of best approximation of the function $\varphi_0(z)$ on B by means of polynomials of degree not exceeding n is equal to zero and the polynomial $\varphi_0(z)$ is itself a polynomial of best approximation. Therefore, in what follows, we shall assume that the set B contains at least $n+2$ distinct points and that the function $\varphi(z)$ either is not a polynomial or is a polynomial of degree at least n. It is then obvious that $\mu > 0$ and conditions (17) take the form

$$\sum_{l=1}^{m} \delta_l e^{-i\theta_l} z_l^j = 0, \quad j = 0, 1, \ldots, n \tag{18}$$

where $\{z_l\}_1^m$ is the characteristic set of the points $z_l \in B$ for the function $\varphi_0(z)$, where $\theta_l = \arg(\varphi_0(z) - \pi_n(z))$ (where $\pi_n(z)$ in turn is a polynomial of best approximation), and where $\delta_l > 0$ (for $l = 1, 2, \ldots, m$). It follows from the exposition given above that m cannot be less than $n+2$ and that $1 \leqslant m \leqslant 2n+3$; that is, $n+2 \leqslant m \leqslant 2n+3$.

Theorem 3 (Vidensky [2, 4] *)* For the numbers $\{\delta_l e^{-i\theta_l}\}_1^m$, where $n+2 \leqslant m \leqslant 2n+3$ and $\delta_l > 0$, to satisfy conditions

(18), it is necessary and sufficient that a polynomial $u(z)$, of degree not exceeding $m-n-2$, exist such that $u(z_l) \neq 0$, for $l=0, 1, \ldots, n$, and

$$\delta_l e^{-i\theta_l} = \frac{u(z_l)}{\omega'(z_l)}, \quad l=1, 2, \ldots, m, \quad \omega(z) = \prod_{k=1}^{m} (z-z_k) \quad \textbf{(19)}$$

PROOF: NECESSITY Suppose that Equations (18) are satisfied. We set

$$u(z) = \sum_{l=1}^{m} \delta_l e^{-i\theta_l} \frac{\omega(z)}{z-z_l} \quad \textbf{(20)}$$

From the identity

$$\frac{1}{z-z_l} = \frac{1}{z}\sum_{j=0}^{k}\left(\frac{z_l}{z}\right)^j + \left(\frac{z_k}{z}\right)^{k+1}\frac{1}{z-z_l}, \quad k=0, 1, \ldots,$$

and Equations (18), we have

$$u(z) = \sum_{j=0}^{k} \frac{\omega(z)}{z^{j+1}} \sum_{l=1}^{m} \delta_l e^{-i\theta_l} z_l^j + \sum_{l=1}^{m} \delta_l e^{-i\theta_l} \frac{\omega(z)}{z-z_l}\left(\frac{z_l}{z}\right)^{k+1}$$

$$= \sum_{l=1}^{m} \delta_l e^{-i\theta_l} \frac{\omega(z)}{z-z_l}\left(\frac{z_l}{z}\right)^{k+1}, \quad k=0, 1, \ldots, n$$

If we set $k=n$ in this equation, we see that the degree of the polynomial $u(z)$ does not exceed $(m-1)-(n+1)=m-n-2$. From (20), we obtain by use of Lagrange's formula

$$u(z_l) = \delta_l e^{-i\theta_l} \omega'(z_l), \quad l=1, 2, \ldots, m$$

Equations (19) follow from this. This proves the necessity.

SUFFICIENCY Suppose that $u(z)$ is an arbitrary polynomial of degree and does not exceed $m-n-2$ such that $u(z_k) \neq 0$ for $k=1, 2, \ldots, m$. From Lagrange's formula,

$$z^j u(z) = \sum_{l=1}^{m} z_l^j u(z_l) \cdot \frac{\omega(z)}{(z-z_l)\,\omega'(z_l)}, \quad j=0, 1, \ldots, n$$

If we equate the coefficients of z^{m-1}, on the left- and right-hand sides of this equation, we obtain

$$0 = \sum_{l=1}^{m} \frac{u(z_l)}{\omega'(z_l)} z_l^j, \quad l=0, 1, \ldots, n$$

If we now set

$$\frac{u(z_k)}{\omega'(z_k)} = \delta_k e^{i\theta_k}, \quad \delta_k > 0, \quad k = 1, 2, \ldots, m$$

we obtain Equations (18). This completes the proof of the theorem.

COROLLARY 1 If the degree of a polynomial $u(z)$ is $r < m-n-2$, then Equations (18) are satisfied for $l = 0, 1, \ldots, m-r-2$. Consequently, the polynomial of best approximation for $f(z)$ on B out of all the polynomials of degree not exceeding n is also the polynomial of best approximation for $f(z)$ on B out of all the polynomials of degree not exceeding $m-r-2 > n$.

COROLLARY 2 A polynomial $u(z)$ such that $u(z_k) \neq 0$, (for $k = 1, 2, \ldots, m$) the degree of which does not exceed $m-n-2$ defines by means of Equations (19) a set of arguments $\{-\theta_k\}_1^m$ with which the best approximation of the function on the set $\{z_k\}_1^m$ can be achieved. Here, to an arbitrary set of arguments $\{-\theta_k\}_1^m$ defined in this manner, there corresponds at least one function for which the best approximation is attained with just these arguments.

The function

$$f(z_k) = e^{-i\theta_k}, \quad k = 1, 2, \ldots, m$$

is, by virtue of Theorem 2, Section 5.3.3 and Equations (19), such a function.

Suppose now that $T_n(z) = z^n + c_1 z^{n-1} + \ldots + c_n$ is Chebychev's polynomial of degree n for the set B (that is, the polynomial deviating least from zero out of all the polynomials of degree n with leading coefficient equal to one). As pointed out at the beginning of the chapter, the problem of finding Chebyshev's polynomial $T_n(z)$ is equivalent to the problem of finding the polynomial of best approximation for the function z^n on B out of all the polynomials of degree not exceeding $n-1$. If the set B holds no more than n points, obviously Chebyshev's polynomial $T_n(z)$ for the set B may be any polynomial that vanishes at each point of the set B. Then,

$$\mu = \max_{z \in B} |T_n(z)| = 0$$

In what follows, we shall assume that the set B holds at

least $n+1$ distinct points. If $B_m=\{z_l\}_1^m$ where $n+1\leqslant m\leqslant 2n+1$, is the characteristic set for the function z^n and the set B, there exist numbers $\delta_l>0$, for $l=1, 2, \ldots, m$, such that

$$\sum_{l=1}^{m}\delta_l e^{-i\theta_l} z_l^j=0, \quad j=0, 1, \ldots, n-1$$

where $\theta_l=\arg T_n(z_l)$, and, by virtue of Theorem 3, there exists a polynomial $u(z)$ of degree not exceeding $m-n-1$ such that

$$\delta_l e^{-i\theta_l}=\frac{u(z_l)}{\omega'(z_l)}, \quad \omega(z)=\prod_{k=1}^{n}(z-z_k)$$

On the basis of Corollary 1 to this same theorem, we may assert that the degree of the polynomial $u(z)$ is equal to $m-n-1$ because otherwise the polynomial $T_n(z)-z^n$ would be a polynomial of best approximation of the function z^n out of all the polynomials of degree n, which is not true. Lagrange's interpolation formula yields

$$T_n(z)\,u(z)=\sum_{l=1}^{m}T_n(z_l)\,u(z_l)\frac{\omega(z)}{(z-z_l)\,\omega'(z_l)}$$

or

$$T_n(z)\,u(z)=\mu\sum_{l=1}^{m}\delta_l\frac{\omega(z)}{z-z_l} \qquad \delta_l=\left|\frac{u(z_l)}{\omega'(z_l)}\right| \qquad \mu=\sup_{z\in B}|T_n(z)|$$

If we equate the coefficients of z^m in this equation, we conclude that the leading coefficient of the polynomial $u(z)$ is positive. We note that if $m=n+1$, the polynomial $u(z)$ is a constant and the preceding formula takes the form

$$T_n(z)=\mu\sum_{l=1}^{n+1}\frac{1}{|\omega'(z_l)|}\,\frac{\omega(z)}{z-z_l}$$

5.3.4 *The uniqueness of the polynomial of best approximation*

Suppose that B is a set of arbitrary elements z, containing at least n distinct elements, and that $\varphi_j(z)$, for $j=1, 2, \ldots, n$ are complex functions defined on B and assuming only finite values. We shall call such a system of functions a Chebyshev system if an arbitrary generalised polynomial

$$p(z, \tilde{X}) = \sum_{j=1}^{n} c_j \varphi_j(z), \quad \tilde{X}(c_1, \ldots, c_n) \in \tilde{\Re}_n$$

where not all the c_j are equal to zero, has no more than $n-1$ distinct roots on B. Obviously, a Chebyshev system of functions is a system of linearly independent functions defined on B.

For what follows, we shall need the following lemma.

Lemma 1 If a system of functions $\varphi_j(z)$, for $j = 1, 2, \ldots, n$ and $z \in B$, is a Chebyshev system, then, for an arbitrary system of n distinct elements z_l, for $l = 1, 2, \ldots, n$, belonging to B, it is possible to construct generalised polynomials $\omega_m(z)$, for $m = 1, 2, \ldots, n$, possessing the property that

$$\omega_m(z_l) = \begin{cases} 1 & \text{for } l = m \\ 0 & \text{for } l \neq m, \end{cases} \quad l, m = 1, 2, \ldots, n$$

and this polynomial is unique.

PROOF We note first that the determinant

$$\Delta = \begin{vmatrix} \varphi_1(z_1), & \varphi_2(z_1), & \ldots, & \varphi_n(z_1) \\ \varphi_1(z_2), & \varphi_2(z_2), & \ldots, & \varphi_n(z_2) \\ \cdot & \cdot & \cdot & \cdot \\ \varphi_1(z_n), & \varphi_2(z_n), & \ldots, & \varphi_n(z_n) \end{vmatrix} \neq 0$$

since there would otherwise exist a point $\tilde{X}(c_1, \ldots, c_n) \in \tilde{\Re}_n$, where not all the c_j are equal to zero, such that

$$c_1\varphi_1(z_l) + c_2\varphi_2(z_l) + \ldots + c_n\varphi_n(z_l) = 0, \quad l = 1, 2, \ldots, n$$

which is impossible in view of the fact that the system of functions is a Chebyshev system.

Let $\Delta_m(z)$ be the determinant obtained from Δ by replacing the m-th row with the row $\varphi_1(z), \varphi_2(z), \ldots, \varphi_n(z)$. Clearly, the polynomial $\omega_m(z) = \frac{\Delta_m(z)}{\Delta}$ is the desired one. The uniqueness of $\omega_m(z)$ follows immediately from the fact that $\varphi_j(z)$, for $j = 1, 2, \ldots, n$, is a Chebyshev system. This proves the lemma.

COROLLARY 1 Suppose that u_l, for $l=1, 2, \ldots, n$, are arbitrary given numbers. The generalised polynomial

$$p(z)=\sum_{l=1}^{n} u_l \omega_l(z) \tag{1}$$

is the only polynomial satisfying the conditions

$$p(z_l)=u_l, \quad l=1, 2, \ldots, n$$

We shall call the polynomial (1) an interpolational polynomial and we shall call the elements z_l, for $l=1, 2, \ldots, n$, the basic points of the interpolation.

COROLLARY 2 If the set $B'\subset B$ holds no more than n elements, then, for every function $\varphi_0(z)$, that is defined on B', the measure $\mu(B')$ of the best approximation for $\varphi_0(z)$ on B' by means of generalised polynomials

$$p(z)=\sum_{j-1}^{n} c_j \varphi_j(z)$$

is equal to zero.

PROOF Suppose that $B'=\{z_l\}_1^k$, where the z_l are distinct elements of B and $k\leqslant n$. If $k<n$, we extend the set B' to include the n distinct elements z_l, for $l=1, 2, \ldots, n$, in B. Let us construct an interpolational generalised polynomial $p(z)$ for the basic points z_l, for $l=1, 2, \ldots, n$, satisfying the conditions $p(z_l)=\varphi_0(z_l)$ for $l=1, 2, \ldots, k$, and $p(z_l)=u_l$, for $l=k+1, \ldots, n$, where the u_l are arbitrary numbers. Then, $|\varphi_0(z)-p(z)|=0$ for $z\in B'$ and, consequently, $\mu(B')=0$.

COROLLARY 3 If the functions $\varphi_j(z)$, for $j=0, 1, \ldots, n$, are linearly independent on a set B containing at least $n+1$ elements and if the system of functions $\varphi_j(z)$, for $j=1, 2, \ldots, n$, is a Chebyshev system, then every characteristic set $B_m^{(0)}=\{z_l\}_1^m$ of elements $z_l\in B$ for the set B and the system $\varphi_j(z)$, for $j=0, 1, \ldots, n$, contains at least $n+1$ elements; consequently, $n+1\leqslant m\leqslant 2n+1$. If in addition the functions $\varphi_j(z)$ are real, $m=n+1$.

To see this, note that in the present case $\mu(B)>0$ and, by virtue of Corollary 2, $m\geqslant n+1$.

Theorem 1 (Tonelli [1]*)* Suppose that $\varphi_j(z)$ for $j=0, 1, \ldots, n$, is a system of continuous functions defined on a compact set B, and that B contains at least $n+1$ elements.

If $\varphi_j(z)$, where $j=1, 2, \ldots, n$, and $z \in B$, is a Chebyshev system of functions, the following assertions hold: (1) the polynomial

$$\pi(z)=\sum_{j=1}^{n} c_j\varphi_j(z)$$

of best approximation for the function $\varphi_0(z)$ on B is unique; (2) the function $|\varphi_0(z)-\pi(z)|$ attains its greatest value on B at at least $n+1$ elements z_l of B, for $l=1, 2, \ldots, n+1$.

PROOF The existence of at least one polynomial of best approximation was shown earlier (see the corollary to Theorem 1, Section 5.3.2). Let us first prove assertion (2) of the theorem. If the functions $\varphi_j(z)$, for $j=0, 1, \ldots, n$, are linearly dependent on B, $\mu(B)$ must be zero (since the $\varphi_j(z)$, for $j=1, 2, \ldots, n$, constitute a Chebyshev system and, consequently, are linearly independent), so that assertion (2) of the theorem is obvious. If the functions $\varphi_j(z)$, for $j=0, 1, \ldots, n$, are linearly independent on B, we obtain assertion (2) of the theorem when we apply Corollary 3 of the lemma and Note 3 following the definition of a characteristic set in Section 5.3.3.

Let us now prove the first assertion. Suppose that, in addition to $\pi(z)$, there is another polynomial $\pi_*(z)\not\equiv\pi(z)$, which is also a polynomial of best approximation for the function $\varphi_0(z)$ on B. We introduce the polynomial

$$P(z)=\frac{1}{2}(\pi(z)+\pi_*(z))$$

For $z \in B$, we have

$$|\varphi_0(z)-P(z)|=\left|\frac{1}{2}(\varphi_0(z)-\pi(z))+\frac{1}{2}(\varphi_0(z)-\pi_*(z))\right|$$

$$\leqslant\frac{1}{2}|\varphi_0(z)-\pi(z)|+\frac{1}{2}|\varphi_0(z)-\pi_*(z)|\leqslant\frac{1}{2}\mu+\frac{1}{2}\mu=\mu$$

that is, $P(z)$ is also a polynomial of best approximation for $\varphi_0(z)$ on B. Consequently, there exist at least $n+1$ elements z_l of B, for $l=1, 2, \ldots, n+1$, at which the left member of the inequality attains its maximum value μ such that equality holds in the inequality for $z=z_l$, for $l=1, 2, \ldots, n+1$. This last is possible only if the quantities $\frac{1}{2}(\varphi_0(z_l)-\pi(z_l))$ and $\frac{1}{2}(\varphi_0(z_l)-\pi_*(z_l))$ have absolute values equal to $\frac{1}{2}\mu(B)$, and identical arguments (if $\mu(B)\neq 0$). From this it follows that

$$\tfrac{1}{2}(\varphi_0(z_l)-\pi(z_l))=\frac{1}{2}(\varphi_0(z_l)-\pi_*(z_l)), \quad l=1, 2, \ldots, n+1$$

and

$$\pi(z_l)=\pi_*(z_l)$$

for $l=1, 2, \ldots, n+1$, thus, the polynomial $\pi(z)-\pi_*(z)\not\equiv 0$ has $n+1$ roots. This contradiction proves the first assertion of the theorem.

Theorem 2 (Haar [1]*)* Let us suppose that $\varphi_j(z)$, for $j=1, 2, \ldots, n$, is a system of continuous functions defined on a compact set B, and that B contains at least n elements. Then, for an arbitrary function $\varphi_0(z)$, that is, continuous on B to have a unique polynomial

$$\pi(z)=\sum_{j=1}^{n} c_j\varphi_j(z)$$

of best approximation on B, it is necessary and sufficient that the system of functions $\varphi_j(z)$, for $j=1, 2, \ldots, n$, be a Chebyshev system.

PROOF The sufficiency of this condition was proven in Tonelli's theorem for the case in which the set B contains at least $n+1$ elements. If B contains n elements z_l, for $l=1, 2, \ldots, n$, an arbitrary function $\varphi_0(z)$ is a polynomial

$$\varphi_0(z)=\sum_{l=1}^{n}\varphi_0(z_l)\,\omega_l(z)$$

and it is its own unique best approximation.

To prove the necessity, suppose that, in contradiction to the assertion of the theorem, the system $\varphi_j(z)$, for $j=1, 2, \ldots, n$, is not a Chebyshev system, that is, suppose that there exists a polynomial

$$p_*(z)=\sum_{j=1}^{n} a_j\varphi_j(z)$$

possessing at least n distinct roots $z_1, z_2, \ldots, z_n$ on B, and that not all the a_j are equal to zero. From this, we conclude that the determinant

$$\Delta=\begin{vmatrix}\varphi_1(z_1), & \ldots, & \varphi_n(z_1)\\ \cdot\;\cdot\;\cdot & \cdot\;\cdot\;\cdot & \cdot\;\cdot\;\cdot\\ \varphi_1(z_n), & \ldots, & \varphi_n(z_n)\end{vmatrix}=0$$

Consequently, there exists a system of numbers $b_1, b_2, \ldots, b_n$, not all zero, such that

$$\sum_{\nu=1}^{n} b_\nu \varphi_j(z_\nu) = 0$$

where $j=1, 2, \ldots, n$, and, for an arbitrary polynomial

$$p(z) = \sum_{j=1}^{n} c_j \varphi_j(z)$$

we have

$$\sum_{\nu=1}^{n} b_\nu p(z_\nu) = \sum_{\nu=1}^{n} b_\nu \sum_{j=1}^{n} c_j \varphi_j(z_\nu) = \sum_{j=1}^{n} c_j \sum_{\nu=1}^{n} b_\nu \varphi_j(z_\nu) = 0 \qquad (2)$$

Let $g(z)$ be an arbitrary function that is continuous on B, such that the following.

1. $|g(z)| \leqslant 1$, for $z \in B$;
2. $g(z_m) = \frac{b_m}{|b_m|}$ for all m such that $b_m \neq 0$.

It is not difficult to construct such a function (see Aleksandrov [1], pp. 284-287).

We introduce the continuous function

$$\varphi_0(z) = g(z)\left(1 - \frac{1}{H+1}|p_*(z)|\right), \quad H = \sup_{z \in B} |p_*(z)|, \quad z \in B$$

which satisfies the condition $|\varphi_0(z)| \leqslant 1$ for $z \in B$. Let us show that every polynomial

$$p_\varepsilon(z) = \frac{\varepsilon}{H+1} p_*(z)$$

where ε is an arbitrary number such that $|\varepsilon| \leqslant 1$ is a polynomial of best approximation for $\varphi_0(z)$ on B. For $z \in B$

$$|\varphi_0(z) - p_\varepsilon(z)| \leqslant |\varphi_0(z)| + |p_\varepsilon(z)| \leqslant 1 - \frac{1}{H+1}|p_*(z)| + \frac{|\varepsilon|}{H+1}|p_*(z)| \leqslant 1$$

Let us suppose that there exists a polynomial $\pi(z)$ such that

$$\sup_{z \in B} |\varphi_0(z) - \pi(z)| < 1$$

For $b_m \neq 0$ we have

$$\operatorname{Re}(b_m \pi(z_m)) = \operatorname{Re}\{b_m \varphi_0(z_m) - b_m(\varphi_0(z_m) - \pi(z_m))\} \geqslant |b_m| - |b_m|\,|\varphi(z_m) - \pi(z_m)| > 0$$

and, consequently,

$$\operatorname{Re} \sum_{\nu=1}^{m} b_\nu \pi(z_\nu) > 0$$

which is impossible since

$$\sum_{\nu=1}^{n} b_\nu \pi(z_\nu) = 0$$

by virtue of Equation (2). Thus, every polynomial $p_\varepsilon(z)$ is a polynomial of best approximation for $\varphi_0(z)$ on B. But this contradicts the hypothesis of the theorem and thus proves the theorem.

5.3.5 *Chebyshev's theorem*

Suppose that $\varphi_j(z)$, for $j = 1, 2, \ldots, n$, is a Chebyshev system of continuous functions defined on a compact set B, and that B contains at least $n+1$ elements. Let us look at the approximation of an arbitrary function $\varphi_0(z)$ that is continuous on B by means of the generalised polynomials

$$p_n(z) = \sum_{j=1}^{n} c_j \varphi_j(z)$$

under the condition that the function $\varphi_0(z)$ is not a generalised polynomial. In this case, for the number m of elements of the characteristic set (see Corollary 3 following the lemma in Section 5.3.4), we have the condition $n+1 \leqslant m \leqslant 2n+1$. We shall examine in detail the case in which there exists a characteristic set $B_{n+1}^{(0)} = \{z_l\}_1^{n+1}$ for the set B and the system $\varphi_j(z)$ (for $j = 0, 1, \ldots, n$), which has $n+1$ elements. This is the case, for example, when the functions $\varphi_j(z)$, for $j = 0, 1, \ldots, n$, are real functions defined on B. Let $\pi(z)$ be an arbitrary generalised polynomial. We define

$$\rho = \sup_{z \in B} |\varphi_0(z) - \pi(z)|$$

Lemma 1 (Vidensky [2] *)* For a generalised polynomial $\pi(z)$ to be a polynomial of best approximation for $\varphi_0(z)$ on B, it is necessary and sufficient that (1) the numbers $w_l = \varphi_0(z_l) - \pi(z_l)$, for $l = 1, 2, \ldots, n+1$, have identical absolute values equal to ρ; (2) the numbers $(-1)^l \frac{w_l}{\Delta_l}$, for $l = 1, 2, \ldots, n+1$ where

$$\Delta_l = \begin{vmatrix} \varphi_1(z_1), & \varphi_2(z_1), & \ldots, & \varphi_n(z_1) \\ \cdot\cdot\cdot & \cdot\cdot\cdot & \cdot\cdot\cdot & \cdot\cdot\cdot \\ \varphi_1(z_{l-1}), & \varphi_2(z_{l-1}), & \ldots, & \varphi_n(z_{l-1}) \\ \varphi_1(z_{l+1}), & \varphi_2(z_{l+1}), & \ldots, & \varphi_n(z_{l+1}) \\ \cdot\cdot\cdot & \cdot\cdot\cdot & \cdot\cdot\cdot & \cdot\cdot\cdot \\ \varphi_1(z_{n+1}), & \varphi_2(z_{n+1}), & \ldots, & \varphi_n(z_{n+1}) \end{vmatrix}, \quad l = 1, 2, \ldots, n+1 \qquad (1)$$

have identical arguments.

PROOF: NECESSITY Since $\pi(z)$ is a polynomial of best approximation, we have $|w_l| = \mu = \rho$, where $l = 1, 2, \ldots, n+1$ (see Note 3, Section 5.3.3, following the definition of a characteristic set), and Remez's conditions are satisfied:

$$\sum_{l=1}^{n+1} \delta_l \overline{w}_l \varphi_j(z_l) = 0, \quad \delta_l > 0, \ j = 1, 2, \ldots, n \qquad (2)$$

From this it follows that, for every generalised polynomial $p(z)$

$$\sum_{l=1}^{n+1} \delta_l \overline{w}_l p(z_l) = 0 \qquad (3)$$

In particular,

$$\sum_{l=1}^{n+1} \delta_l \overline{w}_l \omega_j(z_l) = 0, \quad j = 1, 2, \ldots, n \qquad (4)$$

where $\omega_j(z) = \frac{\Delta_j(z)}{\Delta}$ are generalised polynomials constructed as in the lemma of Section 5.3.4. Since $\omega_j(z_l) = 0$ for $j \neq l$ and $\omega_l(z_l) = 1$, where $j, l = 1, 2, \ldots, n$, conditions (4) take the form

$$\delta_l \overline{w}_l + \delta_{n+1} \overline{w}_{n+1} \omega_l(z_{n+1}) = 0, \quad l = 1, 2, \ldots, n \qquad (5)$$

or, since $\omega_l(z_{n+1}) = \frac{\Delta_l(z_{n+1})}{\Delta}$, we obtain

$$\delta_l \frac{\overline{w}_l}{\Delta_l(z_{n+1})} = -\delta_{n+1} \frac{\overline{w}_{n+1}}{\Delta}, \quad l = 1, 2, \ldots, n \qquad (6)$$

But $\Delta_l(z_{n+1}) = (-1)^{n-l} \Delta_l$ for $l = 1, 2, \ldots, n$, and $\Delta = \Delta_{n+1}$

Therefore, conditions (6) take the form

$$(-1)^l \delta_l \frac{\overline{w_l}}{\Delta_l} = (-1)^{n+1} \delta_{n+1} \frac{\overline{w_{n+1}}}{\Delta_{n+1}}, \quad l = 1, 2, \ldots, n \tag{7}$$

From this we conclude that the numbers $(-1)^l \frac{\overline{w_l}}{\Delta_l}$, for $l = 1, 2, \ldots, n+1$, have identical arguments, which proves the necessity.

SUFFICIENCY By hypothesis, the numbers $(-1)^l \frac{\overline{w_l}}{\Delta_l}$, for $l = 1, 2, \ldots n+1$, have identical arguments. For arbitrary $\delta_{n+1} > 0$, there exist $\delta_l > 0$, for $l = 1, 2, \ldots, n$, such that Equations (7), and consequently Equations (6), (5), and (4), are satisfied. Since every generalised polynomial $p(z)$ can be represented in the form

$$p(z) = \sum_{j=1}^{n} p(z_j)\,\omega_j(z)$$

Equations (4) imply Equations (3) and, in particular, Equations (2); that is, Remez's conditions are satisfied. Furthermore, by hypothesis, $|w_l| = \rho$. Consequently, by virtue of Remez's theorem (Theorem 2, Section 5.3.3), $\pi(z)$ is a polynomial of best approximation for $\varphi_0(z)$ on B. This completes the proof of the lemma.

NOTE 1 In the proof of the sufficiency, we did not use the fact that the set $\{z_l\}_1^{n+1}$ is characteristic.

Theorem 1 (Chebyshev [1]*)* Suppose that $\varphi_j(x)$, for $j = 1, 2, \ldots, n$, is a Chebyshev system of continuous real functions defined on a finite closed interval $[a, b]$ of the real axis, that

$$\pi(x) = \sum_{k=0}^{n} c_k \varphi_k(x)$$

is a polynomial, and that $\varphi_0(x)$ is a non-polynomial continuous real function defined on $[a, b]$. We set

$$\rho = \sup_{x \in [a, b]} |\varphi_0(x) - \pi(x)|$$

For the polynomial $\pi(x)$ to be a polynomial of best approximation for $\varphi_0(x)$ on $[a, b]$, it is necessary and sufficient that there exist $n+1$ points x_l such that

$$a \leqslant x_1 < x_2 < \ldots < x_{n+1} \leqslant b$$

and in addition that

1. $|\varphi_0(x_l) - \pi(x_l)| = \rho$, for $l = 1, 2, \ldots, n+1$ (8)
2. The numbers $w_l = \varphi_0(x_l) - \pi(x_l)$, for $l = 1, 2, \ldots, n+1$, have alternating signs.

PROOF: NECESSITY Suppose that $\pi(x)$ is a polynomial of best approximation for $\varphi_0(x)$ on $[a, b]$ and that the set $B_{n+1}^{(0)} = \{x_l\}_1^{n+1}$ is characteristic. Then, by virtue of the lemma, condition (1) is satisfied and the numbers $(-1)^l \frac{w_l}{\Delta_l}$, for $l = 1, 2, \ldots, n+1$, have identical signs. If we can show that the numbers Δ_l, for $l = 1, 2, \ldots, n+1$, have identical signs, condition (2), and hence the necessity, will be proven. Let us think of the determinant Δ_{n+1} as a function of the variables $x_l, x_{l+1}, \ldots, x_n$ and let us increase these variables continuously in such a way that they remain different and x_k becomes x_{k+1}, for $k = l, l+1, \ldots, n$. The determinant Δ_{n+1} varies continuously without changing signs and it becomes the determinant Δ_l. Consequently, all the determinants Δ_l, for $l = 1, 2, \ldots, n+1$, have identical signs.

SUFFICIENCY Suppose that conditions (1) and (2) are satisfied. Since the determinants Δ_l, for $l = 1, 2, \ldots, n$, have identical signs, conditions (2) are equivalent to the condition that the numbers $(-1)^l \frac{w_l}{\Delta_l}$ have identical signs. On the basis of the lemma (we are using the sufficiency [see also the note following the lemma]), $\pi(x)$ is a polynomial of best approximation for $\varphi_0(x)$ on $[a, b]$. This completes the proof of the theorem.

COROLLARY (de la Vallée Poussin [1]) Suppose that $p(x)$ is a generalised polynomial and that x_l, where $a \leqslant x_1 < x_2 < \ldots < x_{n+1} \leqslant b$, is an arbitrary system of $n+1$ points on $[a, b]$. If the numbers

$$\lambda_l = \varphi_0(x_l) - p(x_l), \quad l = 1, 2, \ldots, n+1$$

have alternating signs, the measure μ of best approximation on $[a, b]$ of the function $\varphi_0(x)$ satisfies the inequality

$$\mu \geqslant \min_{l=1, 2, \ldots, n+1} |\lambda_l|$$

If, under these conditions, $B_{n+1}^{(0)} = \{x_l\}_1^{n+1}$ is the characteristic set for $\varphi_0(x)$ on B, then

$$\min_{l=1,2,\ldots,n+1} |\lambda_l| \leqslant \mu \leqslant \max_{l=1,2,\ldots,n+1} |\lambda_l|.$$

PROOF Consider the approximation of the function $\varphi_0(x)$ on the set $\{x_l\}_1^{n+1}$ and denote by μ' the measure of best approximation for $\varphi_0(x)$ on that set. Then, by virtue of Corollary 3 to Remez's theorem,

$$\min_{l=1,2,\ldots,n+1} |\lambda_l| \leqslant \mu' \leqslant \max_{l=1,2,\ldots,n+1} |\lambda_l|$$

Remembering that $\mu' \leqslant \mu$ and that $\mu' = \mu$, if the set $\{x_l\}_1^{n+1}$ is characteristic for $\varphi_0(x)$ on B, we arrive at the conclusion of the corollary.

NOTE 2 The theory of best uniform approximation of a function by means of generalised polynomials that we have expounded in Sections 5.2 and 3 is of a qualitative nature. This theory makes it possible to check whether a particular polynomial $\pi(z)$ is a polynomial of best approximation for a given function $f(z)$ on a given set B, but it is not of great help in finding a polynomial of best approximation. It is also not clear, even in the case of approximation of a complex function by means of algebraic polynomials, whether (and under what conditions) a sequence of polynomials $\pi_n(z) \in \mathscr{P}_n$, (for $n = 1, 2, \ldots$) of best approximation for a function $f(z)$ on a set B will converge uniformly on B to the function $f(z)$, that is, whether the sequence of numbers

$$\mu_n = \sup_{z \in B} |f(z) - \pi_n(z)|$$

approaches zero as $n \to \infty$. This problem is solved by Mergellian's theorem (see Section 1.7.4).

5.4 GENERALISED POLYNOMIALS THAT DEVIATE LEAST FROM ZERO WITH COEFFICIENTS SATISFYING A LINEAR CONDITION

In this section, we shall apply certain results obtained in Sections 5.2 and 3 to the investigation of the problem formulated below. Also, we shall use these investigations to obtain certain exact bounds.

Suppose that $\tilde{\alpha} = (\alpha_0, \alpha_1, \ldots, \alpha_n) \in \tilde{\Re}_{n+1}$, where the α_j are given numbers not all zero. This last hypothesis will apply to all that follows. Suppose that B is a non-empty set of arbitrary elements z and that $\zeta_j = \varphi_j(z)$, for $j = 0, 1, \ldots, n$, are functions defined on B that assume on B only finite values. For an arbitrary generalised polynomial

$$p(z)=\sum_{j=0}^{n} c_j\varphi_j(z)$$

where $z\in B$, we obtain

$$L_{\tilde{\alpha}}\{p(z)\}=\sum_{j=0}^{n}\alpha_j c_j$$

We denote by $\mathscr{P}(\tilde{\alpha})$ the class of generalised polynomials $p(z)$ that satisfy the condition $L_{\tilde{\alpha}}\{p(z)\}=1$. Consider the problem of finding the polynomials $\pi(z)$ that deviate least from zero on B out of all the polynomials $p(z)\in\mathscr{P}(\tilde{\alpha})$, that is, out of all the polynomials $\pi(z)\in\mathscr{P}(\tilde{\alpha})$ such that*

$$\sup_{z\in B}|\pi(z)|=\mu,\ \text{where}\ \mu=\inf_{p(z)\in\mathscr{P}(\tilde{\alpha})}\ \sup_{z\in B}|p(z)|$$

The problem of best approximation of the function $\varphi_0(z)$ on B by means of the polynomials

$$p(z)=\sum_{j=1}^{n} c_j\varphi_j(z)$$

is obviously equivalent to this problem for $\tilde{\alpha}=(1, 0, 0, \ldots, 0)$. The problem as stated is easily reduced to the problem of best approximation. If, for example, $\alpha_{j_0}\neq 0$, then

$$c_{j_0}=\frac{1}{\alpha_{j_0}}-\sum_{\substack{j=0\\ j\neq j_0}}^{n}\frac{\alpha_j}{\alpha_{j_0}}c_j$$

and, for an arbitrary polynomial

$$p(z)=\sum_{j=0}^{n} c_j\varphi_j(z)\in\mathscr{P}(\tilde{\alpha})$$

we have

$$p(z)=\frac{1}{\alpha_{j_0}}\varphi_{j_0}(z)-\sum_{\substack{j=0\\ j\neq j_0}}^{n} c_j\left(\frac{\alpha_j}{\alpha_{j_0}}\varphi_{j_0}(z)-\varphi_j(z)\right)$$

*The statement and solution of this problem for the case in which $\phi_j(z)=z^j$ and B is a closed interval of the real axis is due to V.A. Markov [1].

where c_j, for $j = 0, 1, \ldots, n$, and $j \neq j_0$, are arbitrary numbers. It follows from this that the problem stated is equivalent to the problem of best approximation of the function $(1/\alpha_{j_0})\,\varphi_{j_0}(z)$ on B by means of the polynomials

$$p^*(z) = \sum_{\substack{j=0 \\ j \neq j_0}}^{n} c_j \left(\frac{\alpha_j}{\alpha_{j_0}} \varphi_{j_0}(z) - \varphi_j(z) \right)$$

We use the usual notation

$$\mu(B') = \inf_{p(z) \in \mathscr{P}(\tilde{\alpha})} \sup_{z \in B'} |p(z)|, \quad B' \subset B$$

We shall refer to the set $B_m^{(0)} = \{z_l\}_1^m$ of distinct elements $z_l \in B$ as the characteristic set for the set B in the class $\mathscr{P}(\tilde{\alpha})$ if $\mu(B_m^{(0)}) = \mu = \mu(B)$ and if $\mu(B^*) < \mu(B_m^{(0)})$ for an arbitrary sub-set $B^* \subset B_m^{(0)}$ (where $B^* \neq B_m^{(0)}$). The characteristic set $B_m^{(0)}$ for B in the class $\mathscr{P}(\tilde{\alpha})$ coincides with the characteristic set for B in the problem of best approximation, to which is reduced our problem. Therefore, $1 \leqslant m \leqslant 2n+1$.

In all that follows, we shall assume that B is a compact set and that the functions $\varphi_j(z)$, for $j = 0, 1, \ldots, n$, are continuous on B. Suppose that

$$\pi(z) = \sum_{j=0}^{n} c_j^{(0)} \varphi_j(z)$$

is a generalised polynomial of the class $\mathscr{P}(\tilde{\alpha})$. Setting

$$\rho = \sup_{z \in B} |\pi(z)|$$

we denote by $B(\rho)$ the set of points $z \in B$ such that $|\pi(z)| = \rho$.

We shall use Theorem 2, Section 5.3.3, and the remarks that we have just made to prove the following theorem:

Theorem 1 (Vidensky [3, 6]*)* Suppose that B is a compact set and that the functions $\varphi_j(z)$, for $j = 0, 1, \ldots, n$, are continuous on B. For a generalised polynomial $\pi(z)$ to be a polynomial deviating least from zero on B out of all the polynomials

$$p(z) = \sum_{j=0}^{n} c_j \varphi_j(z) \in \mathscr{P}(\tilde{\alpha})$$

it is necessary and sufficient that there exist a set $B_m = \{z_l\}_1^m$, where $1 \leqslant m \leqslant 2n+1$, of distinct elements $z_l \in B(\rho)$ and that there exist numbers $\delta_l > 0$ for $l = 1, 2, \ldots, m$, such that

$$\sum_{l=1}^{m} \delta_l \overline{\pi(z_l)} \varphi_j(z_l) = \delta_0 \alpha_j, \quad j = 0, 1, \ldots, n \tag{1}$$

where

$$\delta_0 = \rho^2 \sum_{l=1}^{m} \delta_l$$

PROOF On the basis of Theorem 2, Section 5.3.3, it is necessary and sufficient that

$$\sum_{l=1}^{m} \delta_l \overline{\pi(z_l)} \left(\frac{\alpha_j}{\alpha_{j_0}} \varphi_{j_0}(z_l) - \varphi_j(z_l) \right) = 0, \quad j = 0, 1, \ldots, n, \quad j \neq j_0$$

or

$$\sum_{l=1}^{m} \delta_l \overline{\pi(z_l)} \varphi_j(z_l) = \frac{\alpha_j}{\alpha_{j_0}} \sum_{l=1}^{m} \delta_l \overline{\pi(z_l)} \varphi_{j_0}(z_l)$$

This last sum may be represented in the form $\delta_0 \alpha_{j_0} e^{i\beta}$, where $0 \leqslant \beta < 2\pi$ and $\delta_0 \geqslant 0$. Then, we have

$$\sum_{l=1}^{m} \delta_l \overline{\pi(z_l)} \varphi_j(z_l) = \delta_0 \alpha_j e^{i\beta}, \quad j = 0, 1, \ldots, n$$

If we multiply the j-th equation by c_j for each j and add the resulting equations, we obtain

$$\sum_{l=1}^{m} \delta_l |\pi(z_l)|^2 = \delta_0 e^{i\beta} \sum_{j=0}^{n} \alpha_j c_j = \delta_0 e^{i\beta}$$

Since the left-hand side of this equation is positive, $e^{i\beta} = 1$ and then the last equation takes the form

$$\rho^2 \sum_{l=1}^{m} \delta_l = \delta_0$$

This completes the proof of the theorem.

Notes and corollaries analogous to those following Theorem 2, Section 5.3.3, are applicable to this theorem. In particular, (1) if we set $\pi(z_l) = \rho e^{i\theta_l}$, where $0 \leqslant \theta_l < 2\pi$ and $\delta_0 = \rho$, Equations (1) take the form

$$\sum_{l=1}^{m} \delta_l e^{-i\theta_l} \varphi_j(z_l) = \alpha_j, \quad j = 0, 1, \ldots, n \tag{2}$$

in this case,

$$\rho \sum_{l=1}^{m} \delta_l = 1$$

(2) all the polynomials $\pi(z) \in \mathscr{P}(\tilde{\alpha})$ that deviate least from zero on B out of all the polynomials $p(z) \in \mathscr{P}(\tilde{\alpha})$ have identical values on the elements of each characteristic set $B_m^{(0)}$ for B in the class $\mathscr{P}(\tilde{\alpha})$. We then conclude on the basis of Corollary 1 to the Lemma in Section 5.3.4, that if m, the number of elements of the characteristic set for B, in the class $\mathscr{P}(\tilde{\alpha})$, satisfies the inequality $n+1 \leqslant m \leqslant 2n+1$ and if the system of functions $\varphi_j(z)$, for $j=0, 1, \ldots, n$ and $z \in B$, is a Chebyshev system, then, out of all the polynomials $p(z) \in \mathscr{P}(\tilde{\alpha})$, there exists a unique polynomial $\pi(z)$ that deviates least from zero on B. If $m \leqslant n$, there may not be a unique polynomial $\pi(z)$ that deviates least from zero on B.

Suppose that we know the characteristic set $B_m^{(0)} = \{z_l\}_1^m$ for B in the class $\mathscr{P}(\alpha)$, where $1 \leqslant m \leqslant n+1$. We denote by $\mathscr{P}^*(\tilde{\alpha})$ the class of generalised polynomials that deviate least from zero $B_m^{(0)}$ out of the polynomials $p(z) \in \mathscr{P}(\tilde{\alpha})$. If we assume that the rank of the matrix $\|\varphi_j(z_l)\|$ (for $j=0, 1, \ldots, n$, and $l=1, 2, \ldots, m$), is equal to m, we have a representation of the polynomials $p^*(z) \in \mathscr{P}^*(\tilde{\alpha})$.

Since the numbering of the functions $\varphi_j(z)$ is at our disposal, we shall assume that the determinant

$$\Delta = \begin{vmatrix} \varphi_0(z_1) & \ldots, & \varphi_0(z_m) \\ \cdot & \cdot\ \cdot\ \cdot & \cdot \\ \varphi_{m-1}(z_1), & \ldots, & \varphi_{m-1}(z_m) \end{vmatrix} \neq 0 \tag{3}$$

Suppose that $\pi^*(z) \in \mathscr{P}^*(\tilde{\alpha})$. On the basis of Theorem 1 (see condition (2)),

$$\sum_{l=1}^{m} \delta_l e^{-i\theta_l} \varphi_j(z_l) = \alpha_j, \quad j=0, 1, \ldots, n \tag{4}$$

where

$$\theta_l = \arg \pi^*(z_l), \ \delta_l > 0, \ \sum_{l=1}^{m} \delta_l = \frac{1}{\mu}, \ \mu = |\pi^*(z_l)|, \text{ and} \quad l=1, 2, \ldots, m$$

From the first m of Equations (4), we have

$$\delta_l e^{-i\theta_l} = \frac{D_l}{\Delta}, \quad l=1, 2, \ldots, m$$

where D_l is the determinant obtained from Δ by replacing the l-th column with the column $\alpha_0, \alpha_1, \ldots, \alpha_{m-1}$. Furthermore, we obtain successively

$$\delta_l = \frac{|D_l|}{|\Delta|}, \quad \sum_{k=1}^{m} \frac{|D_k|}{|\Delta|} = \frac{1}{\mu}, \quad \mu = \frac{|\Delta|}{\sum_{k=1}^{m} |D_k|}$$

$$\pi^*(z_l) = \mu e^{i\theta} = \frac{|\Delta|}{\sum_{k=1}^{m} |D_k|} \frac{\Delta}{D_l} \frac{|D_l|}{|\Delta|} = \frac{\Delta}{D_l} \frac{|D_l|}{\sum_{k=1}^{m} |D_k|} \tag{5}$$

$$l = 1, 2, \ldots, m.$$

We define

$$\psi_0(z) = \sum_{l=1}^{m} \pi^*(z_l)\,\omega_l(z), \quad \omega_l(z) = \frac{\Delta_l(z)}{\Delta} \tag{6}$$

where the numbers $\pi^*(z_l)$ are determined by Formulae (5) and $\Delta_l(z)$ is the determinant obtained from Δ by replacing the l-th column with the column $\varphi_0(z), \varphi_1(z), \ldots, \varphi_{m-1}(z)$. Noting that $L_{\tilde{\alpha}}\{\Delta_l(z)\} = D_l$, we have

$$L_{\tilde{\alpha}}\{\psi_0(z)\} = \sum_{l=1}^{m} \pi^*(z_l) \frac{1}{\Delta} L_{\tilde{\alpha}}\{\Delta_l(z)\} = \sum_{l=1}^{m} \frac{\Delta}{D_l} \frac{|D_l|}{\sum_{k=1}^{m} |D_k|} \frac{1}{\Delta} D_l = 1$$

and hence the polynomial $\psi_0(z) \in \mathscr{P}(\tilde{\alpha})$. Since $\psi_0(z_l) = \pi^*(z_l)$, for $l = 1, 2, \ldots, m$, we have $\psi_0(z) \in \mathscr{P}^*(\tilde{\alpha})$. But all the polynomials $p^*(z) \in \mathscr{P}^*(\tilde{\alpha})$ have the same values on elements of the set $B_m^{(0)}$; therefore, every polynomial $p^*(z)$ can be represented in the form $p^*(z) = \psi_0(z) + \psi(z)$, where $\psi(z)$ is a polynomial satisfying the conditions $\psi(z_l) = 0$, for $l = 1, 2, \ldots, m$. But every polynomial $\psi(z)$ of this kind can be represented in the form

$$\psi(z) = \sum_{k=1}^{n+1-m} c_k \psi_k(z)$$

where

$$\psi_k(z) = \begin{vmatrix} \varphi_0(z_1) & , \ldots, & \varphi_0(z_m) & , \varphi_0(z) \\ \cdot & \cdots & \cdot & \cdot \\ \varphi_{m-1}(z_1) & , \ldots, & \varphi_{m-1}(z_m) & , \varphi_{m-1}(z) \\ \varphi_{m-1+k}(z_1), & \ldots, & \varphi_{m-1+k}(z_m), & \varphi_{m-1+k}(z) \end{vmatrix} \tag{7}$$

$$k = 1, 2, \ldots, n-1-m$$

and the c_k are constants. The determinant $L_{\tilde{\alpha}}\{\psi_k(z)\}$ obtained from the determinant (7) by replacing the last column with the column $\alpha_0, \ldots, \alpha_{m-1}, \alpha_{m-1+k}$, is equal to zero since the system (4) has a non-zero solution. From this, we easily conclude that the following theorem is valid.

Theorem 2 If the determinant $\Delta \neq 0$ (see (3)), then every polynomial $p^*(z) \in \mathcal{P}^*(\tilde{\alpha})$ can be represented in the form

$$p^*(z) = \psi_0(z) + \sum_{k=1}^{n+1-m} c_k \psi_k(z) \tag{8}$$

where the c_k are constants. For arbitrary constants c_k, Formula (8) defines a generalised polynomial $p^*(z) \in \mathcal{P}^*(\tilde{\alpha})$.

If $m = n+1$ there exists a unique polynomial $\psi_0(z) \in \mathcal{P}^*(\tilde{\alpha})$. Consequently, $\psi_0(z)$ is the unique polynomial that deviates least from zero on B out of all the $p(z) \in \mathcal{P}(\tilde{\alpha})$. If $1 \leqslant m \leqslant n$, the problem of finding the $\pi(z) \in \mathcal{P}(\tilde{\alpha})$ that deviates least from zero on B has been reduced to the problem of finding the polynomial of best approximation on B for the function $\psi_0(z)$ by means of generalised polynomials

$$\psi(z) = \sum_{k=1}^{n+1-m} c_k \psi_k(z)$$

Here, we know the measure

$$\mu = |\Delta| \left(\sum_{k=1}^{m} |D_k| \right)^{-1}$$

of best approximation of the function $\psi_0(z)$ by means of the polynomials $\psi(z)$. If the system of functions $\psi_l(z)$, for $l = 1, 2, \ldots, n+1-m$, is a Chebyshev system, there exists a unique polynomial $\pi(z) \in \mathcal{P}(\tilde{\alpha})$ that deviates least from zero on B; if it is not, uniqueness is not guaranteed.

Obviously, these considerations do not provide an efficient method of constructing the polynomial $\pi(z)$ that deviates least from zero on B out of all the polynomials $p(z) \in \mathcal{P}(\tilde{\alpha})$. We shall now give some re-wordings of Theorems 1 and 2 that will make it possible to obtain certain exact estimates.

We introduce some notations that we shall use in the statements of the following theorems. Let

$$\Pi(z) = \sum_{j=0}^{n} c_j^{(0)} \varphi_j(z)$$

denote an arbitrary generalised polynomial that is not identically equal to zero on B. Define $\rho = \sup_{z \in B} |\Pi(z)|$. Let $B_{m_0}^{(0)} = \{z_l\}_1^{m_0}$, where $1 \leqslant m_0 < +\infty$, be an arbitrary set of distinct elements $z_l \in B$ such that $|\Pi(z_l)| = \rho$.

Theorem 3 Suppose that δ_l, for $l = 1, 2, \ldots, m_0$, are non-negative numbers such that

$$\rho^2 \sum_{l=1}^{m_0} \delta_l = 1$$

We define

$$\alpha_j = \sum_{l=1}^{m_0} \delta_l \overline{\Pi(z_l)} \varphi_j(z_l), \quad j = 0, 1, \ldots, n \tag{9}$$

Then, the generalised polynomial $\Pi(z)$ is a polynomial that deviates least from zero on the set B out of all the polynomials $p(z) \in \mathscr{P}(\tilde{\alpha})$, where $\tilde{\alpha} = (\alpha_0, \alpha_1, \ldots, \alpha_n)$. Also, $\mu = \rho$.

PROOF We note that the α_j, for $j = 0, 1, \ldots, n$, are not all equal to zero and that $\Pi(z) \in \mathscr{P}(\tilde{\alpha})$ since

$$\sum_{j=0}^{n} \alpha_j c_j^{(0)} = \sum_{l=1}^{m_0} \delta_l \overline{\Pi(z_l)} \sum_{j=0}^{n} c_j^{(0)} \varphi_j(z_l) = \sum_{l=1}^{m_0} \delta_l |\Pi(z_l)|^2 = \rho^2 \sum_{l=1}^{m_0} \delta_l = 1$$

By applying Theorem 1 (the sufficiency) and taking as our set B_m, the set (referred to in the statement of Theorem 1) of these elements $z_l \in B_{m_0}^{(0)}$ to which positive values of δ_l correspond, we arrive at the conclusion of the theorem.

If the set $B(\rho)$ of all elements $z \in B$ such that $|\Pi(z)| = \rho$ consists of a finite number of distinct elements z_l, where $l = 1, 2, \ldots, m$, we may take $B(\rho)$ instead of $B_{m_0}^{(0)}$. It is then easy to prove the following theorem.

Theorem 4 For every generalised polynomial

$$p(z) = \sum_{j=0}^{n} c_j \varphi_j(z)$$

satisfying the conditions $p(z)| \leqslant \rho$, where $z \in B$, to satisfy the inequality

$$|L_{\tilde{\alpha}}\{p(z)\}| \leqslant |L_{\tilde{\alpha}}\{\Pi(z)\}|, \quad L_{\tilde{\alpha}}\{p(z)\} = \sum_{j=0}^{n} \alpha_j c_j \tag{10}$$

it is necessary and sufficient that there exist non-negative numbers δ_l, where $l=1, 2, \ldots, m$, and a number ε such that $|\varepsilon|=1$, such that

$$\alpha_j = \varepsilon \sum_{l=1}^{m} \delta_l \overline{\Pi(z_l)}\, \varphi_j(z_l),$$

$$j=0, 1, \ldots, n$$

Equality holds in (10) for the polynomials $p(z)=\eta\Pi(z)$, where $|\eta|=1$ (possibly only for these polynomials).

In what follows, we shall also assume that $1 \leqslant m \leqslant n+1$ and that the determinant Δ (see (3)) is non-zero. Theorem 3 may then be re-worded as follows:

Theorem 3′ Suppose that the polynomial

$$\Pi(z) = \sum_{j=0}^{n} c_j^{(0)} \varphi_j(z)$$

and the numbers α_j, for $j=0, 1, \ldots, n$, and ε, where $|\varepsilon|=1$, satisfy the conditions

$$\delta_l = \varepsilon \Pi(z_l) L_{\tilde{\alpha}} \{\omega_l(z)\} \geqslant 0, \quad l=1, 2, \ldots, m, \sum_{l=1}^{m} \delta_l > 0 \tag{11}$$

$$L_{\tilde{\alpha}} \{\psi_k(z)\} = 0, \quad k=1, 2, \ldots, n+1-m \tag{12}$$

where the polynomials $\omega_l(z)$ and $\psi_k(z)$ satisfy Formulae (6) and (7) respectively. Then,

$$b = \sum_{j=0}^{n} \alpha_j c_j^{(0)} \neq 0$$

and the generalised polynomial $\Pi(z)$ is one which deviates least from zero on B out of all the polynomials $p(z)$ such that $ap(z) \in \mathscr{P}(\tilde{\alpha})$, where $a=b^{-1}$.

PROOF It will be sufficient for us to show that $b \neq 0$ and that the conditions of Theorem 1 are satisfied for the polynomial $\pi(z)=a\Pi(z)$. If we denote by $\Delta_{\nu, l}$ the co-factor of the element $\varphi_\nu(z_l)$ of the determinant Δ, we have, for $j=0, 1, \ldots, m-1$

$$\sum_{l=1}^{m} \delta_l \overline{\Pi(z_l)} \varphi_j(z_l) = \sum_{l=1}^{m} \varepsilon \Pi(z_l) \frac{D_l}{\Delta} \overline{\Pi(z_l)} \varphi_j(z_l)$$

$$= \rho^2 \varepsilon \frac{1}{\Delta} \sum_{l=1}^{m} \left(\sum_{\nu=0}^{m-1} \alpha_\nu \Delta_{\nu, l} \right) \varphi_j(z_l) = \rho^2 \varepsilon \frac{1}{\Delta} \sum_{\nu=0}^{m-1} \alpha_\nu \sum_{l=1}^{m} \Delta_{\nu, l} \varphi_j(z_l)$$

$$= \rho^2 \varepsilon \frac{1}{\Delta} \alpha_j \sum_{l=1}^{m} \Delta_{j, l} \varphi_j(z_l) = \rho^2 \varepsilon \alpha_j$$

If we denote by $\Delta_\nu^{(k)}$ the co-factor of the element in the last column and ν-th row of the determinant (7), condition (12) may be written in the form

$$\alpha_j \Delta + \sum_{\nu=0}^{m+1} \alpha_\nu \Delta_\nu^{(j)} = 0, \quad j = m, m+1, \ldots, n$$

By analogy with the above, we have, for $j = m, m+1, \ldots, n$

$$\sum_{l=1}^{m} \delta_l \overline{\Pi(z_l)} \varphi_j(z_l) = \rho^2 \varepsilon \frac{1}{\Delta} \sum_{\nu=0}^{m-1} \alpha_\nu \sum_{l=1}^{m} \Delta_{\nu, l} \varphi_j(z_l)$$

$$= \rho^2 \varepsilon \frac{1}{\Delta} \left(- \sum_{\nu=0}^{m-1} \alpha_\nu \Delta_\nu^{(j)} \right) = \rho^2 \varepsilon \alpha_j$$

Thus,

$$\sum_{l=1}^{m} \delta_l \overline{\Pi(z_l)} \varphi_j(z_l) = \rho^2 \varepsilon \alpha_j, \quad j = 0, 1, \ldots, n. \tag{13}$$

If we multiply the jth equation by $c_j^{(0)}$, for $j = 0, 1, \ldots, n$, and add the resulting equations, we have

$$\rho^2 \sum_{l=1}^{m} \delta_l = \rho^2 \varepsilon \sum_{j=0}^{n} \alpha_j c_j^{(0)} = \rho^2 \varepsilon \overline{b} > 0$$

From this we conclude that $b \neq 0$. If we multiply (13) by $\overline{a}$ and set

$$\delta_0 = |a|^2 \rho^2 \sum_{l=1}^{n} \delta_l = \rho^2 \varepsilon \overline{a}$$

we obtain

$$\sum_{l=1}^{m} \delta_l \overline{\pi(z_l)} \varphi_j(z_l) = \delta_0 \alpha_j, \quad j = 0, 1, \ldots, n$$

where $\delta_l \geqslant 0$, for $l = 1, 2, \ldots, m$. In view of Theorem 1, this completes the proof of the present theorem.

If we set $B(\rho)$ of all elements $z \in B$ such that $|\Pi(z)| = \rho$, consists of m, distinct elements z_l, where $1 \leqslant m \leqslant n+1$, and $l = 1, 2, \ldots, m$, we have an analogy with Theorem 4.

Theorem 4' For Equation (10) to be satisfied for every generalised polynomial

$$p(z) = \sum_{j=0}^{n} c_j \varphi_j(z)$$

that satisfies the condition $|p(z)| \leqslant \rho$ for $z \in B$, it is necessary and sufficient that the numbers α_j, for $j = 0, 1, \ldots, n$, satisfy conditions (1) and (12) (without the condition that $\sum_{l=1}^{m} \delta_l > 0$ however) for some ε such that $|\varepsilon| = 1$. Then, equality holds in (10) for the polynomials $p(z) = \eta \Pi(z)$, where $|\eta| = 1$ (possibly not only for these).

Let us consider some special cases of Theorem 4'. Let $P_n(z)$ be an algebraic polynomial of degree $n \geqslant 1$, all the zeros of which lie in the open disc $|z| < 1$ and let $\mathscr{P}_n$ be the class of algebraic polynomials

$$p_n(z) = \sum_{j=0}^{n} c_j z^j$$

of degree not exceeding n that satisfy the condition $|p_n(z)| \leqslant |P_n(z)|$ on the circle $|z| = 1$. We introduce the generalised polynomials

$$p(z) = \frac{p_n(z)}{P_n(z)} = \sum_{j=0}^{n} c_j \frac{z^j}{P_n(z)}$$

and we set

$$\Pi(z) = \frac{(1-\lambda) P_n(z) + \lambda z^n e^{i\gamma} \overline{P_n\left(\frac{1}{\bar{z}}\right)}}{P_n(z)}, \quad 0 < \lambda \leqslant 1, \quad 0 \leqslant \gamma < 2\pi$$

where λ and γ are real constants. For our set B, we take the set of points on the circle $|z| = 1$. It is easy to see that $\rho = \sup_{z \in B} |\Pi(z)| = 1$, and that for the set B_m we should take the set of zeros of the polynomial

$$\omega(z) = P_n(z) - e^{i\gamma} z^n \overline{P_n\left(\frac{1}{\bar{z}}\right)}$$

that lie on the circle $|z|=1$. On the basis of Note 3, Section 5.1.3, all the zeros of $\omega(z)$ lie on the circle $|z|=1$ and they are simple; that is, $B_n=\{z_l\}_1^n$, and $\omega(z_l)=0$.

If we apply this to the case of Theorem 4′, we obtain the following theorem.

Theorem 5 For the inequality

$$|L_{\tilde{\alpha}}\{p_n(z)\}|\leqslant|L_{\tilde{\alpha}}\{P_n(z)\}| \qquad L_{\tilde{\alpha}}\{p_n(z)\}=\sum_{j=0}^{n}\alpha_j c_j \tag{14}$$

to hold for every polynomial

$$p_n(z)=\sum_{j=1}^{n}c_j z^j\in\mathscr{P}_n$$

and for equality to hold in it for some polynomial of the form

$$p_n(z)=\eta\left[(1-\lambda)P_n(z)+\lambda e^{i\gamma}z^n\overline{P_n\left(\frac{1}{\bar{z}}\right)}\right],\ |\eta|=1,\ 0<\lambda\leqslant1, \tag{15}$$

with given λ and γ, where $0\leqslant\gamma<2\pi$, it is necessary and sufficient that there exist a number ε, such that $|\varepsilon|=1$ and

$$\varepsilon P_n(z_l)L_{\tilde{\alpha}}\left\{\frac{\omega(z)}{(z-z_l)\,\omega'(z_l)}\right\}\geqslant0,\quad l=1,2,\ldots,n,\quad L_{\tilde{\alpha}}\{\omega(z)\}=0.$$

Equality will then hold in (14) for the polynomials (15) for all η such that $|\eta|=1$ and all λ such that $0\leqslant\lambda\leqslant1$ and only for these polynomials.

The last assertion is proven by reasoning as in the proof of Theorem 1, Section 5.1.3.

If we set $P_n(z)=Mz^n$ in Theorem 5, where $M>0$, we obtain the following corollary.

COROLLARY 1 For the inequality

$$\left|\sum_{\nu=0}^{n}\alpha_\nu c_\nu\zeta^\nu\right|\leqslant M|\alpha_n|\quad\text{for all } \zeta, \text{ such that } |\zeta|=1 \tag{16}$$

to hold for every polynomial

$$p_n(z)=\sum_{\nu=0}^{n}c_\nu z^\nu$$

of degree not exceeding n that satisfies the condition $|p_n(z)|\leqslant M$ on the circle $|z|=1$ and for equality to hold in it for an arbitrarily given $\zeta=\zeta_0$ (where $|\zeta_0|=1$) for every polynomial $p_n(z)$ of the form

$$p_n(z)=\eta M\left[(1-\lambda)z^n+\lambda e^{i\gamma}\zeta_0\right],\quad |\eta|=1,\quad 0<\lambda\leqslant 1 \tag{17}$$

with given λ and γ, such that $0\leqslant\gamma<2\pi$, it is necessary and sufficient that there exist a number ε, such that $|\varepsilon|=1$ and

$$\varepsilon\sum_{\nu=0}^{n-1}\alpha_\nu z_l^{n-\nu}\geqslant 0,\quad l=1,2,\ldots,n$$

$$z_l=e^{i\theta_l},\quad \theta_l=\frac{\gamma}{n}+\frac{2\pi}{n}l,\quad \alpha_n=\alpha_0 e^{i\gamma}$$

Equality holds in (16) for $\zeta=\zeta_0$ for the polynomials (17) for all η such that $|\eta|=1$ and all λ such that $0\leqslant\lambda\leqslant 1$ and only for these polynomials.

First we take inequality (16) for $\zeta=1$ Then, by recalling that $|p_n(\zeta z)|\leqslant M$ on $|z|=1$ for $|\zeta|=1$, we obtain inequality (16) for all $|\zeta|=1$.

By considering polynomials of even degree, we easily obtain the following corollary.

COROLLARY 2 For the inequality

$$\left|\alpha_0 c_0+\sum_{\nu=1}^{n}\left(\alpha_\nu e^{i\gamma}c_\nu\zeta^\nu+\alpha_{-\nu}e^{-i\gamma}c_{-\nu}\zeta^{-\nu}\right)\right|\leqslant M|\alpha_n| \tag{18}$$

for all ζ, such that $|\zeta|=1$

where γ is a constant in $[0,2\pi]$, to hold for every generalised polynomial

$$p_n(z)=\sum_{\nu=-n}^{n}c_\nu z^\nu$$

that satisfies the condition $|p_n(z)|\leqslant M$ (where $M>0$) on the circle $|z|=1$, and for equality to hold in it for an arbitrarily given $\zeta=\zeta_0$ such that $|\zeta_0|=1$ for some polynomial $p_n(z)$ of the form

$$p_n(z)=\eta M\left[(1-\lambda)(\bar{\zeta}_0 z)^n+\lambda e^{2i\gamma}(\zeta_0 z)^{-n}\right],\quad |\eta|=1,\quad 0<\lambda\leqslant 1 \tag{19}$$

it is necessary and sufficient that there exist a number ε such that $|\varepsilon|=1$ and

$$\varepsilon\left[\alpha_{-n}+\alpha_n+2\alpha_0(-1)^l+2\sum_{\nu=1}^{n-1}\left(\alpha_\nu z_l^{n-\nu}+\alpha_{-\nu}z_l^{-(n-\nu)}\right)\right]\geqslant 0$$

$$l=0,1,\ldots,2n-1$$

$$z_l=e^{i\theta_l},\quad \theta_l=\frac{\gamma}{n}+\frac{\pi}{n}l,\quad \alpha_{-n}=\alpha_n$$

Equality holds in (18) for $\zeta=\zeta_0$ for the polynomials (19) for all η such that $|\eta|=1$ and all λ such that $0\leqslant\lambda\leqslant 1$ and only for such polynomials.

If we set $z=\cos\theta+i\sin\theta$ for $0\leqslant\theta<2\pi$, we easily obtain the following corollary.

COROLLARY 3 For the inequality

$$\left|\sum_{\nu=0}^{n}[(\lambda_\nu a_\nu-\mu_\nu b_\nu)\cos(\nu\theta+\gamma)+(\lambda_\nu b_\nu+\mu_\nu a_\nu)\sin(\nu\theta+\gamma)]\right| \leqslant M|\lambda_n-\mu_n i| \tag{20}$$

where the λ_ν, the μ_ν, and γ are constants, with $0\leqslant\gamma<2\pi$ and $\mu_0=0$, to hold for every trigonometric polynomial

$$s_n(\theta)=a_0+\sum_{\nu=1}^{n}(a_\nu\cos\nu\theta+b_\nu\sin\nu\theta)$$

that satisfies the condition $|s_n(\theta)|\leqslant M$ (where $M>0$), for every θ in $[0, 2\pi]$ and for equality to hold in it for an arbitrarily given $\theta=\theta_0$, $\in[0, 2\pi]$ for some trigonometric polynomial $s_n(\theta)$ of the form

$$s_n(\theta)=\eta M[\cos n(\theta-\theta_0-\gamma)+i(1-2\lambda)\sin n(\theta-\theta_0-\gamma)] \quad |\eta|=1,\ 0<\lambda\leqslant 1 \tag{21}$$

it is necessary and sufficient that there exist a number ε such that $|\varepsilon|=1$ and

$$\varepsilon\left\{\lambda_0\cos n\theta_l+\lambda_n+2\sum_{\nu=1}^{n-1}(\lambda_\nu\cos(n-\nu)\theta_l+\mu_n\sin(n-\nu)\theta_l)\right\}\geqslant 0$$
$$l=0, 1, \ldots, n-1,\quad \theta_l=\frac{\gamma}{n}+\frac{\pi}{n}l,\quad \mu_n=0.$$

Equality then holds in (20) for the trigonometric polynomials (20) for all η such that $|\eta|=1$ and all λ such that $0\leqslant\lambda\leqslant 1$ and only for such polynomials.

If we assume that the numbers a_ν, b_ν, λ_ν and μ_ν in this corollary are real and that $\mu_n=0$, $\lambda=1/2$, and $\eta=1$, we obtain a result due to Bernstein [1], p. 173, which generalises a result of Szegö.

NOTE By analogy with the above exposition, we might examine generalised polynomials that deviate least from zero whose coefficients satisfy two or more linear conditions, but we shall not stop to do this. See also Gol'shtein [1].

SUPPLEMENTARY LITERATURE

In addition to the works cited in the text, we mention the monographs dealing with the material of the present chapter: V.L. Goncharov [1], Chapter 5; I.I. Ibragimov [1] and B.Ya. Levin [1], Chapter 9. We also mention the following: N.I. Akhiezer [2], N.I. Akhiezer and B. Ya. Levin [1], V.S. Vidensky [8], E.V. Voronovskaya [1 to 4], E.I. Zoltarev [1], S.I. Zukhovitskii [1, 2], S.I. Zukhovitskii, R.A. Poliak and M.E. Primak [1], I.I. Ibragimov and R.G. Mamedov [1], N.N. Meiman [2 to 4], I. Yu. Ryzhakov [1] and S. Ya. Khavinson [2].

References

AKHIEZER, N.I.

1 'Weighted approximation of continuous functions by polynomials on the whole of the real axis', *Usp. mat. Nauk,* **11**, 4, 3-43 (1956).

2 *Lecture on the theory of approximation,* Gostekhizdat (1947).

AKHIEZER, N.I. and LEVIN, B.Ya.

1 'Generalised inequalities of S.N. Bernstein for derivatives of entire functions', Collection *Investigations of current problems in the theory of functions of a complex variable,* 111-165 (1960).

ALEKSANDROV, P.S.

1 *Introduction to general theory of sets and functions,* Gostekhizdat (1948).

ALPER, S.Y.

1 'Uniform approximation of functions of a complex variable in a closed domain', *Izv. Akad. Nauk SSSR, Maths series,* **19**, 423-444 (1955).

2 'Convergence of Lagrange interpolation polynomials in a complex domain', *Usp. mat. Nauk,* **11**, 5 (71), 44-50 (1956).

ALPER, S.Y. and IVANOV, V.V.

1 'Approximation of functions by partial sums of series of Faber polynomials', *Dokl. Akad. Nauk SSSR*, **90**, No. 3, 325-328 (1953).

ARONSHAIN, N.

1 'Theory of reproductive kernels', *Periodical collection of translations of foreign articles, Maths,* **7**, 2, 67-130 (1963).

BAZILEVICH, I.E.

1 'Distortion theorems and coefficients of schlicht functions', *Mat. Sb.* **28**, (70), No. 1, 147-164 (1951).

BERGMAN, S

1 'The kernel function and conformal mapping', *Mathl Survs*, **5**, American Mathematical Society, New York (1950).

BERNSTEIN, S.N.

1 *Constructive theory of functions* (1905-1930) in Collected works, **I**, Akademii Nauk SSSR (1952).

2 *Constructive theory of functions* (1932-1953) in Collected works, **II**, Akademii Nauk SSSR (1954).

BOCHNER, S.

1 'Über ortogonale Systeme analytischer Funktionen', *Math. Z.*, **14**, 180-207 (1922).

BRUIJN, N.G.

1 'Inequalities concerning polynomials in the complex domain', *Indag. math.*, **9**, No. 5, 591-598 (1947).

CARLEMAN, T.

1 'Über die Approximation analytischer Funktionen durch lineare Aggregate von vorgegebenen Potenzen', *Ark. Mat. Astr. Fys.*, **17** (1922-23).

CHEBYSHEV, P.L.

1 *Questions of smallest values connected with approximate representation of functions*, in complete works, **II**, Akademii Nauk SSSR (1947).

COURANT, R.

1 *Dirichlet's principle, conformal representations and minimal surfaces*, IL (1953).

DOCHEV, KIRIL

1 'A theorem of S.N. Bernstein', *Dokl. Akad. Nauk SSSR*, **146**, No. 1, 17-19 (1962).

DOLZHENKO, E.P.

1 'Estimates of derivatives of rational functions', *Izv. Akad. Nauk SSSR, Maths series*, **27**, No. 1, 9-28 (1963).

DRAGILEV, M.M.

1 'Stability of the base $\{z^n\}$', *Izv. vuzov, mathematics*, **3**, 67-73 (1958).

2 'Regular convergence of basic expansions of analytic functions', *Nauch. Dokl. vўssh. Shk, fiz.-mat. n.*, **4**, 27-32 (1958).

3 'Basis of regular convergence in the space of analytic functions', *Nauch. Dokl. vўssh. Shk, fiz.-mat. n.*, **6**, 61-70 (1958).

4 'Canonical form of basis of the space of analytic functions', *Usp. mat. Nauk*, **15**, No. 2, 181-188 (1960).

DZHRBASHIAN, M.M.

1 *Metric theorems of completeness and representability of analytic functions*, Dissertation, Erevan, (1948) (see Mergelian, **VIII**).

2 'Questions of the theory of weighted polynomial approximation in the complex domain', *Mat. Sb.*, **36** (78), No. 3, 353-440 (1955).

3 'Best approximation by entire functions in the complex domain', *Izv. Akad. Nauk SSSR, Maths series*, **20**, No. 4, 485-512 (1956).

4 'Expansion of an analytic function into a series of rational functions with a given set of poles', *Izv. Akad. Nauk Armyan. SSR, Maths. series*, **10**, No. 1, 21-29 (1957).

5 'Orthogonal systems of rational functions on a circle with a given set of poles', *Dokl. Akad. Nauk SSSR*, **147**, No. 6, 1278-1281 (1962).

6 'Expansion with respect to systems of rational functions with fixed poles', *Dokl. Akad. Nauk SSSR*, **143**, No. 1, 17-20 (1962).

DZHRBASHIAN, M.M. and KHACHATRIAN, I.O.

1 'Complete system of functions $\{z^k\}$ in the complex domain for weighted quadratic approximation', *Dokl. Akad. Nauk SSSR*, **110**, No. 6, 914-917 (1956).

DZHRBASHIAN, M.M. and TAMADIAN, A.P.

1 'Best approximation with entire functions in the complex domain', *Izv. Akad. Nauk SSSR, Maths series*, **20**, 485-512 (1956).

DZIADYK, V.K.

1 'S.M. Nikolovsky's problem in the complex domain', *Izv. Akad. Nauk SSSR, Maths series*, **23**, 697-736 (1959).

2 'Inverse theorems of approximation of functions in domains with angular singularities', *Dokl. Akad. Nauk SSSR*, **127**, No.3, 505-508 (1959).

EROKHIN, V.

1 'Best approximation of analytic functions by means of rational fractions with free poles', *Dokl. Akad. Nauk SSSR*, **128**, No. 1, 29-32 (1959).

EVGRAFOV, M.A.

1 'Interpolation of entire functions', *Mathematics in USSR over 40 years 1917-1957*, **1**, 398-444 (1959).

EVGRAFOV, M.A. and SOLOVEV, A.D.

1 'A general criterion of a basis', *Dokl. Akad. Nauk SSSR*, **113**, No. 3, 493-496 (1957).

FABER, G.

1 'Über polynomische Entwicklungen', *Math. Ann.*, **57**, 389-408 (1903); **64**, 116-135 (1907).

2 'Über Tschebyscheffsche Polynome', *Grell J.*, 150 (1920).

3 'Über nach Polynomen fortschreitende Reihe', *Münchn. Ber.*, 157-178, (1922).

FEJER, L.

1 'Über Interpolation', *Göttingen Nachr.*, 66-91 (1916).

2 'Interpolation und konforme Abbildung', *Nachr. Ges. Wiss.* Göttingen, 319-331 (1918).

FEKETE, M.

1 'Über die Verteilung der Wurzeln bei gewissen algebraischen Gleichungen mit ganzzahligen Koeffizienten', *Math. Z.*, **17**, 228-249 (1923).

2 'Über Interpolation', *Z. angew. Math. Mech.*, **6**, 410-413 (1926).

3 'Über den transfiniten Durchmesser ebenen Punktmengen', *Math. Z.* **37**, 635-646 (1933).

GAGUA, M.

1 'Complete systems of functions', Collection *Investigations of current problems in the theory of functions of a complex variable*, 208-211 (1961).

GAKHOV, F.D.

1 *Boundary problems*, Fizmatgiz (1963).

GELFOND, A.O.

1 *Calculus of finite differences*, Fizmatgiz (1959).

GERONIMUS, Ya.L.

1 *Theory of orthogonal polynomials*, Gostekhizdat (1950).

2 *Polynomials, orthogonal on a circle and on an interval*, **IV**, 77, Fizmatgiz (1958).

3 'Certain asymptotic properties of polynomials', *Mat. Sb.*, **23** (65), 77-88 (1948).

GO CHU-JOO

1 'Improvement of a theorem of S.N. Mergelian', *Acta math. Sin.*, **9**, No.3, 271-280 (1959).

GOL'SHTEIN, E.G.

1 'Problem of best Chebyshev approximation in the complex domain with additional conditions for the coefficients of the approximating polynomials', *Dokl. Akad. Nauk SSSR*, **141**, No.2, 274-276 (1961).

GOLUZIN, G.M.

1 *Geometric theory of functions of a complex variable*, Gostekhizdat (1952).

2 'Complete systems of functions in the complex domain', *Uchen. Zap. Leningr. gos. Univ.*, **37**, 48-51 (1939).

3 'The Carathéodory-Fejer problem and one analogous problem', *Mat. Sb.* **18** (60), No. 2, 214-226 (1946).

GONCHAR, A.A.

1 'Best approximation of measurable functions by rational functions and polynomials', *Izv. Vuzov, mat.*, **6**, 74-81 (1960).

2 'Inverse theorems of best approximation by rational functions', *Izv. Akad. Nauk SSSR, Maths series*, **25**, 347-356 (1961).

GONCHAROV, V.L.

1 *Theory of interpolation and approximation of functions*, Gostekhizdat (1954).

GROMOV, V.P.

1 'Complete systems of derivatives of analytic functions', *Izv. Akad. Nauk SSSR, Maths series*, **25**, 543-556 (1961).

GRONWALL, T.H.

1 'A sequence of polynomials connected with the roots of unity', *Bull. Amer. math. Soc.*, **27** (1920-21).

GUSEV, V.A.

1 'Functionals of derivatives of algebraic polynomials and the theorem of V.A. Markov', *Izv. Akad. Nauk SSSR, Maths series*, **25**, No. 3, 367-384 (1961).

HAAR, A.

1 'Die Minkowskische Geometrie und die Annaherung an stetige Funktionen', *Math. Ann.*, **78** (1914).

HALL, T.

1 *On polynomials bounded at an infinity of points*, Uppsala (1950).

IBRAGIMOV, I.I.

1 'Extremal properties of entire functions of a finite exponential type', *Izv. Akad. Nauk Azerb. SSR* (1962).

IBRAGIMOV, I.I. and MAMEDOV, R.G.

1 'Inequality of the S.N. Bernstein type for analytic functions', Collection *Investigations of current problems in the theory of functions of a complex variable*, 269-273 (1961).

IGNATEVA, A.V. and VOSTRETSOV, B.A.

1 'On the question of the rate of approximation of functions on arbitrary continuums', *Uchen. Zap. Mosc. obl. ped. in-ta, Trudȳ Kaf. mat.*, **57**, No. 4, 45-50 (1957).

ILIEV, L.

1 'Series of Faber polynomials, the coefficients of which have a finite

number of values', *Dokl. Akad. Nauk SSSR*, **40**, No. 4, 499-502 (1953).

IVANOV, V.K.

1 'Problem of minimax of a system of linear functions', *Mat. Sb.*, 28 (70), No. 3, 685-706 (1951).

2 'Uniform approximation of continuous functions', *Mat. Sb.*, **30** (72), No. 3, 543-558 (1952).

3 Letter to editor, *Mat. Sb.*, **33** (75), No. 3, 676 (1953).

KALASHNIKOV, M.D.

1 'Polynomials of best (quadratic) approximation on a given system of points', *Dokl. Akad. Nauk SSSR*, **105**, No. 4, 634-636 (1955).

KALMAR, L.

1 'Über Interpolation', *Mat. Phys. Lap.*, 120-149 (1926).

KAZMIN, U.A.

1 'Infinite systems of linear equations and bases of analytic functions', *Dokl. Akad. Nauk SSSR*, **106**, No. 2, 179-182 (1956).

2 'Completeness of systems of functions of the form $\{f(z+\alpha_n)\}$ and $\{f^{(n)}(z)\}$', *Usp. mat. Nauk*, 12, No. 2, (74), 151-154 (1957).

3 'Completeness of certain systems of functions in L_2 and C', *Sb. nauchno-tech. Rab. Azovo-Chernomorskovo in-ta mech. elektrifikh.s. kh.*, No. 11, Part 2, 193-199 (1959).

4 'Completeness of a system of analytic functions I and II', *Vest. Moskgos. Univ., Mat. and mechan.*, No. 5, 3-13, No.6, 11-19 (1960).

5 'Order of best approximation of functions, defined on a continuum, by linear combinations of elements of a continuable basis' *Dokl. Akad. Nauk SSSR*, **139**, No. 3, 528-530 (1961).

KELDYSH, M.B.

1 'Sur l'approximation en moyenne quadratique des fonctions analytiques', *Mat. Sb.*, **5** (47), 391-402 (1939).

2 'Sur l'approximation en moyenne par polynômes des fonctions d'une variable complexe', *Mat. Sb.*, **16** (58), 1-20 (1945).

KELDYSH, M.B. and LAVRENTEV, M.A.

1 'Sur la representation conforme des domaines limités par des courbes rectifiables', *Ann. Écl. Norm.*, **54**, 1-38 (1937).

KHAPLANOV, M.G.

1 'The completeness of certain systems of analytic functions', *Uchen. Zap. Rostov. n/D gos. Ped. in-ta*, No. 3, 53-58 (1955).

KHAVIN, B.P.

1 'Analytic continuation of power series and Faber polynomials', *Dokl. Akad. Nauk SSSR*, **118**, No.5, 879-881 (1958).

2 'Problems of moments and analytic continuation of power series', Collection *Investigations of current problems in the theory of functions of a complex variable*, 113-131 (1961).

KHAVINSON, S.Ya.

1 'Analytic capacity of sets, joint non-trivial nature of various classes of analytic functions and Schwartz's lemma in arbitrary domains', *Mat. Sb.*, **54**, (96), 1, 3-50 (1961).

2 'Theory of extremal problems for bounded analytic functions satisfying additional conditions in a domain', *Usp. mat. Nauk*, **18**, 2 (110), 25-98 (1963).

KHELLI

1 'Classes of convex bodies with common points', *Usp. mat. Nauk*, No. 2, 80-81 (1936).

KISH, O.

1 'Notes on interpolation', *Acta math. Hung.*, **11**, No. 1-2, 49-64 (1960).

KOLMOGOROV, A.N.

1 'Remarks concerning P.L. Chebyshev polynomials deviating least from a given function', *Usp. mat. Nauk*, **3**, No. 1, 216-221 (1948).

KOROBEINIK, U.F.

1 'Necessary and sufficient criterion of whether a basis is of a quasi-power nature', *Sb. stud. Rab. Rostov. in-ta*, No. 3, 65-68 (1957).

KOROVKIN, P.P.

1 'Asymptotic expression for polynomials, orthogonal on a rectifiable contour', *Dokl. Akad. Nauk SSSR*, **27**, 521-524 (IV) (1940).

2 'Polynomials, orthogonal on a rectifiable contour in the presence of a weight function', *Mat. Sb.*, **9** (51), No. 3, 469-488 (1941).

3 'Asymptotic representation of polynomials, orthogonal with respect to a surface', *Dokl. Akad. Nauk SSSR*, **58**, 1883-1885 (1947).

KRAPLIN, M.A.

1 'Two examples of a matrix method of investigating quasi-power bases', *Sb. nauchno-tech. Rab. Azovo-Chernomorsk. in-ta mechaniz. s.h.*, No. 10, 313-332 (1957).

KRASNOSHCHEKOVA, T.I.

1 'A theorem on series of polynomials', *Dokl. Akad. Nauk SSSR*, **77**, No. 5, 787-789 (1951).

2 'Some properties of series of Faber polynomials', *Dokl. Akad. Nauk SSSR*, **115**, No. 1, 38-41 (1957).

KRYLOV, V.I.

1 'Determination of the smallest domain of holomorphicity in which

Hermite's interpolation converges for any system of interpolation points', *Dokl. Akad. Nauk SSSR*, **78**, 857-859 (1951).

KUZMINA, A.L.

1 'Asymptotic representation of polynomials, orthogonal on the unit circle', *Dokl. Akad. Nauk SSSR*, **107**, No. 6, 793-795 (1956).

LEBEDEV, N.A.

1 'A method for analytic continuation of power series', *Dokl. Akad. Nauk SSSR*, **125**, No. 4, 730-732 (1959).

2 'Extension of complex functions', *Vest. Leningr. gos. Univ.*, No. 13 (1963).

LEBEDEV, M.A. and MILIN, I.M.

1 'Coefficients of certain classes of analytical functions', *Mat. Sb.*, No. 2, 359-400 (1951).

LEVIN, B. Ya.

1 *Distribution of roots of entire functions*, Gostekhizdat (1956).

LOKHIN, I.F.

1 'On the question of the completeness of systems of functions $f(\lambda_n z)$', *Mat. Sb.*, **35**, (77), No. 2, 215-222 (1954).

2 'On the question of representing analytic functions by Faber polynomials', *Mat. Sb.*, **36**, (78), No. 3, 441-444 (1955).

LOZINSKY, S.M.

1 'Fejer's interpolating process', *Dokl. Akad. Nauk SSSR*, **24**, 318-321 (1939).

MARKOV, A.A.

1 'On a question of D.I. Mendeleev', *Izv. Akad. Nauk SSSR*, **62**, 1-24 (1889), or in the book *Selected works on the theory of continuous functions* (1949).

MARKOV, V.A.

1 *Functions deviating least from zero in a given interval*, St. Petersburg (1892).

MARKUSHEVICH, A.I.

1 *Conformal mapping of domains with variable boundaries with applications to the approximation of functions by polynomials*, Dissertation, MGU (1934).

2 'On Faber polynomials', *Izv. Akad. Nauk SSSR, Maths. series*, **8**, 49-60 (1944).

MEIMAN, N.N.

1 'Differential inequalities and some questions of extending the zeros

of entire and single-valued analytic functions', *Usp. mat. Nauk,* **7**, No. 3 (49), 3-62 (1952).

2 'On the theory of polynomials deviating least from zero', *Dokl. Akad. Nauk SSSR,* **130**, No. 2, 257-260 (1960).

3 'Polynomials deviating least from zero with any number of given coefficients', *Dokl. Akad. Nauk SSSR,* **130**, No. 3, 503-506 (1960).

4 'Solution of basic problems of the theory of polynomials and entire functions deviating least from zero', *Trudȳ Mosk. mat. Obshch.,* **9**, 507-535 (1960).

MERGELIAN, S.N.

1 'Some questions of the constructive theory of functions', *Trudȳ mat. in-ta Akad. Nauk SSSR,* **37** (1951).

2 'Uniform approximation of functions of a complex variable', *Usp. mat. Nauk,* **7**, 2 (48), 31-122 (1952).

3 'On the completeness of systems of analytic functions', *Usp. mat. Nauk,* **8**, 4 (56), 3-63 (1953).

4 'General metrical criteria for the completeness of systems of polynomials', *Dokl. Akad. Nauk SSSR,* 105, No. 5, 901-904 (1955).

5 'Weighted approximations by polynomials', *Usp. mat. Nauk,* **11**, No. 5, 107-152 (1956).

6 'Approximation of functions of a complex variable', in *Mathematics in USSR over 40 years, 1917-1957,* **I**, 383-398 (1959).

NATANSON, I.P.

1 *Constructive theory of functions,* Gostekhizdat (1949).

POLYA, G. and SZEGÖ, G.

1 Aufgaben und dehrsötze aus der Analysis, Berlin, Springer (1925).

POMMERENKE, Ch.

1 'On the derivative of a polynomial', *Mich. math. J.,* **6**, No. 4, 373-375 (1959).

PRIVALOV, I.I.

1 *Boundary properties of analytic functions,* Gostekhizdat (1950).

REMEZ, E.Ya.

1 'On approximations in the complex domain', *Dokl. Akad. Nauk SSSR,* **77**, No. 6, 965-968 (1951).

2 'Certain questions of Chebyshev approximation in the complex plane', *Ukr. mat. Zh.,* **5**, No. 1, 3-49 (1953).

3 'Fixed points of deviation of solutions of the problems of Chebyshev approximation with linearly entering parameters I and II', *Ukr. mat. Zh.,* **9**, No. 1, 44-65; 9, No. 3, 310-328 (1957).

ROSENBLOOM and WARSCHAWSKI, S.E.

1 'Approximation by polynomials'. *Lectures on functions of a complex variable,* Ann Arbor, University of Michigan Press 287-302 (1959).

RYZHAKOV, I.U.

1 'A problem of S.N. Bernstein', *Dokl. Akad. Nauk SSSR,* **153**, No. 2, 282-285 (1963).

SCHAEFFER, A.G. and DUFFIN, R.J.

1 'On some inequalities of S. Bernstein and W. Markov for derivatives of polynomials', *Bull. Amer. math. Soc.,* **44**, No. 4, 289-297 (1938).

SELEZNEV, A.I.

1 'Representation of functions by Faber polynomials', Collection *Investigations of current problems in the theory of functions of a complex variable,* 131-138 (1961).

SHAGINIAN, A.L.

1 *Completeness of orthogonal polynomials in the complex domain,* Dissertation LGU (1939).

2 'On the question of approximation in the mean in the complex domain', *Izv. Akad. Nauk SSSR, Maths. series,* **5**, No. 4-5, 285-296 (1941).

3 'Completeness of families of analytic functions in the complex domain', *Izv. in-ta mat. mekh. Akad Nauk Armyan SSR,* **1**, 3-59 (1947).

4 'Best approximations in non-Jordan domains', *Dokl. Akad. Nauk Armyan SSR,* **11**, 97-100 (1949).

5 'On the question of weighted mean-square approximation in the complex domain', *Dokl. Akad. Nauk Armyan SSR* **16**, No. 1, 3-8 (1953).

6 'Approximation by polynomials', *Izv. Akad. Nauk Armyan. SSR, ser. fiz.- mat. estestv. tech. nauk,* **7 : 4**, 1-21 (1954).

7 'A problem of the theory of approximation in the complex domain', *Sib. mat. Zh.,* **1**, No. 3, 523-543 (1960).

SHNIREL'MAN, L.G.

1 'Uniform approximation', *Izv. Akad. Nauk SSSR, Maths series,* **2** No. 1, 53-60 (1938).

SHUVALOVA, E.Z.

1 'A sufficient condition for the completeness of the system $\{f^{(n)}(z)\}$', *Mat. Sb.,* **44**, No. 1, 131-136 (1958).

SMIRNOV, V.I.

1 *Course of higher mathematics,* **IV**, Gostekhizdat (1953).

2 *Course of higher mathematics,* **V**, Gostekhizdat (1959).

3 'Sur quelques polynômes aux propriétés extremales', *Zap. Khayk. Mat. o-va,* (4), 2, 67-72 (1928).

4 'Sur la théorie des polynômes ortogonaux à une variable complexe',

Zh. Leningr. fiz.-mat. Obshch., 2, 155-179 (1928).

5 'Sur les formules de Cauchy et de Green et quelques problèmes qui s'y rattachent', *Izv. Akad. Nauk SSSR*, 337-372 (1932).

STEFFENSEN, I.F.

1 *Theory of interpolation*, Chelsea Pub. Co., N.Y. (1950).

SUETIN, P.K.

1 'Uniqueness of series of Faber polynomials', *Dokl. Akad. Nauk SSSR*, **105**, No. 3, 423-425 (1955).

2 'On polynomials, orthogonal with a differentiable weight function', *Dokl. Akad. Nauk SSSR*, **106**, No. 5, 789-791 (1956).

3 'Representation of analytical functions by series of orthogonal polynomials', *Dokl. Akad. Nauk SSSR*, **109**, No. 1, 36-39 (1959).

4 'Polynomials, orthogonal with respect to a smooth contour with a differentiable weight function', *Dokl. Akad. Nauk SSSR*, **114**, No. 3, 498-501 (1957).

5 'Polynomials, orthogonal with respect to a surface', *Dokl. Akad. Nauk SSSR*, **126**, No. 5, 943-945 (1959).

6 'Interpolation of analytical functions', *Dokl. Akad. Nauk SSSR*, **128**, No. 6, 1125-1128 (1959).

7 'Certain asymptotic properties of polynomials', *Dokl. Akad. Nauk SSSR*, **129**, No. 1, 30-33 (1959).

8 'On the order of approximation of analytic functions by partial sums of series of orthogonal polynomials', *Izv. Akad. Nauk Armyan SSR, ser. fiz-mat. nauk*, **15**, No. 2, 81-86 (1962).

SZEGÖ, G.

1 *Orthogonal polynomials*, Fizmatgiz (1962).

2 'Beiträge zur Theorie der Toeplitzschen Formen'. II, *Math. Z.*, **9**, 167-190 (1921).

3 'Über orthogonale Polynome, die zu einer gegeben Kurve der komplexen Ebene gehören', *Math. Z.*, **9**, 218-270 (1921).

4 'Über einen Satz von A. Markoff', *Math. Z.*, 23 (1925).

TAMADIAN, E.

1 'A theorem of M.V. Keldysh', *Izv. Akad. Nauk Armyan SSR*, **6**, No. 2, 5-11 (1953).

TITCHMARSH, E.C.

1 *The theory of functions*, (2nd edition) Oxford (1939).

TONELLI, L.

1 'J. polinomi d'approssimazione di Tchebychev', *Annali. Mat.*, **15** (1908).

TONIAN, V.A.

1 'Weighted polynomial approximation of analytical functions in infinite domains', *Dokl. Akad. Nauk SSSR*, **133**, No. 3, 535-536 (1960).

TUMARKIN, G. Ts.

1 'Approximation of a function by rational fractions with given poles', *Dokl. Akad. Nauk SSSR*, **48**, No. 6, 409-912 (1954).

2 'Approximation in the mean of functions on rectifiable curves', *Mat. Sb.*, **42**, No. 1, 79-128 (1957).

3 'Expansion of analytic functions into series of rational fractions with a given set of poles', *Izv. Akad. Nauk Armyan SSR, ser. fiz-mat. nauk*, **14**, No. 1, 9-31 (1961).

4 'A sufficient condition for a domain to belong to class *S*'', *Vestn. Leningr. gos. Univ.*, **13**, 3, 47-55 (1963).

VALLÉE POUSSIN, CH.

1 'Sur les polynômes d'approximation et la représentation approchée d'un angle', *Bull. Acad. Belg.*, **12** (1910).

VARGANOVA, S.V. and KOROBEINIK, U.F.

1 'Upper convergence and non-continuation of functional series', *Dokl. Akad. Nauk SSSR*, **137**, No. 3, 499-501 (1961).

VIDENSKY, V.S.

1 'Estimation of derivatives of a polynomial', *Izv. Akad. Nauk SSSR, Maths series*, **15**, No. 5, 401-420 (1951).

2 'Uniform approximation in the complex plane', *Usp. mat. Nauk*, **11**, 5 (71), 169-175 (1956).

3 'Polynomials deviating least from zero, the coefficients of which satisfy a given linear dependence', *Dokl. Akad. Nauk SSSR*, **126**, No. 2, 248-250 (1959).

4 'Qualitative questions of the theory of best approximation of functions of a complex variable', Collection *Investigations of current problems in the theory of functions of a complex variable*, 258-272 (1960).

5 'Extremal estimates of the derivative of a trigonometrical polynomial on an interval smaller than the period', *Dokl. Akad. Nauk SSSR*, **130**, No. 1, 13-16 (1960).

6 'Second report on polynomials deviating least from zero, the coefficients of which satisfy a given linear dependence', *Izv. Akad. Nauk Armyan SSR*, No. 1, 3-7 (1961) (Ref. 1962 2B100).

7 'Extremal estimates of derivatives of a polynomial on an interval', Collection *Investigations of current problems in the theory of functions of a complex variable*, 98-106 (1961).

8 'Some estimates of derivatives of rational fractions', *Izv. Akad. Nauk*

SSR, Maths series, **26**, 415-426 (1962).

VITUSHKIN, A.G.

1 'Best approximation of differentiable and analytic functions', *Dokl. Akad. Nauk SSSR*, **119**, No. 3, 418-420 (1958).

2 'Necessary and sufficient conditions for a set in order that it should be possible to approximate uniformly every function continuous on this set by means of analytic (or rational) functions', *Dokl. Akad. Nauk SSSR*, **128**, No. 1, 17-20 (1959).

VORONOVSKAYA, E.V.

1 'Application of functional analysis to polynomials of least deviation', *Dokl. Akad. Nauk SSSR*, **99**, No. 1, 5-8 (1954).

2 'Extremal polynomials of some simpler functionals',*Dokl. Akad. Nauk SSSR*, **99**, No.2, 193-196 (1954).

3 'Functional of the first derivative and improvement of the theorem of A.A. Markov', *Izv. Akad. Nauk SSSR, Maths series*, **23**, 951-962 (1959).

4 *Functional method and its applications*, Izd. Leningr. Bonch-Bruevich elektrotekh. inst. svyazi (1963).

WALSCH, J.L.

1 *Interpolation and approximation by rational functions in a complex domain*, IL (1961).

2 'Note on polynomial interpolation to analytic functions', *Proc. Nat. Acad. Sci.*, **19**, 959-963 (1933).

3 'On interpolation by polynomials', *Bull. Amer. math. Soc.*, **40**, Abstract No. 198 (1934).

ZAKHARUTA, V.P.

1 'Bases in the space of functions analytic in the closed disc $|z| \leqslant R$', *Dokl. Akad. Nauk SSSR*, **141**, No. 5, 1023-1025 (1961).

ZOLOTAREV, E.I.

1 *Application of elliptical functions to problems of functions deviating most and least from zero*, Complete Works, **II**, St. Petersburg (1877).

ZUKHOVITSKII, V.P.

1 'An algorithm for constructing the best approximation of continuous functions by polynomials in the complex domain', Collection *Investigations of current problems in the theory of functions of a complex variable*', 201-208 (1961).

2 'Some algorithms for constructing Chebyshev approximations', Collection *Investigations of current problems in the theory of functions of a complex variable*', 9-14 (1961).

ZUKHOVITSKII, S.I., POLIAK, R.A. and PRIMAK, M.E.

1 'An algorithm for solving problems of convex Chebyshev approximation', *Dokl. Akad. Nauk SSSR*, **151**, No. 1, 27-30 (1963).

ZYGMUND, A.

1 *Trigonometrical series*, (1935).

Index